Environmental Science and Engineering

Series Editors

Ulrich Förstner, Buchholz, Germany

Wim H. Rulkens, Department of Environmental Technology, Wageningen University & Research, Wageningen, Gelderland, The Netherlands

The ultimate goal of this series is to contribute to the protection of our environment, which calls for both profound research and the ongoing development of solutions and measurements by experts in the field. Accordingly, the series promotes not only a deeper understanding of environmental processes and the evaluation of management strategies, but also design and technology aimed at improving environmental quality. Books focusing on the former are published in the subseries Environmental Science, those focusing on the latter in the subseries Environmental Engineering.

Matthew Chidozie Ogwu · Sylvester Chibueze Izah
Editors

Evaluating Environmental Processes and Technologies

Springer

Editors
Matthew Chidozie Ogwu
Goodnight Family Sustainable
Development Department
Appalachian State University
Boone, NC, USA

Sylvester Chibueze Izah
Department of Community Medicine
Bayelsa Medical University
Yenagoa, Nigeria

ISSN 1863-5520 ISSN 1863-5539 (electronic)
Environmental Science and Engineering
ISBN 978-3-031-85329-6 ISBN 978-3-031-85327-2 (eBook)
https://doi.org/10.1007/978-3-031-85327-2

This Springer imprint is published by the registered company Springer Nature Switzerland AG
The registered company address is: Gewerbestrasse 11, 6330 Cham, Switzerland

Preface

Evaluating Environmental Processes and Technologies: Sustainable Management Strategies is an in-depth exploration of the diverse challenges, transformative innovations, and sustainable practices shaping modern environmental management. As the world grapples with escalating environmental issues, this book provides a multidisciplinary perspective on navigating the complex nexus of ecological processes, technological advancements, and sustainability imperatives. It is a culmination of collaborative efforts, theoretical insights, and practical approaches aimed at fostering a sustainable future for our planet.

Environmental systems are inherently dynamic, characterized by intricate interconnections between natural, human, and socio-economic factors. These complexities demand innovative and context-sensitive solutions to ensure the long-term health of ecosystems and communities. This book acknowledges these challenges, offering a comprehensive narrative that addresses critical topics such as pollution mitigation, resource management, climate adaptation, and green technological applications. With sustainability as the central theme, the chapters unravel the layers of environmental challenges and present actionable strategies for achieving resilience and balance in environmental systems.

The book is organized into 16 chapters, each of which delves into vital aspects of environmental processes and technologies including the principles of environmental systems and sustainable development, integrated natural resource management and the role of ecology in technological advancements as well as how to tackle pressing issues such as pollution sources and controls, waste-to-energy innovations, and climate change mitigation technologies. The book also emphasizes the role of policy frameworks, regulatory strategies, and economic instruments in driving sustainable environmental practices. Notably, the book bridges theoretical concepts and real-world applications, enriched with case studies and multidisciplinary insights. It underscores the importance of collaboration across sectors and regions, offering globally relevant strategies for diverse environmental contexts. The contributors, drawn from various disciplines, provide a holistic perspective that is both scientifically rigorous and practically relevant. Our goal is to provide readers with a resource that not only informs but also inspires action. By examining current challenges and showcasing innovative solutions, *Evaluating Environmental Processes and Technologies* aims to empower stakeholders at all levels to contribute meaningfully to sustainable development.

This book is a valuable resource for academics, researchers, policymakers, and practitioners in environmental science, sustainability, and technology. It aims to inspire meaningful dialogue and action, equipping readers with the knowledge and tools needed to address environmental challenges effectively. As you explore these pages, we hope you find inspiration and guidance for contributing to a sustainable future, where innovation and ecological stewardship work hand in hand to secure the health of our planet.

Boone, North Carolina, USA
Yenagoa, Bayelsa State, Nigeria
2025

Dr. Matthew Chidozie Ogwu
Dr. Sylvester Chibueze Izah

About This Book

The book, *Evaluating Environmental Processes and Technologies: Sustainable Management Strategies*, is a definitive resource that delves into the complexities of environmental systems, innovative technologies, and sustainable management strategies. It highlights innovative methods, foundational theories, and practical applications to tackle pressing environmental challenges. It also emphasizes sustainable practices and cutting-edge technologies, making it a valuable resource for advancing environmental science and management. This edited volume brings together an international cohort of experts to comprehensively explore the challenges and opportunities within environmental science and sustainability.

This volume bridges theoretical foundations with practical applications, offering readers insights into the science of pollution, green technologies, waste-to-energy innovations, and climate change adaptation. It extends beyond technical discussions to examine the socio-economic and policy dimensions critical for fostering sustainable practices. The chapters are carefully curated to provide a holistic understanding of diverse topics, including carbon capture, sustainable urban planning, and the valuation of ecosystem services. Each contribution integrates cutting-edge research with actionable insights, making the content equally valuable for academic inquiry and real-world applications.

The book's interdisciplinary nature ensures a multidimensional perspective, with contributors from fields such as ecology, environmental engineering, urban planning, and environmental economics. By addressing environmental challenges through a global lens, the book offers solutions adaptable to developed and developing regions, underscoring its relevance in varied contexts. Furthermore, it emphasizes the role of innovative technologies and strategic policy frameworks in driving sustainable environmental practices.

Designed for academics, researchers, policymakers, and practitioners, *Evaluating Environmental Processes and Technologies* is vital for advancing knowledge and promoting sustainable development. Its rich content and accessible presentation make it a valuable resource for higher education, professional training programs, and policy formulation. This book is not only an academic contribution but also a call to integrate innovation, science, and governance to address some of our time's most pressing environmental issues.

Boone, USA
Yenagoa, Nigeria
2025

Matthew Chidozie Ogwu, Ph.D.
Sylvester Chibueze Izah, Ph.D.

Contents

About the Editors

Dr. Matthew Chidozie Ogwu is an assistant professor of Integrated Ecology and Sustainable Development at Appalachian State University, USA. He is an interdisciplinary academic with transdisciplinary skills and diverse convergence research interests (One Health and Eco Health) pertinent to the assessment of coupled human, natural, and socio-ecological systems. His work reflects a commitment to bridging diverse fields to promote sustainability and resilience in both human and ecological contexts. He has numerous awards, research grants, and scholarships to his name. Dr. Ogwu serves on the board of and as a reviewer for many peer-reviewed journals. He continues to volunteer his time and skills to promote sustainable development.

Dr. Sylvester Chibueze Izah is a lecturer at Bayelsa Medical University in Yenagoa, Nigeria, where he also serves as the assistant director of Academic Planning, Research, and Innovations. Dr. Izah is a licensed environmental health specialist in Nigeria. He is a multidisciplinary academic with multifaceted abilities relevant to Sustainable Human-Environmental Health Interactions (covering air, soil, and water quality; toxicology; hygiene and sanitation; food science; waste management; biodiversity and their sustainability and risk assessment). He has an impressive track record of over 280 peer-reviewed publications, including journal articles, book chapters, and edited books. Dr. Izah has also collaborated on research projects with colleagues worldwide and actively contributes as an editorial and review board member for several esteemed journals.

Chapter 1
Introduction to Environmental Systems and Processes

Matthew Chidozie Ogwu, Sylvester Chibueze Izah, and Adams Ovie Iyiola

Abstract This chapter reviews environmental systems and processes to highlight the complex relationships between natural and man-made systems. An environment's abiotic (non-living) and biotic (life) elements interact dynamically to form environmental systems and emergent processes and functions. These systems operate across various scales, from local ecosystems to the global biosphere, and are driven by energy flow, nutrient cycling, and ecosystem regulation. Key processes like photosynthesis, respiration, and the water and carbon cycles are fundamental to maintaining ecological balance and supporting life. Human activities, including resource extraction, industrialization, and urbanization, significantly impact these systems, often disrupting natural processes and leading to challenges such as climate change. Understanding these interactions is crucial for promoting sustainability in the face of twenty-first-century challenges. Disruptions caused by human activities stress the urgent need for conservation efforts and the maintenance of ecosystem functions to ensure the sustainability of both ecological and human systems. The critical role of feedback loops, thresholds, and tipping points in ecosystem resilience illustrates how these factors interconnect within environmental interactions. Integrating sustainability thinking into educational curricula and adopting robust analytical frameworks

---

M. C. Ogwu
Goodnight Family Department of Sustainable Development, Appalachian State University, Boone, NC, USA
e-mail: ogwumc@appstate.edu

S. C. Izah (✉)
Department of Community Medicine, Faculty of Clinical Sciences, Bayelsa Medical University, Yenagoa, Bayelsa State, Nigeria
e-mail: sylvester.izah@bmu.edu.ng

Department of Microbiology, Faculty of Science, Bayelsa Medical University, Yenagoa, Bayelsa State, Nigeria

A. O. Iyiola
Department of Fisheries and Aquatic Resources Management, Faculty of Renewable Natural Resources Management, College of Agriculture and Renewable Resources, Ejigbo Campus, Osun State University, Ejigbo, Osun State, Nigeria
e-mail: adams.iyiola@uniosun.edu.ng

M. C. Ogwu and S. Chibueze Izah (eds.), *Evaluating Environmental Processes and Technologies*, Environmental Science and Engineering,
https://doi.org/10.1007/978-3-031-85327-2_1

can enhance understanding of these processes. Furthermore, the importance of Environmental Impact Assessments is highlighted as a tool for evaluating potential environmental impacts, promoting sustainable development, and incorporating stakeholder perspectives into decision-making. Because of these complexities of ecological systems, interdisciplinary collaboration, and holistic management approaches are essential. Therefore, future research should focus on the interdependencies within ecosystems, particularly microbial communities' role and the impacts of anthropogenic activities. Society may strive toward practical solutions that support sustainability, resilience, and just futures for future generations by cultivating a deeper awareness of these linked systems.

Keywords Environmental systems · Ecosystems · Biogeosociochemical cycles · Sustainability · Anthropogenic processes · Biodiversity · Environmental impact assessment · Ecosystem resilience

1.1 Introduction

Environmental systems, which include a wide range of interactions and processes that characterize ecological and anthropogenic landscapes, are based on the complex interactions between natural and man-made systems. These systems must be understood to address today's environmental issues, such as resource depletion, biodiversity loss, and climate change. Environmental systems are complex networks of interactions among biotic and abiotic components, where both natural phenomena and human activities converge. These systems encompass various scales, from local ecosystems to global biogeosociochemical cycles, and include the interactions of physical, chemical, and biological processes that sustain life on Earth (Sundsbø et al. 2015). The scope of environmental systems extends beyond mere ecological considerations; it incorporates social, economic, and political dimensions, reflecting the multifaceted nature of sustainability (Spencer et al. 2013). For instance, integrating sustainability thinking in science and engineering curricula highlights the necessity of a holistic approach to environmental management, which is often overlooked in traditional educational frameworks (Sundsbø et al. 2015). The complexity of environmental systems necessitates a comprehensive understanding of their components and interactions. For example, the NEIMiner framework illustrates how nanomaterials interact with environmental systems, emphasizing the need for a thorough understanding of their physical and chemical properties and their ecological impacts (Tang et al. 2013). Additionally, the life-cycle assessment (LCA) methodology is a critical tool for evaluating the environmental impacts of products and processes, reinforcing the importance of defining system boundaries and scopes in environmental analysis (Đjekić et al. 2019; Akudugu and Ogwu 2024). This multifaceted approach is essential for addressing the challenges posed by unsustainable practices and for fostering resilience in environmental systems

The interaction between natural and human-made systems is a defining characteristic of environmental systems. Human activities, such as agriculture, urbanization, and industrialization, exert significant pressures on natural ecosystems, leading to alterations in biodiversity, habitat loss, and climate change (Gardner et al. 2023). Conversely, natural systems also influence human activities, providing essential ecosystem services supporting human well-being, including clean air, water, and fertile soil. This reciprocal relationship underscores the necessity for integrated management approaches considering ecological and anthropogenic factors. For instance, the Capacity-Building in Sustainability and Environmental Management Model (CapSEM-Model) illustrates how organizations can transition towards sustainability by recognizing the interdependencies between various systemic levels (Fet and Knudson 2021). This model emphasizes the importance of capacity-building initiatives that foster collaboration among stakeholders, thereby enhancing the overall sustainability of environmental systems. Furthermore, the application of system dynamics modeling in environmental health decision-making highlights the complexities inherent in these interactions, as it allows for exploring feedback loops and causal relationships within environmental systems (Currie et al. 2018). To effectively create policies and practices aimed at minimizing environmental deterioration, it is imperative to comprehend these relationships. Organizations with a strong commitment to sustainability are more likely to implement comprehensive environmental management practices, as evidenced by the mediating effects of environmental information systems on management commitment and performance (Spencer et al. 2013). This research emphasizes how crucial it is to incorporate environmental factors into organizational decision-making procedures to promote a sustainable culture that cuts over conventional lines.

Addressing the numerous issues the world faces today requires an understanding of environmental dynamics. Strong analytical frameworks and procedures are essential for evaluating environmental systems' dynamics as they become more complex. For instance, life-cycle assessment offers a methodical way to examine how processes and products affect the environment at every stage of their life cycle, from the extraction of raw materials to disposal (Song et al. 2020). This all-encompassing viewpoint is crucial for spotting areas that could use development and for encouraging sustainable practices in various industries. Furthermore, it has become clear that incorporating environmental management accounting within organizational structures is essential to sustainability initiatives. By extending the spatial scope of ecological assessments beyond individual organizations to encompass supply chains and circular systems, businesses can better understand their environmental footprints and implement strategies to mitigate negative impacts (Burritt et al. 2023). This approach aligns with the United Nations Sustainable Development Goals (SDGs), which emphasize the need for holistic and integrated solutions to global challenges (Burritt et al. 2023). The need for interdisciplinary collaboration further underscores the significance of understanding environmental processes. For instance, the scoping review on children's environmental health advocates for a systems approach incorporating material, biological, cultural, and social dimensions. This holistic perspective is essential for developing effective interventions that address the root causes of environmental health disparities and promote equity in environmental decision-making. The conceptual framework proposed in Peng and Jia (2023) highlights innovative behavior as the intersection of the environmental systems and innovation processes, emphasizing the dynamic interplay between these components (Fig. 1.1). At their intersection lies the symbolism of space, representing the synergy of these environments to create a conducive atmosphere for innovation. This framework underscores the importance of integrating social dynamics, physical design, and structured innovation processes to foster creativity and innovative behavior effectively.

The chapter aims to present an overview of environmental systems and processes to promote understanding the intricate interactions within natural and anthropogenic systems, emphasizing their relevance in addressing ecological challenges. It explores ecosystems' structures and functions, detailing energy flows, trophic levels, and the essential role of biodiversity in ensuring stability and resilience. The chapter delves into biogeosociochemical cycles, distinguishing natural processes from anthropogenic disruptions, and underscores the importance of Environmental Impact Assessments (EIAs) for promoting sustainable development. The chapter advocates for integrating sustainability thinking into education and decision-making by highlighting the interconnectedness of ecological components and processes. This approach fosters interdisciplinary collaboration, equipping society with practical solutions to mitigate environmental challenges and supporting a sustainable future.

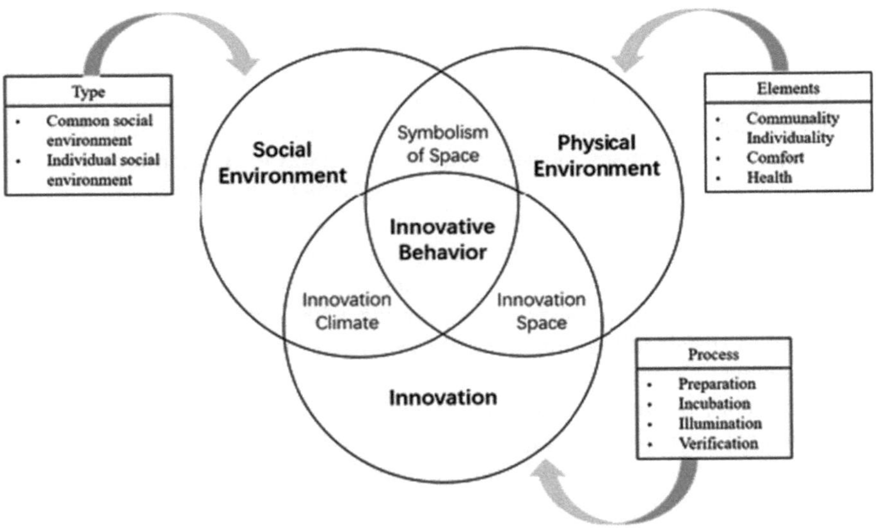

Fig. 1.1 Conceptual framework for innovative behavior. *Source* Peng and Jia (2023)

1.2 Environmental Systems and Their Components

Environmental systems are dynamic frameworks where energy, nutrients, and organisms interact to sustain ecological processes. These systems operate across various scales, from local ecosystems to global biogeosociochemical cycles, and integrate biological, physical, and chemical processes. They are not only ecological constructs but also incorporate socio-economic and cultural dimensions, reflecting the complexity of sustainability (Sundsbø et al. 2015). Environmental systems consist of biotic and abiotic components that interact to sustain life.

- **Biotic Components**: These include all living organisms, such as plants, animals, and microorganisms. They contribute to processes like photosynthesis, decomposition, and nutrient cycling, essential for ecosystem health (Cardinale et al. 2011).
- **Abiotic Components**: These include non-living factors like air, water, soil, and climate. Abiotic elements regulate temperature, precipitation, and nutrient availability, shaping ecosystem functionality (Lin 2014).

Ecosystems are complex networks of interactions among living organisms and their physical environment, characterized by a multitude of components that work together to sustain life. The structure of an ecosystem refers to its physical arrangement, including the types and numbers of organisms, their spatial distribution, and the interactions among them. Function, on the other hand, encompasses the processes that occur within the ecosystem, such as energy flow, nutrient cycling, and the various ecological roles played by different species. Understanding the interplay between structure and function is crucial for ecological research and conservation efforts, as it can reveal how ecosystems respond to environmental changes and disturbances (Inostroza et al. 2020). Figure 1.1 illustrates the concept of complex adaptive ecosystems, emphasizing the interplay between biodiversity and ecosystem functioning. Biodiversity is a key factor that increases ecosystem productivity, nutrient cycling, network dynamics, and stability. The energy and matter mediate these processes' flow within the ecosystem, which is central to sustaining its dynamics. Figure 1.1 highlights how environmental context (spatial and ecological scales) and temporal scales (evolutionary processes) influence these interactions. As biodiversity increases, the ecosystem's ability to adapt and maintain functional stability improves, underscoring its resilience to environmental changes.

The structure of ecosystems (Table 1.1) can be analyzed through various frameworks, such as trait-based approaches that consider the functional traits of species within a community. For instance, Fahey et al. proposed a multivariate framework to describe canopy structure variation across vegetated ecosystems, emphasizing the importance of functional traits in understanding ecosystem dynamics (Fahey et al. 2019). This approach allows researchers to evaluate how specific traits, such as primary productivity and disturbance resilience, influence ecosystem functions. Additionally, the structural diversity of ecosystems can be assessed using remote sensing technologies, which can provide insights into how structural metrics relate to ecosystem functions across large spatial scales (LaRue et al. 2019). Ecosystems, as subsets of environmental systems, are defined by their structure (physical arrangement) and function (processes that sustain them). The structure of an ecosystem includes the spatial distribution of organisms, trophic levels, and interactions among species (Fahey et al. 2019). Functions include energy flow, nutrient cycling, and ecological services like pollination and carbon sequestration, which underpin human and environmental health (Izah 2018; Ogwu et al. 2014; 2024a, b).

Energy flow is a fundamental aspect of ecosystem function, primarily described through trophic levels. Trophic levels categorize organisms based on their feeding relationships, ranging from primary producers, such as plants, to various levels of consumers, including herbivores and carnivores. Energy transfer efficiency between these levels is critical for understanding ecosystem productivity and stability. For example, studies have shown that different food web structures can lead to distinct ecological functions, indicating that the organization of trophic interactions is a key driver of ecosystem dynamics (Yen et al. 2016). Moreover, temporal variability in primary production has been linked to marine ecosystem structure and function, highlighting the importance of energy flow in shaping ecological communities (Schlenger et al. 2019).

Biodiversity plays a vital role in maintaining ecosystem stability and resilience (Fig. 1.2; Izah 2018; Izah et al. 2018, 2023; Izah and Seiyaboh 2018a, b; Izah and Aigberua 2023; Ogwu et al. 2023a, b). High levels of biodiversity can enhance ecosystem functioning by promoting a variety of ecological processes, such as nutrient cycling and primary production. Cardinale et al. (2013) demonstrated that biodiversity simultaneously enhances the production and stability of community biomass, suggesting that diverse ecosystems are better equipped to withstand environmental fluctuations. Furthermore, the evolutionary relationships among species, or phylogenetic diversity, have been shown to explain variations in ecosystem function and stability more effectively than species richness alone (Cadotte et al. 2012). This indicates that the functional roles of species, rather than their mere abundance, are crucial for ecosystem health. The relationship between biodiversity and ecosystem stability is complex and can vary across spatial and temporal scales. For instance, research has indicated that biodiversity is a good predictor of stability, explaining a significant portion of the variation in stability at both local and regional scales. However, the effects of biodiversity on stability may differ depending on the specific ecological context and the types of interactions present within the community. For example, dominant species can significantly influence ecosystem stability, particularly in dryland ecosystems, where their presence can stabilize community dynamics (García-Palacios et al. 2018; Chen et al. 2021). This highlights the need for a nuanced understanding of how different components of biodiversity contribute to ecosystem resilience.

In addition to species diversity, functional diversity—referring to the range of functional traits present within a community—also plays a critical role in ecosystem stability. Studies have shown that functional diversity can independently predict ecosystem processes, providing additional explanatory power beyond taxonomic

Table 1.1 Ecosystems and their components

Ecosystem component	Characteristics	Key processes and functions	Ecological importance	Examples
Structure of ecosystems	The physical arrangement of biotic (living) and abiotic (non-living) components	Provides habitat for organisms and influences interactions	Determines ecosystem types and their functionality	Forests, wetlands, grasslands
Function of ecosystems	Processes and interactions among components that maintain ecosystem health and productivity	Nutrient cycling, energy flow, and habitat provision	Supports life processes and ecosystem services	Pollination, decomposition, water purification
Energy flow	The transfer of energy through food webs and trophic levels in an ecosystem	Energy is captured by producers and transferred to consumers	Essential for sustaining life and dictates ecosystem productivity	Plants capture solar energy, and energy is lost through trophic levels
Trophic levels	Hierarchical levels in an ecosystem based on energy transfer	Producers → Primary consumers → Secondary consumers → Decomposers	Affects population dynamics and energy availability	Grass (producer), rabbit (primary consumer), fox (secondary consumer)
Biodiversity	An ecosystem's variety of life forms, including species richness and genetic diversity	Enhances resilience and stability of ecosystems	Contributes to ecosystem services, such as pollination and nutrient cycling	Coral reefs (high biodiversity), agricultural fields (lower biodiversity)
Role of biodiversity in stability	Biodiversity contributes to ecosystem resilience against disturbances and changes	Diverse species can adapt to changes and stabilize ecosystems	Ensures sustainable ecosystem functions over time	Diverse plant and animal species in forests support various animal life

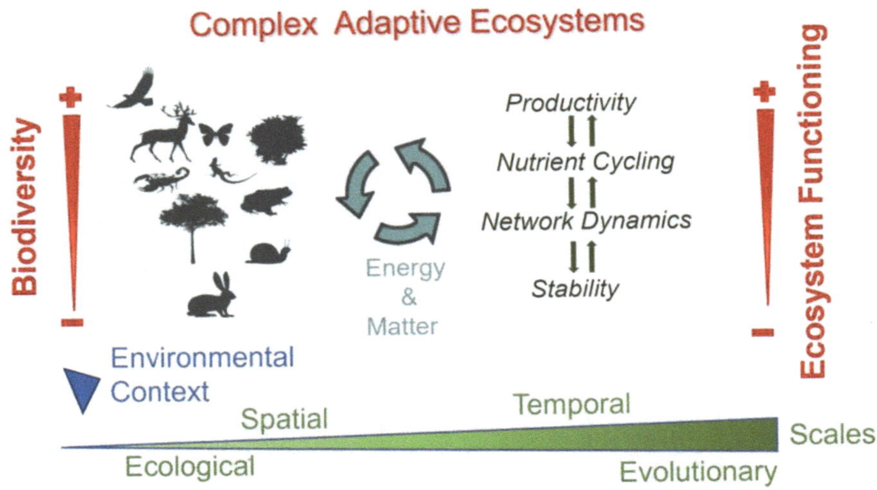

Fig. 1.2 Complex adaptive environmental system. *Source* Correia and Lopes (2023)

diversity (Mouillot et al. 2011; Balvanera et al. 2005). This suggests that maintaining diverse functional traits within ecosystems is essential for sustaining ecological processes and enhancing resilience to disturbances. The interaction between aboveground and belowground biodiversity is another important factor influencing ecosystem stability. Research has demonstrated that both components jointly drive ecosystem stability, with aboveground biodiversity having a more pronounced effect than belowground diversity in specific contexts (Chen et al. 2021). This underscores the interconnectedness of different ecological compartments and the need to consider aboveground and belowground dynamics when assessing ecosystem health and stability. Ecosystem functions, such as carbon storage, nutrient cycling, and pollination, are critical for human well-being and environmental sustainability. The loss of biodiversity can disrupt these functions, leading to declining ecosystem services vital for human survival (Cardinale et al. 2011; Ogwu 2019). Therefore, understanding how biodiversity influences ecosystem functioning is essential for effective conservation and management strategies. For instance, developing methods to analyze ecosystems' functional structure can help assess biodiversity loss's impacts on ecosystem services (Balvanera et al. 2005).

Moreover, the resilience of ecosystems to external disturbances, such as climate change or habitat degradation, is closely linked to their structural and functional attributes. Ecosystems with high structural diversity tend to exhibit greater resilience, as they can better absorb shocks and recover from disturbances (Kareksela et al. 2015). This resilience is often mediated by the presence of diverse species that fulfill various ecological roles, thereby enhancing the overall stability of the ecosystem.

1.3 Biogeosociochemical Cycles

Biogeosociochemical cycles represent an integrated framework that encapsulates the movement of matter and energy through biological, geological, chemical, and sociological domains. These cycles are foundational to understanding ecosystem processes, global sustainability, and human impacts on Earth's systems. These cycles are interconnected, and their functioning is crucial for sustaining life on Earth. The primary biogeosociochemical cycles include the carbon, nitrogen, and water cycles, each significant in maintaining ecological balance and supporting various life forms. The traditional framework of biogeochemical cycles, focusing on the interaction of biological, geological, and chemical processes, has evolved to incorporate sociological dimensions, reflecting humanity's role in altering these cycles. Early works by Vernadsky introduced the biosphere as a dynamic system where life interacts with Earth's physical and chemical components. Over time, researchers recognized that anthropogenic activities, such as agriculture, industrialization, and urbanization, have disrupted the natural cycles of carbon, nitrogen, phosphorus, and water (Galloway et al. 2008; Steffen et al. 2015). Studying biogeosociochemical cycles is essential for understanding the intricate interactions between biological, geological, and chemical processes within ecosystems. Various factors, including anthropogenic activities, climate change, and ecological dynamics, influence these cycles. Core components of biogeosociochemical cycles include.

- **Biological Processes**: Biological entities, from microorganisms to plants and animals, mediate key processes like decomposition, photosynthesis, and respiration. For instance, the microbial breakdown of organic matter drives the nitrogen cycle by converting atmospheric nitrogen into bioavailable forms (Vitousek et al. 1997).
- **Geological Interactions**: Geological processes, such as rock weathering, sedimentation, and tectonic activity, influence the availability of nutrients. These processes are crucial for the long-term storage and release of elements like phosphorus and sulfur.
- **Chemical Transformations**: Chemical reactions facilitate the transfer of elements across different reservoirs. Redox reactions in soil and water govern nutrient cycling and trace gas fluxes, with implications for atmospheric chemistry.
- **Sociological Dimensions**: Human activities have significantly altered these natural cycles. Industrial processes, fossil fuel combustion, deforestation, and urbanization contribute to imbalances, such as eutrophication, greenhouse gas emissions, and biodiversity loss.

In urban systems, these cycles are significantly altered due to human activities. Mitchell (2023) highlights that urbanization leads to changes in hydrological dynamics and surface cover, affecting the cycling of nutrients, contaminants, and greenhouse gases. The proximity of urban areas to anthropogenic emissions creates unique biogeochemical dynamics that require tailored management strategies. As urban populations grow, understanding these dynamics becomes increasingly important for sustainable urban planning and environmental protection. The role of microorganisms in biogeochemical cycles is critical, particularly in extreme environments. Martínez-Espinosa (2020) discusses how microorganisms contribute to the nitrogen cycle in extreme conditions, emphasizing their metabolic capabilities and adaptations. These microorganisms are vital in nutrient cycling and can significantly influence ecosystem functions. Similarly, Shen et al. (2023) review the microbial community structures in coastal wetland sediments, highlighting their importance in maintaining biogeochemical cycles and the impacts of human activities and climate change on these ecosystems. The resilience and adaptability of microbial communities are crucial for sustaining biogeochemical processes in changing environments. Interactions between different components of ecosystems also play a significant role in biogeochemical cycles. For example, Devlin et al. (2015) demonstrate that the presence of top consumers, such as fish, can influence methane efflux in lakes through trophic cascades that affect microbial communities. This finding underscores the interconnectedness of trophic levels and their collective impact on biogeochemical cycling. Similarly, Goldman et al. (2017) explore the biogeochemical cycling at the aquatic-terrestrial interface, emphasizing the importance of hyporheic zones in mediating nutrient and chemical exchanges between terrestrial and aquatic systems. Understanding these interactions is essential for developing effective management strategies for ecosystems. The impact of environmental changes on biogeochemical cycles is a growing concern. Gruber (2011) discusses how rising temperatures, ocean acidification, and deoxygenation are expected to alter ocean biogeochemistry significantly. These changes can disrupt nutrient cycling and affect marine ecosystems, leading to broader implications for global biogeochemical processes. Lanning et al. (2019) reveal that acid deposition from fossil fuel combustion affects soil biogeochemistry and vegetation water use in freshwater systems, highlighting the cascading effects of anthropogenic activities on ecosystem functions. Such studies emphasize the need for integrated approaches to assess and mitigate the impacts of environmental changes on biogeochemical cycles. Furthermore, the coupling of biogeochemical processes with socio-economic factors is crucial for sustainable management. Ma et al. illustrate how soil biogeochemical cycles are foundational to ecosystem function and are influenced by human activities, such as agriculture and land use changes (Ma et al. 2021). Effective soil microbial community management can help address food security, biodiversity loss, and climate change challenges. Similarly, Peña et al. discuss modeling dissolved oxygen dynamics and hypoxia, emphasizing the importance of understanding biogeochemical processes in managing aquatic ecosystems (Peña et al. 2010).

The carbon cycle is one of the most critical biogeosociochemical cycles, involving the continuous movement of carbon among the atmosphere, oceans, soil, and living organisms. Carbon has various forms, including carbon dioxide (CO_2), organic matter, and carbonate minerals. Photosynthesis by plants and phytoplankton captures atmospheric CO_2, converting it into organic carbon, which consumers then utilize in the food web. Respiration, decomposition, and combustion release carbon into the atmosphere, completing the cycle. The balance of this cycle is vital for regulating Earth's climate, as excess CO_2 contributes to global warming and climate change (Raimi et al. 2021). The nitrogen cycle is another essential biogeosociochemical cycle that involves the transformation of nitrogen through various chemical forms and biological processes. Nitrogen is a critical nutrient for all living organisms, primarily in the atmosphere as inert N_2. Biological nitrogen fixation, carried out by certain bacteria and archaea, converts atmospheric nitrogen into reactive forms, such as ammonia, which plants can take up (Kuypers et al. 2018). The cycle also includes nitrification, denitrification, and ammonification, which further transform nitrogen compounds, ensuring their availability for biological uptake and maintaining ecosystem productivity (Singh and Bakshi 2013). Disruptions to the nitrogen cycle, often due to agricultural practices and industrial activities, can lead to environmental issues such as eutrophication and biodiversity loss (Leach et al. 2012). The water cycle, or hydrological cycle, describes the continuous movement of water on, above, and below the surface of the Earth. This cycle involves evaporation, condensation, precipitation, infiltration, and runoff. Water is essential for all life forms, and its availability directly influences ecosystem health and productivity. The water cycle is intricately linked to other biogeosociochemical cycles, particularly the carbon and nitrogen cycles, as water facilitates the transport and transformation of these elements (Mitchell 2023). Changes in precipitation patterns and water availability due to climate change can significantly impact biogeosociochemical processes and ecosystem functions (Raimi et al. 2021).

Human activities have profoundly impacted biogeosociochemical cycles, significantly altering their natural functioning. Industrialization, urbanization, and agricultural intensification have introduced excess nutrients, pollutants, and greenhouse gases into the environment, disrupting the delicate balance of these cycles. For instance, the excessive use of nitrogen fertilizers has led to increased nitrogen runoff into water bodies, causing eutrophication and the formation of dead zones, where aquatic life cannot survive (Leach et al. 2012). Similarly, deforestation and land-use changes have altered the carbon cycle by reducing carbon sequestration capacity and increasing carbon emissions (Raimi et al. 2021). The interplay between human activities and biogeosociochemical cycles is complex and multifaceted. Urbanization, for example, modifies the natural water cycle through increased impervious surfaces, leading to altered runoff patterns and reduced groundwater recharge (Mitchell 2023). Additionally, urban areas often experience higher concentrations of pollutants, affecting local biogeosociochemical processes and ecosystem health. Understanding these interactions is crucial for developing sustainable practices that mitigate negative impacts on biogeosociochemical cycles and promote environmental resilience (Mitchell 2023). Microbial communities play a vital role in mediating biogeosociochemical cycles, influencing the transformation and transport of essential elements. Microorganisms are involved in various processes, including nitrogen fixation, decomposition, and organic matter mineralization, which are vital for nutrient cycling (Griffiths and Philippot 2013). The dynamics of microbial populations can significantly affect the rates of biogeosociochemical processes, highlighting the importance of microbial ecology in understanding ecosystem functioning (Buchkowski et al. 2017).

Moreover, shifts in microbial community composition due to environmental changes can lead to decoupling of biogeosociochemical cycles, further complicating the response of ecosystems to disturbances (Mooshammer et al. 2017). Extremophilic microorganisms thrive in extreme environments and contribute to biogeosociochemical cycles, particularly in nutrient-poor ecosystems. These organisms have unique metabolic pathways that enable them to participate in the cycling of carbon, nitrogen, and sulfur, thereby influencing overall ecosystem productivity (González and Terrón 2021). The study of extremophiles provides insights into the resilience of biogeosociochemical processes under changing environmental conditions, emphasizing the need for further research in this area (González and Terrón 2021). The coupling of biogeosociochemical cycles is an emerging area of research that seeks to understand how different cycles interact and influence one another. For instance, carbon cycling is closely linked to nitrogen and phosphorus cycles, as these elements are often co-limited in ecosystems (Finzi et al. 2011). Understanding these interactions is crucial for addressing environmental challenges, as disruptions in one cycle can have cascading effects on others. Research into coupled biogeosociochemical cycles can inform management strategies to mitigate human activities' environmental impacts (Finzi et al. 2011).

1.4 Natural Versus Anthropogenic Processes

The distinction between natural and anthropogenic processes is crucial for understanding environmental dynamics and the implications of human activities on ecological systems (Table 1.2). Natural processes like erosion and weathering are fundamental to shaping landscapes and ecosystems. Erosion, for instance, is a natural phenomenon that involves the removal of soil and rock from one location and its deposition in another, driven by forces such as water, wind, and ice. This process plays a vital role in soil formation and nutrient cycling, which is essential for sustaining plant life and maintaining biodiversity (Lin 2014). Weathering refers to the breakdown of rocks and minerals through physical, chemical, and biological processes, contributing to soil formation and the geochemical cycles governing ecosystem health (Lin 2014).

Anthropogenic processes, in contrast, refer to human-induced changes that significantly alter the environment. These activities include pollution, deforestation, and urbanization, which profoundly affect natural ecosystems. Pollution can be categorized into various types, including biological, chemical, physical, and radiological. Each type of pollution poses unique threats to environmental health and biodiversity. Biological pollution, often resulting from introducing of invasive species, disrupts local ecosystems and can lead to the extinction of native species (Shahzad

Table 1.2 Natural versus anthropogenic processes

Process type	Focal approaches	Examples	Environmental consequences	Mitigation strategies
Natural processes	Natural mechanisms that shape the environment without human intervention	Erosion, weathering, and natural disasters (floods, earthquakes)	Maintains ecosystem balance and shapes landscapes over time	Minimal mitigation is needed, and natural processes are self-regulating
Erosion	The wearing away of soil and rock by wind, water, or ice	Riverbank erosion, coastal erosion	Loss of topsoil, sedimentation in water bodies, and habitat loss	Vegetative cover, contour plowing, and buffer zones
Weathering	The breakdown of rocks into smaller particles by physical, chemical, or biological means	Frost weathering, chemical weathering of limestone	Contributes to soil formation and nutrient cycling	Conservation of natural landscapes and soil management
Human activities	Actions taken by humans that impact the environment	Pollution, deforestation and urbanization	Degradation of ecosystems, loss of biodiversity, climate change	Regulations, sustainable practices, and restoration efforts
Pollution	Introduction of harmful substances into the environment	Air pollution from vehicles, and water pollution from runoff	It Impairs water and air quality affects health and ecosystems	Implementation of cleaner technologies and waste management
Deforestation	The large-scale removal of forests for agriculture, urban development, or logging	Amazon rainforest clearing logging activities	Habitat loss, increased carbon emissions, and soil degradation	Reforestation, sustainable forestry practices, and conservation policies
Urbanization	The growth of urban areas is due to population migration and development	Expansion of cities, and construction of infrastructure	Loss of habitats, increased pollution, and resource depletion	Smart growth planning and green infrastructure development

et al. 2021). Chemical pollution, primarily from industrial activities, contaminates air, water, and soil, leading to health risks for humans and wildlife (Shahzad et al. 2021). Physical pollution, such as plastic waste, contributes to habitat degradation and directly threatens marine and terrestrial organisms (Shahzad et al. 2021). Radiological pollution, while less common, can devastate ecosystems and human health, as evidenced by incidents such as the Chernobyl disaster (Shahzad et al. 2021). Deforestation is another significant anthropogenic process that has far-reaching environmental consequences. Clearing forests for agriculture, urban development, and logging reduces biodiversity and disrupts carbon storage and water cycles (Murugan et al. 2021). Forests play a critical role in sequestering carbon dioxide, and their removal contributes to increased greenhouse gas emissions, exacerbating climate change (Murugan et al. 2021). Furthermore, deforestation leads to soil erosion, loss of habitat for countless species, and disruption of local weather patterns, which can have cascading effects on regional and global climates (Murugan et al. 2021; Ogwu et al. 2024b). The interplay between natural and anthropogenic processes highlights the need for sustainable management practices considering ecological integrity and human development.

The consequences of human activities extend beyond immediate environmental degradation; they also impact social structures and human health. For instance, air pollution from industrial emissions and vehicular traffic has been linked to respiratory diseases and other health issues (Shahzad et al. 2021). Studies have shown that populations living in areas with high levels of air pollution experience increased rates of asthma, lung cancer, and cardiovascular diseases (Shahzad et al. 2021). Moreover, the socio-economic implications of environmental degradation are profound, as marginalized communities often bear the brunt of pollution and resource depletion, leading to disparities in health and quality of life. This highlights the importance of integrating environmental justice into discussions about anthropogenic impacts, ensuring that vulnerable populations are considered in policy-making and environmental management. In response to the challenges posed by anthropogenic processes, there has been a growing emphasis on environmental education and awareness. Educational initiatives that foster a connection to nature and promote pro-environmental behaviors are essential for cultivating a culture of sustainability (Vladova 2023). Research indicates that individuals with positive experiences in natural settings are more likely to engage in environmentally responsible behaviors (Rosa et al. 2018). This underscores the importance of early exposure to nature, as it can shape attitudes and behaviors towards environmental stewardship throughout a person's life (Rosa et al. 2018). Furthermore, ecological education can empower individuals and communities to advocate for sustainable practices and policies, thereby mitigating the negative impacts of human activities on the environment (Vladova 2023).

The role of businesses in addressing environmental challenges cannot be overlooked. Companies increasingly recognize the importance of adopting sustainable practices and innovations to minimize their ecological footprint (Moktadir et al. 2020). Environmental innovations, or eco-innovations, encompass many techniques, from developing cleaner production processes to implementing waste reduction strategies (Galliano and Nadel 2018). These innovations contribute to environmental sustainability and enhance a company's competitiveness in the market (Moktadir et al. 2020). Integrating environmental considerations into business strategies is essential for fostering a circular economy, where resources are reused and recycled, reducing the demand for new raw materials and minimizing waste (Moktadir et al. 2020). Legislation is critical in shaping the relationship between human activities and environmental outcomes. Strong regulatory frameworks can incentivize businesses to adopt sustainable practices and hold them accountable for their environmental impacts (Moktadir et al. 2020). For instance, policies that promote renewable energy sources and penalize pollution can drive industries toward more sustainable operations (Moktadir et al. 2020). Public awareness and advocacy can also influence policy decisions, leading to more stringent environmental regulations that protect ecosystems and public health (Vladova 2023). The interplay between legislation, business practices, and public engagement is vital for creating a sustainable future.

1.5 Environmental Impact Assessments (EIA)

Environmental Impact Assessments (EIAs) are a critical tool in the decision-making process for projects that may significantly affect the environment. The primary purpose of an EIA is to evaluate the potential environmental impacts of proposed projects before they are carried out, thus facilitating informed decision-making that considers ecological sustainability. EIAs are essential for identifying adverse environmental effects and proposing measures to mitigate these impacts, promoting sustainable development practices (Vanclay 2015; Mahmoudi et al. 2018; Thamir et al. 2019). The importance of EIAs extends beyond mere compliance with regulations; they also foster public participation and transparency, allowing stakeholders to voice their concerns and contribute to the planning process (Stradere and Didenko 2020; Bishoge 2020).

The steps in conducting an EIA are systematic and typically include screening, scoping, assessment, reporting, review, decision-making, and monitoring (Table 1.3; Fig. 1.3). Screening determines whether a project requires a full EIA based on its potential environmental impact. Scoping identifies the key issues and impacts that need to be assessed. The assessment phase involves a detailed analysis of the possible ecological effects, including direct and indirect impacts. A report is compiled, summarizing the findings and proposed mitigation measures. This report undergoes a review process, where stakeholders can provide feedback before making a final decision. Finally, monitoring ensures compliance with the proposed mitigation measures and assesses the actual impacts of the project post-implementation (Matemilola and Salami 2020; Wilson et al. 2017).

Studies across various sectors illustrate the diverse applications and outcomes of EIAs. For instance, in the renewable energy sector, EIAs have been instrumental in assessing the environmental impacts of wind and solar energy projects. Basit et al. (2022) highlighted that integrating various assessment methods has improved the

Table 1.3 Steps in conducting an EIA

Step/stage	Focus	Key activities	Outputs	Importance
Screening	Determine if an EIA is necessary for the proposed project	Review project details and assess potential impacts	Decision on whether an EIA is required	Prevents unnecessary assessments for low-impact projects
Scoping	Identify key issues, stakeholders, and the scope of the EIA study	Consult with stakeholders and define assessment boundaries	A scoping report outlining key environmental issues	Focuses the assessment on significant impacts
Impact assessment	Evaluate the potential environmental effects of the project	Conduct field studies, analyze data, and predict impacts	Impact assessment report detailing findings	Identifying and quantifying has implications for informed decision-making
Mitigation	Propose measures to avoid, minimize, or offset adverse impacts identified	Develop mitigation strategies and assess feasibility	Mitigation plan outlining measures to reduce impacts	Ensures that negative impacts are addressed and minimized
Reporting	Compile the findings and recommendations into a comprehensive EIA report	Draft the EIA report, and include public comments and responses	EIA report submitted to decision-makers	Provides a transparent record of the assessment process
Review and decision-making	Authorities review the EIA report and decide on project approval	Evaluate compliance with regulations and make recommendations	Decision on whether to approve or reject the project	Ensures that all significant impacts have been considered
Monitoring	Implement a monitoring plan to ensure compliance with mitigation measures post-approval	Establish monitoring protocols and conduct regular assessments	Monitoring reports detailing compliance and effectiveness	Ensures ongoing protection of the environment during project execution

effectiveness of EIAs in this sector, ensuring that renewable energy developments align with environmental sustainability goals. Similarly, the EIA for the Halfaya Oilfield Project in Iraq demonstrated the necessity of evaluating both the ecological and socio-economic impacts of oil extraction activities, emphasizing the importance of comprehensive assessments in resource extraction projects (Thamir et al. 2019; Pazoki et al. 2015; Izah et al. 2024).

In the construction sector, applying life cycle impact assessment (LCIA) methodologies within EIAs has gained traction. Liu et al. (2015) argued that incorporating LCIA allows for a more holistic view of environmental impacts throughout a project's

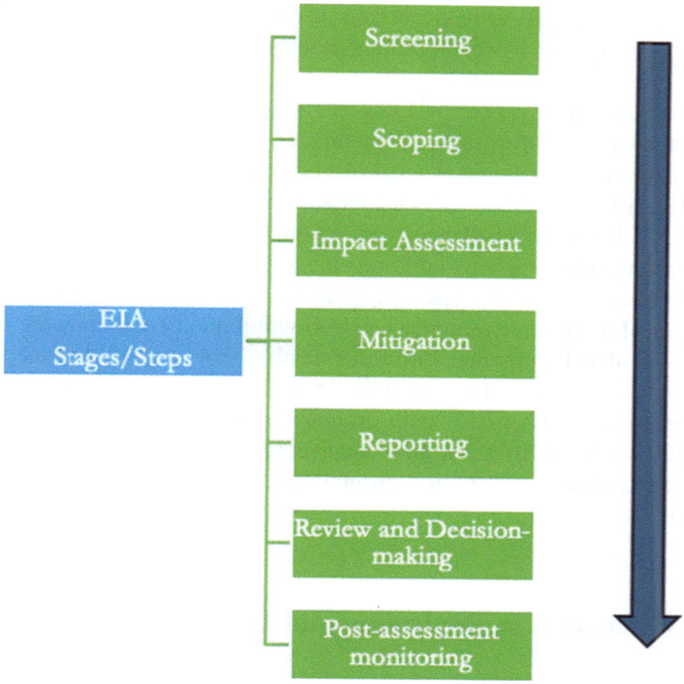

Fig. 1.3 Environmental impact assessment steps

life cycle, from planning to decommissioning. This approach not only aids in regulatory compliance but also enhances the understanding of long-term ecological consequences, thereby informing better project design and management practices (Hong et al. 2013; Bond et al. 2012). The construction industry's reliance on EIAs underscores the need for robust frameworks that can adapt to the complexities of modern development projects.

The agricultural sector also benefits from EIAs, particularly in assessing the impacts of large-scale agrarian projects on local ecosystems and communities. The integration of resilience assessments into EIAs, as discussed by Mahmoudi et al. (2018), can enhance the ability to adapt to climate change and other environmental stressors, ensuring that agricultural practices are sustainable and resilient. This approach aligns with global sustainability goals and highlights the role of EIAs in promoting environmentally responsible farming practices. In urban planning, the role of EIAs is equally significant. Stradere and Didenko (2020) emphasized that EIAs contribute to territorial planning by assessing the potential impacts of urban development projects on local environments and communities. This is particularly relevant in rapidly urbanizing regions, where the pressure on natural resources and ecosystems is intense. EIAs help mitigate negative impacts and promote sustainable urban development by incorporating environmental considerations into urban planning.

Moreover, the effectiveness of EIAs can vary significantly based on regional and national contexts. For instance, Vietnam's EIA system has undergone numerous reforms to improve environmental performance, reflecting the government's commitment to integrating environmental considerations into development planning (Clarke and Vu 2021). However, challenges remain, including the need for greater stakeholder engagement and capacity building among local authorities to ensure EIAs are conducted effectively and transparently (Hamid and Long 2017; Machaka 2020). Integrating health impact assessments (HIAs) within the EIA framework is another important area that is growing. Linzalone et al. (2014) argued that incorporating health considerations into EIAs can enhance public health outcomes by identifying potential health risks associated with proposed projects. This integration not only broadens the scope of EIAs but also underscores the interconnectedness of environmental and public health issues, promoting a more holistic approach to impact assessment.

1.6 Key Concepts in Environmental Interactions

Environmental interactions are complex and multifaceted, encompassing various key concepts essential for understanding ecosystem dynamics (Table 1.4). Among these concepts, feedback loops, thresholds and tipping points, and the sustainability and resilience of ecosystems stand out as critical components. Feedback loops, both positive and negative, play a significant role in regulating ecosystem processes. Positive feedback loops can amplify changes, leading to rapid shifts in ecosystem states, while negative feedback loops tend to stabilize systems by counteracting changes. For instance, in climate change, the melting of polar ice reduces the Earth's albedo, leading to increased absorption of solar radiation and further warming—a classic example of a positive feedback loop (Dell et al. 2019). Conversely, negative feedback mechanisms, such as regulating predator–prey dynamics, help maintain ecological balance, ensuring that populations do not exceed their carrying capacity (Sasaki et al. 2015).

The concept of thresholds and tipping points is closely related to feedback loops, as it describes critical junctures where small changes can lead to significant and often irreversible shifts in ecosystem structure and function. Tipping points are particularly relevant in climate change, where gradual temperature increases can trigger abrupt changes in ecosystems, such as the dieback of forests or the collapse of coral reefs (Chapin et al. 2010). Understanding these thresholds is crucial for effective ecosystem management, as it allows for the identification of critical limits beyond which recovery may be impossible (Huang and Xia 2019). For example, biodiversity loss can lower an ecosystem's resilience, making it more susceptible to crossing tipping points (Tang et al. 2022). This interplay between feedback loops and

Table 1.4 Key concepts in environmental interactions

Concept	Key features	Types	Impacts on ecosystems	Examples
Feedback loops	Processes where the output of a system influences its operation	Positive feedback amplifies change while negative feedback dampens change	Stabilizes or destabilizes ecosystems; influences climate and species dynamics	Positive: melting ice reduces albedo, increasing warming Negative: Plant growth absorbs CO_2 and stabilizes the climate
Thresholds	Points at which a relatively small change in conditions can have significant effects	Ecological thresholds: focuses on shifts in species composition whereas social-ecological thresholds: seek to address changes in human behavior and policies	Crossing thresholds can lead to irreversible changes in ecosystems	Coral reef bleaching at higher temperature thresholds; biodiversity loss beyond habitat loss
Tipping points	Specific thresholds that, once crossed, lead to abrupt changes in system behavior	Social tipping points addresses shift in societal norms and ecological tipping points focuses on changes in species dominance	It can lead to sudden ecosystem shifts, altering functions and services	Amazon rainforest turning into savanna after deforestation; Arctic Sea ice loss leading to further warming
Sustainability	The capacity to endure in a relatively constant state over time; balance of social, economic, and environmental needs	Environmental sustainability is focused on maintaining ecosystem health whereas economic sustainability ensures the security of resources for future generations	Promotes long-term health of ecosystems and human societies	Sustainable agriculture practices that maintain soil health and productivity
Resilience	The ability of an ecosystem to absorb disturbances and still maintain its basic structure and functions	Ecological resilience focuses on the ability to recover from disturbances while social resilience addresses the ability of communities to cope with change	Enhances an ecosystem's capacity to withstand shocks, ensuring stability	Wetlands restored after floods; forests regrowing after wildfires

tipping points underscores the importance of monitoring and managing ecosystems to prevent undesirable shifts.

Sustainability and resilience are foundational concepts in environmental science, reflecting the capacity of ecosystems to withstand disturbances while maintaining their essential functions and services. Resilience refers to the ability of an ecosystem to absorb shocks and recover from disturbances, while sustainability encompasses the long-term maintenance of ecosystem health and functionality (Parlee et al. 2012). The resilience of ecosystems is influenced by various factors, including biodiversity, ecological interactions, and feedback mechanisms (McColl-Kennedy 2023). For instance, ecosystems with high biodiversity tend to exhibit greater resilience, as diverse species can fulfill multiple ecological roles, thereby enhancing overall stability (Zampieri et al. 2021). Furthermore, the concept of adaptive management emphasizes the need for flexible and responsive strategies that can accommodate changing environmental conditions and uncertainties (Hogan et al. 2021). In addition to resilience, the sustainability of ecosystems is increasingly recognized as a critical goal in the face of global environmental changes. Sustainable practices balance ecological health with human needs, ensuring that ecosystems can continue providing essential services, such as clean water, food production, and climate regulation (Sasaki et al. 2015). Integrating sustainability principles into ecosystem management involves recognizing the interconnectedness of social, economic, and environmental systems and fostering collaborative approaches that engage multiple stakeholders (Selkoe et al. 2015). For example, community-based resource management initiatives have shown promise in promoting sustainable practices while enhancing local resilience (Hogan et al. 2021).

Feedback loops, thresholds, and resilience are not isolated concepts but are interconnected within the broader framework of ecosystem dynamics. The interactions between these elements can lead to complex behaviors that challenge traditional management approaches. For instance, positive feedback loops can push ecosystems toward tipping points, while effective management strategies can enhance resilience by fostering negative feedback mechanisms (Liu et al. 2023). This complexity necessitates a systems-thinking approach considering the interplay of various ecological, social, and economic factors (Green et al. 2014). By understanding these interactions, policymakers and managers can develop more effective strategies for maintaining ecosystem health and resilience in the face of ongoing environmental changes. The role of traditional ecological knowledge (TEK) in understanding thresholds and resilience is also significant. Indigenous communities often possess deep insights into local ecosystems and their dynamics, which can inform contemporary management practices (Williams et al. 2014). TEK can provide valuable information about historical ecosystem changes, helping identify potential tipping points and adaptive strategies. Integrating TEK with scientific research can enhance the resilience of ecosystems by fostering a more holistic understanding of ecological dynamics and promoting culturally relevant management practices (DeClerck et al. 2016).

Moreover, ecological thresholds are crucial for understanding the potential for regime shifts in ecosystems. Regime shifts occur when ecosystems transition from one stable state to another, often due to cumulative stressors that exceed critical thresholds (Ramaano et al. 2021). These shifts can have profound implications for ecosystem services and human well-being, highlighting the need for proactive management strategies that prevent crossing these thresholds (Winn and Pogutz

2013). For example, the degradation of coral reefs due to climate change and pollution can shift from a coral-dominated ecosystem to an algal-dominated state, with significant consequences for marine biodiversity and fisheries (Admiraal et al. 2013).

In the climate change context, ecosystems' resilience is increasingly tested by extreme weather events and shifting climatic conditions. Research indicates that highly resilient ecosystems are better equipped to withstand such disturbances, maintaining their functionality and services (Asmamaw et al. 2015). For instance, studies have shown that resilient forests can recover more quickly from droughts and wildfires, highlighting the importance of biodiversity and ecological interactions in promoting resilience (Hunt and Batelaan 2016). As climate change continues to pose challenges to ecosystems worldwide, understanding the factors contributing to resilience will be essential for effective management and conservation efforts. Integrating sustainability and resilience into ecosystem management frameworks is vital for addressing the complex challenges posed by environmental change. This involves adopting adaptive management practices that allow flexibility and responsiveness to changing conditions (Tavoni and İriş 2020). Ecosystems can better cope with disturbances and maintain their essential functions by fostering resilience through biodiversity conservation, habitat restoration, and sustainable resource management (Biggs et al. 2012). Furthermore, promoting stakeholder engagement and collaboration can enhance the effectiveness of management strategies, ensuring that diverse perspectives and knowledge are incorporated into decision-making processes (Fabricius et al. 2022).

1.7 Conclusion

The concept of environmental systems encompasses the intricate interplay between natural and human-made systems, highlighting the significance of understanding their complexities and interactions. Environmental systems are defined by their components—such as ecosystems and biogeosociochemical cycles—and the dynamic processes that shape them. This understanding is essential, as it provides a framework for analyzing how human activities influence these systems. The confrontation of unprecedented environmental challenges, such as climate change, pollution, and biodiversity loss, makes the need for a comprehensive approach to studying ecological processes increasingly apparent. Integrating sustainability thinking into educational curricula, robust analytical frameworks, and interdisciplinary collaboration will enhance the ability to address these challenges and foster a more sustainable future. Ecosystems play a crucial role in this discussion, serving as the foundational units of environmental systems. Their structure and function are governed by the flow of energy through trophic levels and the diversity of species that inhabit them. Biodiversity is not merely a collection of species but a critical component that underpins ecosystem stability and resilience. As anthropogenic pressures continue to mount, it is vital to prioritize biodiversity conservation and maintaining ecosystem functions. Future research should focus on developing holistic approaches to ecosystem management and restoration, emphasizing the interdependencies that sustain ecological health. By understanding the intrinsic value of ecosystems and their services, the complexities of human–environment interactions and implementing effective conservation strategies can be better assessed.

Biogeosociochemical cycles, including the carbon, nitrogen, and water cycles, are integral to sustaining life on Earth. These cycles regulate essential processes that support biodiversity and ecosystem functions. However, human activities have disrupted these natural processes, leading to significant environmental challenges. For instance, the excessive release of carbon into the atmosphere has intensified climate change, while pollution has adversely affected water quality and ecosystem health. Understanding the delicate balance of these cycles and the role of microbial communities is crucial for developing effective strategies to mitigate human impacts and promote environmental resilience. By recognizing the interconnectedness of biogeosociochemical cycles and human activities, more informed approaches to environmental sustainability can be devised. The contrast between natural and anthropogenic processes underscores the complexities inherent in environmental systems. While natural processes are vital for maintaining ecological balance, human activities often threaten this equilibrium. A multifaceted approach is necessary to address these challenges, involving education, innovation, and effective policymaking. EIAs exemplify such an approach, serving as a vital mechanism for evaluating potential environmental impacts of projects. They promote sustainable development and ensure stakeholder participation in decision-making processes. As ecological challenges evolve, the role of EIAs in fostering sustainable practices and safeguarding ecosystems remains essential. By deepening the understanding of the interactions between natural and human systems, a more sustainable and equitable future that ensures the health of the planet for generations to come will be achieved.

References

Admiraal JF, Wossink A, de Groot WT, de Snoo GR (2013) More than total economic value: how to combine economic valuation of biodiversity with ecological resilience. Ecol Econ 89:115–122. https://doi.org/10.1016/j.ecolecon.2013.02.009

Akudugu MA, Ogwu MC (2024) Sustainable development policies and interventions: a bibliometric analysis of the contributions of the academic community. J Clean Prod (Article # 139919). https://doi.org/10.1016/j.jclepro.2023.139919

Asmamaw M, Ambellu A, Tiku S (2015) Resilience of Ecosystems to climate change. Am J Environ Protect 12(1):325–333. https://doi.org/10.11648/j.ajep.20150406.19

Balvanera P, Kremen C, Martínez-Ramos M (2005) Applying community structure analysis to ecosystem function: examples from pollination and carbon storage. Ecol Appl 15(1):360–375. https://doi.org/10.1890/03-5192

Basit A, Hassan M, Kanwal S, Anwar M, Kazmi SAA, Ansari AA (2022) Environmental impact assessments of the renewable energy technologies adaptation. Pak J Eng Technol 5(2):100–103. https://doi.org/10.51846/vol5iss2pp100-103

Biggs R, Schlüter M, Biggs D, Bohensky EL, BurnSilver S, Cundill G, Dakos V, Daw TM, Evans LS, Kotschy K, Leitch AM, Quinlan A, Raudsepp-Hearne C, Robards MD, Schoon ML, Schultz

L, West PC (2012) Toward principles for enhancing the resilience of ecosystem services. Annu Rev Environ Resour 37(1):421–448. https://doi.org/10.1146/annurev-environ-051211-123836

Bishoge OB (2020). The significance of environmental agencies in the environmental impact assessment process. Earth Environ Sci Res Rev 3(4). https://doi.org/10.33140/eesrr.03.04.05

Bond A, Morrison-Saunders A, Pope J (2012) Sustainability assessment: the state of the art. Impact Assess Project Appraisal 30(1):53–62. https://doi.org/10.1080/14615517.2012.661974

Buchkowski RW, Bradford MA, Grandy AS, Schmitz OJ, Wieder WR (2017) Applying population and community ecology theory to advance understanding of belowground biogeochemistry. Ecol Lett 20(2):231–245. https://doi.org/10.1111/ele.12712

Burritt RL, Schaltegger S, Christ KL (2023) Environmental management accounting–developments over the last 20 years from a framework perspective. Aust Account Rev 33(4):336–351. https://doi.org/10.1111/auar.12407

Cadotte MW, Dinnage R, Tilman D (2012) Phylogenetic diversity promotes ecosystem stability. Ecology 93(sp8):S223–S233. https://doi.org/10.1890/11-0426.1

Cardinale BJ, Gross K, Fritschie K, Flombaum P, Fox JW, Rixen C, Van Ruijven J, Reich PB, Scherer-Lorenzen M, Wilsey BJ (2013) Biodiversity simultaneously enhances the production and stability of community biomass, but the effects are independent. Ecology 94(8):1697–1707. https://doi.org/10.1890/12-1334.1

Cardinale BJ, Matulich KL, Hooper DU, Byrnes JE, Duffy E, Gamfeldt L, Balvanera P, O'connor MI, Gonzalez A (2011) The functional role of producer diversity in ecosystems. Am J Bot 98(3):572–592. https://doi.org/10.3732/ajb.1000364

Chapin FS, Carpenter SR, Kofinas GP, Folke C, Abel N, Clark WC, Olsson P, Smith DMS, Walker B, Young OR, Berkes F, Biggs R, Grove JM, Naylor RL, Pinkerton E, Steffen W, Swanson FJ (2010) Ecosystem stewardship: sustainability strategies for a rapidly changing planet. Trends Ecol Evol 25(4):241–249. https://doi.org/10.1016/j.tree.2009.10.008

Chen L, Jiang L, Jing X, Wang J, Shi Y, Chu H, He JS (2021) Above-and belowground biodiversity jointly drive ecosystem stability in natural alpine grasslands on the Tibetan Plateau. Glob Ecol Biogeogr 30(7):1418–1429. https://doi.org/10.1111/geb.13307

Clarke BD, Vu CC (2021) EIA effectiveness in Vietnam: key stakeholder perceptions. Heliyon 7(2). https://doi.org/10.1016/j.heliyon.2021.e06157

Correia AM, Lopes LF (2023) Revisiting biodiversity and ecosystem functioning through the lens of complex adaptive systems. Diversity 15(8):895. https://doi.org/10.3390/d15080895

Currie DJ, Smith C, Jagals P (2018) The application of system dynamics modelling to environmental health decision-making and policy-a scoping review. BMC Public Health 18:1–11. https://doi.org/10.1186/s12889-018-5318-8

DeClerck FA, Jones SK, Attwood S, Bossio D, Girvetz E, Chaplin-Kramer B, Enfors E, Fremier AK, Gordon LJ, Kizito F, Noriega IL, Matthews N, McCartney M, Meacham M, Noble A, Quintero M, Remans R, Soppe R, Willemen L, Wood SLR, Zhang W (2016) Agricultural ecosystems and their services: the vanguard of sustainability? Curr Opin Environ Sustain 23:92–99. https://doi.org/10.1016/j.cosust.2016.11.016

Dell JE, Salcido DM, Lumpkin W, Richards LA, Pokswinski SM, Loudermilk EL, O'Brien JJ, Dyer LA (2019) Interaction diversity maintains resiliency in a frequently disturbed ecosystem. Front Ecol Evol 7:145. https://doi.org/10.3389/fevo.2019.00145

Devlin S, Saarenheimo J, Syväranta J, Jones R (2015) Top consumer abundance influences lake methane efflux. Nat Commun 6(1). https://doi.org/10.1038/ncomms9787

Djekic I, Pojić M, Tonda A, Putnik P, Bursać Kovačević D, Režek-Jambrak A, Tomasevic I (2019) Scientific challenges in performing life-cycle assessment in the food supply chain. Foods 8(8):301. https://doi.org/10.3390/foods8080301

Fabricius C, Novellie P, Ringler C, Uhlenbrook S, Wright D (2022) Resilience in agro-ecological landscapes: process principles and outcome indicators, vol 4. International Water Management Institute (IWMI).; CGIAR Research Program on Water, Land and Ecosystems (WLE). https://doi.org/10.5337/2022.206

Fahey RT, Atkins JW, Gough CM, Hardiman BS, Nave LE, Tallant JM, Nadehoffer KJ, Vogel C, Scheuermann CM, Stuart-Haëntjens E, Haber LT, Fotis AL, Ricart R, Curtis PS (2019) Defining a spectrum of integrative trait-based vegetation canopy structural types. Ecol Lett 22(12):2049–2059. https://doi.org/10.1111/ele.13388

Fet AM, Knudson H (2021) An approach to sustainability management across systemic levels: the capacity-building in sustainability and environmental management model (CapSEM-model). Sustainability 13(9):4910. https://doi.org/10.3390/su13094910

Finzi AC, Cole JJ, Doney SC, Holland EA, Jackson RB (2011) Research frontiers in the analysis of coupled biogeosociochemical cycles. Front Ecol Environ 9(1):74–80. https://doi.org/10.1890/100137

Galliano D, Nadel S (2018) Environmental innovations and firms' organizational changes: what kind of complementarity? Evidence from French industrial firms. Rev Econ Ind 164:37–71. https://doi.org/10.4000/rei.7600

Galloway JN, Townsend AR, Erisman JW, Bekunda M, Cai Z, Freney JR, Martinelli LA, Seitzinger SP, Sutton MA (2008) Transformation of the nitrogen cycle: recent trends, questions, and potential solutions. Science, 320(5878):889–892. https://doi.org/10.1126/science.1136674

García-Palacios P, Gross N, Gaitán J, Maestre FT (2018) Climate mediates the biodiversity–ecosystem stability relationship globally. Proc Natl Acad Sci 115(33):8400–8405. https://doi.org/10.1073/pnas.1800425115

Gardner G, Burton W, Sinclair M, Bryant M (2023) Interventions to strengthen environmental sustainability of school food systems: narrative scoping review. Int J Environ Res Public Health 20(11):5916. https://doi.org/10.3390/ijerph20115916

Goldman A, Graham E, Crump A, Kennedy D, Romero E, Anderson C et al (2017) Biogeochemical cycling at the aquatic–terrestrial interface is linked to parafluvial hyporheic zone inundation history. Biogeosciences 14(18):4229–4241. https://doi.org/10.5194/bg-14-4229-2017

González AG, y Terrón RP (2021) Importance of extremophilic microorganisms in biogeosociochemical cycles. GSC Adv Res Rev 9(1):082–093. https://doi.org/10.30574/gscarr.2021.9.1.0229

Green OO, Garmestani AS, Hopton ME, Heberling MT (2014) A multi-scalar examination of law for sustainable ecosystems. Sustainability 6(6):3534–3551. https://doi.org/10.3390/su6063534

Griffiths BS, Philippot L (2013) Insights into the resistance and resilience of the soil microbial community. FEMS Microbiol Rev 37(2):112–129. https://doi.org/10.1111/j.1574-6976.2012.00343.x

Gruber N (2011) Warming up, turning sour, losing breath: ocean biogeochemistry under global change. Philos Trans Royal Soc Math Phys Eng Sci 369(1943):1980–1996. https://doi.org/10.1098/rsta.2011.0003

Hamid MA, Long KQ (2017) An assessment of environmental impacts assessment (EIA) in Malaysia. In: SHS web of conferences, vol 36. EDP Sciences, p 00018. https://doi.org/10.1051/shsconf/20173600018

Hogan KF, Nash KL, Bennett E (2021) Adaptive management of ecosystem services for multisystemic resilience. Multisyst Resil 725-43. https://doi.org/10.1093/oso/9780190095888.003.0038

Hong T, Ji C, Jeong K, Park J (2013) Development of the assessment framework for the environmental impacts in construction. J Construct Eng Project Manage 3(3):1–9. https://doi.org/10.6106/jcepm.2013.3.3.001

Huang K, Xia J (2019) High ecosystem stability of evergreen broadleaf forests under severe droughts. Glob Change Biol 25(10):3494–3503. https://doi.org/10.1111/gcb.14748

Hunt RJ, Hayashi M, Batelaan O (2016) Ecohydrology and its relation to integrated groundwater management. Integr Groundw Manage Concepts Approaches Challenges 297–312. https://doi.org/10.1007/978-3-319-23576-9_12

Inostroza L, Zepp H, Pickett S, de Groot R (2020) Ecosystem function. In: Life on land. Encyclopedia of the UN sustainable development goals, pp 282–289. https://doi.org/10.1007/978-3-319-71065-5_21-1

Izah SC (2018) Ecosystem of the Niger-Delta region of Nigeria: potentials and threats. Biodiv Int J 2(4):338–345

Izah SC, Aigberua AO (2023) Potential threats and possible conservation strategies of biodiversity in Niger Delta Region of Nigeria. In: Sustainable utilization and conservation of Africa's biological resources and environment. Springer Nature Singapore, Singapore, pp 151–176. https://doi.org/10.1007/978-981-19-6974-4_6

Izah SC, Seiyaboh EI (2018a) Changes in the protected areas of Bayelsa state, Nigeria. Int J Mol Evol Biodiv 8(1):1–11

Izah SC, Seiyaboh EI (2018b) Challenges of wildlife with therapeutic properties in Nigeria; a conservation perspective. Int J Avian Wildl Biol 3(4):252–257

Izah SC, Iyiola AO, Yarkwan B, Richard G (2023) Impact of air quality as a component of climate change on biodiversity-based ecosystem services. In: Visualization techniques for climate change with machine learning and artificial intelligence. Elsevier, pp 123–148. ISBN: 978-0-323-99714-0

Izah SC, Ogwu MC, Hamidifar H (2024) Biomonitoring of Pollutants in the Global South. Springer Nature, Switzerland. https://doi.org/10.1007/978-981-97-1658-6

Kareksela S, Haapalehto T, Juutinen R, Matilainen R, Tahvanainen T, Kotiaho JS (2015) Fighting carbon loss of degraded peatlands by jump-starting ecosystem functioning with ecological restoration. Sci Total Environ 537:268–276. https://doi.org/10.1016/j.scitotenv.2015.07.094

Kuypers MM, Marchant HK, Kartal B (2018) The microbial nitrogen-cycling network. Nat Rev Microbiol 16(5):263–276. https://doi.org/10.1038/nrmicro.2018.9

Lanning M, Wang L, Scanlon T, Vadeboncoeur M, Adams M, Epstein H et al (2019) Intensified vegetation water use under acid deposition. Sci Adv 5(7). https://doi.org/10.1126/sciadv.aav5168

LaRue EA, Hardiman BS, Elliott JM, Fei S (2019) Structural diversity as a predictor of ecosystem function. Environ Res Lett 14(11):114011. https://doi.org/10.1088/1748-9326/ab49bb

Leach AM, Galloway JN, Bleeker A, Erisman JW, Kohn R, Kitzes J (2012) A nitrogen footprint model to help consumers understand their role in nitrogen losses to the environment. Environ Dev 1(1):40–66. https://doi.org/10.1016/j.envdev.2011.12.005

Lin YP (2014) Environments: a new cutting-edge international and interdisciplinary scholarly open access journal. Environments 1(1):1–3. https://doi.org/10.3390/environments1010001

Linzalone N, Assennato G, Ballarini A, Cadum E, Cirillo M, Cori L, De Maio F, Musmeci L, Natali M, Rieti S, Soggiu ME, Bianchi F (2014) Health Impact Assessment practice and potential for integration within environmental impact and strategic environmental assessments in Italy. Int J Environ Res Public Health 11(12):12683–12699. https://doi.org/10.3390/ijerph111212683

Liu KFR, Chiu SY, Yeh PC, Kuo JY (2015) Case study of using life cycle impact assessment in environmental impact assessment. Int J Environ Sci Dev 6(11):849. https://doi.org/10.7763/ijesd.2015.v6.711

Liu X, Li Z, Wang T, Cheng S, Zheng X, Jia L, Liu B, Li J, Miao Z (2023) The change of ecosystem resilience and its response to economic factors in Yulin, China. Environ Res Commun 5(4):045006. https://doi.org/10.1088/2515-7620/acc5dc

Ma B, Stirling E, Liu Y, Zhao K, Zhou J, Singh B et al (2021) Soil biogeochemical cycle couplings inferred from a function-taxon network. Research 2021. https://doi.org/10.34133/2021/7102769

Machaka RK (2020) The improved model of the method, rights, and resources (MRR) for the evaluation of the EIA system: Revising the sustainability indicators. IntechOpen. https://doi.org/10.5772/intechopen.88747

Mahmoudi H, Sayahnia R, Esmaeilzadeh H, Azadi H (2018) Integrating resilience assessment in environmental impact assessment. Integr Environ Assess Manag 14(5):567–570. https://doi.org/10.1002/ieam.4075

Martínez-Espinosa R (2020) Microorganisms and their metabolic capabilities in the context of the biogeochemical nitrogen cycle at extreme environments. Int J Mol Sci 21(12):4228. https://doi.org/10.3390/ijms21124228

Matemilola S, Salami H (2020) Environmental assessment 1–4. https://doi.org/10.1007/978-3-030-02006-4_521-1

McColl-Kennedy JR, Breidbach CF, Green T, Zaki M, Gain AM, van Driel ML (2023) Cultivating resilience for sustainable service ecosystems in turbulent times: evidence from primary health care. J Serv Mark 37(9):1167–1185. https://doi.org/10.1108/jsm-03-2023-0100

Mitchell C (2023) Editorial: biogeochemical dynamics in urban systems: interactions, feedbacks and cumulative effects. Front Environ Sci 11. https://doi.org/10.3389/fenvs.2023.1338537

Moktadir MA, Kumar A, Ali SM, Paul SK, Sultana R, Rezaei J (2020) Critical success factors for a circular economy: implications for business strategy and the environment. Bus Strateg Environ 29(8):3611–3635. https://doi.org/10.1002/bse.2600

Mooshammer M, Hofhansl F, Frank AH, Wanek W, Hämmerle I, Leitner S, Schnecker J, Wild B, Watzka M, Keiblinger KM, Zechmeister-Boltenstern S, Richter A (2017) Decoupling of microbial carbon, nitrogen, and phosphorus cycling in response to extreme temperature events. Sci Adv 3(5):e1602781. https://doi.org/10.1126/sciadv.1602781

Mouillot D, Villéger S, Scherer-Lorenzen M, Mason NW (2011) Functional structure of biological communities predicts ecosystem multifunctionality. PLoS ONE 6(3):e17476. https://doi.org/10.1371/journal.pone.0017476

Murugan M, Tee L, Oh K (2021) Evaluation of the environment impact of extraction of bioactive compounds from darcyodes rostrata using deep eutectic solvent (des) using life cycle assessment (lca). J Phys Conf Ser 2120(1):012005. https://doi.org/10.1088/1742-6596/2120/1/012005

Ogwu MC (2019) Towards sustainable development in Africa: the challenge of urbanization and climate change adaptation. In: Cobbinah PB, Addaney M (eds) The geography of climate change adaptation in urban Africa. Springer Nature, Switzerland, pp 29–55. https://doi.org/10.1007/978-3-030-04873-0_2

Ogwu MC, Osawaru ME, Ahana CM (2014) Challenges in conserving and utilizing plant genetic resources (PGR). Int J Genet Mol Biol 6(2):16–22. https://doi.org/10.5897/IJGMB2013.0083

Ogwu MC, Iyiola AO, Izah SC (2023a) Overview of African biological resources and environment. In: Sustainable utilization and conservation of Africa's biological resources and environment. Springer Nature Singapore, Singapore, pp 1–31. https://doi.org/10.1007/978-981-19-6974-4_1

Ogwu MC, Osawaru ME, Owie MO (2023b) Effects of storage at room temperature on the food components of three Cocoyam species (*Colocasia esculenta, Xanthosoma atrovirens*, and *X. sagittifolium*). Food Stud Interdisc J 13(2):59–83. https://doi.org/10.18848/2160-1933/CGP/v13i02/59-83

Ogwu MC, Izah SC, Dessureault-Rompré J, Gasparatos D (2024a) Sustainable soil systems in global south. Springer Nature, Switzerland. https://doi.org/10.1007/978-981-97-5276-8

Ogwu MC, Onah TP, Kosoe EA, Mumuni E, Akolgo-Azupogo H, Obahiagbon EG, Izah SC, Imarhiagbe O (2024b) Factors driving the acceptance of genetically modified food crops in Ghana. Food Safety Health. https://doi.org/10.1002/fsh3.12031

Parlee BL, Geertsema K, Willier A (2012) Social-ecological thresholds in a changing boreal landscape: insights from Cree knowledge of the Lesser Slave Lake region of Alberta, Canada. Ecol Soc 17(2). https://doi.org/10.5751/es-04410-170220

Pazoki M, Pari MA, Dalaei P, Ghasemzadeh R (2015) Environmental impact assessment of a water transfer project. Jundishapur J Health Sci 7(3). https://doi.org/10.5812/jjhs.27238v2

Peña M, Katsev S, Oğuz T, Gilbert D (2010) Modeling dissolved oxygen dynamics and hypoxia. Biogeosciences 7(3):933–957. https://doi.org/10.5194/bg-7-933-2010

Peng L, Jia R (2023) Exploring the interplay of the physical environment and organizational climate in innovation. Sustainability 15(20):15013. https://doi.org/10.3390/su152015013

Raimi MO, Abiola I, Alima O, Omini DE (2021) Exploring how human activities disturb the balance of biogeosociochemical cycles: evidence from the carbon, nitrogen and hydrologic cycles. Nitrogen Hydrol Cycles. https://doi.org/10.36956/rwae.v2i3.426

Ramaano R, Moyo H, Mugwedi L, Tsvuura Z (2021) Survive or perish: resprouting ability of repeatedly cut *Chromolaena odorata* in a reforestation site. Afr J Ecol 59(3). https://doi.org/10.1111/aje.12870

Rosa CD, Profice CC, Collado S (2018) Nature experiences and adults' self-reported pro-environmental behaviors: the role of connectedness to nature and childhood nature experiences. Front Psychol 9:1055. https://doi.org/10.3389/fpsyg.2018.01055

Sasaki T, Furukawa T, Iwasaki Y, Seto M, Mori AS (2015) Perspectives for ecosystem management based on ecosystem resilience and ecological thresholds against multiple and stochastic disturbances. Ecol Ind 57:395–408. https://doi.org/10.1016/j.ecolind.2015.05.019

Schlenger AJ, Libralato S, Ballance LT (2019) Temporal variability of primary production explains marine ecosystem structure and function. Ecosystems 22:331–345. https://doi.org/10.1007/s10021-018-0272-y

Selkoe KA, Blenckner T, Caldwell MR, Crowder LB, Erickson AL, Essington TE, Estes JA, Fujita RM, Halpern BS, Hunsicker ME, Kappel CV, Kelly PK, Kittinger JN, Levin PS, Lynham JM, Mach ME, Martone RG, Mease LA, Salomon AK, Samhouri JF, Scarborough C, Stier AC, White C, Zedler J (2015) Principles for managing marine ecosystems prone to tipping points. Ecosyst Health Sustain 1(5):1–18. https://doi.org/10.1890/ehs14-0024.1

Shahzad A, Sabiha ZUA, Haq M, Ullah H, Zafar MH, Khalid S (2021) Air pollution and respiratory health among men exposed to two different environs as occupational exposure. Polish J Environ Stud 30(1):833. https://doi.org/10.15244/pjoes/122502

Shen L, Li H, Wu H, Yan B, Song A (2023) Microorganisms in coastal wetland sediments: a review on microbial community structure, functional gene, and environmental potential. Front Microbiol 14. https://doi.org/10.3389/fmicb.2023.1163896

Singh S, Bakshi BR (2013) Accounting for the biogeosociochemical cycle of nitrogen in input-output life cycle assessment. Environ Sci Technol 47(16):9388–9396. https://doi.org/10.1021/es4009757

Song X, Carlsson C, Kiilsgaard R, Bendz D, Kennedy H (2020) Life cycle assessment of geotechnical works in building construction: a review and recommendations. Sustainability 12(20):8442. https://doi.org/10.3390/su12208442

Spencer SY, Adams C, Yapa PW (2013) The mediating effects of the adoption of an environmental information system on top management's commitment and environmental performance. Sustain Account Manage Policy J 4(1):75–102. https://doi.org/10.1108/sampj-10-2011-0030

Steffen W, Richardson K, Rockström J, Cornell SE, Fetzer I, Bennett EM, Biggs R, Carpenter SR, De Vries W, De Wit CA, Folke C, Gerten D, Heinke J, Mace GM, Persson LM, Ramanathan V, Reyers B, Sörlin S (2015) Planetary boundaries: guiding human development on a changing planet. Science 347(6223). https://doi.org/10.1126/science.1259855

Stradere M, Didenko K (2020) The role of the environmental impact assessment in territorial planning. https://doi.org/10.3846/bm.2020.640

Sundsbø AO, Runkle BR, McMonagle S, Jantke K, Lottermoser F, Gottschick M, Häseler S, Rodriguez Lopez JM, Scheele M (2015) One metaphor—several meanings: an interdisciplinary approach to sustainable development. In: Integrating sustainability thinking in science and engineering curricula: innovative approaches, methods and tools, pp 197–213 https://doi.org/10.1007/978-3-319-09474-8

Tang J, Xiong K, Chen Y, Wang Q, Ying B, Zhou J (2022) A review of village ecosystem vulnerability and resilience: implications for the rocky desertification control. Int J Environ Res Public Health 19(11):6664. https://doi.org/10.3390/ijerph19116664

Tang K, Liu X, Harper SL, Steevens JA, Xu R (2013) NEIMiner: nanomaterial environmental impact data miner. Int J Nanomed 8(sup1):15–29

Tavoni A, İriş D (2020) Tipping and reference points in climate change games. In: Handbook on the economics of climate change. Edward Elgar Publishing, pp 239–258. https://doi.org/10.4337/9780857939067.00018

Thamir AD, Lafta SH, Mohammed FQ, Hanon MM (2019) Environmental impact assessment of Halfaya Oilfield project. Eng Technol J 37(2):258–267. https://doi.org/10.30684/etj.37.2c.10

Vanclay F (2015) Changes in the impact assessment family 2003–2014: implications for considering achievements, gaps and future directions. JEAPM 17(01):1550003. https://doi.org/10.1142/s1464333215500039

Vitousek PM, Aber JD, Howarth RW, Likens GE, Matson PA, Schindler DW, Schlesinger WH, Tilman DG (1997) Human alteration of the global nitrogen cycle: sources and consequences. Ecol Appl 7(3):737–750. https://doi.org/10.1890/1051-0761(1997)007[0737:HAOTGN]2.0.CO;2

Vladova I (2023) Towards a more sustainable future: the importance of environmental education in developing attitudes towards environmental protection. In: SHS web of conferences, vol 176. EDP Sciencesm, p 01009. https://doi.org/10.1051/shsconf/202317601009

Williams IN, Torn MS, Riley WJ, Wehner MF (2014) Impacts of climate extremes on gross primary production under global warming. Environ Res Lett 9(9):094011. https://doi.org/10.1088/1748-9326/9/9/094011

Wilson J, Hinz S, Coston-Guarini J, Mazé C, Guarini JM, Chauvaud L (2017) System-based assessments—improving the confidence in the EIA process. Environments 4(4):95. https://doi.org/10.3390/environments4040095

Winn MI, Pogutz S (2013) Business, ecosystems, and biodiversity: New horizons for management research. Organ Environ 26(2):203–229. https://doi.org/10.1177/1086026613490173

Yen JD, Cabral RB, Cantor M, Hatton I, Kortsch S, Patrício J, Yamamichi M (2016) Linking structure and function in food webs: maximization of different ecological functions generates distinct food web structures. J Anim Ecol 85(2):537–547. https://doi.org/10.1111/1365-2656.12484

Zampieri M, Grizzetti B, Toreti A, De Palma P, Collalti A (2021) Rise and fall of vegetation primary production resilience to climate variability anticipated by a large ensemble of Earth System Models' simulations. https://doi.org/10.31223/x5zw36

Chapter 2
Principles of Sustainable Development and Environmental Management

Sylvester Chibueze Izah, Matthew Chidozie Ogwu, and Saoban Sunkanmi Salimon

Abstract Sustainable development has become a cornerstone for addressing global challenges such as environmental degradation, climate change, and social inequality. Rooted in the Brundtland Report's definition, sustainable development emphasizes balancing economic growth, social equity, and environmental protection to meet the needs of current and future generations. The interconnected principles of intergenerational equity, environmental integrity, economic viability, and social inclusion are vital for achieving sustainable outcomes. Environmental management is pivotal in operationalizing these principles, offering tools such as Environmental Impact Assessments, Life Cycle Assessments, and ecosystem-based management strategies to ensure responsible resource use and conservation. Economic sustainability fosters innovation and transitioning to circular economies, reduces resource dependency, and aligns corporate strategies with environmental goals. Simultaneously, social sustainability promotes community involvement, equitable access to resources, and the preservation of cultural heritage, ensuring inclusive decision-making and equitable development. The adoption of international frameworks such as the Sustainable Development Goals (SDGs) and the Paris Agreement underscores

---

S. C. Izah
Department of Community Medicine, Faculty of Clinical Sciences, Bayelsa Medical University, Yenagoa, Bayelsa, Nigeria

Department of Microbiology, Faculty of Science, Bayelsa Medical University, Yenagoa, Bayelsa, Nigeria

S. C. Izah
e-mail: sylvester.izah@bmu.edu.ng

M. C. Ogwu
Goodnight Family Department of Sustainable Development, Appalachian State University, Boone, NC, USA
e-mail: ogwumc@appstate.edu

S. S. Salimon (✉)
School of Biological and Medical Physics, Moscow Institute of Physics and Technology, Dolgoprudny, Moscow Region, Russian Federation
e-mail: salimon.s@phystech.edu

M. C. Ogwu and S. Chibueze Izah (eds.), *Evaluating Environmental Processes and Technologies*, Environmental Science and Engineering,
https://doi.org/10.1007/978-3-031-85327-2_2

the importance of integrating sustainability into policy and governance. These frameworks drive collaborative efforts among governments, businesses, and civil society, promoting accountability and transparency in achieving sustainability objectives. Despite progress, challenges such as conflicting priorities, resource depletion, and biodiversity loss persist. Innovative approaches, including renewable energy adoption, green infrastructure integration, and participatory governance, demonstrate pathways for overcoming these obstacles while addressing economic, environmental, and social dimensions. This chapter examines the core principles of sustainable development and explores practical applications in environmental management, financial planning, and social equity. It highlights the synergies between governance frameworks, stakeholder collaboration, and technological innovations in fostering sustainable practices. Through an examination of international case studies and tactics, the study emphasizes the need for comprehensive, inclusive approaches to sustainable development. In an increasingly linked world, attaining long-term sustainability, equity, and resilience requires bridging policy gaps, encouraging innovation, and coordinating local and global goals.

Keywords Sustainable development · Environmental management · Intergenerational equity · Climate change · Governance · Social inclusion

2.1 Introduction

Sustainable development is often defined as "development that meets the needs of the present without compromising the ability of future generations to meet their own needs" (Shrestha 2019). This definition, articulated in the Brundtland Report, that emphasized the interconnectedness of environmental integrity, economic viability, and social equity. However, the term is not without its complexities and contestations. Various scholars have pointed out that the ambiguity inherent in the definition allows for diverse interpretations, which can lead to misapplication in practice (Imran et al. 2014; Barua 2015). For instance, while some view sustainable development as a pathway to economic growth, others argue it should prioritize ecological preservation and social justice (Giddings et al. 2002; Murphy 2012). The concept of sustainable development has evolved to involve a broader understanding of sustainability, including environmental considerations and social and economic dimensions. This tripartite framework is often visualized as three overlapping circles representing the economy, society, and environment, highlighting the need for a holistic approach to development (Giddings et al. 2002; Murphy 2012). The challenge lies in balancing these dimensions, as they often present conflicting interests and priorities.

The historical roots of sustainable development can be traced back to the early twentieth century, but it gained significant traction in the latter half due to increasing awareness of environmental degradation and social inequalities (Pisani 2006; Dobers and Zilahy 2023; Akudugu and Ogwu 2024). The 1972 United Nations Conference on the Human Environment in Stockholm marked a critical turning point, as it brought global attention to environmental issues and laid the groundwork for future international agreements (Pisani 2006). However, the Brundtland Report, published in 1987 by the World Commission on Environment and Development (WCED), indeed catalyzed the sustainable development movement by linking environmental concerns with economic and social development (Sardain et al. 2016). The Brundtland Report's influence extended beyond its immediate context, inspiring subsequent global frameworks such as Agenda 21, adopted at the 1992 Earth Summit in Rio de Janeiro. Agenda 21 provided a comprehensive blueprint for sustainable development, outlining actions for governments, businesses, and civil society to promote local, national, and global sustainability (Sardain et al. 2016; Oloyede 2009; Ogwu 2019). The report emphasized the importance of integrating environmental considerations into economic planning and decision-making, a principle that remains central to sustainable development discourse today (Bebbington and Unerman 2018). In the years following the Earth Summit, the concept of sustainable development continued to evolve, leading to establishing of the Sustainable Development Goals (SDGs) in 2015. The SDGs consist of 17 interconnected goals to address global challenges such as poverty, inequality, climate change, and environmental degradation (Bebbington and Unerman 2018; Takala and Häyry 2023). These goals reflect a shift towards a more inclusive and participatory approach to development, recognizing the need for collaboration among various stakeholders to achieve sustainability (Vannadil and Prasanna 2023; Obahiagbon and Ogwu 2024).

The Brundtland Report is often regarded as a seminal document in sustainable development, as it defined the concept and framed the discourse around it. By articulating the need for a balance between economic growth and environmental protection, the report set the stage for a new paradigm of development that prioritizes sustainability (Shrestha 2019; Dobers and Zilahy 2023). However, the report has also faced criticism for its perceived ambiguity and for promoting "sustainable economic growth," which some argue undermines the core principles of sustainability (Zovanyi 2008; Izah and Ogwu 2023). Critics contend that the emphasis on economic growth within the Brundtland Report has led to a narrow interpretation of sustainable development, focusing primarily on economic indicators while neglecting social and environmental dimensions (Zovanyi 2008; Giddings et al. 2002). This has resulted in a disconnect between the theoretical framework of sustainable development and its practical implementation, as policymakers often prioritize short-term economic gains over long-term sustainability (Giddings et al. 2002; Murphy 2012). The challenge, therefore, lies in reconciling these competing interests and ensuring that sustainable development is understood and applied holistically. The SDGs represent a significant advancement in the global commitment to sustainable development, building on the foundations laid by the Brundtland Report and Agenda 21. Adopted by all United Nations member states in 2015, the SDGs provide a comprehensive framework for addressing the world's most pressing challenges through a sustainable lens (Bebbington and Unerman 2018; Takala and Häyry 2023). Each goal is interconnected, emphasizing the need for integrated approaches to development that consider the complex relationships between economic, social, and environmental factors (Vannadil and Prasanna 2023).

The SDGs also reflect a shift towards inclusivity and equity, recognizing that sustainable development must benefit all people, particularly marginalized and vulnerable communities (Vannadil and Prasanna 2023). This focus on social justice is crucial, as it acknowledges that environmental degradation and economic inequality are often intertwined and that solutions must address both issues simultaneously (Murphy 2012; Takala and Häyry 2023). Furthermore, the SDGs encourage collaboration among governments, businesses, and civil society, fostering partnerships essential for achieving sustainability at all levels (Vannadil and Prasanna 2023). Despite the progress in defining and promoting sustainable development, significant challenges remain in its implementation. One of the primary obstacles is the lack of a universally accepted framework for measuring sustainability, which complicates efforts to assess progress and hold stakeholders accountable (Boron and Murray 2004). The diversity of interpretations and applications of sustainable development further exacerbates this issue, as different contexts may require tailored approaches that are not easily comparable (Giddings et al. 2002; Murphy 2012). Additionally, the tension between economic growth and environmental protection continues to challenge policymakers. Many governments prioritize short-term economic gains over long-term sustainability, often leading to decisions that exacerbate environmental degradation and social inequalities (Giddings et al. 2002; Murphy 2012). This conflict highlights the need for a paradigm shift in how development is conceptualized and pursued, moving away from a growth-centric model towards one that prioritizes sustainability as a fundamental goal (Zovanyi 2008).

Implementing sustainable development should also consider local contexts and the unique challenges faced by different communities. While global frameworks such as the SDGs provide a valuable roadmap, local adaptations are essential for ensuring that sustainability efforts are relevant and practical (Vannadil and Prasanna 2023). This requires engaging local stakeholders in decision-making and recognizing the diverse cultural, social, and economic factors that shape sustainability in different regions (Oloyede 2009). For instance, in developing countries, sustainable development initiatives may need to prioritize poverty alleviation and social equity, while in developed nations, the focus may be on reducing consumption and promoting environmental stewardship (Giddings et al. 2002; Murphy 2012). By tailoring approaches to local contexts, stakeholders can better address their communities' specific needs and challenges, ultimately leading to more effective and sustainable outcomes (Vannadil and Prasanna 2023). Public engagement and participation are also essential for fostering a sense of ownership and responsibility towards sustainability initiatives. By involving communities in the planning and implementing of sustainable development projects, stakeholders can ensure that local perspectives and needs are considered, ultimately leading to more successful outcomes (Vannadil and Prasanna 2023). This participatory approach enhances the effectiveness of sustainability efforts, builds social capital, and strengthens community resilience (Giddings et al. 2002; Murphy 2012). Looking ahead, the future of sustainable development will depend on the ability of stakeholders to navigate the complexities and challenges associated with its implementation. As global challenges such as climate change, resource depletion, and social inequality continue to escalate, the urgency for effective sustainable development strategies becomes increasingly apparent (Vannadil and Prasanna 2023). This necessitates a collaborative approach that transcends traditional boundaries, fostering partnerships among governments, businesses, civil society, and local communities (Vannadil and Prasanna 2023). Moreover, integrating technology and innovation into sustainable development practices can enhance efficiency and effectiveness. For instance, advancements in renewable energy, sustainable agriculture, and waste management can provide practical solutions to pressing sustainability challenges (Giddings et al. 2002; Murphy 2012). By leveraging technology, stakeholders can develop more sustainable practices that address immediate concerns and contribute to long-term sustainability goals (Vannadil and Prasanna 2023).

The principles of environmental management and sustainable development are examined in this chapter, focusing on the necessity of striking a balance between social justice, environmental preservation, and economic progress. By looking at fundamental ideas like environmental integrity and intergenerational equality, as well as pragmatic aspects like resource management and pollution control, this study emphasizes how crucial it is to incorporate sustainability into national and international policies. In addition, it discusses the possibility of creativity and collaboration in accomplishing long-term sustainability objectives, the importance of governance, and the difficulties of putting sustainable principles into practice.

2.2 Core Principles of Sustainable Development

Sustainable development is a multifaceted concept that balances economic growth, social equity, and environmental integrity (Fig. 2.1 and Table 2.1). Sustainable development is grounded in key principles, including intergenerational equity, ecological integrity, economic viability, social inclusion, etc. (Fig. 2.1). These principles play a crucial role in shaping policies and practices to foster a sustainable future for all. Figure 2.2 illustrates the alignment of key principles and impact categories with SDGs, emphasizing a people-centered approach to sustainable development. It categorizes contributions across economic growth, human development, food production, energy supply, sanitation, and ecosystem management. Economic, social, and environmental principles support these categories. Strategies like experimental games to enhance workforce participation, fair consultation processes, and legal community registration mechanisms are emphasized on the economic and social front. Environmentally focused initiatives include small-scale bioenergy systems, agrovoltaic systems, land-sharing techniques, and carbon capture solutions. The integration of these approaches underlines the necessity of collaborative and inclusive frameworks to address the multidimensional challenges of sustainability, fostering equitable growth while safeguarding ecosystems. This figure effectively connects actionable strategies to the broader SDGs framework, promoting holistic development through equitable and sustainable practices.

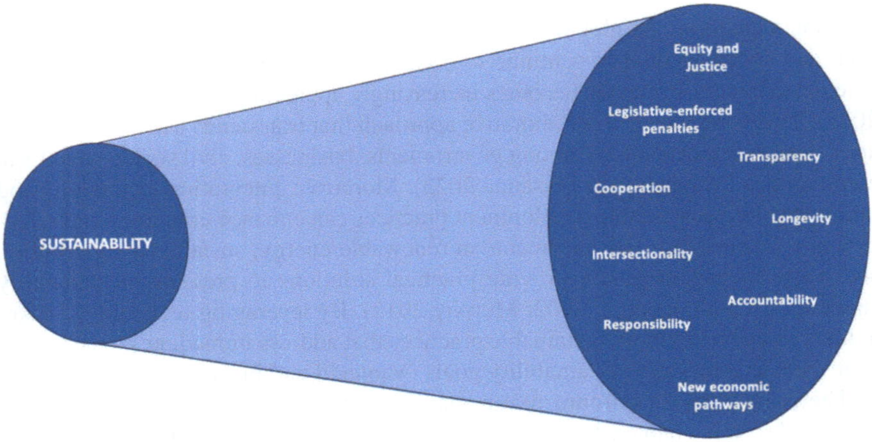

Fig. 2.1 Some core principles of sustainability. *Source* Axon (2024)

Table 2.1 Core principles of sustainable development and their interconnections

Core principle	Focus	Interconnections and linkages	References
Intergenerational equity	Ensuring that the current generation's use of resources does not compromise the ability of future generations to meet their own needs	Interconnected with *environmental integrity*, protecting ecosystems ensures resources for future generations. *Economic Viability* is also key to sustaining resources for long-term use	Holdgate (1987) and United Nations (2015)
Environmental integrity	Preserving biodiversity and ecosystem functions to maintain the planet's life-support systems for present and future generations	Closely linked with *Intergenerational Equity*, ensuring the environment is preserved for future use. It also supports *Social Inclusion* by maintaining ecosystem services for vulnerable populations	Finlayson et al. (2005) and UNEP (2019)
Economic viability	Promoting economic growth that is sustained over the long term while integrating environmental protection and resource conservation	It integrates with *environmental integrity* by fostering economic systems that do not degrade ecosystems. *Social Inclusion is* achieved by promoting equitable access to financial resources	Pearce et al. (1990) and Sachs (2015)
Social inclusion	Addressing inequality, poverty alleviation, and ensuring access to resources and opportunities for everyone, regardless of social class, race, or gender	Strongly linked to *Economic Viability* as equitable resource distribution improves economic opportunities. *Environmental Integrity* ensures that marginalized groups can access essential ecosystem services	Sen (1999) and UNEP (2020)

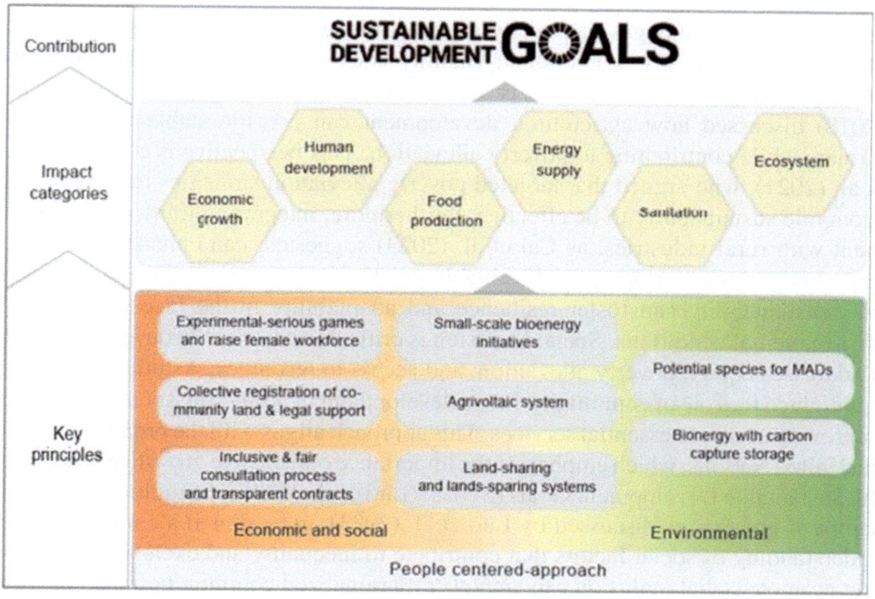

Fig. 2.2 Integrative framework for achieving sustainable development goals through economic, social, and environmental principles. *Source* Varela Pérez et al. (2022)

Intergenerational equity emphasizes the responsibility of the current generation to manage resources in a way that does not compromise the ability of future generations to meet their own needs. This principle is particularly relevant in discussions about resource management and environmental conservation. Wang and Peng (2023) highlighted the importance of sustainable practices in poverty-stricken areas, arguing that effective resource management can create a foundation for future generations to thrive. Similarly, You et al. (2020) asserted that inclusive growth and sustainable development are essential for improving living standards, ensuring that future generations inherit a viable environment and economy. The United Nations' 2030 Agenda for Sustainable Development further underscores this principle by calling for actions that promote sustainability across generations (You et al. 2020).

Environmental integrity is another fundamental principle of sustainable development, focusing on preserving biodiversity and ecosystems. The degradation of natural habitats and biodiversity loss pose significant threats to the planet's ecological balance. Garnett et al. (2016) emphasized the need for organizations to incorporate sustainability principles into their mission statements, which can influence their operational practices and contribute to environmental conservation. Additionally, the work of Burns et al. (2021) revealed alarming trends in biodiversity loss across Europe, underscoring the urgent need for transformative changes to protect ecosystems. Integrating biodiversity and ecosystem services into development agendas is crucial for achieving long-term sustainability and addressing the interconnected challenges of poverty and environmental degradation (Lucas et al. 2013; Ogwu et al. 2014). Economic viability is essential for sustainable development, ensuring economic growth is aligned with sustainable practices. Liu et al. (2018) discussed how agricultural development can provide stable incomes for farmers while contributing to poverty alleviation. This perspective is echoed by Xu et al. (2021), who argued that targeted poverty alleviation strategies must consider economic sustainability to be effective. Furthermore, integrating transport development with rural industries, as Cai et al. (2023) suggested, can enhance economic opportunities while promoting sustainable practices. Economic policies that prioritize sustainability can foster resilience and adaptability in the face of changing environmental conditions. Social inclusion is critical to sustainable development, as it addresses equity, poverty alleviation, and access to resources. Astuti et al. (2015) highlighted the role of community-based development in promoting social equity and improving access to essential services. This approach aligns with the broader goals of the United Nations, which emphasize the importance of inclusive growth in achieving the Sustainable Development Goals (SDGs) (Luo et al. 2023). The multidimensional nature of poverty, as discussed by Luo et al. (2023), necessitated a comprehensive understanding of social factors that contribute to inequality and exclusion. Policies that promote social inclusion can empower marginalized communities and enhance their capacity to participate in sustainable development initiatives.

The interplay between these core principles of sustainable development is evident in various case studies and research findings. For example, Baffoe et al. (2021) explored urban–rural linkages in Ghana, demonstrating how effectively integrating these areas can lead to sustainable development outcomes. Their findings suggest that addressing the disparities between urban and rural regions is essential for achieving social equity and economic viability. Similarly, the research conducted by Zeng et al. (2021) on agricultural fiscal expenditures highlights the importance of targeted investments in poverty reduction strategies that consider environmental sustainability. These studies illustrate the interconnectedness of economic, social, and ecological dimensions in pursuing sustainable development.

Moreover, the role of innovation in promoting sustainable development cannot be overlooked. Fahrudi (2020) discussed how social innovation can create opportunities for businesses to address social issues while achieving profitability. This perspective aligns with the need for creative solutions that integrate economic viability with social inclusion, ultimately contributing to sustainable development. The emphasis on innovation is also reflected in the work of Yin et al. (2022), who advocate for rural innovation systems to revitalize countryside areas and promote sustainable practices. By fostering innovation, communities can develop adaptive strategies that enhance their resilience to environmental and economic challenges. The complexities of global interdependencies compound the challenges of achieving sustainable development. The impact of poverty on sustainable economic growth is a pressing concern, as highlighted by various studies that emphasize the need for comprehensive strategies to address this issue (Utuk 2022). The recognition that poverty alleviation is a principal yardstick for measuring sustainable development underscores the importance of integrating social equity into economic policies (Utuk 2022). As nations strive to meet the SDGs, the interconnectedness of these challenges necessitates a holistic approach that considers the diverse needs of communities.

2.3 Dimensions of Sustainable Development

Sustainable development involves a multifaceted approach that integrates environmental, economic, and social dimensions (Table 2.2). Each dimension plays a critical role in ensuring that development meets the needs of the present without compromising the ability of future generations to meet their own needs. This comprehensive discussion will explore these dimensions in detail, drawing on relevant literature to substantiate the claims made.

2.3.1 Environmental Sustainability

Environmental sustainability is fundamentally concerned with the responsible management of natural resources, pollution control, and the conservation of ecosystems. Effective resource management is essential for maintaining the health of ecosystems and ensuring that natural resources are available for future generations. According to Sharma (2019), sustainable natural resource management practices are crucial for enhancing the qualitative livelihood of communities, particularly in regions rich in biodiversity, such as forests and pastures. These practices include conserving and regenerating natural resources, which are vital for sustaining ecological balance and supporting local economies.

Pollution control is another critical aspect of environmental sustainability. The increasing pollution levels from industrial activities and urbanization pose significant threats to human health and the environment. Chen (2023) highlighted the relationship between sustainable development and economic growth, emphasizing that economic activities must be aligned with environmental protection to foster sustainable outcomes. This alignment requires implementing stringent pollution control measures and adopting cleaner production technologies that minimize waste and emissions. Ecosystem conservation is integral to maintaining biodiversity and the services that ecosystems provide. The Communal Areas Management Programme for Indigenous Resources (CAMPFIRE) in Zimbabwe exemplifies a successful initiative that balances conservation with community needs by incentivizing local communities to engage in sustainable practices (Gandiwa et al. 2013). Such programs demonstrate that when communities are empowered to manage their natural resources, they can contribute significantly to conservation efforts and economic well-being.

Table 2.2 Dimensions of sustainable development with some model approaches

Dimension	Focus	Key areas	Model approaches	Interconnections	References
Environmental sustainability	It focuses on the responsible management of natural resources to ensure long-term ecosystem health, pollution control, and biodiversity conservation	Resource management, pollution control, and ecosystem conservation	Circular economy model: promotes recycling, reducing waste, and reusing resources to minimize environmental degradation	Interlinked with *Economic Sustainability* by promoting sustainable consumption and production. *Social Sustainability* depends on healthy ecosystems for community livelihoods	Rockström et al. (2009a, b), United Nations (2015) and Ellen MacArthur Foundation (2013)
Economic sustainability	Ensures long-term economic growth by promoting sustainable production, consumption patterns, and circular economy models that minimize waste and maximize resource use	Sustainable production and consumption, circular economy, and resource efficiency	Sustainable development goals (SDG) model: aligns economic activities with UN SDGs to support sustainable economic growth	Relates to *Environmental Sustainability* as the circular economy minimizes environmental impact. *Social Sustainability* benefits from economic equity and sustainable job creation	Pearce et al. (1989), Raworth (2017) and Sachs (2015)
Social sustainability	Emphasizes equity, social justice, cultural preservation, and active community participation to ensure that all individuals and groups benefit from sustainable development	Community involvement, social justice and cultural preservation	Social innovation model: focuses on solutions that integrate community empowerment, equity, and social justice	Strongly linked to *Economic Sustainability* through equitable access to resources. *Environmental Sustainability* ensures all access to clean air, water, and other ecosystem services	Sen (1999), Leach et al. (2010) and UNDP (2020)

2.3.2 Economic Sustainability

Economic sustainability focuses on the capacity of an economy to support a defined level of economic production indefinitely. It emphasizes sustainable production and consumption patterns, essential for reducing resource depletion and environmental degradation. The concept of a circular economy, which promotes the reuse and recycling of materials, is a key strategy for achieving economic sustainability. This approach minimizes waste, fosters innovation, and creates new economic opportunities (Jeronen 2020). As Klitgaard (2020) noted, economic sustainability requires a shift in how resources are allocated and utilized, emphasizing efficiency and long-term viability over short-term gains. Moreover, the relationship between economic growth and sustainable development is complex. While economic growth can provide the resources necessary for sustainable development, it can also lead to environmental degradation if not appropriately managed. As Uysal (2015) highlighted, integratingf human resource management practices into economic strategies can enhance overall economic performance while ensuring that social and environmental considerations are considered. This holistic approach is essential for fostering sustainable economic growth that benefits all segments of society. The role of technology in promoting economic sustainability cannot be overlooked. Luo and Bu (2016) discussed how advancements in information and communication technology (ICT) can enhance resource management and improve operational efficiency in businesses, particularly in emerging economies. By leveraging technology, organizations can optimize resource use, reduce waste, and contribute to more sustainable economic practices.

2.3.3 Social Sustainability

Social sustainability is concerned with the social dimensions of sustainability, including community involvement, social justice, and cultural preservation. Community involvement is crucial for ensuring that development initiatives are inclusive and equitable. Eneji and Ering (2018) emphasized the importance of community-based natural resource management, which empowers local communities to participate in decision-making processes regarding resource use. This participatory approach enhances resource management's effectiveness and fosters a sense of ownership and responsibility among community members. Social justice is another critical aspect of social sustainability. It involves ensuring all individuals have equal access to resources and opportunities, regardless of their socio-economic status. The concept of socioeconomic sustainability, as discussed by Sadykova (2022), highlighted the need for policies that promote equitable economic growth and address disparities within communities. By prioritizing social justice, societies can create a more inclusive environment that supports the well-being of all individuals. Cultural preservation is also vital for social sustainability. The loss of cultural heritage can undermine community identity and cohesion. Initiatives that promote the conservation of cultural practices and traditions contribute to social sustainability by fostering a sense of belonging and continuity within communities. Integrating cultural considerations into conservation efforts can enhance the effectiveness of these initiatives and promote community engagement.

2.3.4 Interconnections Among the Dimensions

Environmental, economic, and social sustainability interconnections are critical for achieving comprehensive sustainable development. For instance, effective resource management supports environmental sustainability and enhances economic opportunities for communities. As Huang et al. (2014) highlighted, integrating intelligent resource management systems can optimize resource use and improve overall sustainability outcomes. This interconnectedness underscores the importance of adopting a holistic approach to sustainable development, considering the interplay between different dimensions. Furthermore, achieving sustainable development goals (SDGs) requires collaboration among various stakeholders, including governments, businesses, and civil society. The PESTLE analysis conducted by Chen (2023) illustrated how political, economic, social, technological, legal, and environmental factors influence sustainable development outcomes. By understanding these interactions, stakeholders can develop more effective strategies that address the complexities of sustainable development (Fig. 2.3).

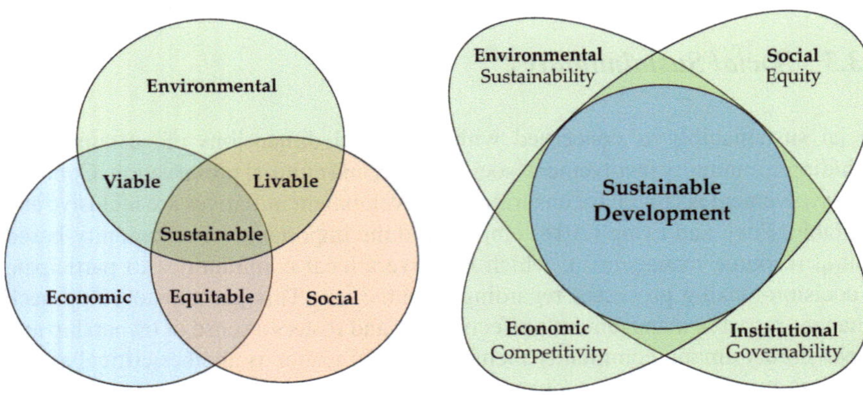

Fig. 2.3 Three (**a**) and four (**b**) dimensions of Sustainable development. *Source*: Londoño-Pineda et al. (2021)

2.4 Environmental Management: Concepts and Approaches

Environmental management has emerged as a critical field of study and practice, aiming to address global environmental challenges societies face. At its core, ecological management involves a set of strategies, policies, and procedures designed to minimize environmental impacts while promoting sustainable development (Table 2.3). Environmental management's primary goals include preserving natural resources, reducing pollution, and enhancing ecosystem services, all of which are essential for achieving the United Nations Sustainable Development Goals (SDGs) (Bruce and Ohlsson 2020). Integrating environmental considerations into organizational decision-making processes is crucial for fostering a culture of sustainability and ensuring long-term ecological health (Eya et al. 2022).

One of the foundational tools in environmental management is the Environmental Impact Assessment (EIA), which systematically evaluates the potential environmental effects of proposed projects before they are carried out. EIAs serve as a critical mechanism for informed decision-making, allowing stakeholders to weigh the benefits of development against potential environmental costs (Ladyman et al. 2022). Life Cycle Analysis (LCA) is another essential technique that assesses the environmental impacts associated with all stages of a product's life, from raw material extraction to disposal. By providing a comprehensive view of a product's ecological footprint, LCA enables organizations to identify opportunities for improvement and innovation in their processes (Ghosh et al. 2022).

Ecosystem-based management (EBM) represents a holistic approach to environmental management that considers the interconnectedness of ecological systems and the services they provide. Unlike traditional management strategies that often focus on single species or specific issues, EBM emphasizes the importance of understanding and managing entire ecosystems (Sandström et al. 2013). This approach is particularly relevant in natural resource management, where the sustainable use of resources is essential for maintaining ecological balance and supporting human livelihoods (Zheng 2020). Adaptive management, closely related to EBM, involves a systematic process of learning and adjusting management strategies based on the outcomes of previous actions, thus fostering resilience in the face of environmental change (Curtin and Prellezo 2010). Implementing effective environmental management practices requires the active involvement of all organizational levels, particularly in fostering employee awareness and engagement. Research indicates that organizations are more likely to achieve their environmental goals when employees are informed and involved in environmental initiatives (Ivashchenko 2023). Moreover, the commitment of senior management to environmental issues plays a pivotal role in shaping organizational culture and driving positive environmental performance (Chen 2013). This top-down approach, combined with grassroots involvement, creates a synergistic effect that enhances the overall effectiveness of environmental management systems. Green human resource management (GHRM) has

Table 2.3 Diverse Environmental management tools, techniques, and approaches for sustainable societal development

Category	Tools, techniques, and approaches	Application	Key focus areas	Model approaches	References
Environmental assessment	Environmental impact assessment (EIA)	A systematic process for evaluating the environmental consequences of proposed projects before decisions are made	Predicting environmental impacts and public participation as well as mitigation strategies	EIA framework: focuses on assessing potential environmental risks and benefits of development projects	Petts (2005) and Glasson and Therivel (2013)
Sustainability auditing	Life cycle assessment (LCA)	Assesses the environmental impacts of all stages of a product's life, from raw material extraction to disposal	Resource use, energy consumption, and waste generation	Cradle-to-grave approach: evaluate the total environmental cost of product life cycles	Curran (2012) abd ISO (2006)
Geospatial analysis	Geographic information systems (GIS)	A tool for mapping and analyzing spatial data related to natural resources, land use, and environmental hazards	Spatial analysis, resource management, and environmental planning	GIS-based planning: helps design and implement sustainable development projects by mapping critical areas	Goodchild et al. (2005) and Burrough et al. (2015)
Risk assessment and management	Environmental risk assessment (ERA)	Evaluates the likelihood and consequences of environmental hazards, enabling better decision-making and mitigation measures	Hazard identification, risk estimation, and mitigation measures	Probabilistic risk assessment: combines environmental science and risk analysis to assess potential impacts	Suter (2006) and Burgman (2005)

(continued)

Table 2.3 (continued)

Category	Tools, techniques, and approaches	Application	Key focus areas	Model approaches	References
Sustainability metrics	Sustainable development goals (SDG) monitoring	Tools to measure progress toward achieving sustainability goals across social, economic, and environmental pillars	Monitoring sustainability, international collaboration, and data collection	SDG framework: used globally for policy alignment and tracking progress in sustainable development initiatives	Sachs (2015) and United Nations (2015)
Ecosystem management	Ecosystem-based management (EBM)	Integrates ecological, social, and economic goals to sustainably manage ecosystems and their services	Biodiversity conservation, resource management, and ecosystem health	Adaptive ecosystem management: a flexible management approach that adjusts based on ecosystem feedback	Grumbine (1994) and Berkes et al. (2000)
Pollution prevention	Pollution prevention (P2) programs	Aimed at reducing or eliminating waste at the source through changes in production, operation, and raw materials	Waste reduction, cleaner (green) production and emission control	Source reduction strategy: focuses on reducing the pollutants and emissions produced by industrial activities	Freeman (1995) and EPA (1992)
Circular economy approaches	Closed-loop recycling	A process that minimizes waste by recovering and reusing materials in production cycles	Resource efficiency, waste reduction and economic sustainability	Circular economy model: aims for sustainable economic growth by maintaining products and materials at their highest utility	Ghisellini et al. (2016)

(continued)

Table 2.3 (continued)

Category	Tools, techniques, and approaches	Application	Key focus areas	Model approaches	References
Integrated resource management	Integrated water resource management (IWRM)	Promotes coordinated development and management of water, land, and related resources to maximize economic and social welfare	Water conservation, integrated resource use, and stakeholder engagement	IWRM approach: facilitates the equitable allocation of water resources while protecting ecosystems and improving water quality	Agarwal et al. (2000) and Biswas (2004)
Renewable energy and green technologies	Renewable energy technologies (RETs)	Promotes solar, wind, geothermal, and bioenergy use to reduce reliance on fossil fuels and mitigate climate change	Carbon reduction, energy security, and sustainable energy production	Low-carbon energy transition: focuses on shifting from fossil fuels to renewable energy sources to achieve sustainability	REN21 (2019) and IRENA (2018)

emerged as a vital component of environmental management, focusing on integrating sustainability into human resource practices. GHRM involves various strategies, including recruitment, training, and performance evaluation, to foster pro-environmental behaviors among employees (Esen and Çalışkan 2019). By aligning human resource policies with environmental objectives, organizations can cultivate a workforce aware of environmental issues and actively engaging in sustainability initiatives (Saeed et al. 2019). This alignment is crucial for ensuring that environmental management goals are met and organizations can adapt to evolving regulatory and societal expectations.

In urban environments, sustainable waste management has become a priority for many municipalities, reflecting the growing recognition of the need for integrated approaches to urban sustainability (Khosravani et al. 2023). The 2030 Agenda for Sustainable Development emphasizes the importance of creating inclusive, safe, resilient, and sustainable cities, with waste management being a critical aspect of this agenda (Khosravani et al. 2023). Effective waste management systems reduce environmental impacts and contribute to public health and community well-being, highlighting the interconnectedness of environmental and social dimensions of sustainability. The role of policy frameworks in shaping environmental management practices cannot be overstated. Governments and international organizations are crucial in establishing regulations and guidelines that promote sustainable practices across various sectors (Zhou 2023). For instance, China's environmental constraint target policy exemplifies how clear goal-setting and management can drive organizations toward achieving specific environmental outcomes (Zhou 2023). Such policies enable organizations to implement effective environmental management systems and contribute to broader sustainability objectives.

Barriers to the successful implementation of environmental management systems (EMS) often stem from a lack of resources, inadequate training, and insufficient support from management (Seifert 2018). Addressing these barriers requires a comprehensive understanding of the organizational context and the specific challenges different sectors face. For example, hospitals implementing the Eco-Management and Audit Scheme (EMAS) must balance environmental goals with the primary objective of providing high-quality healthcare services (Seifert and Guenther 2019). This necessitates a tailored approach that considers the unique operational realities of healthcare institutions while striving for improved environmental performance. Integrating environmental management into organizational practices also has implications for financial performance. Studies have shown that organizations prioritizing environmental sustainability often experience enhanced financial outcomes, as they can reduce costs associated with waste and inefficiency while improving their public image (Jiménez et al. 2013). This relationship underscores the importance of viewing environmental management as a compliance obligation and a strategic opportunity for value creation. Stakeholder engagement is another critical aspect of effective environmental management. By involving various stakeholders, including local communities, businesses, and government agencies, organizations can foster collaboration and build trust, which is essential for the successful implementation of environmental initiatives (Zheng 2020). This participatory approach enhances the legitimacy of environmental management efforts and ensures that diverse perspectives are considered in decision-making processes.

2.5 Policy and Governance for Sustainability

The discourse surrounding policy and governance for sustainability has gained significant traction in recent years, particularly considering pressing global challenges such as climate change, biodiversity loss, and social inequality. International treaties and agreements, such as the Paris Agreement and the Convention on Biological Diversity, play a pivotal role in shaping national and local policies to foster sustainable development. These agreements establish a framework for international cooperation and serve as catalysts for domestic policy reforms that align with global sustainability objectives. The Paris Agreement, for instance, sets forth a collective commitment to limit global warming to well below 2 °C, thereby necessitating robust national strategies to reduce greenhouse gas emissions and promote sustainable practices across various sectors (Iacobuţă et al. 2021; Dondisch and Montanez 2023; Markkanen and Anger-Kraavi 2019).

Integrating sustainability indicators into national policy frameworks is crucial for tracking progress and ensuring accountability in implementing sustainable development goals (SDGs). Research indicates that while the direct policy use of these indicators has been limited, they have proven effective in facilitating communication among stakeholders and breaking down siloed approaches within governmental structures (Lepenies et al. 2023). This is particularly relevant in the context of the SDGs, which encompass a broad spectrum of social, economic, and environmental objectives that require coordinated action across multiple sectors and levels of governance (Desa et al. 2021; Dzebo et al. 2023). The successful implementation of these goals hinges on the ability of national governments to translate international commitments into actionable policies that resonate with local contexts and priorities. At the national level, policies supporting sustainable development often manifest through legislative frameworks, regulatory measures, and financial incentives to promote environmentally responsible practices. For instance, countries increasingly adopt carbon pricing mechanisms, renewable energy mandates, and sustainable land-use policies as part of their climate action strategies (Von Stechow et al. 2015; Soergel et al. 2021). These measures contribute to reducing greenhouse gas emissions and foster economic resilience and social equity by creating green jobs and promoting sustainable livelihoods (Munguia et al. 2017; Doelman et al. 2022). Moreover, aligning national policies with international agreements such as the Paris Agreement enhances the credibility of countries on the global stage and facilitates access to climate finance and technology transfer (Davide et al. 2018; Tanaka et al. 2022).

Local policies, on the other hand, play a critical role in addressing sustainability challenges at the community level. Municipalities are often at the forefront of implementing sustainability initiatives, ranging from urban planning and waste management to public transportation and energy efficiency programs (Von Stechow et al. 2015; Rumagia et al. 2020). The effectiveness of these local policies is often contingent upon the active engagement of stakeholders, including citizens, businesses, and civil society organizations, in the decision-making process. This participatory approach enhances local policies' legitimacy and fosters a sense of ownership and accountability among stakeholders (Markkanen and Anger-Kraavi 2019; Honegger et al. 2021). Furthermore, local governments increasingly recognize the importance of integrating climate adaptation strategies into their planning processes to enhance community resilience to climate impacts (Soergel et al. 2021). Corporate social responsibility (CSR) has emerged as a vital component of environmental governance as businesses are increasingly held accountable for their environmental and social impacts. The Paris Agreement has intensified the demand for corporate transparency and accountability in climate-related disclosures, compelling firms to adopt more sustainable practices and integrate sustainability into their strategic planning (Munguia et al. 2017; Issa and Hanaysha 2023). Research suggests that companies that proactively engage in CSR initiatives are better positioned to mitigate risks associated with climate change and enhance their long-term competitiveness (Čadež and Czerny 2016; Ferreira et al. 2019). Moreover, aligning corporate sustainability strategies with international agreements fosters collaboration between the private sector and governments, facilitating the mobilization of resources and expertise needed to achieve sustainability objectives (Mashari et al. 2023; Migliorelli 2021).

The role of international treaties and agreements in promoting sustainable development cannot be overstated. For example, the Convention on Biological Diversity emphasizes the need for the conservation and sustainable use of biodiversity, which is essential for maintaining ecosystem services and supporting human well-being (Oktay et al. 2021; Dzebo et al. 2023). This treaty has spurred national governments to develop policies that protect natural habitats, promote sustainable agriculture, and enhance the resilience of ecosystems to climate change (Trancossi et al. 2023). Similarly, the Paris Agreement has catalyzed a global movement towards decarbonization, prompting countries to set ambitious targets for reducing emissions and transitioning to low-carbon economies (Dondisch and Montanez 2023; Tanaka et al. 2022). The interconnectedness of these agreements underscores the importance of a holistic approach to sustainability that recognizes the interdependence of environmental, social, and economic dimensions. In addition to international agreements, national and local policies must be underpinned by robust governance frameworks that promote transparency, accountability, and stakeholder engagement. Effective governance structures are essential for ensuring that sustainability initiatives are implemented effectively and that progress is monitored and reported transparently (Von Stechow et al. 2015; Soergel et al. 2021). This includes establishing mechanisms for public participation in decision-making processes and creating platforms for collaboration among government agencies, civil society, and the private sector (Markkanen and Anger-Kraavi 2019; Honegger et al. 2021). By fostering a culture of inclusivity and collaboration, governance frameworks can enhance the legitimacy of sustainability policies and facilitate the achievement of shared goals.

The concept of sustainable finance has gained prominence as a critical enabler of sustainable development. Financial institutions increasingly recognize the importance of aligning their investment strategies with sustainability objectives, thereby contributing to the mobilization of resources needed to support the transition to a low-carbon economy (Mashari et al. 2023; Migliorelli 2021). Integrating environmental, social, and governance (ESG) criteria into investment decision-making processes enhances financial portfolio resilience and promotes responsible corporate behavior (Oktay et al. 2021; Migliorelli 2021). Furthermore, sustainable finance frameworks aligned with international agreements, such as the Paris Agreement and the SDGs, can facilitate the flow of capital towards projects that deliver positive environmental and social outcomes (Davide et al. 2018; Migliorelli 2021). The interplay between corporate governance and sustainability is also evident in the growing emphasis on executive compensation linked to sustainability performance. Research indicates that companies that adopt sustainability-linked compensation structures are more likely to achieve their sustainability targets and enhance their overall performance (Issa and Hanaysha 2023; Soergel et al. 2021). This trend reflects a broader recognition of the need for businesses to align their strategic objectives with societal expectations and environmental imperatives. By incentivizing sustainable practices at the executive level, companies can drive meaningful change and contribute to achieving global sustainability goals (Munguia et al. 2017; Issa and Hanaysha 2023). The importance of international cooperation and collaboration cannot be overstated as the global community grapples with the multifaceted challenges of climate change and sustainable development. The Paris Agreement and the SDGs represent a collective commitment to addressing these challenges through coordinated action at all levels of governance (Dondisch and Montanez 2023; Dzebo et al. 2023). However, successfully implementing these agreements requires sustained political will, adequate financial resources, and robust institutional frameworks that facilitate collaboration among diverse stakeholders (Tanaka et al. 2022; Schneider and Tarawali 2021). By fostering a culture of cooperation and shared responsibility, the global community can achieve a more sustainable and equitable future for all.

2.6 Sustainable Development in Practice

Sustainable development is a multifaceted concept that seeks to balance economic growth, environmental stewardship, and social equity. In practice, it manifests through various case studies across agriculture, industry, and urban planning, each demonstrating innovative approaches to sustainability. Integrating technology and innovation is pivotal in these practices, particularly in renewable energy and green infrastructure. This discussion synthesizes relevant literature to explore these themes in depth. In agriculture, sustainable practices are increasingly adopted to enhance productivity while minimizing environmental impact. For instance, precision agriculture utilizes technology to optimize field-level management regarding crop farming. This approach employs GPS, IoT sensors, and data analytics to monitor crop health and soil conditions, thereby reducing resource waste and increasing yield efficiency (Usman et al. 2023; Izah et al. 2024). Moreover, agroecological practices, which emphasize biodiversity and ecosystem services, have improved resilience against climate change while maintaining soil health (Asnor et al. 2022; Ogwu et al. 2024). These practices contribute to food security and promote ecological balance, demonstrating a successful intersection of technology and sustainable agriculture (Ogwu 2020; 2023).

In the industrial sector, the transition towards sustainability is marked by adopting circular economy principles. This model focuses on resource efficiency, waste reduction, and the recycling of materials, thereby minimizing environmental degradation (Koval et al. 2023). Innovations in manufacturing processes, such as using biodegradable materials and energy-efficient technologies, are critical in this transition. For instance, companies are increasingly investing in renewable energy sources to power their operations, reducing carbon footprints and enhancing long-term economic viability (Maradin 2021). Integrating sustainability into corporate strategies is not merely a regulatory compliance measure; it is increasingly viewed as a competitive advantage in the marketplace.

Urban planning is another critical area where sustainable development practices are being implemented. The concept of green infrastructure (GI) has gained traction as cities seek to integrate natural systems into urban environments. GI includes parks, green roofs, and permeable pavements, which provide multiple benefits such as stormwater management, urban heat reduction, and enhanced biodiversity (Tayouga and Gagné 2016). Studies have shown that urban green spaces significantly contribute to public health by promoting physical activity and social cohesion (Jennings and Bamkole 2019). Furthermore, the incorporation of health impact assessments in urban planning can ensure that developments align with broader health promotion policies, thereby fostering healthier communities (Lee et al. 2015). The role of technology and innovation in promoting sustainability cannot be overstated. Renewable energy technologies, such as solar and wind power, are at the forefront of this transition. These technologies not only provide cleaner energy alternatives but also stimulate economic growth through job creation in new sectors (Paramati et al. 2020). The development of energy storage solutions, such as advanced battery systems, is crucial for addressing the intermittency of renewable energy sources, thereby enhancing their reliability and integration into existing power grids (Bernal and Ablian 2023). Moreover, innovations in energy efficiency technologies, such as smart grids and energy management systems, are essential for optimizing energy consumption and reducing waste (Usman et al. 2021).

In the context of urban planning, the integration of smart technologies can enhance the effectiveness of green infrastructure. For example, smart water management systems can optimize water usage in urban landscapes, reducing waste and promoting sustainability (Palliwoda et al. 2020). Additionally, the use of data analytics in urban design can facilitate better decision-making processes, ensuring that developments are environmentally sustainable and socially equitable (Arsaeva et al. 2019). The synergy between technology and urban planning is vital for creating resilient cities that can adapt to the challenges posed by climate change. The economic implications of sustainable practices are significant. Investments in renewable energy and green technologies are not only essential for environmental sustainability but also present lucrative opportunities for economic growth. Policymakers are encouraged to provide financial and regulatory support for research and development in these sectors, as this can lead to significant innovations that drive down costs and improve efficiency (Paramati et al. 2020). Furthermore, fostering a culture of innovation within industries can lead to the development of new business models that prioritize sustainability, thereby enhancing competitiveness in the global market (Li et al. 2022).

Challenges remain in the implementation of sustainable practices across sectors. In agriculture, for instance, adopting sustainable practices can be hindered by economic constraints and a lack of access to technology (Asnor et al. 2022). Similarly, in the industrial sector, the transition to a circular economy requires significant changes in production processes and supply chain management, which can be met with resistance from established practices (Koval et al. 2023). Urban planners face challenges integrating green infrastructure into existing urban landscapes, particularly in areas with high population density and limited space (Faisal et al. 2022). Addressing these challenges requires collaborative efforts among stakeholders, including governments, businesses, and communities. The importance of stakeholder engagement in promoting sustainable practices cannot be overlooked. Community involvement in urban planning processes can lead to more inclusive and equitable outcomes as local knowledge and preferences are considered (Muzdalifah and Maryono 2018). Similarly, engaging farmers in developing sustainable practices can enhance their adoption and effectiveness (Asnor et al. 2022). Collaborative approaches that involve multiple stakeholders can foster innovation and drive the implementation of sustainable solutions across sectors. Moreover, the global nature of sustainability challenges necessitates international cooperation and knowledge sharing. Countries can learn from each other's experiences in implementing sustainable practices, particularly in areas such as renewable energy deployment and urban planning (Liu et al. 2021). Collaborative research initiatives can facilitate the exchange of best practices and innovative solutions, thereby accelerating progress towards sustainability goals on a global scale.

2.7 Challenges and Future Directions

The interplay between development and environmental protection presents a complex challenge requiring a multifaceted approach (Table 2.4). As nations pursue economic growth, they often encounter the detrimental consequences of resource depletion, climate change, and biodiversity loss. These issues are interconnected and require integrated strategies harmonizing development goals with ecological sustainability. For instance, Leclère et al. (2020) emphasized that without a comprehensive plan that aligns with the Sustainable Development agenda, habitat losses will persist, and biodiversity will continue to decline, potentially leading to irreversible losses. This highlights the urgency of adopting integrated approaches that simultaneously consider ecological, social, and economic dimensions.

Urbanization is a significant driver of biodiversity loss, particularly in rapidly developing regions. Zhang et al. (2023) illustrated how the Hukou system in China has led to rural built-up land expansion that threatens biodiversity as urban–rural dynamics shift without adequate ecological assessments. This situation underscores the need for systematic evaluations of habitat changes due to urbanization, which are often overlooked in biodiversity conservation efforts. The failure to recognize urban expansion as a critical factor in biodiversity loss complicates the development

Table 2.4 Balancing development with environmental protection

Category	Approaches	Model approaches	References
Balancing development with environmental protection	The challenge of pursuing economic development while ensuring environmental protection and sustainability	Integrated land-use planning, green infrastructure, circular economy	Sachs (2015) and UNEP (2019)
Addressing climate change	Tackling the global crisis of rising temperatures through mitigation strategies such as reducing emissions and adaptation strategies to minimize impacts	Renewable energy transition, carbon capture and storage (CCS), climate-resilient infrastructure, reforestation, and promoting energy efficiency	IPCC (2014) and Stern (2008)
Resource depletion	Managing the over-exploitation of natural resources for sustainable use, including water, soil, and minerals	Resource conservation policies, water efficiency techniques, sustainable mining practices, agroecology, and regenerative agriculture	Rockström et al. (2009a, b) and WEF (2020)
Biodiversity loss	Preventing the rapid loss of biodiversity due to habitat destruction, pollution, climate change, and overuse of resources	Biodiversity conservation strategies, habitat restoration, species protection laws, the creation of biodiversity corridors	WWF (2020) and IPBES (2019)
Global cooperation	International partnerships are important in addressing cross-borders environmental issues, such as climate change and ocean pollution	International treaties (e.g., Paris Agreement), global partnerships (e.g., UN SDGs), multilateral environmental agreements, transboundary water cooperation, and biodiversity frameworks	UN (2015) and Paris Agreement (2015)
Public participation	Engaging communities in sustainable development through education, awareness, and participatory decision-making processes	Participatory environmental governance, stakeholder engagement frameworks, public consultations, grassroots movements, and local sustainability initiatives	Arnstein (1969) and Pretty (1995)

of effective conservation policies and strategies. The Belt and Road Initiative (BRI) exemplifies the challenges of balancing development with environmental sustainability on a global scale. Coenen et al. (2021) argued that the environmental governance of the BRI requires a deeper understanding of how non-participating countries, such as those in the EU and the US, influence sustainability outcomes through their consumption patterns. This highlights the need for global cooperation to address transboundary environmental impacts and ensure that development initiatives do not exacerbate ecological degradation.

Biodiversity governance is further complicated by the low levels of policy integration across various sectors, particularly agriculture. van der Hoff and Anyango (2022) posit that a more integrated land-sharing approach is essential to bend the curve of biodiversity loss while ensuring food security and climate resilience (van der Hoff and Anyango 2022). This perspective emphasizes that agricultural policies must align with biodiversity conservation efforts to mitigate the adverse effects of land-use changes on ecosystems. The concept of social-ecological systems (SES) is crucial in understanding the long-term sustainability of human–environment interactions. Lafuite et al. (2017) discussed how delayed behavioral shifts in response to biodiversity loss can undermine the resilience of SES, potentially leading to social-ecological traps. This indicates that immediate actions are necessary to address the root causes of biodiversity loss and to foster adaptive management practices that enhance the resilience of ecosystems and communities alike. In the context of global supply chains, Cabernard et al. (2022) highlighted the importance of sustainability assessments encompassing climate change, air pollution, and biodiversity loss. The interconnections between these environmental challenges necessitate a holistic approach to sustainability that integrates economic growth with ecological health. By promoting responsible consumption and production patterns, nations can work towards achieving the Sustainable Development Goals (SDGs) while mitigating environmental impacts.

The built environment is pivotal in supporting biodiversity and achieving the SDGs. Opoku (2019) emphasized that a built environment designed with biodiversity in mind can enhance ecosystem services and contribute to human well-being. This highlights the importance of incorporating ecological considerations into urban planning and development processes to foster sustainable communities that thrive alongside natural ecosystems. The urgency of addressing biodiversity loss is further underscored by the alarming rates of species extinction, as Gibassier et al. (2019) noted. The United Nations has recognized the critical need to halt biodiversity loss by 2030, yet current trends indicate that these goals may not be met without significant intervention. This situation calls for immediate and concerted efforts from all sectors of society, including businesses, governments, and civil society, to implement effective biodiversity conservation strategies. The evolution of corporate commitments to biodiversity, as explored by De Silva et al. (2019), illustrates the growing recognition of businesses as key players in biodiversity conservation. However, challenges remain in translating these commitments into actionable strategies that yield tangible outcomes. Companies must navigate the complexities of biodiversity impacts and develop frameworks that align their operations with sustainability objectives. Cuckston (2019) argued for the necessity of ecologically defensible calculations of biodiversity loss and gain, emphasizing the importance of transparent and accountable reporting mechanisms. This is particularly relevant in corporate sustainability initiatives, where the potential for greenwashing poses a significant risk to genuine conservation efforts. By establishing robust metrics for biodiversity assessment, organizations can better understand their ecological footprints and implement strategies to mitigate negative impacts.

The loss of forest fauna, as discussed by Krause and Tilker (2022), has profound implications for achieving the SDGs, particularly with nutrition, health, and climate action. The interlinkages between biodiversity and human well-being necessitate a comprehensive approach that recognizes the vital role of ecosystems in supporting sustainable development. Addressing defaunation and its consequences is critical for maintaining ecosystem functions that underpin food security and health outcomes. The Ecological Footprint is a valuable tool for tracking human-induced pressures on biodiversity and ecosystems. Lazarus et al. highlight its potential in informing biodiversity-related policies and guiding sustainable practices (Lazarus et al. 2015). By breaking down the components of the Ecological Footprint, policymakers can identify key drivers of biodiversity loss and develop targeted interventions to address these challenges. China's biodiversity crisis, exacerbated by economic trade dynamics, underscores the urgent need for public awareness and action to halt species loss (Liu et al. 2022). Liu et al. (2022) emphasized that effective governance and civil society engagement are essential for implementing strategies promoting sustainable biodiversity use. This highlights the importance of fostering a culture of environmental stewardship that empowers communities to take an active role in conservation efforts. The concept of social-ecological resilience, as articulated by Folke et al. (2016), provides a framework for understanding the dynamics of human–environment interactions. By recognizing the interconnectedness of social and ecological systems, stakeholders can develop adaptive management strategies that enhance resilience and sustainability. This approach is crucial for navigating the complexities of environmental challenges in diverse contexts.

Achieving ecosystem health is a multifaceted challenge that requires integration across various dimensions, including socioeconomic and biophysical factors. Lü et al. (2015) emphasized the need for effective strategies that maintain ecosystem health while addressing human needs. This calls for collaborative efforts among stakeholders to devise solutions that balance ecological integrity with economic development. Spatial variations in biodiversity loss across regions, as highlighted by Lü et al. (2020), illustrated the diverse impacts of environmental stressors on ecosystems. Understanding these spatial dynamics is essential for developing targeted conservation strategies that address specific regional challenges. Policymakers must consider local contexts when designing interventions to mitigate biodiversity loss and promote sustainable land-use practices.

Human expansion has been a significant driver of biodiversity loss across Asia, as Yang (2023) demonstrated. The need for ambitious conservation targets and protected area coverage underscores the urgency of halting this trend. Addressing habitat loss and fragmentation requires coordinated efforts among governments, NGOs, and local communities to implement effective conservation measures. The competition between development initiatives, such as those in China and Japan, raises questions about the sustainability of infrastructure projects (Mao 2023). As both countries strive for quality and sustainability, the potential for triangular cooperation presents an opportunity to enhance environmental governance and promote sustainable development practices. Collaborative approaches can facilitate knowledge sharing and resource mobilization to address common challenges. Australia's biodiversity offset legislation exemplifies the complexities of balancing development with environmental protection (Abdo and Kemp 2019). While offsets aim to mitigate the impacts of development on biodiversity, their effectiveness in achieving sustainable outcomes remains a topic of debate. A critical assessment of offset policies is necessary to ensure they align with broader sustainability principles and contribute to genuine conservation efforts.

The post-2000 biodiversity agendas, as discussed by Moranta et al. (2022), highlight the need to transcend growth-oriented strategies that contribute to biodiversity loss. Recognizing the economic drivers of biodiversity decline is essential for formulating effective conservation policies prioritizing ecological health and economic development. This shift in perspective is crucial for fostering a sustainable future that respects the intrinsic value of biodiversity. Balancing development with environmental protection.

2.8 Conclusion

Sustainable development is an essential framework for addressing the global challenges of economic growth, environmental protection, and social equity. Rooted in the principles outlined by the Brundtland Report and further developed through the United Nations' Sustainable Development Goals (SDGs), sustainable development promotes the integration of these three dimensions to create a balanced approach to development that ensures the well-being of current and future generations. The interconnectedness of these goals—environmental sustainability, economic prosperity, and social inclusion—highlights the need for a holistic perspective when implementing policies and strategies to achieve sustainability. The core principles of sustainable development—intergenerational equity, environmental integrity, economic viability, and social inclusion—are vital in ensuring that future generations inherit a livable and equitable world. Intergenerational equity emphasizes the need to safeguard resources and ecosystems for future generations, while environmental integrity underscores the protection of biodiversity and ecosystem services. Economic viability focuses on creating economic systems that are both profitable and sustainable in the long term without degrading natural resources. Social inclusion, an equally important component, ensures that all members of society benefit from sustainable practices, addressing inequality and access to resources. These principles are profoundly interconnected and mutually reinforcing, forming the foundation of sustainable development. The practical application of sustainable development involves addressing three interrelated dimensions: environmental, economic, and social sustainability. Environmental sustainability requires effective resource management, pollution control, and the conservation of ecosystems. In this context, environmental management plays a critical role, providing tools such as Environmental Impact Assessments (EIAs), Life Cycle Assessments (LCAs), and ecosystem-based management strategies. These tools help identify and mitigate potential environmental risks, ensuring that development projects minimize harm to natural systems. In turn, adaptive management strategies enable continuous learning and adjustment as new information becomes available, promoting resilience in the face of environmental changes.

Economic sustainability, on the other hand, calls for a shift towards sustainable production and consumption patterns, supported by technological advancements. The growing emphasis on a circular economy, where resources are reused, recycled, and repurposed, is an example of how innovation can lead to more sustainable economic practices. This transition also demands that businesses and industries adopt corporate social responsibility (CSR) principles, incorporating sustainability into their operations and supply chains. Such practices not only contribute to environmental goals but also create economic opportunities that benefit communities. Social sustainability emphasizes the importance of community engagement, social justice, and cultural preservation. Ensuring that all individuals have access to the resources they need to live healthy, fulfilling lives is essential to sustainable development. Furthermore, inclusive decision-making processes that involve local communities, indigenous peoples, and marginalized groups are critical in promoting equitable outcomes. This collaborative approach fosters a sense of ownership and responsibility among stakeholders, enhancing the long-term success of sustainability initiatives. Effective governance and policy frameworks are critical to the success of sustainable development. International agreements such as the Paris Agreement on climate change and the Convention on Biological Diversity have set the stage for coordinated global action. At the national and local levels, policies must align with sustainability goals, integrating environmental, social, and economic considerations into planning and decision-making processes. The role of transparency, accountability, and collaboration among governments, businesses, and civil society cannot be overstated. These governance mechanisms ensure that sustainability efforts are monitored, evaluated, and improved over time.

References

Abdo L, Kemp A (2019) Apples for oranges: disparities in offset legislation and policy among jurisdictions and its implications for environmental protection and sustainable development in Australia. Environ Manage Sustain Dev 8(1):172. https://doi.org/10.5296/emsd.v8i1.14081

Agarwal A, de los Angeles MS, Bhatia R, Chéret I, Davila-Poblete S, Falkenmark M et al (2000) Integrated water resources management. Global Water Partnership, Stockholm, pp 1–67

Akudugu MA, Ogwu MC (2024) Sustainable development policies and interventions: a bibliometric analysis of the contributions of the academic community. J Clean Prod (Article # 139919). https://doi.org/10.1016/j.jclepro.2023.139919

Arnstein SR (1969) A ladder of citizen participation. J Am Inst Plann 35(4):216–224. https://doi.org/10.1080/01944366908977225

Arsaeva IL, Mashyanova EE, Sekerin VD, Gorokhova AE, Dolgov SI (2019) Infrastructure and ecology as important systems for sustainable urban development. Int J Innov Technol Explor Eng 9(2):4219–4223. https://doi.org/10.35940/ijitee.b7581.129219

Asnor AS, Al-Mohammad MS, Wan Ahmad S, Almutairi S, Rahman RA (2022) Challenges for implementing environmental management plans in construction projects: the case of Malaysia. Sustainability 14(10):6231. https://doi.org/10.3390/su14106231

Astuti W, Santosa BH, Hardiana A (2015) Contribution of Bedah Kampung and relocaton program toward poverty alleviation–best practises of surakarta. J Arch Environ 14(1):1–18. https://doi.org/10.12962/j2355262x.v14i1.a883

Axon S (2024) Unveiling understandings of the Rio declaration's sustainability principles: a case of alternative concepts, misaligned (dis)connections, and terminological evolution. Sustainability 16(6):2600. https://doi.org/10.3390/su16062600

Baffoe G, Zhou X, Moinuddin M, Somanje A, Kuriyama A, Mohan G, Saito O, Takeuchi K (2021) Urban–rural linkages: effective solutions for achieving sustainable development in Ghana from an SDG interlinkage perspective. Sustain Sci 16(4):1341–1362. https://doi.org/10.1007/s11625-021-00929-8

Barua A (2015) Towards a philosophy of sustainability: the Gandhian way. Sociol Anthropol 3(2):136–143. https://doi.org/10.13189/sa.2015.030208

Bebbington J, Unerman J (2018) Achieving the United Nations sustainable development goals. Account Audit Account J 31(1):2–24. https://doi.org/10.1108/aaaj-05-2017-2929

Berkes F, Colding J, Folke C (2000) Rediscovery of traditional ecological knowledge as adaptive management. Ecol Appl 10(5):1251–1262. https://doi.org/10.1890/1051-0761(2000)010[1251:ROTEKA]2.0.CO;2

Bernal AS, Ablian JD (2023). A review on the innovation of renewable energy system. Int J Comput Eng Res Trends 10(2):45–53. https://doi.org/10.22362/ijcert/2023/v10/i02/v10i0201

Biswas AK (2004) Integrated water resources management: a reassessment: a water forum contribution. Water Int 29(2):248–256

Boron S, Murray K (2004) Bridging the unsustainability gap: a framework for sustainable development. Sustain Dev 12(2):65–73. https://doi.org/10.1002/sd.231

Bruce P, Ohlsson Y (2020) Environmental goals addressed in assessments of contaminated sediments. Integr Environ Assess Manag 16(1):128–139. https://doi.org/10.1002/ieam.4223

Burgman M (2005) Risks and decisions for conservation and environmental management. Cambridge University Press

Burns F, Eaton MA, Burfield IJ, Klvaňová A, Šilarová E, Staneva A, Gregory RD (2021) Abundance decline in the avifauna of the European Union reveals cross-continental similarities in biodiversity change. Ecol Evol 11(23):16647–16660. https://doi.org/10.1002/ece3.8282

Burrough PA, McDonnell RA, Lloyd CD (2015) Principles of geographical information systems. Oxford University Press, USA

Cabernard L, Pfister S, Hellweg S (2022) Improved sustainability assessment of the G20's supply chains of materials, fuels, and food. Environ Res Lett 17(3):034027. https://doi.org/10.1088/1748-9326/ac52c7

Čadež S, Czerny A (2016) Climate change mitigation strategies in carbon-intensive firms. J Clean Prod 112:4132–4143. https://doi.org/10.1016/j.jclepro.2015.07.099

Cai J, Huang C, Deng Z, Li L (2023) Transport accessibility and poverty alleviation in Guizhou province of China: spatiotemporal pattern and impact analysis. Sustainability 15(4):3143. https://doi.org/10.3390/su15043143

Chen F (2013) Managers' views on environmental management: an examination of the Taiwanese airline industry. J Sustain Dev 6(1):65. https://doi.org/10.5539/jsd.v6n1p65

Chen Y (2023) Using the pestle method to analyze the relationship between sustainable development and economic growth: a comparison of Chengdu, China and Sheffield, UK. Adv Econ Manage Polit Sci 62(1):45–52. https://doi.org/10.54254/2754-1169/62/20231311

Coenen J, Bager S, Meyfroidt P, Newig J, Challies E (2021) Environmental governance of china's belt and road initiative. Environ Policy Gov 31(1):3–17. https://doi.org/10.1002/eet.1901

Cuckston T (2019) Seeking an ecologically defensible calculation of net loss/gain of biodiversity. Account Audit Account J 32(5):1358–1383. https://doi.org/10.1108/aaaj-01-2018-3339

Curran MA (2012) Life cycle assessment handbook: a guide for environmentally sustainable products. Wiley

Curtin R, Prellezo R (2010) Understanding marine ecosystem based management: a literature review. Mar Policy 34(5):821–830. https://doi.org/10.1016/j.marpol.2010.01.003

Davide M, De Cian E, Bernigaud A (2018) Energy for adaptation: connecting the paris agreement with the sustainable development goals. SSRN Electron J. https://doi.org/10.2139/ssrn.3290449

De Silva GC, Regan EC, Pollard EHB, Addison PFE (2019) The evolution of corporate no net loss and net positive impact biodiversity commitments: understanding appetite and addressing challenges. Bus Strateg Environ 28(7):1481–1495. https://doi.org/10.1002/bse.2379

Desa S, Abdullah M, Mutalib N, Mansor R (2021) The readiness of integrating sustainable development into biology teacher education program. Cakrawala Pendidikan 40(2):305–315. https://doi.org/10.21831/cp.v40i2.37116

Dobers P, Zilahy G (2023) Sustainability in transforming societies. Sustain Dev 31(6):3965–3967. https://doi.org/10.1002/sd.2594

Doelman J, Beier F, Stehfest E, Bodirsky B, Beusen A, Humpenöder F et al (2022) Quantifying synergies and trade-offs in the global water-land-food-climate nexus using a multi-model scenario approach. Environ Res Lett 17(4):045004. https://doi.org/10.1088/1748-9326/ac5766

Dondisch L, Montanez S (2023) What impact can financial engineering have on hastening the transition to sustainable energy? J Stud Res 12(4):5445. https://doi.org/10.47611/jsrhs.v12i4.5445

Du Pisani JA (2006) Sustainable development—historical roots of the concept. Environ Sci 3(2):83–96. https://doi.org/10.1080/15693430600688831

Dzebo A, Iacobuţă GI, Beaussart R (2023) The Paris agreement and the sustainable development goals: evolving connections. Stockholm, Sweden. https://doi.org/10.51414/sei2023.036

Ellen MacArthur Foundation (2013) Towards the circular economy: economic and business rationale for an accelerated transition

Eneji RI, Ering SO (2018) Community-based natural resources management and rural development in Nigeria: a theoretical review. Int J Human Social Sci 8(11):111–116. https://doi.org/10.30845/ijhss.v8n11p13

EPA (1992) Pollution prevention act of 1990. Environmental Protection Agency, U.S

Esen E, Çalışkan AO (2019) Green human resource management (GHRM) and environmental sustainability. Press Acad Proc 9(1):58–60. https://doi.org/10.17261/pressacademia.2019.1065

Eya AM, Sinniah GK, Junaidu AM, Zubairu M (2022) Comparing environmental management and cities sustainability as a basis for sustainable development in Nigeria. Plan Malay 20. https://doi.org/10.21837/pm.v20i21.1119

Fahrudi AN (2020) Alleviating poverty through social innovation. Aust Account Bus Finance J 14(1):71–78. https://doi.org/10.14453/aabfj.v14i1.7

Faisal B, Dahlan MZ, Chaeriyah S, Hutriani IW, Amelia M (2022) Analysis of green infrastructure development policy in Indonesia: an adaptive strategy for sustainable landscape development.

IOP Conf Ser Earth Environ Sci 1092(1):012013. https://doi.org/10.1088/1755-1315/1092/1/012013
Ferreira A, Pinheiro MD, de Brito J, Mateus R (2019) Decarbonizing strategies of the retail sector following the Paris agreement. Energy Policy 135:110999. https://doi.org/10.1016/j.enpol.2019.110999
Finlayson M, Cruz RD, Davidson N, Alder J, Cork S, De Groot RS, et al (2005) Millennium ecosystem assessment: ecosystems and human well-being: wetlands and water synthesis. Island Press
Folke C, Biggs R, Norström AV, Reyers B, Rockström J (2016) Social-ecological resilience and biosphere-based sustainability science. Ecol Soc 21(3):41. https://doi.org/10.5751/es-08748-210341
Freeman HM (1995) Industrial pollution prevention handbook. McGraw-Hill
Gandiwa E, Heitkönig IM, Lokhorst AM, Prins HH, Leeuwis C (2013) Campfire and human-wildlife conflicts in local communities bordering northern Gonarezhou National Park, Zimbabwe. Ecol Soc 18(4):7. https://doi.org/10.5751/es-05817-180407
Garnett ST, Lawes MJ, James R, Bigland K, Zander KK (2016) Portrayal of sustainability principles in the mission statements and on home pages of the world's largest organizations. Conserv Biol 30(2):297–307. https://doi.org/10.1111/cobi.12617
Ghisellini P, Cialani C, Ulgiati S (2016) A review on circular economy: the expected transition to a balanced interplay of environmental and economic systems. J Clean Prod 114:11–32. https://doi.org/10.1016/j.jclepro.2015.09.007
Ghosh BK, Mekhilef S, Ahmad S, Ghosh SK (2022) A review on global emissions by e-products based waste: technical management for reduced effects and achieving sustainable development goals. Sustainability 14(7):4036. https://doi.org/10.3390/su14074036
Gibassier D, Maas K, Schaltegger S (2019) Special issue of business, strategy, and the environment call for papers business, society, biodiversity, and natural capital. Bus Strateg Environ 28(5):921–924. https://doi.org/10.1002/bse.2348
Giddings B, Hopwood B, O'Brien G (2002) Environment, economy and society: fitting them together into sustainable development. Sustain Dev 10(4):187–196. https://doi.org/10.1002/sd.199
Glasson J, Therivel R (2013) Introduction to environmental impact assessment, 4th edn. Routledge, London. https://doi.org/10.4324/9781315881218
Goodchild MF, Longley PA, Maguire DJ, Rhind DW (2005) Geographic information systems and science, vol 17. Wiley, West Sussex, p 517
Grumbine RE (1994) What is ecosystem management? Conserv Biol 8(1):27–38
Holdgate MW (1987) Our common future: report of the world commission on environment and development. Oxford University Press. Environ Conserv 14(3):280–282
Honegger M, Michaelowa A, Roy J (2021) Potential implications of carbon dioxide removal for the sustainable development goals. Clim Policy 21(5):678–698. https://doi.org/10.1080/14693062.2020.1843388
Huang GQ, Guo SX, Xiong W (2014) Model design of self-service intelligent resource management system based on cloud platform. Adv Mater Res 846–847:1491–1495. https://doi.org/10.4028/www.scientific.net/amr.846-847.1491
Iacobuţă GI, Höhne N, van Soest HL, Leemans R (2021) Transitioning to low-carbon economies under the 2030 agenda: Minimizing trade-offs and enhancing co-benefits of climate-change action for the sdgs. Sustainability 13(19):10774. https://doi.org/10.3390/su131910774
Imran S, Alam K, Beaumont N (2014) Reinterpreting the definition of sustainable development for a more ecocentric reorientation. Sustain Dev 22(2):134–144. https://doi.org/10.1002/sd.537
Intergovernmental Panel on Climate Change (IPCC) (2014) Climate change 2014: synthesis report. IPCC. https://www.ipcc.ch/report/ar5/syr/
IPBES (2019) Global assessment report on biodiversity and ecosystem services. Intergovernmental science-policy platform on biodiversity and ecosystem services. https://doi.org/10.5281/zenodo.3831673

IRENA (2018) Global energy transformation: a roadmap to 2050. Available online: https://www.irena.org/-/media/Files/IRENA/Agency/Publication/2018/Apr/IRENA_Report_GET_2018.pdf

ISO (2006) Environmental management—life cycle assessment—principles and framework. International Organization for Standardization

Issa A, Hanaysha JR (2023) Sustainability performance, executive compensation, market value, and say-on-pay voting adoption: evidence from the STOXX Europe 600. Corp Govern 24(2):418–438. https://doi.org/10.1108/cg-02-2023-0072

Ivashchenko N (2023) Factor system and framework approach to measuring the environmental potential of textile enterprises. E3S Web Conf 460:02018. https://doi.org/10.1051/e3sconf/202346002018

Izah SC, Ogwu MC, Hamidifar H (2024) Biomonitoring of pollutants in the Global South. Springer Nature, Switzerland. https://doi.org/10.1007/978-981-97-1658-6

Jennings V, Bamkole O (2019) The relationship between social cohesion and urban green space: an avenue for health promotion. Int J Environ Res Public Health 16(3):452. https://doi.org/10.3390/ijerph16030452

Jeronen E (2020) Economic sustainability. In: Idowu S, Schmidpeter R, Capaldi N, Zu L, Del Baldo M, Abreu R (eds) Encyclopedia of sustainable management. Springer, Cham. https://doi.org/10.1007/978-3-030-02006-4_197-1

Jiménez J, Vázquez-Brust D, Plaza-Úbeda J, Dijkshoorn J (2013) Environmental protection and financial performance: an empirical analysis in Wales. Int J Oper Prod Manag 33(8):981–1018. https://doi.org/10.1108/ijopm-11-2010-0374

Khosravani F, Abbasi E, Choobchian S, Jalili Ghazizade M (2023) A comprehensive study on criteria of sustainable urban waste management system: using content analysis. Sci Rep 13(1):22526. https://doi.org/10.1038/s41598-023-49187-x

Klitgaard K (2020) Sustainability as an economic issue: a biophysical economic perspective. Sustainability 12(1):364. https://doi.org/10.3390/su12010364

Koval V, Ostapenko O, Halushchak O, Olczak P, Dobrovolska K, Kaptalan S (2023) Economic and environmental dimensions of energy production with the use of renewable technologies. Energy Policy J 28(1):5–22. https://doi.org/10.33223/epj/161355

Krause T, Tilker A (2022) How the loss of forest fauna undermines the achievement of the SDGs. Ambio 51(1):103–113. https://doi.org/10.1007/s13280-021-01547-5

Ladyman M, Gutierrez-Carazo E, Persico F, Temple T, Coulon F (2022) Assessing the performance of environmental management in academic research laboratories. Heliyon 8(3):e09135. https://doi.org/10.1016/j.heliyon.2022.e09135

Lafuite AS, de Mazancourt C, Loreau M (2017) Delayed behavioural shifts undermine the sustainability of social–ecological systems. Proc Royal Soc B Biol Sci 284(1868):20171192. https://doi.org/10.1098/rspb.2017.1192

Lazarus E, Lin D, Martindill J, Hardiman J, Pitney L, Galli A (2015) Biodiversity loss and the ecological footprint of trade. Diversity 7(2):170–191. https://doi.org/10.3390/d7020170

Leach M, Scoones I, Stirling A (2010) Dynamic sustainabilities: technology, environment, social justice. Taylor & Francis, Routledge, p 232

Leclère D, Obersteiner M, Barrett M, Butchart SH, Chaudhary A, De Palma A et al (2020) Bending the curve of terrestrial biodiversity needs an integrated strategy. Nature 585(7826):551–556. https://doi.org/10.1038/s41586-020-2705-y

Lee ACK, Jordan HC, Horsley J (2015) Value of urban green spaces in promoting healthy living and wellbeing: prospects for planning. Risk Manage Healthcare Policy 131–137. https://doi.org/10.2147/rmhp.s61654

Lepenies R, Büttner L, Bärlund I, Jax K, Lyytimäki J, Pedersen A, Richard D (2023) The politics of national SDG indicator systems: a comparison of four European countries. Ambio 52(4):743–756. https://doi.org/10.1007/s13280-022-01809-w

Li W, Yüksel S, Dınçer H (2022) Understanding the financial innovation priorities for renewable energy investors via QFD-based picture fuzzy and rough numbers. Financial Innov 8(1):67. https://doi.org/10.1186/s40854-022-00372-3

Liu B, Liu Y, Zhang A (2021) Heterogeneous impact of CO_2 emissions on renewable energy technology innovation between oil importers and exporters. Energy Environ 32(2):281–294. https://doi.org/10.1177/0958305x20928934

Liu X, Zhang J, Zhang H, Tang D, Hu G, Li X (2022) China's mismatch of public awareness and biodiversity threats under economic trade. Environ Sci Technol 56(13):9784–9796. https://doi.org/10.1021/acs.est.2c00844

Liu Y, Guo Y, Zhou Y (2018) Poverty alleviation in rural china: policy changes, future challenges and policy implications. China Agric Econ Rev 10(2):241–259. https://doi.org/10.1108/caer-10-2017-0192

Londoño-Pineda A, Cano JA, Gómez-Montoya R (2021) Application of AHP for the weighting of sustainable development indicators at the subnational level. Economies 9(4):169. https://doi.org/10.3390/economies9040169

Lü Y, Wang R, Zhang Y, Su H, Wang P, Jenkins A et al (2015) Ecosystem health towards sustainability. Ecosyst Health Sustain 1(1):1–15. https://doi.org/10.1890/ehs14-0013.1

Lü Y, Yang Y, Sun B, Yuan J, Yu M, Stenseth N et al (2020) Spatial variation in biodiversity loss across china under multiple environmental stressors. Sci Adv 6(47):eabd0952. https://doi.org/10.1126/sciadv.abd0952

Lucas PL, Kok MT, Nilsson M, Alkemade R (2013) Integrating biodiversity and ecosystem services in the post-2015 development agenda: goal structure, target areas and means of implementation. Sustainability 6(1):193–216. https://doi.org/10.3390/su6010193

Luo H, Shu Y, Cai Z (2023) Investigating the multidimensional relative poverty in China: evidence from Nanling Yao ethnic group area. Environ Dev Sustain 25(11):12357–12370. https://doi.org/10.1007/s10668-022-02570-6

Luo Y, Bu J (2016) How valuable is information and communication technology? A study of emerging economy enterprises. J World Bus 51(2):200–211. https://doi.org/10.1016/j.jwb.2015.06.001

Mao C (2023) North-South development competition and the quality and sustainability of Chinese and Japanese trans-national infrastructure initiatives. J Int Dev 35(2):331–346. https://doi.org/10.1002/jid.3694

Maradin D (2021) Advantages and disadvantages of renewable energy sources utilization. Int J Energy Econ Policy 11(3):176–183. https://doi.org/10.32479/ijeep.11027

Markkanen S, Anger-Kraavi A (2019) Social impacts of climate change mitigation policies and their implications for inequality. Clim Policy 19(7):827–844. https://doi.org/10.1080/14693062.2019.1596873

Mashari DPS, Zagloel TYM, Soesilo TEB, Maftuchah I (2023) A bibliometric and literature review: alignment of green finance and carbon trading. Sustainability 15(10):7877. https://doi.org/10.3390/su15107877

Migliorelli M (2021) What do we mean by sustainable finance? Assessing existing frameworks and policy risks. Sustainability 13(2):975. https://doi.org/10.3390/su13020975

Moranta J, Torres C, Murray I, Hidalgo M, Hinz H, Gouraguine A (2022) Transcending capitalism growth strategies for biodiversity conservation. Conserv Biol 36(2):e13821. https://doi.org/10.1111/cobi.13821

Munguia N, Varela A, Esquexr J, Velazquez LE (2017) Fostering corporate sustainability in the Mexican coffee industry. PSU Res Rev 1(1):51–62. https://doi.org/10.1108/prr-10-2016-0002

Murphy K (2012) The social pillar of sustainable development: a literature review and framework for policy analysis. Sustain Sci Pract Policy 8(1):15–29. https://doi.org/10.1080/15487733.2012.11908081

Muzdalifah A, Maryono M (2018) Preliminary identification of urban park infrastructure resilience in Semarang Central Java. E3S Web Conf 31:07006. https://doi.org/10.1051/e3sconf/20183107006

Obahiagbon EG, Ogwu MC (2024) The Nexus of business, sustainability, and herbal medicine. In: Izah SC, Ogwu MC, Akram M (eds) Herbal medicine phytochemistry. Reference Series in Phytochemistry. Springer, Cham. https://doi.org/10.1007/978-3-031-21973-3_67-1

Ogwu MC (2019) Towards sustainable development in Africa: the challenge of urbanization and climate change adaptation. In: Cobbinah PB, Addaney M (eds) The geography of climate change adaptation in urban Africa. Springer Nature, Switzerland, pp 29–55. https://doi.org/10.1007/978-3-030-04873-0_2

Ogwu MC (2020) Value of *Amaranthus* [L.] species in Nigeria. In: Waisundara V (ed). Nutritional value of Amaranth. IntechOpen, UK, pp 1–21. https://doi.org/10.5772/intechopen.86990

Ogwu MC (2023) Plants as monitors and managers of pollution. In: Srivastav AL, Grewal AS, Tiwari M, Pham TD (eds) Role of green chemistry in ecosystem restoration to achieve environmental sustainability. Elsevier, Amsterdam, Netherlands. pp 51–60. ISBN: 978-0-443-15291-7. https://doi.org/10.1016/B978-0-443-15291-7.00022-5

Ogwu MC, Izah SC, Dessureault-Rompré J, Gasparatos D (2024b) Sustainable soil systems in global south. Springer Nature, Switzerland. https://doi.org/10.1007/978-981-97-5276-8

Ogwu MC, Osawaru ME, Ahana CM (2014) Challenges in conserving and utilizing plant genetic resources (PGR). Int J Genet Mol Biol 6(2):16–22. https://doi.org/10.5897/IJGMB2013.0083

Oktay S, Bozkurt S, Yazici K (2021) The relationship between carbon disclosure project scores and global 500 companies: a perspective from national culture. SAGE Open 11(2):215824402110145. https://doi.org/10.1177/21582440211014521

Oloyede O (2009) Developing sustainable communities in Africa: components for a framework. Identity Cult Polit 10(2):56–64. https://doi.org/10.57054/icp.v10i2.5119

Opoku A (2019) Biodiversity and the built environment: implications for the sustainable development goals (SDGs). Resour Conserv Recycl 141:1–7. https://doi.org/10.1016/j.resconrec.2018.10.011

Palliwoda J, Banzhaf E, Priess JA (2020) How do the green components of urban green infrastructure influence the use of ecosystem services? Examples from Leipzig, Germany. Landsc Ecol 35(5):1127–1142. https://doi.org/10.1007/s10980-020-01004-w

Paramati SR, Alam MS, Hammoudeh S, Hafeez K (2020) Long-run relationship between R&D investment and environmental sustainability: evidence from the European Union member countries. Int J Financ Econ 26(4):5775–5792. https://doi.org/10.1002/ijfe.2093

Paris Agreement (2015) Paris agreement to the United Nations framework convention on climate change. https://unfccc.int/sites/default/files/english_paris_agreement.pdf

Pearce DW, Barbier EB, Markandya A (1990) Sustainable development: economics and environment in the third world. Edward Elgar Publishing

Pearce DW, Markandya A, Barbier EB (1989) Blueprint for a green economy. Earthscan Publications Ltd

Petts J (2005) Handbook of environmental impact assessment. Blackwell Publishing

Pretty JN (1995) Participatory learning for sustainable agriculture. World Dev 23(8):1247–1263. https://doi.org/10.1016/0305-750X(95)00046-F

Raworth K (2017) Doughnut economics: seven ways to think like a 21st-century economist. Chelsea Green Publishing

REN21 (2019) Asia and the pacific renewable energy status report (Paris: REN21 Secretariat)

Rockström J, Steffen W, Noone K, Persson Å, Chapin FS, Lambin EF et al (2009a) A safe operating space for humanity. Nature 461(7263):472–475. https://doi.org/10.1038/461472a

Rockström J, Steffen W, Noone K, Persson Å, Chapin FS, Lambin EF et al (2009b) Planetary boundaries: exploring the safe operating space for humanity. Ecol Soc 14(2):32

Rumagia F, Boer M, Kurnia R, Kamal M (2020) Sustainability window approach for fisheries management at the coastal area of Ternate Island, North Maluku Province. Indones J Urban Environ Technol 3(2):136–148. https://doi.org/10.25105/urbanenvirotech.v3i2.5476

Sachs JD (2015) The age of sustainable development. Columbia University Press. https://doi.org/10.7312/sach17314

Sadykova G (2022) Essence and factors of socio-economic sustainability of the region. Socialis Ser Social Sci 2:13–28. https://doi.org/10.20319/socv2.1328
Saeed BB, Afsar B, Hafeez S, Khan I, Tahir M, Afridi M (2019) Promoting employee's proenvironmental behavior through green human resource management practices. Corp Soc Responsib Environ Manag 26(2):424–438. https://doi.org/10.1002/csr.1694
Sandström C, Gasper S, Öhman K (2013) Conflict resolution through ecosystem-based management: the case of Swedish moose management. Int J Commons 7(2):349. https://doi.org/10.18352/ijc.349
Sardain A, Tang C, Potvin C (2016) Towards a dashboard of sustainability indicators for panama: a participatory approach. Ecol Ind 70:545–556. https://doi.org/10.1016/j.ecolind.2016.06.038
Schneider F, Tarawali S (2021) Sustainable development goals and livestock systems. Revue Scientifique et Technique (International Office of Epizootics) 40(2):585–595. https://doi.org/10.20506/rst.40.2.3247
Seifert C (2018) The barriers for voluntary environmental management systems—the case of EMAS in hospitals. Sustainability 10(5):1420. https://doi.org/10.3390/su10051420
Seifert C, Guenther E (2019) Prevention is better than cure—environmental management measures in hospitals. Corp Soc Responsib Environ Manag 26(4):781–790. https://doi.org/10.1002/csr.1720
Sen A (1999) Development as freedom. Oxford University Press
Sharma, M. (2019). A review of scenario and status of natural resource management practices in Nepal. Acta Sci Agric 3(10):79–84. https://doi.org/10.31080/asag.2019.03.0651
Shrestha P (2019) Mainstreaming the 'brown' agenda. Sustainability 11(23):6660. https://doi.org/10.3390/su11236660
Soergel B, Kriegler E, Weindl I, Rauner S, Dirnaichner A, Ruhe C et al (2021) A sustainable development pathway for climate action within the un 2030 agenda. Nat Clim Chang 11(8):656–664. https://doi.org/10.1038/s41558-021-01098-3
Stern N (2008) The economics of climate change. Am Econ Rev 98(2):1–37. https://doi.org/10.1017/CBO9780511817434
Suter GW (2006) Ecological risk assessment, 2nd edn. CRC Press
Takala T, Häyry M (2023). Justainability. Cambridge Quarterly of Healthcare Ethics, pp 1–11. https://doi.org/10.1017/s0963180123000361
Tanaka K, Azar C, Boucher O, Ciais P, Gaucher Y, Johansson DJ (2022) Paris agreement requires substantial, broad, and sustained policy efforts beyond covid-19 public stimulus packages. Clim Change 172(1):1. https://doi.org/10.1007/s10584-022-03355-6
Tayouga S, Gagné S (2016) The socio-ecological factors that influence the adoption of green infrastructure. Sustainability 8(12):1277. https://doi.org/10.3390/su8121277
Trancossi M, Pascoa J, Catellani T (2023) Exergy, ecology and democracy - concepts of a vital society or a proposal for an exergy tax 30 years after—part 1: generalities. Thermal Sci 27(2 Part B):1337–1353. https://doi.org/10.2298/tsci220907019t
UNEP (2020) Human development report 2020: the next frontier—human development and the anthropocene
United Nations (2015) Transforming our world: the 2030 agenda for sustainable development. UN General Assembly. https://www.un.org/ga/search/view_doc.asp?symbol=A/RES/70/1&Lang=E
United Nations Environment Programme (UNEP) (2019) Global environment outlook—GEO-6: Healthy planet, healthy people. Cambridge University Press. https://doi.org/10.1017/9781108627146
United Nations Development Programme (UNDP) (2020) Human development report 2019: beyond income, beyond averages, beyond today: empowered lives. Resilient nations. Inequalities in human development in the 21st century. https://hdr.undp.org/sites/default/files/hdr2019.pdf
Usman M, Rădulescu M, Balsalobre-Lorente D, Rehman A (2023) Investigation on the causality relationship between environmental innovation and energy consumption: empirical evidence

from EU countries. Energy Environ 34(8):3130–3159. https://doi.org/10.1177/0958305x221120931
Usman O, Iorember PT, Jelilov G, Işık A, Ike GN, Sarkodıe SA (2021) Towards mitigating ecological degradation in g-7 countries: accounting for economic effect dynamics, renewable energy consumption, and innovation. Heliyon 7(12):e08592. https://doi.org/10.1016/j.heliyon.2021.e08592
Utuk IO (2022) The impact of poverty on sustainable economic development in Nigeria. AKSU J Manage Sci 7(1&2):80–95. https://doi.org/10.61090/aksujomas.2022.005
Uysal G (2015) Venn relations of human resource: impact of human resource management on economy. Chin Bus Rev 14(12):583–589. https://doi.org/10.17265/1537-1506/2015.12.003
van der Hoff R, Anyango MN (2022) Transforming biodiversity governance: biodiversity finance and transformative governance: the limitations of innovative financial instruments. In: Transforming biodiversity governance. Cambridge University Press, pp 115–136. https://doi.org/10.1017/9781108856348
Vannadil P, Prasanna C (2023) Integrating sustainable development goals into environment impact assessment in India: a conceptual analysis. Sustain Dev 31(6):3993–4006. https://doi.org/10.1002/sd.2579
Varela Pérez P, Greiner BE, von Cossel M (2022) Socio-economic and environmental implications of bioenergy crop cultivation on marginal African drylands and key principles for a sustainable development. Earth 3(2):652–682. https://doi.org/10.3390/earth3020038
Von Stechow C, McCollum D, Riahi K, Minx JC, Kriegler E, Van Vuuren DP et al (2015) Integrating global climate change mitigation goals with other sustainability objectives: a synthesis. Annu Rev Environ Resour 40(1):363–394. https://doi.org/10.1146/annurev-environ-021113-095626
Wang M, Peng X (2023) How to develop sustainably after poverty alleviation in poverty-stricken areas under paired assistance: a quantitative assessment framework based on system dynamics model. Sustainability 15(6):4764. https://doi.org/10.3390/su15064764
World Economic Forum (WEF) (2020) The global risks report 2020. World Econ Forum. https://www.weforum.org/reports/the-global-risks-report-2020
World Wide Fund for Nature (WWF) (2020) Living planet report 2020: bending the curve of biodiversity loss. WWF. https://www.worldwildlife.org/publications/living-planet-report-2020
Xu L, Deng X, Jiang Q, Ma F (2021) Identification and alleviation pathways of multidimensional poverty and relative poverty in counties of China. J Geog Sci 31(12):1715–1736. https://doi.org/10.1007/s11442-021-1919-8
Yang C, Li Q, Wang X, Cui A, Chen J, Liu H et al (2023) Human expansion-induced biodiversity crisis over Asia from 2000 to 2020. Research 6:0226. https://doi.org/10.34133/research.0226
Yin X, Chen J, Li J (2022) Rural innovation system: revitalize the countryside for a sustainable development. J Rural Stud 93:471–478. https://doi.org/10.1016/j.jrurstud.2019.10.014
You K, Bianco S, Amankwah-Amoah J (2020) Closing technological gaps to alleviate poverty: evidence from 17 Sub-Saharan African countries. Technol Forec Soc Chang 157:120055. https://doi.org/10.1016/j.techfore.2020.120055
Zeng G, Zhang C, Li S, Sun H (2021) The dynamic impact of agricultural fiscal expenditures and gross agricultural output on poverty reduction: a VAR model analysis. Sustainability 13(11):5766. https://doi.org/10.3390/su13115766
Zhang L, Geng H, Qu Y, Peng S, Feng X, Liu Y et al (2023) The hukou system-induced rural built-up land outpaces urban expansion in threatening biodiversity in China. https://doi.org/10.21203/rs.3.rs-3097025/v1
Zheng S (2020) EBM–based innovation strategy for integrated management of marine ecological protection. Geol Behav 3(2):15–19. https://doi.org/10.26480/gbr.02.2019.15.19
Zhou L (2023) Improving the sustainable development of firms: evidence from the environmental constraint target policy in China. Sustain Dev 32(1):1217–1225. https://doi.org/10.1002/sd.2730
Zovanyi G (2008) Confronting the 'sustainable-growth' fallacy impeding the realization of sustainable development and sustainable cities. WIT Trans Ecol Environ 117:411–420. https://doi.org/10.2495/sc080391

Chapter 3
Integrated and Collaborative Approaches to Managing Natural Resources

Kurotimipa Frank Ovuru, Prasann Kumar, Debjani Choudhury, and Matthew Chidozie Ogwu

Abstract Natural resource management (NRM) is central to achieving sustainable development, requiring integrated and collaborative approaches that involve diverse stakeholders. This chapter focuses on the role of Integrated Natural Resource Management (INRM) in promoting sustainable outcomes through the interconnection of ecological, social, and economic factors. Over time, the transition from traditional to modern management strategies has highlighted the critical role of local knowledge and community participation in ensuring effective resource use. INRM creates sustainable results for present and future generations by encouraging inclusion, adaptive management, and stakeholder participation. Integrating top-down and bottom-up approaches in INRM enhances stakeholder engagement and impact by aligning policies and resources from higher authorities with community-driven insights and initiatives. This dynamic feedback loop fosters trust, boosts interest, and ensures sustainable and inclusive resource management. Interests and influence are central to natural resource management, shaping decisions and stakeholder dynamics. Interests reflect the goals stakeholders seek, such as economic benefits or ecological sustainability, while influence denotes their power to impact decisions. Balancing diverse interests and leveraging influence fosters inclusive decision-making, reduces conflicts, and promotes sustainable resource use for long-term ecological and socio-economic benefits. Collaborative models, such as Community-Based Natural

K. F. Ovuru (✉)
Department of Disease Control and Immunization, Bayelsa State Primary Health Care Board, Yenagoa, Bayelsa State, Nigeria
e-mail: kurotimipaovuru@yahoo.com

P. Kumar
Department of Agronomy, School of Agriculture, Lovely Professional University, Phagwara, Punjab, India

D. Choudhury
Department of Plant Pathology, School of Agriculture, Lovely Professional University, Phagwara, Punjab, India

M. C. Ogwu
Goodnight Family Department of Sustainable Development, 212 Living Learning Center, Appalachian State University, Boone, NC, USA

M. C. Ogwu and S. Chibueze Izah (eds.), *Evaluating Environmental Processes and Technologies*, Environmental Science and Engineering,
https://doi.org/10.1007/978-3-031-85327-2_3

Resource Management, Public–Private Partnerships, and transboundary cooperation, have effectively addressed the complexities of resource management. These frameworks improve resource governance, enhance conflict resolution, and reduce the burden on individual stakeholders. However, despite the potential benefits, several challenges remain. Institutional and policy barriers, conflicting stakeholder interests, and resource constraints impede the implementation of integrated and collaborative approaches. To overcome these barriers, lessons from past initiatives emphasize the importance of effective governance and adaptive strategies. The implications for sustainable development are significant, with integrated management contributing to advancing the United Nations Sustainable Development Goals, environmental sustainability, climate change resilience, and poverty reduction. Ultimately, integrated and collaborative approaches to managing natural resources are essential for fostering ecological balance, supporting livelihoods, and ensuring the wellbeing of future generations.

Keywords Natural resource management · Integrated natural resource management · Collaborative governance · Sustainable development · Stakeholder collaboration · Environmental sustainability · Climate resilience

3.1 Introduction

Natural resource management (NRM) is a critical field that encompasses the sustainable use and conservation of natural resources, including land, water, soil, plants, and animals. Its importance lies in the fact that effective NRM is essential for maintaining ecological balance, supporting livelihoods, and ensuring the sustainability of resources for future generations. The increasing pressures from population growth, urbanization, and climate change have made adopting integrated and collaborative approaches to NRM imperative. These approaches enhance resource use efficiency and promote stakeholder engagement, which is crucial for addressing the complex challenges associated with resource management (Scott 2015; Behnken et al. 2016; Said and Kusumasari 2022). Integrated and collaborative approaches to NRM emphasize the need for cooperation among various stakeholders, including government agencies, local communities, and private sector actors. Such collaboration can improve environmental outcomes by encouraging the sharing of knowledge, resources, and responsibilities. For instance, collaborative governance has been shown to enhance the effectiveness of resource management by fostering trust and communication among stakeholders, which can lead to better decision-making and conflict resolution (Scott 2015; Hossu et al. 2018; Scott and Thomas 2017). Furthermore, these approaches often incorporate adaptive management strategies, allowing flexibility and responsiveness to changing environmental conditions and stakeholder needs (Tengberg and Valencia 2018).

Historically, top-down approaches have often characterized traditional resource management strategies, where centralized authorities make decisions without significant input from local communities. This has led to inefficiencies and conflicts, as local knowledge and needs are frequently overlooked (Behnken et al. 2016; Berkes 2010). In contrast, modern resource management strategies advocate for decentralized governance and community-based management, recognizing local knowledge's value and stakeholder participation's importance (Eneji and Ering 2018). The shift from traditional to modern strategies reflects a growing understanding of the interconnectedness of ecological systems and the need for holistic approaches that consider social, economic, and environmental factors (Said and Kusumasari 2022; Koontz and Newig 2014). Exploring various approaches to NRM reveals a range of benefits and challenges associated with their implementation. Integrated approaches, such as Integrated Water Resource Management (IWRM) and Integrated Coastal Zone Management (ICZM), aim to harmonize the management of land and water resources, addressing conflicts and promoting sustainable use (Sabai 2023; Qi and Altinakar 2011). These approaches can lead to improved resource efficiency, enhanced ecosystem services, and increased resilience to climate change. However, challenges remain, including the need for effective coordination among diverse stakeholders, the complexity of integrating multiple objectives, and the potential for institutional resistance to change (Behnken et al. 2016; Sabai 2023; Scott and Thomas 2017). The implications of effective NRM for sustainable development are profound. By promoting sustainable practices, NRM can contribute to poverty alleviation, food security, and the preservation of biodiversity (Izah and Ogwu 2023; Ogwu 2019). For instance, community-based natural resource management (CBNRM) initiatives have successfully empowered local communities to enhance their livelihoods while ensuring the sustainable use of resources (Eneji and Ering 2018; Gandiwa et al. 2013). However, achieving sustainable development through NRM requires addressing underlying socio-economic inequalities and ensuring all stakeholders have a voice in decision-making processes (Said and Kusumasari 2022; Haeril and Purnomo 2019).

Moreover, the role of technology and innovation in enhancing NRM practices must be considered. Using internet-based collaboration tools and data-driven decision-making can facilitate better stakeholder communication and coordination, leading to more effective resource management outcomes (Wahl and Kitchel 2017; Diamond and Vangen 2016). Additionally, integrating traditional knowledge with modern scientific approaches can enhance the resilience of resource management systems, particularly in the face of climate change (Mohd Arzaman et al. 2022; Ikhajiagbe et al. 2021). However, successfully implementing these technologies requires adequate training and stakeholder capacity building to ensure effective use (Wahl and Kitchel 2017).

This chapter explores the significance of integrated and collaborative approaches in managing natural resources to achieve sustainability. It examines the principles of Integrated Natural Resource Management (INRM), highlighting the importance of ecological, social, and economic considerations, stakeholder inclusivity, and adaptive management. The chapter further discusses various collaborative frameworks, their benefits, and the challenges involved in their implementation while emphasizing the implications for sustainable development, including contributions to the United Nations Sustainable Development Goals (SDGs) and long-term resilience to climate change and biodiversity loss.

3.2 Principles of Integrated Natural Resource Management

Integrated Natural Resource Management (INRM) is a holistic approach that seeks to balance ecological, social, and economic factors in managing natural resources (Fig. 3.1). This approach recognizes the interdependence of ecosystems, communities, and markets, essential for sustainable development (Obahiagbon and Ogwu 2023, 2024a, b). The principles of INRM emphasize the need for a comprehensive understanding of the interactions between various components of the environment and human society. For instance, Ali-Nakyea et al. (2019) highlighted the importance of defining responsibilities in natural resource revenue management to ensure that resources are utilized effectively and transparently, benefiting current and future generations. This principle aligns with the notion that sustainable management must consider the long-term implications of resource use on both ecological integrity and community wellbeing. Table 3.1 shows the core principles of integrated natural resource management.

Natural Resource Management Systems (NRMS) encompass several key components essential for sustainable and equitable management. Resource inventory and assessment involve identifying and quantifying available resources while monitoring their health and usage patterns to ensure sustainable utilization. Policy and governance establish legal and institutional frameworks, emphasizing transparency, accountability, and enforcement to promote sustainable practices. Stakeholder engagement is crucial, involving governments, communities, industries, and NGOs in collaborative decision-making processes that encourage participatory governance. Technological tools such as Geographic Information Systems (GIS), remote sensing, data analytics, and predictive modeling are vital in mapping, monitoring, and optimizing resource management. Sustainability goals aim to balance ecological conservation with economic development while reducing environmental impacts and mitigating climate change. Conflict resolution mechanisms address competing interests among stakeholders, ensuring equitable benefit distribution and fostering harmonious resource use. Capacity building enhances the skills and knowledge of stakeholders through education, training, and awareness programs, empowering them to contribute effectively to sustainable resource management initiatives. These components collectively form the backbone of effective NRMS, ensuring ecosystems' long-term health and communities' well-being. Table 3.2 outlines the key elements of NRMS, highlighting their essential roles in achieving sustainable resource use and conservation. Each component is described briefly and illustrated with examples to demonstrate its practical application in managing resources such as land, water, forests, and biodiversity. These components collectively ensure

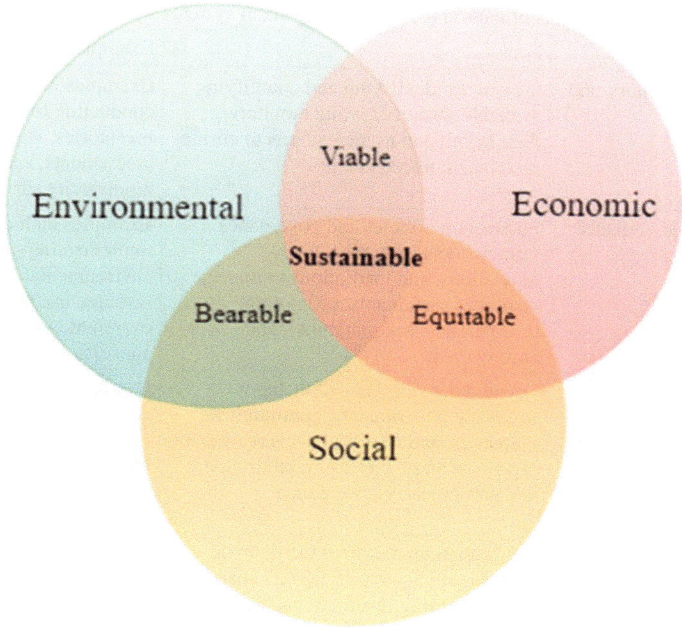

Fig. 3.1 Key components of integration. *Source* Dobrovolskienė et al. (2021)

Table 3.1 Core principles of integrated natural resource management

Principle	Key points	Details
Key components of integration	Ecological, social, and economic factors	Ensures holistic resource management by balancing ecosystem health, community wellbeing, and markets
Stakeholder inclusivity and participatory decision-making	Involving diverse stakeholders, including local communities, government, private sector, and NGOs	Promotes transparency, shared responsibility, and empowerment of marginalized groups
Adaptive management and feedback loops	Flexible strategies that evolve with changing conditions	It relies on monitoring, evaluation, and learning to refine and improve resource management practices

ecological balance, stakeholder inclusivity, and socio-economic benefits, forming a comprehensive framework for effective natural resource management.

Integrating ecological, social, and economic factors requires a nuanced understanding of how these interact. Enukwa et al. (2022) provide evidence from Cameroon, where community forestry initiatives have successfully improved livelihoods while protecting the environment. This case illustrates the potential for local

Table 3.2 Key components of natural resource management systems and their focus

Core component	Focus	Examples
Resource inventory and assessment	It involves identifying and quantifying available resources while monitoring their health and usage patterns to ensure sustainable management	Examples include conducting forest inventories, water quality assessments, and biodiversity surveys
Policy and governance	Focuses on policies and governance frameworks to establish laws, regulations, and institutions to manage resources sustainably, promoting transparency, accountability, and enforcement	Examples include implementing environmental laws, resource use permits, and oversight by regulatory agencies
Stakeholder engagement	Stakeholder engagement includes involving governments, communities, industries, and NGOs in decision-making processes and fostering collaboration and participatory governance	Examples include community forest management programs and resource-sharing agreements facilitated by NGOs
Technological tools	This component utilizes Geographic Information Systems (GIS) for mapping and monitoring resources and remote sensing, data analytics, and predictive modeling to enhance decision-making	Examples include using drones for forest mapping, satellite imagery for monitoring land-use changes, and predictive climate models
Sustainability goals	Focus on balancing ecological conservation with economic development while reducing environmental impacts and mitigating climate change	Examples include integrated watershed management, renewable energy projects, and sustainable agriculture practices
Conflict resolution mechanisms	Conflict resolution mechanisms address competing interests among stakeholders and ensure the equitable distribution of benefits derived from resource management	Examples include mediating disputes over water allocation and ensuring fair sharing of mining royalties
Capacity building	Capacity building enhances the skills and knowledge of stakeholders through education, training, and awareness programs, empowering them to contribute to sustainable resource management	Examples include workshops on sustainable farming techniques, training in GIS tools, and public awareness campaigns

communities to manage resources sustainably when they are empowered and given ownership over their natural resources. Nhem and Lee (2019) further emphasize the interdependence of ecosystems and communities and discuss the significant contributions of forest resources to rural household incomes in Cambodia, underscoring the need for sustainable forestry practices that benefit local populations.

Thus, effective INRM must incorporate local knowledge and techniques to ensure that resource management strategies are relevant and practical. For example, Fig. 3.2 illustrates a dynamic framework for sustainable land use, emphasizing the interdependence between environmental problems, resource supply, human demand, and socio-economic pressures. Policies act as the intermediary, guiding the balance between supply and demand while addressing environmental issues. This cyclical model underscores the importance of aligning human activities with ecological constraints to achieve long-term sustainability and mitigate environmental challenges.

Figure 3.3 illustrates two complementary perspectives of Integrated Water Resources Management (IWRM)—the framework and process views. The framework view of IWRM identifies four key components essential for sustainable water management: enabling environment, institutions and participation, management instruments, and financing. These elements provide the foundation for creating policies, engaging stakeholders, applying practical tools, and ensuring financial support for IWRM initiatives. The process view outlines a collaborative problem-solving model emphasizing a sequential approach to addressing water resource challenges. It begins with issue identification and progresses through community capacity-building, consensus-building, and forming multi-stakeholder partnerships. Constructive engagement, sound management and implementation, and iterative evaluation and replication complete the cycle, ensuring adaptive and inclusive management

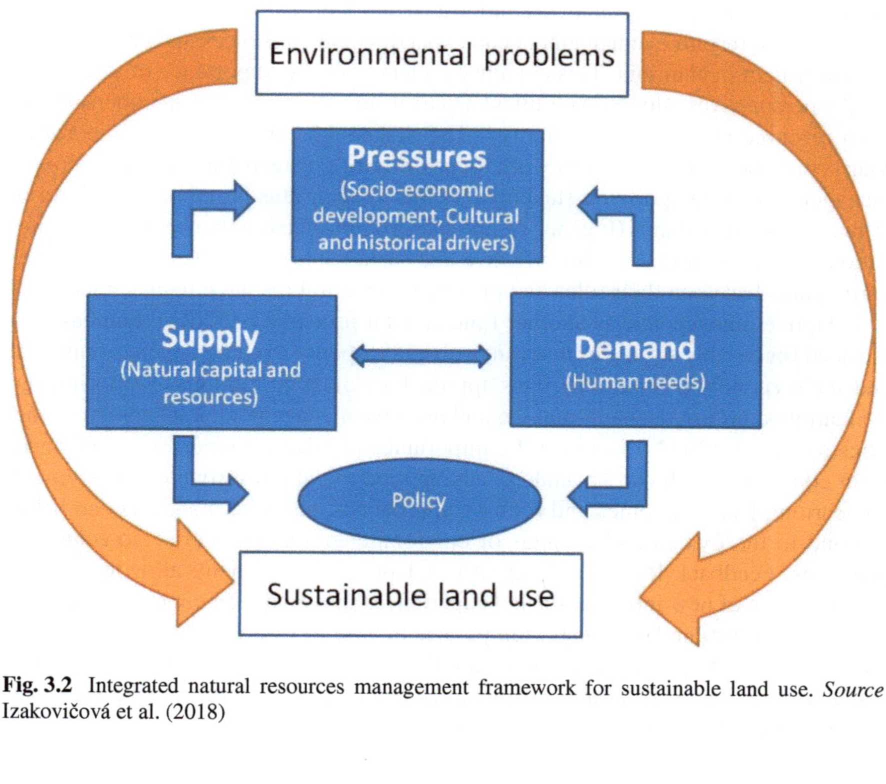

Fig. 3.2 Integrated natural resources management framework for sustainable land use. *Source* Izakovičová et al. (2018)

practices. These models highlight the strategic and participatory processes necessary for effective and sustainable water resource management.

Stakeholder inclusivity and participatory decision-making are critical components of INRM. Engaging diverse stakeholders—including local communities, government entities, the private sector, and non-governmental organizations (NGOs)—is essential for fostering transparency and shared responsibility in resource management. Ismail et al. (2023) emphasize the importance of open communication channels that allow marginalized communities to express their concerns and ideas, which enhances their sense of ownership and agency in resource management decisions. This participatory approach is supported by the findings of Tamang and Nepali (2023), who argue that social workers play a vital role in empowering marginalized groups to engage in resource management processes. By promoting inclusivity, INRM can ensure that the voices of all stakeholders are heard, leading to more equitable and effective resource management outcomes. Empowering marginalized groups is particularly important in natural resource management, as these groups often bear the brunt of resource degradation and environmental change. Ababu et al. (2023) highlighted the significance of women's participation in community land rehabilitation programs, noting that their involvement leads to better governance and resource management outcomes. This empowerment is crucial for achieving sustainable development goals, as women are often key resource users and managers within their communities. The need for gender-sensitive approaches in resource management is further supported by the work of Khumalo and Freimund (2014), who explored how community-based natural resource management can expand women's choices and enhance their empowerment in Namibia. This evidence underscores the importance of integrating gender perspectives into INRM frameworks to ensure all community members benefit from resource management initiatives. The Stakeholder Influence and Importance Matrix provides a strategic framework for prioritizing engagement in natural resource management by categorizing stakeholders into four groups based on their level of influence and importance (Fig. 3.4). Application of the matrix enhances resource allocation and decision-making for effective and sustainable outcomes as it categorizes participants based on their roles and priorities in natural resource management

Adaptive management is another fundamental principle of INRM, emphasizing the need for flexible strategies that can evolve in response to changing environmental and socio-economic conditions. This approach involves continuous monitoring and evaluation to inform decision-making and improve management practices over time. Engström et al. (2022) discussed the importance of adaptive strategies in management contexts, which can be analogously applied to natural resource management. By learning from outcomes and refining approaches, resource managers can better respond to the dynamic challenges of environmental change and socio-economic pressures. Feedback loops are integral to adaptive management, allowing for the incorporation of new information and experiences into management practices. Zhao et al. (2018) illustrated how spatio-temporal analyses can inform sustainable development strategies by providing insights into the ecological footprint and resource use patterns. This data-driven approach enables resource managers to make informed decisions that align with sustainability goals.

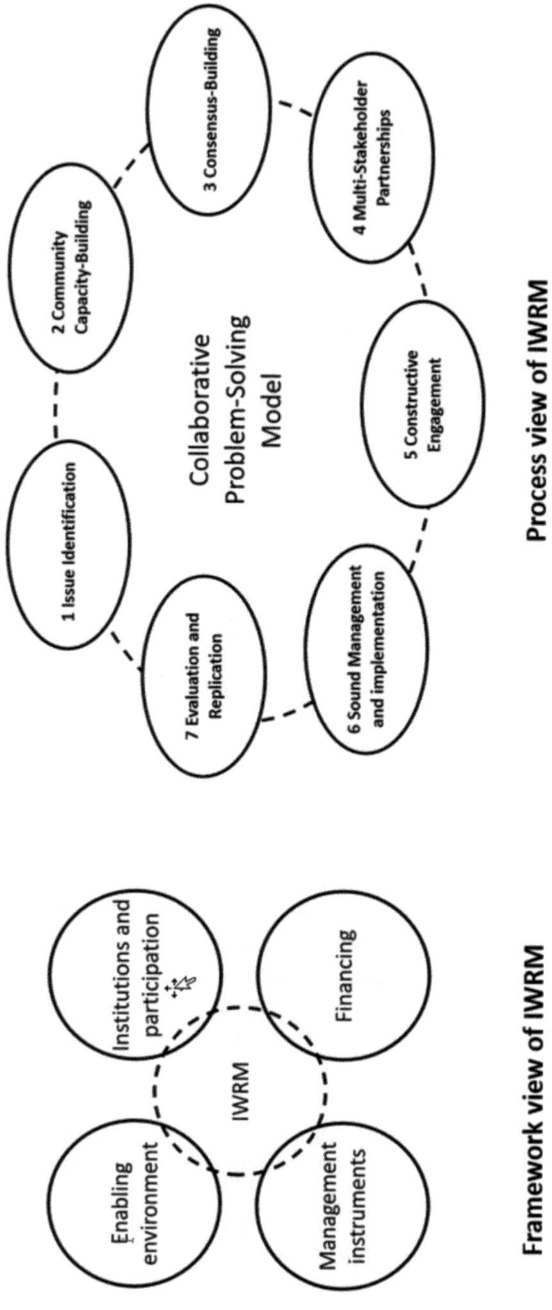

Fig. 3.3 Integrated water resources management: framework and collaborative problem-solving model. *Source* Grigg (2024)

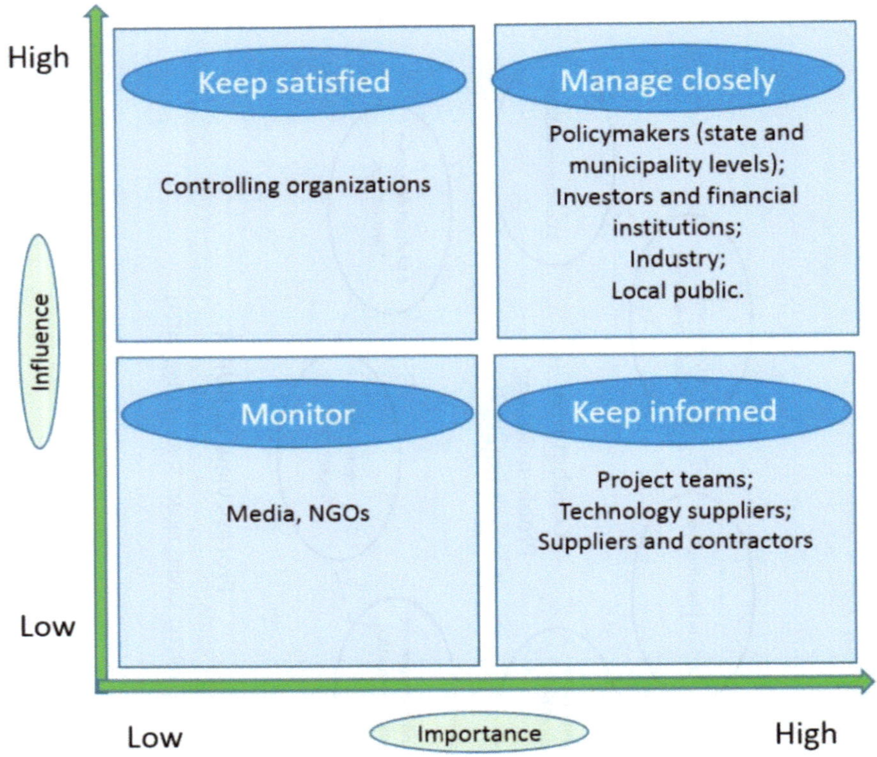

Fig. 3.4 Stakeholder influence and importance matrix for integrated natural resource management. *Sources* Ilinova et al. (2018)

Moreover, the role of leadership in fostering adaptive management practices is emphasized by Marjanović et al. (2019), who argued that effective leadership is essential for promoting sustainable development and efficient resource management. Thus, integrating adaptive management principles into INRM frameworks can enhance resilience and sustainability in resource management practices. In addition to adaptive management, combining ecological, social, and economic factors necessitates focusing on sustainability principles. Pasha et al. (2023) discusses the importance of maintaining sustainability in water resource management by considering social status, economic development, and energy costs. This holistic perspective is critical for ensuring that resource management practices do not compromise the ecological integrity of ecosystems while meeting the needs of local communities. Furthermore, integrating sustainability principles into INRM can help mitigate the impacts of climate change and promote resilience among vulnerable populations.

The challenges of implementing INRM are multifaceted, often requiring a shift in governance structures and practices. Zhou et al. (2022) highlighted the complexities of natural resource management in American national parks, emphasizing the need for comprehensive management systems that address ecological and social dimensions. This complexity is mirrored in the challenges faced by community-based natural resource management initiatives, as noted by Leisher et al. (2016) discuss the importance of gender composition in resource governance and conservation outcomes. Effective INRM must, therefore, navigate these complexities by fostering collaboration among stakeholders and promoting inclusive governance structures. Ultimately, INRM's success hinges on its ability to balance ecological, social, and economic factors while promoting stakeholder inclusivity and adaptive management practices. This requires a commitment to transparency, shared responsibility, and collaboration among all stakeholders involved in resource management. INRM can create more equitable and sustainable resource management outcomes by empowering marginalized groups and fostering participatory decision-making processes. The integration of adaptive management principles further enhances the resilience of resource management strategies, allowing them to evolve in response to changing conditions.

3.3 Collaborative Frameworks in Resource Management

Collaborative frameworks in resource management have gained significant attention in recent years due to their potential to enhance sustainability and efficiency in managing natural resources (Table 3.3). These frameworks encompass various collaboration models, each designed to address specific challenges and leverage the strengths of different stakeholders. Among these models, Community-Based Natural Resource Management (CBNRM) stands out as a powerful approach that empowers local communities to take an active role in decision-making and stewardship of their resources. CBNRM emphasizes the importance of local knowledge and participation, which can lead to more effective and sustainable management outcomes (Kenney et al. 2015). By involving local stakeholders, CBNRM fosters a sense of ownership and responsibility, resulting in better conservation practices and resource utilization (Wilson et al. 2023).

Public–private partnerships (PPP) represent another collaboration model that combines government oversight's strengths with the efficiency of the private sector. These partnerships can facilitate the implementation of resource projects by leveraging private investment and expertise while ensuring that public interests are safeguarded (McDonald 2011). The effectiveness of PPPs in resource management has been demonstrated in various sectors, including water supply, energy, and waste management, where they have led to improved service delivery and infrastructure development (Mason et al. 2020). However, the success of PPPs relies heavily on transparent governance structures and the alignment of interests among stakeholders, which can sometimes take time to achieve (Guillaume and Loufrani-Fedida 2023). Transboundary cooperation is a critical aspect of collaborative frameworks, particularly in managing resources that span national borders, such as rivers and forests.

Table 3.3 Key aspects of collaborative frameworks in resource management

Aspect	Details	Examples
Models of collaboration	Community-based natural resource management (CBNRM)	Local community forest management initiatives
	Public–private partnerships (PPP)	Private investment in sustainable agriculture projects
	Transboundary cooperation	Joint management of shared river basins (e.g., Nile Basin Initiative)
Examples of initiatives	Watershed management programs	Integrated water resource management for flood control and irrigation
	Forest conservation partnerships	REDD + programs for reducing deforestation and supporting livelihoods
	Marine protected areas (MPAs)	Regional efforts to conserve fisheries and protect biodiversity in coral reefs
Benefits of collaboration	Enhanced outcomes through shared expertise and resources	Improved resource sustainability and stakeholder satisfaction
	Shared responsibilities reduce the burden on individuals or groups	Multi-stakeholder support in managing natural disasters
	Conflict resolution through dialogue and consensus-building	Successful mediation in disputes over resource access and usage

Effective transboundary management requires establishing joint governance structures facilitating communication and collaboration among countries (Dumont et al. 2019). Successful examples of transboundary cooperation, such as managing the Nile River Basin, illustrate the potential for shared benefits through collaborative approaches (Cantino et al. 2017). These initiatives often involve complex negotiations and balancing diverse interests, but they can lead to more sustainable resource management and conflict resolution (Nkuruziza et al. 2016).

Several successful collaborative initiatives exemplify the effectiveness of these frameworks in resource management. Watershed management programs, for instance, have emerged as integrated efforts to protect water sources, improve irrigation practices, and prevent erosion. These programs often involve multiple stakeholders, including local communities, government agencies, and non-governmental organizations, working together to develop and implement sustainable practices (Lestari et al. 2020; Adom and Simatele 2022). The collaborative nature of these initiatives allows for the pooling of resources and expertise, leading to more comprehensive and effective management strategies (Mott Lacroix and Megdal 2016). The framework presented in Fig. 3.5 outlines a multifaceted approach to fostering sustainable and transformative collaboration across various sectors. It highlights the importance of collective action, where stakeholders come together to align their visions and implement strategies grounded in mutual understanding and accountability.

Emphasis is placed on fostering empowering relationships that allow all parties to benefit equitably, ensuring that collaboration leads to shared growth and innovation. Additionally, the framework underscores the role of disruptive yet constructive strategies in driving cultural and systemic change, promoting fresh perspectives, and reimagining traditional practices (Osawaru and Ogwu 2013; Ogwu et al. 2017). It advocates for cooperative efforts that prioritize resilience and sustainability, enabling communities and organizations to adapt effectively to environmental and social challenges.

Forest conservation partnerships, such as the REDD + (Reducing Emissions from Deforestation and Forest Degradation) initiative, showcase the potential of multi-stakeholder collaboration in addressing environmental challenges. REDD + emphasizes the importance of integrating local knowledge and values into conservation efforts, thereby enhancing the effectiveness of forest management (Qian 2023; Freeman et al. 2018). By involving local communities in decision-making processes, REDD + initiatives can address carbon sequestration and other critical issues such as water quality and erosion control (Johnson et al. 2015). This holistic approach to forest conservation highlights the benefits of collaboration in achieving sustainable outcomes. Marine protected areas (MPAs) represent another successful example of collaborative governance for sustainable fisheries and biodiversity conservation. The establishment and management of MPAs often require the engagement of various

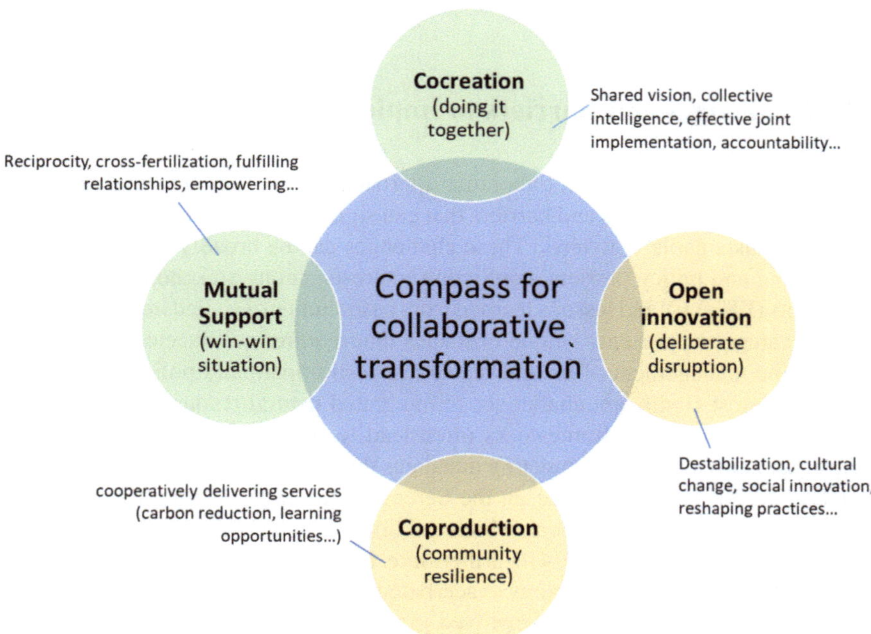

Fig. 3.5 Compass for collaborative transformation and management of natural resources by fostering resilient and inclusive change. *Source* Macedo et al. (2020)

stakeholders, including fishers, conservationists, and government agencies (Said and Kusumasari 2022; Carmona et al. 2013). Collaborative governance in MPAs has been shown to enhance compliance with regulations and improve the overall health of marine ecosystems (Al-Haddad and Al-Abed 2021). By fostering dialogue and consensus-building among diverse interest groups, MPAs can effectively address conflicts and promote sustainable resource use (Goethel et al. 2019).

The benefits of collaboration in resource management are manifold. Enhanced resource management outcomes are often achieved by sharing expertise and resources among stakeholders (Pollard and Bennun 2016; Carr et al. 2012). This collaborative approach improves the effectiveness of management strategies and fosters innovation and adaptive learning, which are crucial in addressing the dynamic challenges of resource management (Tan and Yang 2023; Gutenson et al. 2020). Furthermore, the distribution of responsibilities among stakeholders reduces the burden on individual entities, allowing for a more equitable and efficient allocation of resources. Collaboration promotes conflict resolution by fostering dialogue and consensus-building among diverse interest groups. Engaging stakeholders in decision-making ensures that multiple perspectives are considered, leading to more balanced and inclusive outcomes. This participatory approach can mitigate tensions and build stakeholder trust, ultimately contributing to more sustainable resource management practices. The importance of stakeholder engagement in resource management is increasingly recognized, with many agencies mandating documented interactions at various stages of the decision-making process (Talley et al. 2016).

3.4 Challenges and Barriers to Implementation

Implementing integrated and collaborative approaches to managing natural resources faces numerous challenges and barriers that can significantly hinder effective governance and sustainable outcomes. These challenges can be broadly categorized into institutional and policy barriers, conflicting interests among stakeholders, resource constraints (Fig. 3.6), and lessons learned from case studies of stalled initiatives. Each of these categories presents unique obstacles that require multifaceted solutions to enhance the management of natural resources. Institutional and policy barriers are among the most significant challenges to integrated natural resource management. Fragmented institutional frameworks often lead to overlaps and inefficiencies, as various agencies may have competing mandates that complicate collaborative efforts (Nuño et al. 2014). For instance, in the Serengeti, multiple institutional actors are involved in resource management, which can dilute accountability and hinder effective governance (Nuño et al. 2014). Furthermore, inconsistent policies and regulations across different sectors and regions exacerbate these challenges, creating confusion and uncertainty for stakeholders involved in resource management (Olvera-Garcia and Neil 2020). Weak enforcement mechanisms for existing environmental laws further undermine efforts to protect natural resources, as compliance is often not adequately monitored or enforced (Armitage et al. 2012).

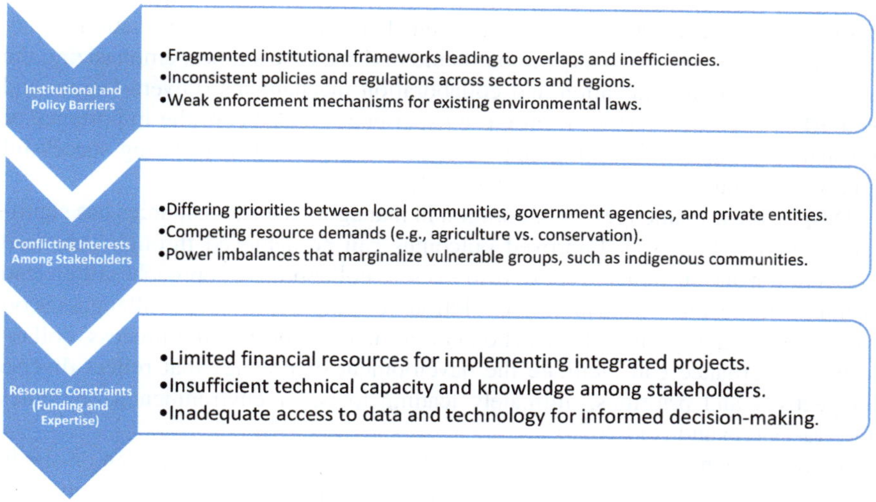

Fig. 3.6 Challenges and barriers to implementation of integrated and collaborative approaches to managing natural resources

Conflicting interests among stakeholders present another significant barrier to effective natural resource management. Local communities, government agencies, and private entities often have differing priorities, leading to tensions and disputes over resource use (Austin et al. 2010). For example, the competing demands of agriculture versus conservation can create significant challenges in decision-making processes, as stakeholders may prioritize short-term economic gains over long-term sustainability (Kamoto et al. 2013). Additionally, power imbalances can marginalize vulnerable groups, such as Indigenous communities, who may lack a voice in decision-making processes (Kamoto et al. 2013). This marginalization can lead to a lack of trust and cooperation among stakeholders, further complicating collaborative efforts.

Resource constraints, particularly in funding and expertise, also hinder the implementation of integrated approaches to natural resource management. More financial resources are needed to allow organizations to undertake comprehensive projects that require significant investment (Kusnadi 2021). Moreover, insufficient technical capacity and knowledge among stakeholders can impede effective collaboration, as parties may need more skills to engage in complex decision-making processes (Moore et al. 2013). Inadequate access to data and technology further complicates informed decision-making, as stakeholders may need more information to evaluate the impacts of their actions on natural resources (Michaud 2013).

Studies of failed or stalled initiatives provide valuable insights into the challenges of implementing integrated approaches to natural resource management. Many projects have been derailed by a lack of stakeholder buy-in or insufficient funding, highlighting the importance of securing commitment from all parties involved (Austin et al. 2010). Mismanagement due to inadequate coordination among stakeholders has also been a common theme in these case studies, emphasizing the need for clear communication and collaboration mechanisms (Olvera-Garcia and Neil 2020). Lessons learned from these experiences underscore the importance of transparency, capacity building, and sustained engagement in fostering successful collaborative initiatives (Armitage et al. 2012).

Despite these challenges, collaboration in managing natural resources has significant benefits (Fig. 3.7). Improved outcomes can be achieved through enhanced conservation and sustainable use of resources, as collaborative approaches often lead to more comprehensive and inclusive solutions to complex resource challenges (Scott 2015). By engaging multiple stakeholders in the decision-making process, collaborative governance can facilitate the development of policies that reflect diverse perspectives and priorities, ultimately leading to better environmental outcomes (Olvera-Garcia and Neil 2020).

Shared responsibilities among stakeholders can also reduce the burden on single entities, fostering a sense of mutual accountability and long-term commitment to resource management (Austin et al. 2010). This distribution of roles encourages stakeholders to work together towards common goals, enhancing the effectiveness of management strategies (Scott 2015). Furthermore, collaborative platforms can serve as venues for dialogue and conflict resolution, helping mitigate resource-related disputes and build stakeholder trust (Austin et al. 2010).

The role of social learning in collaborative natural resource management must be considered. Social learning processes involving knowledge sharing and deliberation among stakeholders can establish a foundation for collective action and improve decision-making. By fostering an environment where stakeholders can learn from one

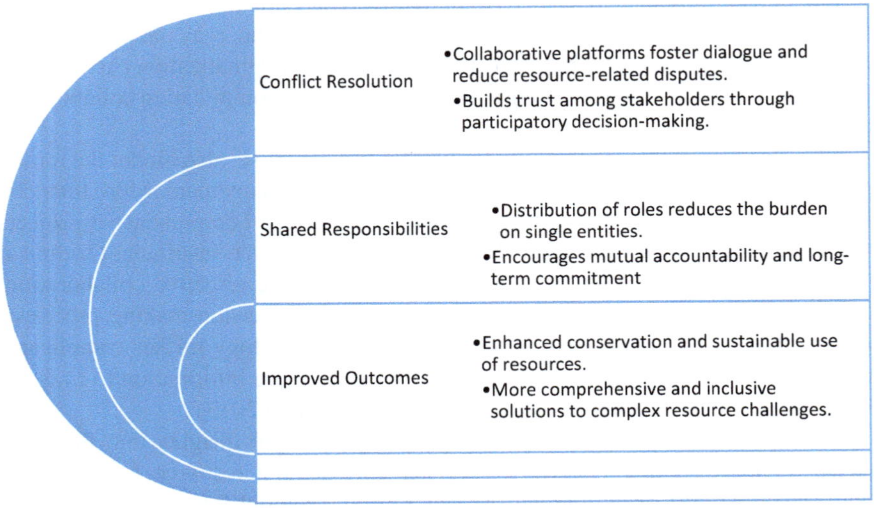

Fig. 3.7 Benefits of collaboration in managing natural resources

another, collaborative governance can enhance the adaptive capacity of communities to respond to changing environmental conditions (Cundill and Rodela 2012). This adaptability is crucial for addressing the complexities and uncertainties inherent in natural resource management.

3.5 Implications for Sustainable Development

Integrated and collaborative resource management is pivotal in supporting several SDGs, notably SDG 6 (Clean Water and Sanitation), SDG 13 (Climate Action), and SDG 15 (Life on Land). By aligning resource use with global sustainability targets, we ensure intergenerational equity, which is essential for maintaining the health of our planet and its resources for future generations. This alignment is increasingly recognized as a critical component of sustainable development, as it fosters a holistic approach to managing resources that transcend mere economic growth and prioritizes ecological integrity and social wellbeing (Leal Filho et al. 2021). Promoting environmental sustainability is another significant implication of integrated resource management. This approach combines ecosystem-based management with human needs, thereby protecting critical habitats and preventing overexploitation of resources. Restoring degraded ecosystems is a vital aspect of this process, as it enhances biodiversity and strengthens natural resource governance. Effective governance minimizes environmental degradation and ensures that natural resources are managed sustainably, thereby contributing to the overall health of ecosystems (Costanza 2012; Ananda et al. 2023). Recognizing the interconnectedness of ecological health and human welfare underscores the importance of sustainable practices in resource management (Brezinskaya and Khartanovich 2020).

Driving social and economic sustainability is equally important. Empowering local communities through participatory governance fosters a sense of ownership and stewardship over natural resources. This empowerment is crucial for balancing economic development with ecological health, as it encourages sustainable industries such as ecotourism, agroforestry, and renewable energy. These industries provide economic opportunities and promote environmental conservation, demonstrating that economic growth and ecological sustainability coexist (Rezaeinejad and Khaniwadekar 2021; Rumagia et al. 2023). Integrating social and economic dimensions into resource management strategies is essential for achieving long-term sustainability (Oliveira et al. 2023). Enhancing resilience to climate change and biodiversity loss is a critical focus of sustainable development initiatives. Collaborative frameworks that support climate adaptation strategies, such as integrated water resource management and forest conservation, are vital for building ecosystem resilience. By maintaining biodiversity and ecosystem services, these strategies ensure the health of both human and environmental systems. The ability of ecosystems to adapt to changing conditions is paramount for sustaining the services they provide, which are essential for human survival and wellbeing (Döring et al. 2015). This resilience is beneficial for the environment and communities that rely on these ecosystems for their livelihoods (Cowx and Ogutu-Owhayo 2019).

Reducing poverty and ensuring equity are fundamental aspects of sustainable development. Integrated approaches that prioritize fair resource distribution are essential for addressing the needs of marginalized groups and reducing inequalities. By supporting livelihood diversification for vulnerable communities, these approaches help reduce poverty while safeguarding natural assets. Recognizing that social equity is intertwined with environmental sustainability is crucial for developing effective policies promoting inclusive growth (Rumagia et al. 2023; Chipriyanova et al. 2022). Ensuring all community members have access to resources and opportunities is essential for fostering a sustainable future (Taha and Taha 2023). The role of education in advancing sustainable development cannot be overstated. Incorporating sustainability into educational curricula equips future generations with the knowledge and skills to address complex environmental challenges. This educational approach fosters a deeper understanding of the interconnectedness of social, economic, and ecological systems, thereby promoting a culture of sustainability (Hutaibat 2019). By preparing students to engage with sustainability issues, educational institutions play a critical role in shaping a more sustainable future (Ivan et al. 2023).

Moreover, integrating technology and innovation in resource management practices is vital for achieving sustainable development goals. The adoption of innovative technologies can enhance the efficiency of resource use and minimize environmental impacts. For instance, advancements in renewable energy technologies can significantly reduce reliance on fossil fuels, contributing to climate action efforts (Shields et al. 2013; Litvinenko et al. 2023). The role of technology in facilitating sustainable practices is increasingly recognized as a key driver of change in various sectors, including agriculture, fisheries, and forestry (Oliveira et al. 2023). The importance of stakeholder engagement in sustainable development initiatives is another critical aspect. Collaborative efforts that involve multiple stakeholders, including governments, businesses, and local communities, are essential for developing effective resource management strategies. Engaging stakeholders in decision-making ensures that diverse perspectives are considered, leading to more equitable and effective outcomes (Barclay 2012; Gisladottir et al. 2021). This collaborative approach fosters a sense of shared responsibility and accountability, which is crucial for the success of sustainability initiatives (Leal Filho et al. 2021; Hidayat and Dewi 2022).

Furthermore, assessing and monitoring sustainability indicators are essential for tracking progress toward achieving the SDGs. The development of reliable indicators allows for evaluating resource management practices and their impacts on social, economic, and environmental dimensions (Pires et al. 2017). These indicators serve as valuable tools for policymakers and practitioners, enabling them to make informed decisions and adjust strategies to enhance sustainability outcomes (Leal Filho et al. 2021; Gisladottir et al. 2021). Continually evaluating sustainability efforts is crucial for ensuring that goals are met and resources are managed effectively. In addition, the interdependence of various sustainability dimensions highlights the need for integrated approaches to resource management. Economic, social, and environmental factors are interconnected, and addressing one dimension without considering the others can lead to unintended consequences (Brezinskaya and Khartanovich 2020; Oliveira et al. 2023). For example, policies aimed solely at economic growth may result in environmental degradation if ecological considerations are overlooked. Therefore, a holistic approach that considers the interplay between these dimensions is essential for achieving sustainable development (Rumagia et al. 2023; Barclay 2012).

The implications of sustainable development extend beyond local contexts to global challenges. Climate change, biodiversity loss, and resource depletion are interconnected and require coordinated global responses. International collaboration is essential for addressing these challenges effectively, as they transcend national boundaries and affect communities worldwide (Shields et al. 2013; Döring et al. 2015). The global nature of these issues underscores the importance of collective action and shared responsibility in pursuing sustainable development goals (Leal Filho et al. 2021; Gisladottir et al. 2021). The future of sustainable development hinges on the commitment of all stakeholders to embrace sustainability principles in their practices. This commitment involves not only adopting sustainable practices but also advocating for policies that promote sustainability at all levels of governance. Integrating sustainability into business practices, community initiatives, and government policies is essential for creating a more sustainable world (Oliveira et al. 2023; Chipriyanova et al. 2022). By fostering a culture of sustainability, we can ensure that future generations inherit a planet capable of sustaining life and providing for their needs.

3.6 Conclusion

Integrated and collaborative approaches to managing natural resources are essential for sustainable development. These approaches promote stakeholder inclusivity and adaptive management practices by addressing the complex interconnections between ecological, social, and economic factors. This framework helps ensure the long-term sustainability of natural resources, advancing key global goals such as environmental protection, poverty reduction, and climate resilience. Despite challenges like institutional barriers, conflicting interests, and resource constraints, the evidence strongly supports the effectiveness of collaboration in fostering ecological balance, supporting livelihoods, and reducing inequality. To scale up these approaches, policymakers, stakeholders, and resource managers must focus on overcoming these challenges by fostering cooperation and enhancing governance structures. Successful models such as community-based natural resource management, public–private partnerships, and transboundary cooperation offer valuable insights for effective resource management. For the future of sustainable natural resource management, it is crucial to engage diverse groups in decision-making, ensure equitable resource distribution, and align management practices with the broader goals of the United Nations Sustainable Development Goals. Ultimately, collective action and shared responsibility will be key to securing a sustainable future for both people and the planet.

References

Ababu T, Seyoum B, Berhanu D, Furo G (2023) Evaluation of women's participation and empowerment in com-munity land rehabilitation programs: lesson drawn from Wera District, Southern Ethiopia. J For Sci 69:158–171. https://doi.org/10.17221/165/2022-jfs

Adom RK, Simatele MD (2022) The role of stakeholder engagement in sustainable water resource management in South Africa. In: Natural resources forum, vol 46, no 4. Blackwell Publishing Ltd., Oxford, UK, pp 410–427. https://doi.org/10.1111/1477-8947.12264

Al-Haddad MA, Al-Abed MS (2021) Stakeholders' engagement and performance efficiency at oil and gas industry in Yemen. J Impact 2(1):1–15. https://doi.org/10.48110/joi.v2i1.25

Ali-Nakyea A, Amoh JK, Mohammed NA (2019) Are sub-Saharan African countries losing it on oil and gas revenue management too?: evidence from Ghana. Int J Energy Econ Policy 9(2):89–97. https://doi.org/10.32479/ijeep.7285

Ananda F, Nofrizal N, Yoswaty D, Farida L, Hamidy R (2023) Integrating environmentally friendly management of water resources toward ecological and socio-economic sustainability. Int J Energy Econ Policy 13(3):73–79. https://doi.org/10.32479/ijeep.14310

Armitage D, De Loë R, Plummer R (2012) Environmental governance and its implications for conservation practice. Conserv Lett 5(4):245–255. https://doi.org/10.1111/j.1755-263x.2012.00238.x

Austin Z, Smart JC, Yearley S, Irvine RJ, White PC (2010) Identifying conflicts and opportunities for collaboration in the management of a wildlife resource: a mixed-methods approach. Wildl Res 37(8):647–657. https://doi.org/10.1071/wr10057

Barclay K (2012) The social in assessing for sustainability: fisheries in Australia. Cosmopolitan Civ Soc: Interdiscip J 4(3):38–53. https://doi.org/10.5130/ccs.v4i3.2655

Behnken JA, Groninger JW, Akamani K (2016) Institutional constraints to collaborative ecosystem management within a wetlands conservation partnership. J Contemp Water Res Educ 158(1):19–33. https://doi.org/10.1111/j.1936-704x.2016.03216.x

Berkes F (2010) Devolution of environment and resources governance: trends and future. Environ Conserv 37(4):489–500. https://doi.org/10.1017/s037689291000072x

Brezinskaya LV, Khartanovich EA (2020) Sustainable forest management as the basis of economic growth of the Krasnoyarsk territory. In: "New Silk Road: business cooperation and prospective of economic development" (NSRBCPED 2019). Atlantis Press, pp 320–324. https://doi.org/10.2991/aebmr.k.200324.060

Cantino V, Devalle A, Cortese D, Ricciardi F, Longo M (2017) Place-based network organizations and embedded entrepreneurial learning: emerging paths to sustainability. Int J Entrep Behav Res 23(3):504–523. https://doi.org/10.1108/ijebr-12-2015-0303

Carmona G, Varela-Ortega C, Bromley J (2013) Participatory modelling to support decision making in water management under uncertainty: two comparative case studies in the Guadiana river basin, Spain. J Environ Manag 128:400–412. https://doi.org/10.1016/j.jenvman.2013.05.019

Carr G, Blöschl G, Loucks DP (2012) Evaluating participation in water resource management: a review. Water Resour Res 48(11). https://doi.org/10.1029/2011wr011662

Chipriyanova G, Krasteva-Hristova R, Kussainova A (2022) Contemporary higher accounting education for social responsibility. Econ Ecol Socium 6(4):27–36. https://doi.org/10.31520/2616-7107/2022.6.4-3

Costanza R (2012) Ecosystem health and ecological engineering. Ecol Eng 45:24–29. https://doi.org/10.1016/j.ecoleng.2012.03.023

Cowx IG, Ogutu-Owhayo R (2019) Towards sustainable fisheries and aquaculture management in the African Great Lakes. Fish Manag Ecol 26(5):397–405. https://doi.org/10.1111/fme.12391
Cundill G, Rodela R (2012) A review of assertions about the processes and outcomes of social learning in natural resource management. J Environ Manag 113:7–14. https://doi.org/10.1016/j.jenvman.2012.08.021
Diamond J, Vangen S (2016) Coping with austerity: innovation via collaboration or retreat to the known? Public Money Manag 37(1):47–54. https://doi.org/10.1080/09540962.2016.1249231
Dobrovolskienė N, Pozniak A, Tvaronavičienė M (2021) Assessment of the sustainability of a real estate project using multi-criteria decision making. Sustainability 13(8):4352
Döring TF, Vieweger A, Pautasso M, Vaarst M, Finckh MR, Wolfe MS (2015) Resilience as a universal criterion of health. J Sci Food Agric 95(3):455–465. https://doi.org/10.1002/jsfa.6539
Dumont ES, Bonhomme S, Pagella TF, Sinclair FL (2019) Structured stakeholder engagement leads to development of more diverse and inclusive agroforestry options. Exp Agric 55(S1):252–274. https://doi.org/10.1017/s0014479716000788
Eneji RI, Ering SO (2018) Community-based natural resources management and rural development in Nigeria: a theoretical review. Int J Hum Soc Sci 8(11):111–116. https://doi.org/10.30845/ijhss.v8n11p13
Engström M, Mårtensson G, Pålsson Y, Strömberg A (2022) What relationships can be found between nurses' working life and turnover? A mixed-methods approach. J Nurs Manag 30(1):288–297. https://doi.org/10.1111/jonm.13494
Enukwa EH, Ngang FD, Jany WNP (2022) The contribution of community forestry on livelihood improvement and environmental protection in Cameroon. Asian J Res Agric for 8(2):14–22. https://doi.org/10.9734/ajraf/2022/v8i230150
Freeman ER, Civera C, Cortese D, Fiandrino S (2018) Strategising stakeholder empowerment for effective co-management within fishery-based commons. British Food J 120(11):2631–2644. https://doi.org/10.1108/bfj-01-2018-0041
Gandiwa E, Heitkönig IM, Lokhorst AM, Prins HH, Leeuwis C (2013) CAMPFIRE and human-wildlife conflicts in local communities bordering northern Gonarezhou National Park, Zimbabwe. Ecol Soc 18(4). https://doi.org/10.5751/es-05817-180407
Gisladottir J, Sigurgeirsdottir S, Ragnarsdóttir KV, Stjernquist I (2021) Economies of scale and perceived corruption in natural resource management: a comparative study between Ukraine, Romania, and Iceland. Sustainability 13(13):7363. https://doi.org/10.3390/su13137363
Goethel DR, Lucey SM, Berger AM, Gaichas SK, Karp MA, Lynch PD, Walter JF III, Deroba JJ, Miller S, Wilberg MJ (2019) Closing the feedback loop: on stakeholder participation in management strategy evaluation. Can J Fish Aquat Sci 76(10):1895–1913. https://doi.org/10.1139/cjfas-2018-0162
Grigg NS (2024) Framework and function of integrated water resources management in support of sustainable development. Sustainability 16(13):5441. https://doi.org/10.3390/su16135441
Guillaume M, Loufrani-Fedida S (2023) Stakeholder engagement in inclusive employability management for employees whose health at work is impaired: empirical evidence from a French public organisation. Pers Rev 52(1):121–144. https://doi.org/10.1108/pr-06-2021-0404
Gutenson JL, Ernest ANS, Bearden BL, Fuller C, Guerrero J (2020) Integrating societal and scientific elements into sustainable and effective water resource policy development. J Environ Informatics Lett. https://doi.org/10.3808/jeil.202000048
Haeril H, Purnomo EP (2019) Management of small-sustainable coastal and island areas based on collaborative management (case study in Bima Regency, West Nusa Tenggara). J Local Gov Issues 2(1):18–37. https://doi.org/10.22219/logos.vol2.no1.18-37
Hidayat F, Dewi OC (2022) Strengthening sustainable context of water management in educational building Universitas Indonesia. In: IOP conference series: earth and environmental science, vol 1058, no 1. IOP Publishing, p 012009. https://doi.org/10.1088/1755-1315/1058/1/012009
Hossu CA, Ioja IC, Susskind LE, Badiu DL, Hersperger AM (2018) Factors driving collaboration in natural resource conflict management: evidence from Romania. Ambio 47(7):816–830. https://doi.org/10.1007/s13280-018-1016-0

Hutaibat K (2019) Incorporating practical sustainability and managerial and financial reporting in accounting education. J Int Educ Bus 12(2):181–197. https://doi.org/10.1108/jieb-10-2018-0047

Ikhajiagbe B, Ogochukwu OF, Ogwu MC (2021) Shelf life, fruit quality and safety of banana fruits ripened through traditional ripening techniques in Nigeria. Int J Fruit Sci. https://doi.org/10.1080/15538362.2020.1858469

Ilinova A, Cherepovitsyn A, Evseeva O (2018) Stakeholder management: an approach in CCS projects. Resources 7(4):83. https://doi.org/10.3390/resources7040083

Ismail IR, Idris R, Kesuma AI (2023) Pa'belle fishermen empowerment: a communicative action perspective in an industrialized society. Tamaddun 22(2):213–224. https://doi.org/10.33096/tamaddun.v22i2.565

Ivan O, Ruslan T, Oleg T, Oleksandr S, Yanina S, Anatoliy R, Pavlo V, Ivan Z, Oleksandr C, Yevheniia K, Anastasiia H (2023) Improvement of economic mechanism of rational use of forest resources using discrete mathematics method. In: 22nd international scientific conference" engineering for rural development": proceedings: [Jelgava, Latvia], 24–26 May 2023, pp 544–552

Izah SC, Ogwu MC (2023) Sustainable utilization and conservation of Africa's biological resources and environment (First Edition). In: Ramawat KG (ed) Sustainable development and biodiversity series, vol 32. Springer, Singapore, 691p. Hardcover. ISBN: 978-981-19-6973-7. https://doi.org/10.1007/978-981-19-6974-4

Izakovičová Z, Špulerová J, Petrovič F (2018) Integrated approach to sustainable land use management. Environments 5(3):37. https://doi.org/10.3390/environments5030037

Johnson TR, Jansujwicz JS, Zydlewski G (2015) Tidal power development in Maine: stakeholder identification and perceptions of engagement. Estuaries Coasts 38:266–278. https://doi.org/10.1007/s12237-013-9703-3

Kamoto J, Clarkson G, Dorward P, Shepherd D (2013) Doing more harm than good? Community based natural resource management and the neglect of local institutions in policy development. Land Use Policy 35:293–301. https://doi.org/10.1016/j.landusepol.2013.06.002

Kenney L, Arvai J, Vardhan M, Catacutan D (2015) Bringing stakeholder values into climate risk management programs: decision aiding for REDD in Vietnam. Soc Nat Resour 28(3):261–279. https://doi.org/10.1080/08941920.2014.918233

Khumalo KE, Freimund WA (2014) Expanding women's choices through employment? Community-based Natural Resource Management and women's empowerment in Kwandu Conservancy, Namibia. Soc Nat Resour 27(10):1024–1039. https://doi.org/10.1080/08941920.2014.905893

Koontz TM, Newig J (2014) From planning to implementation: top-down and bottom-up approaches for collaborative watershed management. Policy Stud J 42(3):416–442. https://doi.org/10.1111/psj.12067

Kusnadi R (2021) Management of the natural attractions in the area Simarjarunjung Simalungun North Sumatra Province. Int J Res Rev 8(11):450–462. https://doi.org/10.52403/ijrr.20211157

Leal Filho W, Frankenberger F, Salvia AL, Azeiteiro U, Alves F, Castro P, Will M, Platje J, Lovren VO, Brandli L, Price E (2021) A framework for the implementation of the Sustainable Development Goals in university programmes. J Clean Prod 299:126915. https://doi.org/10.1016/j.jclepro.2021.126915

Leisher C, Temsah G, Booker F, Day M, Samberg L, Prosnitz D, Agarwal B, Matthews E, Roe D, Russell D et al (2016) Does the gender composition of forest and fishery management groups affect resource governance and conservation outcomes? A systematic map. Environ Evid 5:1–10. https://doi.org/10.1186/s13750-016-0057-8

Lestari S, Winarno B, Premono BT (2020) Increasing stakeholder engagement for sustainable natural resource management in Southern Sumatra, Indonesia. In: E3S web of conferences, vol 153. EDP Sciences, p 03010. https://doi.org/10.1051/e3sconf/202015303010

Litvinenko VS, Petrov EI, Vasilevskaya DV, Yakovenko AV, Naumov IA, Ratnikov MA (2023) Assessment of the role of the state in the management of mineral resources. Записки Горного института (259 (eng)):95–111. https://doi.org/10.31897/pmi.2022.100

Macedo P, Huertas A, Bottone C, del Río J, Hillary N, Brazzini T, Wittmayer JM, Penha-Lopes G (2020) Learnings from local collaborative transformations: setting a basis for a sustainability framework. Sustainability 12(3):795. https://doi.org/10.3390/su12030795

Marjanović N, Jovanović V, Ratknić T, Paunković D (2019) The role of leadership in natural resource conservation and sustainable development a case study of local self-government of Eastern Serbia. Економика Пољопривреде 66(3):889–903. https://doi.org/10.5937/ekopolj1903889m

Mason ET, Kellum AN, Chiu JA, Waltz GT, Murray S, Wendt DE, Starr RM, Semmens BX (2020) Long-term participation in collaborative fisheries research improves angler opinions on marine protected areas. PeerJ 8:e10146

McDonald H (2011) Understanding the antecedents to public interest and engagement with heritage. Eur J Mark 45(5):780–804. https://doi.org/10.1108/03090561111120037

Michaud WR (2013) Evaluating the outcomes of collaborative modeling for decision support. JAWRA J Am Water Resour Assoc 49(3):693–699. https://doi.org/10.1111/jawr.12066

Mohd Arzaman AF, Salleh HS, Nik Mat NH, Asma Wan Talaat WI (2022) Traditional knowledge of marine resources and their relations to the coastal community's economic wellbeing in Peninsular Malaysia. Int J Sustain Dev Plan 17(6):1973–1982. https://doi.org/10.18280/ijsdp.170633

Moore CT, Shaffer TL, Gannon JJ (2013) Spatial education: improving conservation delivery through space-structured decision making. J Fish Wildl Manag 4(1):199–210. https://doi.org/10.3996/082012-jfwm-069

Mott Lacroix KE, Megdal SB (2016) Explore, synthesize, and repeat: unraveling complex water management issues through the stakeholder engagement wheel. Water 8(4):118. https://doi.org/10.3390/w8040118

Nhem S, Lee YJ (2019) Women's participation and the gender perspective in sustainable forestry in Cambodia: local perceptions and the context of forestry research. For Sci Technol 15(3):93–110. https://doi.org/10.1080/21580103.2019.1595174

Nkuruziza G, Kasekende F, Otengei S, Mujabi S, Ntayi J (2016) An investigation of key predictors of performance of agricultural projects in Sub-Saharan Africa. Int J Soc Econ 43(7):676–691. https://doi.org/10.1108/ijse-09-2014-0181

Nuno A, Bunnefeld N, Milner-Gulland EJ (2014) Managing social–ecological systems under uncertainty: implementation in the real world. Ecol Soc 19(2). https://doi.org/10.5751/es-06490-190252

Obahiagbon EG, Ogwu MC (2023) Consumer perception and demand for sustainable herbal medicine products and market. In: Izah SC, Ogwu MC, Akram M (eds) Herbal medicine phytochemistry. Reference series in phytochemistry. Springer, Cham. https://doi.org/10.1007/978-3-031-21973-3_65-1

Obahiagbon EG, Ogwu MC (2024a) Organic food preservatives: the shift towards natural alternatives and sustainability in the global south's markets. In: Ogwu MC, Izah SC, Ntuli NR (eds) Food safety and quality in the global south. Springer, Singapore, pp 299–329. https://doi.org/10.1007/978-981-97-2428-4_10

Obahiagbon EG, Ogwu MC (2024b) The nexus of business, sustainability, and herbal medicine. In: Izah SC, Ogwu MC, Akram M (eds) Herbal medicine phytochemistry. Reference series in phytochemistry. Springer, Cham. https://doi.org/10.1007/978-3-031-21973-3_67-1

Ogwu MC (2019) Towards sustainable development in Africa: the challenge of urbanization and climate change adaptation. In: Cobbinah PB, Addaney M (eds) The geography of climate change adaptation in urban Africa. Springer, Switzerland, pp 29–55. https://doi.org/10.1007/978-3-030-04873-0_2

Ogwu MC, Osawaru ME, Obahiagbon GE (2017) Ethnobotanical survey of medicinal plants used for traditional reproductive care by Usen people of Edo State, Nigeria. Malaya J Biosci 4(1):17–29

Oliveira DCF, Albergaria FC, Martins LT, do Amaral Santos AC, de Sousa Gomes ME, de Souza Ramos AL (2023) The fish industry and the pillars of sustainable development. Caderno de ANAIS HOME. https://doi.org/10.56238/seveniiimulti2023-260

Olvera-Garcia J, Neil S (2020) Examining how collaborative governance facilitates the implementation of natural resource planning policies: a water planning policy case from the Great Barrier Reef. Environ Policy Gov 30(3):115–127. https://doi.org/10.1002/eet.1875

Osawaru ME, Ogwu MC (2013) Collecting West African Okra (*Abelmoschus caillei* (A. Chev.) Stevels) germplasm from traditional agriculture in parts of Southwestern Nigeria. Bioscientist 1(2):171–181

Pasha NH, Zengin E, Baylarli AT (2023) Sustainability of water resources of Azerbaijan, future prospects and problems. J Manag Econ Ind Organ 7(2). https://doi.org/10.31039/jomeino.2023.7.2.2

Pires A, Morató J, Peixoto H, Botero V, Zuluaga L, Casas A (2017) Sustainability assessment of indicators for integrated water resources management. Sci Total Environ 578:139–147. https://doi.org/10.1016/j.scitotenv.2016.10.217

Pollard E, Bennun L (2016) Who are biodiversity and ecosystem services stakeholders? In: SPE international conference and exhibition on health, safety, environment, and sustainability?. SPE, p D032S068R004. https://doi.org/10.2118/179458-ms

Qi H, Altinakar MS (2011) A conceptual framework of agricultural land use planning with BMP for integrated watershed management. J Environ Manag 92(1):149–155. https://doi.org/10.1016/j.jenvman.2010.08.023

Qian J (2023) A new evolution model for collaborative management of knowledge resources. Int J Emerg Technol Learn 18(16). https://doi.org/10.3991/ijet.v18i16.42703

Rezaeinejad I, Khaniwadekar A (2021) The role of Eco-tourism in sustainable development: case study eco-tourism challenges in Iran. In: E3S web of conferences, vol 311. EDP Sciences, p 02004. https://doi.org/10.1051/e3sconf/202131102004

Rumagia F, Boer M, Kurnia R, Kamal MM (2023) Sustainability development index of reef fisheries in Indonesia: a case study of reef fisheries at Ternate Island, North Maluku Province. Omni-Akuatika 19(1):15–26. https://doi.org/10.20884/1.oa.2023.19.1.731

Sabai D (2023) The role of integrated coastal management approach in the protection of coastal and marine resources in the eastern coast of Tanzania. Transylvanian Rev Syst Ecol Res 25(2):77–86. https://doi.org/10.2478/trser-2023-0014

Said M, Kusumasari B (2022) Network structure and collaborative management in natural resource and environmental management: a literature review. Int J Soc Ecol Sustain Dev (IJSESD) 13(1):1–16. https://doi.org/10.4018/ijsesd.287122

Scott T (2015) Does collaboration make any difference? Linking collaborative governance to environmental outcomes. J Policy Anal Manag 34(3):537–566. https://doi.org/10.1002/pam.21836

Scott TA, Thomas CW (2017) Unpacking the collaborative toolbox: why and when do public managers choose collaborative governance strategies? Policy Stud J 45(1):191–214. https://doi.org/10.1111/psj.12162

Shields D, Verga F, Blengini GA (2013) Sustainability versus sustainable development: the case of shale gas. In: SPE Europec featured at EAGE conference and exhibition?. SPE, p SPE-164822. https://doi.org/10.2118/164822-ms

Taha R, Taha N (2023) The role of human resources management in enhancing the economic sustainability of Jordanian banks. J Bus Socio-Econ Dev 3(2):180–193. https://doi.org/10.1108/jbsed-04-2022-0045

Talley JL, Schneider J, Lindquist E (2016) A simplified approach to stakeholder engagement in natural resource management: the Five-Feature Framework. Ecol Soc 21(4). https://doi.org/10.5751/es-08830-210438

Tamang R, Nepali S (2023) Role of social workers vs. farmers training in the context of Nepal. Int J Soc Serv Res 3(6):1346–1351. https://doi.org/10.46799/ijssr.v3i6.419

Tan F, Yang P (2023) Use of a knowledge network for management and sharing of teaching resources for building construction simulations. Int J Emerg Technol Learn 18(8):222–237. https://doi.org/10.3991/ijet.v18i08.39245

Tengberg A, Valencia S (2018) Integrated approaches to natural resources management—theory and practice. Land Degrad Dev 29(6):1845–1857. https://doi.org/10.1002/ldr.2946

Wahl L, Kitchel A (2017) Internet based collaboration tools. In: Remote work and collaboration: breakthroughs in research and practice. IGI Global, pp 105–123

Wilson AR, Miller SK, Galland GR (2023) Management procedure development in RFMOs offer lessons for strategic and impactful stakeholder engagement and collaboration. Front Mar Sci 10:1112236. https://doi.org/10.3389/fmars.2023.1112236

Zhao J, Ma C, Zhao X, Wang X (2018) Spatio-temporal dynamic analysis of sustainable development in China based on the footprint family. Int J Environ Res Public Health 15(2):246. https://doi.org/10.3390/ijerph15020246

Zhou K, Wang L, Li X, Zhang D, Liu T, Lin J, Liu P (2022) Natural resource management in American national parks: principles, problems and enlightenment. Nat Resour Conserv Res 5(2):39–53. https://doi.org/10.24294/nrcr.v5i2.1578

Chapter 4
The Role of Ecology in Environmental Technology Development

Sylvester Chibueze Izah and Kurotimipa Frank Ovuru

Abstract This chapter focuses on the role of ecology in environmental technology development. The study showed that the interplay between ecology and technology is a dynamic and evolving field that plays a crucial role in sustainable development. Integrating ecological principles into technological innovation is essential for addressing the multifaceted challenges of sustainability, from energy flow and nutrient cycling to resilience in environmental systems. This integration not only enhances technological efficacy but also supports ecological integrity. Biomimicry draws inspiration from natural processes and exemplifies how emulating nature can lead to more efficient and sustainable technologies across various sectors, including energy production and healthcare. Cases involving wetland-based water treatment, urban green spaces, and soil microbial applications illustrate the potential benefits of aligning technological interventions with ecological processes. However, achieving this integration requires a holistic approach considering socio-cultural dimensions and ecosystem services. Challenges such as economic constraints, cultural barriers, and regulatory hurdles must be addressed to foster interdisciplinary collaboration and education. As stakeholders, humanity also needs to prioritize ecological insights in technology design. These resilient systems that support human prosperity and environmental integrity can be created. Ultimately, the future of sustainable development hinges on our ability to harness technology in ways that respect and enhance ecological systems, ensuring a viable planet for generations to come.

S. C. Izah
Department of Community Medicine, Faculty of Clinical Sciences, Bayelsa Medical University, Yenagoa, Bayelsa State, Nigeria

Department of Microbiology, Faculty of Science, Bayelsa Medical University, Yenagoa, Bayelsa State, Nigeria

S. C. Izah
e-mail: sylvester.izah@bmu.edu.ng

K. F. Ovuru (✉)
Department of Disease Control and Immunization, Bayelsa State Primary Health Care Board, Yenagoa, Bayelsa State, Nigeria
e-mail: kurotimipaovuru@yahoo.com

M. C. Ogwu and S. Chibueze Izah (eds.), *Evaluating Environmental Processes and Technologies*, Environmental Science and Engineering,
https://doi.org/10.1007/978-3-031-85327-2_4

Keywords Ecology · Sustainable development · Ecosystem services · Interdisciplinary collaboration · Environmental technology · Systems and holistic approach

4.1 Introduction

The intricate relationship between ecology and technology is increasingly recognized as a pivotal factor in addressing the pressing challenges of sustainable development. Technology development must incorporate ecological principles as humanity struggles with the effects of climate change, resource depletion, and biodiversity loss. This relationship is characterized by a reciprocal influence where ecological considerations inform technological innovation, while technology, in turn, provides tools and methodologies that enhance environmental sustainability. For instance, Jones et al. (2017) highlighted those technological innovations, particularly in information and communication technologies (ICT), are crucial in promoting sustainable growth by improving efficiency across various sectors. This notion is echoed by Shrivastava et al. (2016), who asserted that strategic technological innovation must align with ecological sustainability to foster a systems and holistic approach to development. The importance of environmental principles in sustainable development cannot be overstated. Sustainable development is fundamentally about meeting the needs of the present without compromising the ability of future generations to meet their own needs. This requires a deep understanding of ecological systems and the implementation of technologies that respect and enhance these systems. In the recent times, environmental, and technological innovations are designed to mitigate pollution and ensure that development does not exceed the ecological recovery capacity of our planet. This perspective is vital as it underscores the necessity of integrating ecological limits into technological planning and implementation.

Moreover, the role of education in fostering a culture of sustainability is critical. Semenets-Orlova et al. (2021) argued that the development of education for sustainable development necessitates the adoption of new educational technologies and pedagogical approaches that emphasize ecological awareness. This educational shift is essential for preparing future generations to engage with the complexities of sustainability and technology, ensuring that they are equipped to innovate responsibly. In this context, the infusion of sustainability concepts into technology education is paramount, as highlighted by Malik et al. (2019), who noted that a lack of awareness regarding sustainability can hinder students' ability to incorporate these principles into technological development. The concept of "smart cities" exemplifies how technology can be harnessed to promote sustainable urban development. Semyachkov (2021) discussed how innovative city initiatives leverage technological innovations to create economically competitive and environmentally sustainable urban environments. These programs frequently incorporate ICT solutions that promote waste reduction, resource management, and the standard of living for locals. These technological applications show how technology can promote ecological sustainability and align urban development with ecological principles.

Furthermore, the relationship between environmental technological innovation and sustainable development is complex and multifaceted. Yusuf et al. (2018) emphasized that industries must adopt environmental technological innovations to achieve sustainability, highlighting the interconnectedness of eco-innovation and broader sustainable development goals. This relationship is critical as it illustrates that technological advancements must be purposefully directed towards enhancing ecological outcomes rather than merely focusing on economic growth. The challenge of assessing the sustainability of technologies is another critical aspect of this discourse. Park and Jun (2017) pointed out that while sustainable development heavily relies on technological advancements, there is a significant gap in the methodologies available for evaluating the sustainability of these technologies. This gap presents a barrier to effectively integrating ecological principles into technological innovation, as stakeholders lack the necessary tools to measure and compare the sustainability of different technological options. In addition to the challenges, significant opportunities are presented by the intersection of technology and sustainability. Furthermore, scientific and technological innovation drives the sustainable development of regional economies, particularly in the global south. This highlights the potential for technology to address ecological challenges and stimulate economic growth and social equity if it is aligned with sustainable development principles.

The role of renewable energy technologies in achieving sustainable development is particularly noteworthy. Tabrizian (2019) asserted that the rapid growth and implementation of renewable energy technologies are essential for mitigating the adverse effects of traditional energy systems. This assertion is supported by the broader consensus that transitioning to renewable energy sources is a crucial step towards achieving sustainability, as it addresses both environmental concerns and energy security. Furthermore, incorporating sustainability into business operations is becoming more widely acknowledged as a competitive advantage. The significance of comprehending the connection between technology procedures and sustainability aspects in business environments was emphasized by Drejeris and Oželienė (2019). This knowledge is essential for businesses looking to innovate responsibly and match their operations with sustainable development objectives.

This chapter assesses the critical interplay between ecological principles and the advancement of environmental technology, emphasizing sustainable development. It highlights how key concepts like energy flow and nutrient cycling inform innovative designs, including biomimicry and ecosystem service enhancement. The chapter demonstrates successful applications of ecological insights in technology through various cases while addressing the challenges and opportunities for integrating these principles into future innovations.

4.2 Ecological Principles and Their Application in Technology

Ecological principles are crucial in shaping sustainable technologies and practices that align with environmental conservation and resource management. The key concepts such as energy flow, nutrient cycling, and resilience are foundational to understanding ecosystems and their interactions with human activities. These principles inform ecological science and serve as guiding frameworks for designing and implementing technologies to promote sustainability. For instance, energy flow within ecosystems illustrates how energy is transferred through various trophic levels, which can inform the development of energy-efficient technologies that mimic natural processes (Pretty 2007). Similarly, nutrient cycling emphasizes the importance of recycling materials within ecosystems, which can inspire technological solutions that minimize waste and enhance resource recovery (Pretty 2008). Table 4.1 summarizes the ecological principles and their application in technology.

The application of ecological principles in technology design is evident in various sectors, including agriculture, urban planning, and industrial processes. Integrating

Table 4.1 Ecological principles and their application in technology

Ecological principle	Focus	Application in technology
Diversity	The presence of a wide variety of species, genes, and ecosystems	Enhancing resilience in networks and systems through redundancy and diversity of functions (e.g., cloud computing)
Interdependence	Organisms and systems rely on each other for survival and function	Integrated systems design, where components (e.g., IoT devices) interact to improve efficiency and functionality
Energy flow	Energy flows through ecosystems, driving ecological processes	Optimizing energy efficiency in technologies (e.g., renewable energy systems, data centers)
Cycling of materials	Recycling of nutrients and materials within ecosystems	Circular economy models in technology (e.g., recycling electronic waste, using sustainable materials in production)
Adaptation	Organisms evolve and adapt to changing environments	Adaptive algorithms and machine learning that respond to changing data patterns or user needs
Equilibrium	Ecosystems maintain a balance between competing forces, promoting stability	Balancing load in computing networks and maintaining stability in software systems through failover mechanisms
Succession	A gradual process of change and development in ecosystems over time	Iterative development in software engineering (e.g., Agile methodologies) and systems scaling
Resilience	The ability of an ecosystem to recover from disturbances	Building fault-tolerant systems that can quickly recover from failures (e.g., disaster recovery in cloud computing)

environmental management practices in agriculture has improved productivity while reducing reliance on chemical inputs. For example, adopting ecological technologies through farmer field schools has reduced pesticide use and increased vegetable production in Indonesia, demonstrating the potential for environmental principles to enhance economic sustainability in agricultural systems (Mariyono et al. 2020). This approach aligns with the broader concept of sustainable development, which seeks to balance economic growth with ecological integrity de la Vega et al. 2023). Technologies supporting food security and advancing environmental health can be developed by integrating ecological ideas into agricultural activities (Izah et al. 2024a, b). Urban environments also benefit from the application of environmental principles in technology design. The concept of techno-ecological synergy, which emphasizes the integration of ecological and technological systems, has been proposed to enhance the sustainability of residential systems in North America (Urban and Bakshi 2013). This approach advocates for designs that mimic natural ecosystems, optimizing resource use and minimizing environmental impacts. By leveraging ecological processes, urban technologies can be developed that enhance resilience to disturbances, such as climate change, while promoting sustainable living practices among residents (Urban and Bakshi 2013). Recognizing ecological principles in urban planning highlights the necessity of interdisciplinary approaches combining environmental science with engineering and design.

In the industrial sector, the principles of industrial ecology provide a framework for rethinking production processes and resource utilization. By viewing industrial systems as part of more extensive ecological networks, companies can identify resource-sharing opportunities and waste reduction, ultimately leading to more sustainable practices (Wells and Orsato 2005; Ogwu et al. 2023). Integrating eco-efficiency into industrial processes has yielded significant improvements in resource utilization, thereby contributing to ecological sustainability (Wells and Orsato 2005). This perspective encourages industries to adopt circular economy principles, where waste is minimized and materials are continuously reused, reflecting the nutrient cycling concept inherent in ecological systems (Pretty 2008).

Resilience is another critical ecological principle that informs technology design, particularly in environmental management. Resilience refers to the capacity of an ecosystem to absorb disturbances while maintaining its essential functions and structures. Technologies designed with resilience can help communities adapt to environmental changes and recover from disruptions, such as natural disasters or economic shocks (Jørgensen 2006). For example, applying ecological engineering principles in lake management has demonstrated how resilient systems can be developed to enhance water quality and biodiversity while supporting human activities (Jørgensen 2006; Izah et al. 2023a, b, c). This approach underscores the importance of designing technologies that address immediate needs and contribute to long-term ecological stability. The role of innovation in achieving social-ecological sustainability cannot be overstated. Innovations incorporating ecological considerations into their design and implementation processes are essential for fostering sustainable development (Degato 2017). This includes integrating technology, governance, and ecosystem stewardship, collectively contributing to the successful transformation of socio-ecological systems (Degato 2017). By fostering innovation that prioritizes ecological health, societies can develop efficient but also equitable and sustainable technologies, addressing the needs of current and future generations.

Furthermore, sustainable technology involves many applications, from renewable energy systems to waste management solutions. The development of green technologies, such as solar energy and electric vehicles, exemplifies how ecological principles can be harnessed to create sustainable solutions that mitigate environmental impacts (Berawi 2019). These technologies reduce carbon emissions and promote efficient use of resources, aligning with the principles of energy flow and nutrient cycling (Pretty 2008). The transition to sustainable technologies is crucial for achieving global climate targets and fostering a more resilient ecological future (Wang et al. 2023).

In the context of circular economies, integrating ecological principles into business models is increasingly recognized as essential for sustainability. Digital technologies can facilitate the development of sustainable business models that prioritize ecological considerations, enabling companies to align their operations with environmental goals (Böttcher et al. 2024). By leveraging data and technology, businesses can optimize resource use, reduce waste, and enhance their overall sustainability performance (Böttcher et al. 2024). This approach highlights the importance of incorporating ecological principles into the core of business strategies, ensuring that economic activities contribute positively to environmental health. The challenges of implementing sustainable technologies are multifaceted and require a comprehensive understanding of ecological principles. For instance, the rebound effect, where increased efficiency leads to higher overall consumption, poses significant challenges to achieving true sustainability (Wang et al. 2023). Addressing these challenges necessitates ongoing research and the development of policies that promote sustainable practices while mitigating potential negative impacts (Wang et al. 2023). By fostering a deeper understanding of ecological principles, stakeholders can better navigate the complexities of sustainability and develop technologies that genuinely contribute to environmental well-being.

4.3 Evaluating Biomimicry: Learning from Nature

Biomimicry, a term popularized by Janine Benyus in the early 1990s, refers to drawing inspiration from nature to solve human challenges. This approach emphasizes learning from the natural world's strategies and designs that have evolved over millions of years. Biomimicry is not merely about imitation; it involves understanding the principles behind nature's designs and applying them to create innovative solutions in technology, architecture, and various other fields. The core idea is to observe and emulate nature's time-tested patterns and strategies to develop sustainable and efficient human-made systems (Somoza-Norton et al. 2023; Appio et al. 2017). One prominent example of biomimicry in technology is the design of wind turbines.

Traditional wind turbine blades have been enhanced through biomimetic designs inspired by the flippers of humpback whales. The unique tubercle structure on these flippers allows for improved lift and reduced drag, resulting in more efficient energy production. Studies have shown that incorporating these biomimetic features can increase the annual electrical output of wind turbines by approximately 20% (Gu et al. 2016; Carré et al. 2022). This application of biomimicry enhances performance and contributes to the sustainability of energy production methods. A model for evaluating biomimicry is presented in Fig. 4.1. The Biomimicry Design Spiral visualizes a structured approach to innovation by drawing inspiration from nature. It begins with Defining the design challenge and understanding how nature addresses similar problems. In the Biologizephase, nature's principles are framed within the context of the design. The Discover phase focuses on identifying biological strategies that could serve as inspiration. These strategies are further analyzed and abstracted in the Abstractphase to distill their functional essence. The insights are then Emulated to create innovative designs or products based on nature's mechanisms. Finally, the evaluation stage assesses the solution's effectiveness and suitability, ensuring alignment with the initial goals and natural principles. This iterative process fosters sustainable and efficient solutions by leveraging nature's time-tested strategies.

Another significant case of biomimicry is in the field of wastewater treatment. Nature has developed intricate systems for purifying water, such as the self-cleaning properties of certain plant leaves. Researchers have mimicked these natural processes to create bio-inspired filtration systems that enhance the efficiency of wastewater treatment plants. For instance, the design of these systems often incorporates materials and structures that facilitate the natural filtration processes observed in wetlands (Jamei and Vrcelj 2021; Salim and Hassan 2022). By utilizing these nature-inspired designs, engineers can develop more sustainable and effective methods for managing water resources. The construction industry has also embraced biomimicry, particularly in the design of building envelopes. Architects have studied how certain organisms adapt to their environments and have applied these principles to create more energy-efficient buildings and responsive to their surroundings. For example, the self-cleaning characteristics of palm leaves have inspired the design of building facades that require less maintenance and are more resilient to environmental factors (Salim and Hassan 2022). This appropriately reduces the ecological footprint of buildings and enhances their longevity and functionality.

In addition to wind turbines and wastewater treatment, biomimicry has found applications in various other technologies, including materials science and product design. For example, the development of bio-inspired materials has led to composites that mimic the toughness and resilience of natural structures, such as nacre found in mollusks. These materials are being explored in construction, packaging, and even medical applications, where biocompatibility and strength are crucial (Singer et al. 2023; Gu et al. 2016). The interdisciplinary nature of biomimicry allows for integrating concepts from biology, engineering, and design, fostering innovation across multiple sectors.

Globally, numerous case studies illustrate the diverse applications of biomimicry. In Japan, the design of the Shinkansen bullet train was inspired by the beak of

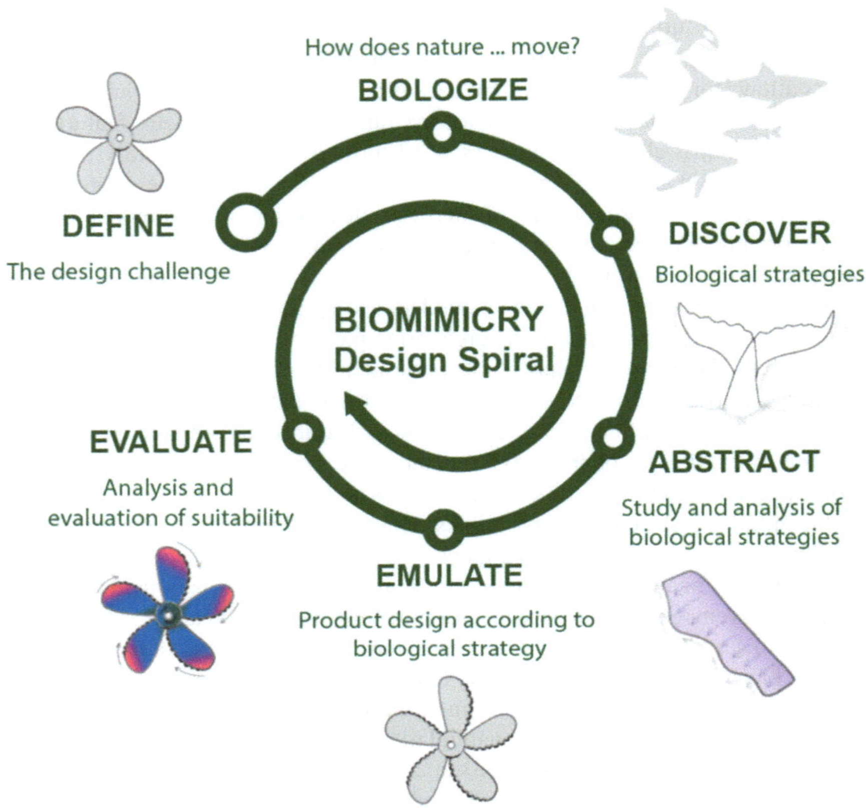

Fig. 4.1 Biomimicry system evaluation design. *Source* Aguilar-Planet and Peralta (2024)

the kingfisher, which allows for smooth transitions between air and water. This biomimetic design improved the train's aerodynamics and reduced noise pollution, demonstrating how nature-inspired solutions can address multiple challenges simultaneously (Zhang and Gruber 2020; Izah et al. 2024a, b). Similarly, in Zimbabwe, the Eastgate Centre's design was influenced by the natural ventilation systems of termite mounds, resulting in a building that maintains comfortable temperatures without relying heavily on air conditioning (Appio et al. 2017; Volstad and Boks 2012). Biomimicry has led to advancements in regenerative medicine and dental materials in healthcare. For instance, the study of shark skin has inspired the development of antimicrobial surfaces that can reduce the risk of infections in medical settings. This application of biomimicry highlights the potential for nature-inspired designs to improve human health and safety (Singer et al. 2023; Kapoor et al. 2022). Furthermore, the exploration of bio-inspired materials in dentistry has opened new avenues for creating biocompatible and durable dental restorations, enhancing patient outcomes (Singer et al. 2023; Carré et al. 2022). The educational sector has also recognized the value of biomimicry as a teaching tool. Educators can foster a deeper understanding of biology and engineering among students by incorporating biomimetic principles into curricula. This approach encourages learners to think critically about how natural systems function and how these principles can be applied to solve real-world problems (Fried et al. 2020; Perricone et al. 2022). Engaging students in biomimicry projects enhances their learning experience and inspires the next generation of innovators to pursue sustainable design practices.

As the world faces increasing environmental challenges, the importance of biomimicry in promoting sustainability cannot be overstated. By learning from nature, we can develop solutions that are not only effective but also environmentally friendly. The principles of biomimicry encourage a shift away from traditional linear models of production and consumption towards more circular and regenerative systems (Ilieva et al. 2022; Regassa et al. 2023). This paradigm shift is essential for addressing climate change, resource depletion, and biodiversity loss. Integrating innovative technologies with biomimicry further enhances the potential for sustainable innovation. For example, artificial intelligence and machine learning can optimize biomimetic designs for more efficient energy conversion and resource management (Regassa et al. 2023). This convergence of technology and nature-inspired design holds promise for creating innovative and sustainable solutions, paving the way for a greener future.

4.4 Ecosystem Services and Technology Development

Ecosystem services (ES) are defined as the benefits humans derive from ecosystems, encompassing a wide range of functions that contribute to human well-being and the sustainability of the environment. The Millennium Ecosystem Assessment (MA) categorizes these services into four main types: provisioning, regulating, cultural, and supporting services. Provisioning services include the production of food, fresh water, and raw materials; regulating services involve climate regulation, water purification, and disease control; cultural services involve recreational, aesthetic, and spiritual benefits; and supporting services are necessary to produce all other ecosystem services, such as nutrient cycling and soil formation (Payne and Itterbeeck 2017; Kaiser et al. 2021; Harrington et al. 2010). This classification highlights the multifaceted nature of ecosystem services and their integral role in maintaining ecological balance and human health. Table 4.2 summarizes ecosystem services and technology development.

The concept of ecosystem services has gained traction in environmental policy and management, as it provides a framework for valuing natural resources and understanding their contributions to economic and social systems. However, the commodification of these services raises ethical concerns, as it can lead to the prioritization of economic gain over ecological integrity (Gómez-Baggethun and Ruiz-Pérez 2011). The valuation of ecosystem services often involves complex methodologies that aim to quantify their benefits, but there remains significant debate about the

Table 4.2 Ecosystem services and technology development

Ecosystem service	Description	Technology development application	References
Provisioning services	Ecosystems provide goods like food, water, timber, and fiber	Development of sustainable agriculture technologies, renewable energy systems, and water purification tech	(Millennium Ecosystem Assessment 2005; FAO 2020)
Regulating services	Natural processes that regulate climate, floods, disease, and water quality	Climate control systems, AI for weather prediction, flood defense technologies, and carbon capture innovations	(IPCC 2014; UNEP 2019)
Cultural services	Non-material benefits such as recreation, aesthetic value, and education	Virtual reality for nature experiences, bio-inspired art and design, and educational apps for environmental awareness	(Chan et al. 2012; Raymond et al. 2009)
Supporting services	Services necessary to produce other ecosystem services, such as nutrient cycling, soil formation, and biodiversity	Biotechnology for soil regeneration, AI for biodiversity conservation, and technologies for ecosystem monitoring	(Rockström et al. 2009; WWF 2020)

appropriateness of these methods and the implications of assigning monetary values to nature (Notte and Rhodes 2020; Morse-Jones et al. 2011; Morelli and Møller 2015). This discourse underscores the need for a nuanced understanding of ecosystem services that transcend mere economic valuation and consider the intrinsic value of biodiversity and ecosystems.

Technological advancements have the potential to mimic or enhance ecosystem services, particularly in urban environments where natural systems are often disrupted. For instance, green infrastructure, such as green roofs and permeable pavements, can improve urban water filtration and reduce runoff, thereby mimicking the natural water regulation services provided by wetlands and forests (Layman et al. 2014; Pettorelli et al. 2017). Additionally, artificial structures, such as constructed wetlands and biofilters, can enhance carbon sequestration and improve air quality, thereby providing regulating services that contribute to climate change mitigation (Layman et al. 2014). These technologies offer practical solutions to urban environmental challenges and highlight the importance of integrating ecological principles into urban planning and development. The integration of technology into ecosystem service provision raises questions about the effectiveness and sustainability of these interventions. While engineered solutions can provide immediate benefits, they may not fully replicate the complexity and resilience of natural ecosystems (Schulp et al.

2014). For example, using bioremediation techniques to clean contaminated sites can enhance soil health and restore ecosystem functions. Still, the long-term success of such interventions depends on the surrounding ecological context and the maintenance of biodiversity (Rendon et al. 2020). Therefore, it is crucial to adopt a holistic approach that considers the interplay between technological solutions and natural processes to ensure the sustainability of ecosystem services.

Moreover, the role of technology in enhancing ecosystem services extends beyond urban environments to agricultural systems. Innovations such as precision agriculture and agroecological practices can improve soil health, enhance biodiversity, and increase the efficiency of resource use, thereby supporting provisioning services such as food production (Payne and Itterbeeck 2017; Rendon et al. 2020). The integration of technology in agriculture not only boosts productivity but also contributes to the resilience of food systems in the face of climate change and environmental degradation. This underscores the potential for technology to serve as a tool for sustainable development, aligning economic interests with ecological health. Despite the promising potential of technology to enhance ecosystem services, challenges remain in effectively implementing these solutions. The complexity of ecosystems and the variability of environmental conditions necessitate adaptive management approaches that can respond to changing circumstances (Caprioli et al. 2020). Furthermore, stakeholder engagement and interdisciplinary collaboration are essential for successfully integrating technological solutions into ecosystem management practices. By fostering partnerships between ecologists, engineers, policymakers, and local communities, it is possible to develop innovative strategies that leverage technology while respecting ecological integrity.

The assessment and quantification of ecosystem services are critical for informing policy decisions and guiding resource management. Various frameworks and methodologies have been developed to evaluate ecosystem services, including spatial mapping and modeling techniques that can visualize service distribution and identify areas of high ecological value (Schulp et al. 2014; Burkhard et al. 2014). These tools enable decision-makers to prioritize conservation efforts and allocate resources effectively, ensuring that ecosystem services are maintained and enhanced over time. However, the effectiveness of these assessments depends on the availability of accurate data and the consideration of socio-cultural values that influence perceptions of ecosystem services (Breyne et al. 2021; Chan et al. 2012). Incorporating socio-cultural values into ecosystem service evaluations is essential for comprehensively understanding their significance to human well-being. Research has shown that communities often prioritize aesthetic and recreational values, highlighting the need for a value pluralism perspective in ecosystem service assessments (Breyne et al. 2021; Chan et al. 2012). By recognizing the diverse ways in which people interact with and benefit from ecosystems, policymakers can develop more inclusive strategies that address the needs and preferences of various stakeholders. This approach enhances the legitimacy of ecosystem service assessments and fosters a sense of stewardship and responsibility towards natural resources. Integrating ecosystem services into spatial planning is another critical aspect of sustainable development. By incorporating ecosystem service considerations into land-use planning and development processes, it is possible to minimize negative impacts on natural systems and enhance the resilience of urban and rural environments (Kowalczyk and Sudra 2014). This requires a shift in perspective, where ecosystems are viewed as integral components of the socio-economic landscape rather than mere resources to be exploited. Such a paradigm shift can lead to more sustainable land-use practices prioritizing ecological health and economic growth.

Furthermore, the role of remote sensing technology in monitoring ecosystem services cannot be overlooked. Advances in satellite imagery and data analytics have enabled researchers to track changes in land cover, assess ecosystem health, and evaluate the effectiveness of management interventions (Pettorelli et al. 2017). These technologies provide valuable insights into the dynamics of ecosystem services, allowing for more informed decision-making and adaptive management strategies. However, the successful application of remote sensing requires collaboration between ecologists, technologists, and policymakers to ensure that data is effectively translated into actionable insights. As the global community grapples with climate change and biodiversity loss, the need for innovative solutions that enhance ecosystem services becomes increasingly urgent. The interplay between technology and ecosystem services presents opportunities and challenges that must be navigated carefully. By fostering interdisciplinary collaboration, integrating socio-cultural values, and leveraging technological advancements, it is possible to develop sustainable strategies that enhance ecosystem services while promoting human well-being.

4.5 Ecological Insights in Technology Design

Integrating ecological insights into technology design has become increasingly relevant as societies grapple with environmental challenges. Table 4.3 summarizes the ecological insights in technology design.

Wetland-based water treatment systems represent a significant advancement in water quality management. These systems utilize the natural processes of wetlands to filter pollutants and improve water quality. For instance, Ackerman et al. developed a wetland management strategy to reduce mercury concentrations in water and fish populations. Their research demonstrated that significantly reduced methylmercury levels could be achieved by modifying seasonal wetlands to promote natural mercury removal processes, such as particulate settling and photodemethylation (Ackerman et al. 2019). This approach enhances water quality and supports biodiversity by providing habitats for various aquatic species.

Moreover, constructed wetlands have emerged as a popular technology for wastewater treatment. These systems mimic the natural filtration processes of wetlands, utilizing vegetation and microbial communities to treat contaminated water. Research by Vymazal (2011) highlighted the evolution of constructed wetlands

Table 4.3 Ecological insights in technology design

Ecological insight	Focus	Application in technology design	References
Biomimicry	Using nature's models to solve human challenges by emulating designs and processes	Development of energy-efficient buildings, robotics, and transportation systems inspired by biological designs	(Benyus 1997; Vincent et al. 2006)
Systems thinking	Understanding the interconnectedness of different components in ecosystems	Designing integrated technological systems that function in harmony, such as smart cities and IoT systems	(Capra 1997; Meadows 2008)
Resilience and adaptation	Ecosystems adapt to changes and recover from disturbances	Developing fault-tolerant and resilient systems in computing, networking, and infrastructure	(Holling 1973; Walker et al. 2004)
Energy efficiency	Ecosystems optimize energy use to maintain balance	Designing low-energy consumption technologies, such as energy-efficient processors and renewable energy systems	(Odum 1993; Smil 2005)
Material cycling	Natural ecosystems recycle nutrients and materials efficiently	The circular economy in technology design focuses on recycling and reusing materials in production and waste management	(McDonough and Braungart 2010; Stahel 2016)
Biodiversity as stability	Diverse ecosystems are more stable and resilient to changes	Implementing diversity in software and hardware systems to increase robustness and prevent failure	(Elmqvist et al. 2003; Tilman 1999)
Ecosystem services framework	Ecosystems provide services that benefit humanity, from clean water to pollination	Applying this framework to create technologies that protect or mimic ecosystem services (e.g., air filtration, water purification)	(Costanza et al. 1997; Daily 1998)

over the past five decades, emphasizing their effectiveness in removing pollutants such as nitrogen and phosphorus from wastewater. The integration of hybrid-constructed wetlands, which combine different flow patterns, has further improved treatment efficiency, showcasing the adaptability of this technology to various environmental contexts (Zhang et al. 2015).

In addition to their role in water treatment, wetlands also serve as critical ecosystems for carbon sequestration and biodiversity conservation. The ecological design of these systems must consider the intricate relationships between water quality, plant communities, and microbial activity. By fostering techno-ecological synergies, as suggested by Urban and Bakshi (2013), we can enhance the sustainability of wetland management practices, ensuring that technological interventions align with ecological processes. This holistic approach is essential for addressing the complex challenges of water pollution and climate change.

Urban green spaces have gained recognition for their potential to improve air quality and enhance urban livability. Incorporating green infrastructure, such as parks and green roofs, can mitigate the urban heat island effect and reduce air pollution levels. Research indicates that urban vegetation is crucial in filtering airborne pollutants, and improving overall air quality (Hussain et al. 2023). For example, implementing urban green spaces has been shown to reduce particulate matter concentrations linked to respiratory diseases and other health issues. The design of urban green spaces must prioritize biodiversity and ecological health. Creating habitats for various species can support urban ecosystems and enhance resilience to environmental stressors. As Feng and Xia (2014) discussed, the concept of landscape ecological design emphasizes the importance of integrating ecological principles into urban planning to create sustainable and functional green spaces. This approach benefits human health and fosters a sense of community and connection to nature. Figure 4.2 illustrates the ongoing evolution of urban ecological worldviews model from Toner et al. (2023), transitioning from the Mechanistic Worldview (Ecology IN the City), focused on conventional sustainability through data and ecological information, to the Transitional Worldview (Ecology OF the City), which integrates ecosystem services and nature-based solutions, reflecting post-modernist approaches. Finally, it progresses to the Ecological Worldview (Ecology FOR the City), emphasizing regenerative sustainability through community engagement, co-creation, and ecological wisdom, aligned with reflexive modernism. This evolution highlights a shift from a segregated understanding of ecology in urban areas to a holistic and collaborative approach prioritizing sustainable and integrated urban ecosystems.

Furthermore, using technology to monitor and manage urban green spaces can enhance their effectiveness. Advanced analytics and sensor networks can provide real-time data on air quality, vegetation health, and biodiversity, enabling informed decision-making for urban planners (Hussain et al. 2023). By leveraging digital technology, cities can optimize the design and maintenance of green infrastructure, ensuring that these spaces continue to provide ecological and social benefits. Soil microbial applications in pollution remediation represent another innovative intersection of technology and ecology. Phytoremediation, which utilizes plants

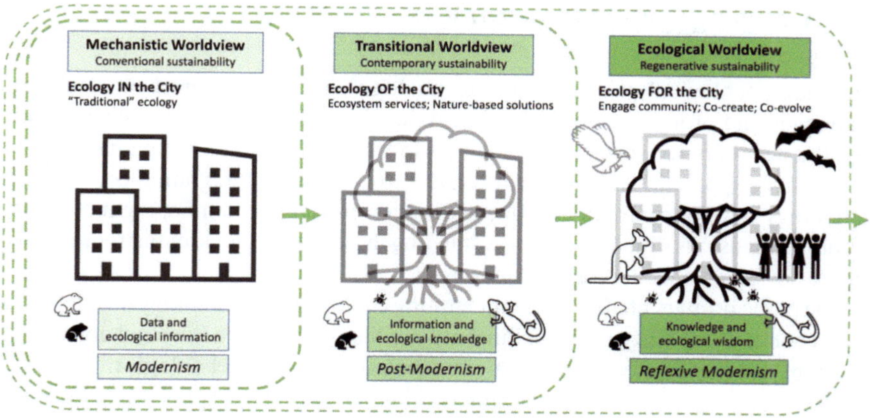

Fig. 4.2 Evolution of urban ecological worldviews model. *Source* Toner et al. (2023)

and their associated microbial communities to degrade or immobilize contaminants, has emerged as a promising strategy for addressing soil and water pollution (Truu et al. 2015). This approach capitalizes on the natural processes of soil ecosystems, harnessing the capabilities of microorganisms to break down toxic compounds and restore ecological health. Research has shown that specific plant species can enhance microbial activity in contaminated soils, improving remediation outcomes. For instance, studies have demonstrated that certain plants can stimulate the growth of beneficial microbes that degrade pollutants such as heavy metals and organic contaminants (Truu et al. 2015). This synergy between plants and microbes underscores the importance of considering ecological interactions in designing pollution remediation strategies.

Moreover, applying soil microbial technologies can be complemented by digital tools that monitor microbial communities and their responses to environmental changes. By employing high-throughput sequencing and bioinformatics, researchers can gain insights into the dynamics of soil microbial populations and their roles in pollutant degradation (Navas-Molina et al. 2013; Ogwu et al. 2024). This information can inform the development of targeted remediation strategies that leverage the natural capabilities of soil ecosystems. The integration of ecological insights into technology design is not without challenges. As highlighted by Oord and Witteloostuijn (2018), the evolution of technology often occurs in isolation from ecological considerations, leading to a disconnect between technological innovation and environmental sustainability. To overcome this barrier, fostering interdisciplinary collaboration among ecologists, engineers, and urban planners is essential, ensuring that ecological principles inform technological solutions.

4.6 Challenges and Opportunities

Integrating ecological principles into technology development presents significant challenges and promising opportunities. As environmental concerns escalate globally, the necessity for sustainable practices in technological innovation has become increasingly apparent. However, the barriers to effectively integrating ecology into technology development are multifaceted, encompassing economic, regulatory, and cultural dimensions (Table 4.4). Understanding these barriers is essential for fostering a more sustainable technological landscape.

One of the primary challenges in integrating ecology into technology development is the economic viability of green innovations. Many firms face financial constraints that limit their ability to invest in environmentally friendly technologies. Empirical evidence suggests that while environmental innovations can drive down toxic emissions and enhance sustainability, the initial costs associated with developing and implementing these technologies can be prohibitive for many businesses (Bönte and Dienes 2013; Obahiagbon and Ogwu 2024). Furthermore, the market often lacks sufficient incentives for companies to prioritize ecological considerations in their product development processes, leading to a reliance on traditional, less sustainable practices (Driessen et al. 2013). This economic barrier is compounded by the lack of comprehensive regulatory frameworks that support green innovation, as many policies still favor established, non-ecological technologies (Dariyabayevichc 2023; Akudugu and Ogwu 2024).

In addition to economic barriers, significant cultural and organizational challenges impede the integration of ecological principles into technology development. Many organizations operate under a paradigm that prioritizes short-term profits over long-term sustainability, which can stifle innovative ecological approaches (Wei et al. 2022). The prevailing corporate culture often emphasizes technological advancement without considering ecological impacts, leading to a disconnect between technological development and environmental stewardship. This cultural inertia can be difficult to overcome, as it requires a fundamental shift in organizational values and priorities (Wang et al. 2022). Moreover, the lack of interdisciplinary collaboration between ecologists and technologists further exacerbates this issue, as both fields often operate in silos, limiting the potential for innovative solutions that incorporate ecological insights into technological development (Angelstam et al. 2013).

Despite these challenges, there are significant opportunities for integrating ecology into technology development. The rise of green technology innovations presents a pathway for sustainable economic growth. Research indicates that green technology can facilitate urbanization while minimizing environmental degradation, addressing the dual challenges of economic development and ecological preservation (Xu et al. 2022; Ogwu et al. 2014). Moreover, advancements in vertical greening systems and other sustainable technologies highlight the potential for innovative solutions that harmonize ecological principles with technological progress (Wang et al. 2022). These innovations contribute to environmental sustainability and enhance the

Table 4.4 Barriers to integrating ecology into technology development

Barrier	Effect	Impact on technology development	References
Lack of interdisciplinary collaboration	Limited collaboration between ecologists and technologists due to disciplinary silos	Slows the integration of ecological principles into technology, hindering innovative sustainable solutions	(Ostrom 2009; Fischer et al. 2015)
Cost and economic incentives	Sustainable technology development may require higher upfront costs	Discourages investment in eco-friendly technologies due to short-term financial priorities	(Loorbach et al. 2010; Porter and Kramer 2011)
Technological determinism	Overreliance on technology as the sole solution to environmental challenges	This leads to the neglect of ecological solutions or hybrid approaches that blend technology with natural processes	(Pitt 2011; Smith and Stirling 2010)
Limited public awareness and engagement	General lack of understanding about the importance of ecology in technology design	The public and policymakers may prioritize convenience and cost over long-term sustainability	(Meadows 2008; Agyeman et al. 2003)
Regulatory and policy gaps	Inconsistent or outdated regulations that fail to incentivize eco-friendly technology	Lack of policy support for innovation in integrating ecological principles into technological development	(Geels 2004; Jänicke and Jacob 2005)

(continued)

Table 4.4 (continued)

Barrier	Effect	Impact on technology development	References
Market competition and profit pressures	Companies often prioritize profitability over ecological sustainability	Results in a focus on short-term financial gains rather than long-term sustainability in tech development	(Schumpeter 2013; Stahel 2016)
Knowledge gaps in ecological science	Incomplete understanding of complex ecological processes and systems	Limits the ability to apply ecological principles effectively in technological innovation	(Loreau et al. 2002; Levin 1998)
Technological lock-in	Entrenched use of existing technologies and infrastructure that are unsustainable	Hinders the adoption of ecologically informed technologies due to the cost and effort required to shift	(Unruh 2000; Arthur 1989)

aesthetic and functional aspects of urban environments, creating a win–win scenario for both ecological and technological advancement.

Future directions in the role of ecology in environmental technology development are promising. The increasing recognition of the interconnectedness of social, ecological, and technological systems underscores the need for a holistic approach to innovation (McPhearson et al. 2021). This perspective advocates for transformative adaptation strategies that integrate ecological considerations into the core of technological development processes (Hellin et al. 2021; Ogwu 2019, 2020, 2023). By fostering collaboration between ecologists, engineers, and policymakers, it is possible to create a more integrated framework for addressing environmental challenges through technology. Moreover, developing regulatory frameworks that incentivize green innovations can significantly enhance the integration of ecological principles into technology development. Policymakers can play a crucial role in shaping the landscape for sustainable technologies by implementing environmental tax policies that encourage the adoption of green technologies (Sayar 2023). Such policies can create a more favorable economic environment for businesses to invest in sustainable practices, ultimately leading to a more robust green economy. Additionally, establishing clear guidelines and standards for ecological performance can help drive innovation in environmentally friendly technologies, ensuring that environmental considerations are embedded in the development process from the outset.

The role of education and training in fostering ecological literacy among future technologists cannot be overstated. By integrating ecological principles into educational curricula, future engineers and technologists can be better equipped to consider environmental impacts in their work (Hlukhaniuk et al. 2023). This educational shift is essential for cultivating a new generation of innovators prioritizing sustainability and ecological integrity in technological solutions. Furthermore, fostering a culture of environmental awareness within organizations can enhance employee engagement and commitment to sustainable practices, ultimately leading to more successful integration of ecological principles into technology development (Driessen et al. 2013).

4.7 Conclusion

A vital component of sustainable development, the nexus of ecology and technology is a dynamic and developing topic. Incorporating ecological principles into technology innovation is crucial as we confront more complicated environmental concerns. We can create technologies that satisfy human needs and ecological integrity and comprehend ideas like energy flow, nutrient cycling, and ecosystem resilience. This environmental and technological synergy is not only advantageous, but also necessary to achieve sustainable results in a variety of fields. Biomimicry, which takes cues from nature's tried-and-true methods, is one of the most promising solutions in this paradigm. By emulating the efficiencies and resilience found in natural ecosystems, we can develop innovative solutions that address pressing global challenges. The potential applications of biomimicry are vast, from renewable energy systems modeled after photosynthesis to healthcare innovations inspired by biological processes. These advancements demonstrate that technology can be developed harmoniously with ecological systems, leading to more sustainable practices that benefit humanity and the environment.

Moreover, the relationship between ecosystem services and technology development is multifaceted. While technology has the potential to enhance and replicate these services, it is crucial to consider the ecological context and socio-cultural dimensions that influence their provision. A holistic and inclusive approach to ecosystem management allows us to create resilient systems supporting environmental health and human prosperity. The wetland-based water treatment case studies, urban green spaces, and soil microbial applications exemplify how aligning technological interventions with ecological processes can yield sustainable benefits. These examples illustrate that integrating ecological insights into technology design can enhance environmental quality and public health. Despite the significant challenges in merging ecological considerations with technological advancements—such as economic, cultural, and regulatory barriers—the opportunities for innovation and sustainability are substantial. Interdisciplinary collaboration among ecologists, engineers, policymakers, and communities is vital for overcoming these obstacles. By fostering a culture of sustainability in educational practices and technological development, we can ensure that future innovations prioritize ecological integrity. As we look ahead, the path to a sustainable future will depend on our ability to leverage technology to respect and enhance ecological systems. The continuous evolution of the relationship between ecology and technology offers a promising avenue for creating solutions that are not only effective but also regenerative. Embracing a techno-ecological approach will be critical in navigating the complexities of environmental challenges, fostering resilience, and promoting sustainable development in our communities.

References

Ackerman JT, Fleck JA, Eagles-Smith CA, Marvin-DiPasquale M, Windham-Myers L, Herzog MP, McQuillen HL (2019) Wetland management strategy to reduce mercury in water and bioaccumulation in fish. Environ Toxicol Chem 38(10):2178–2196. https://doi.org/10.1002/etc.4535

Aguilar-Planet T, Peralta E (2024) Innovation inspired by nature: applications of biomimicry in engineering design. Biomimetics 9(9):523. https://doi.org/10.3390/biomimetics9090523

Agyeman J, Bullard RD, Evans B (eds) (2003) Just sustainabilities: development in an unequal world. MIT Press

Akudugu MA, Ogwu MC (2024) Sustainable development policies and interventions: a bibliometric analysis of the contributions of the academic community. J Clean Prod Article # 139919. https://doi.org/10.1016/j.jclepro.2023.139919

Angelstam P, Andersson K, Annerstedt M, Axelsson R, Elbakidze M, Garrido P, Grahn P, Jönsson KI, Pedersen S, Schlyter P, Skärbäck E, Smith M, Stjernquist I (2013) Solving problems in social–ecological systems: definition, practice and barriers of transdisciplinary research. Ambio 42:254–265. https://doi.org/10.1007/s13280-012-0372-4

Appio FP, Achiche S, Martini A, Beaudry C (2017) On designers' use of biomimicry tools during the new product development process: an empirical investigation. Technol Anal Strateg Manag 29(7):775–789. https://doi.org/10.1080/09537325.2016.1236190

Arthur WB (1989) Competing technologies, increasing returns, and lock-in by historical events. Econ J 99(394):116–131. https://doi.org/10.2307/2234208

Benyus JM (1997) Biomimicry: innovation inspired by nature. Harper Perennial

Berawi MA (2019) The role of industry 4.0 in achieving Sustainable Development Goals. Int J Technol 10(4):644–647. https://doi.org/10.14716/ijtech.v10i4.3341

Bönte W, Dienes C (2013) Environmental innovations and strategies for the development of new production technologies: empirical evidence from Europe. Bus Strateg Environ 22(8):501–516. https://doi.org/10.1002/bse.1753

Böttcher TP, Empelmann S, Weking J, Hein A, Krcmar H (2024) Digital sustainable business models: using digital technology to integrate ecological sustainability into the core of business models. Inf Syst J 34(3):736–761. https://doi.org/10.1111/isj.12436

Breyne J, Dufrêne M, Maréchal K (2021) How integrating 'socio-cultural values' into ecosystem services evaluations can give meaning to value indicators. Ecosyst Serv 49:101278. https://doi.org/10.1016/j.ecoser.2021.101278

Burkhard B, Kandziora M, Hou Y, Müller F (2014) Ecosystem service potentials, flows and demands-concepts for spatial localisation, indication and quantification. Landscape Online 1–34. https://doi.org/10.3097/lo.201434

Capra F (1997) The web of life: a new scientific understanding of living systems. Anchor Books

Caprioli C, Bottero M, Mondini G (2020) Urban ecosystem services: a review of definitions and classifications for the identification of future research perspectives. In: International conference on computational science and its applications. Springer, Cham, pp 332–344. https://doi.org/10.1007/978-3-030-58814-4_23

Carré A, Gasnier P, Roux É, Tabourot L (2022) Extending the operating limits and performances of centimetre-scale wind turbines through biomimicry. Appl Energy 326:119996. https://doi.org/10.1016/j.apenergy.2022.119996

Chan KM, Satterfield T, Goldstein J (2012) Rethinking ecosystem services to better address and navigate cultural values. Ecol Econ 74:8–18. https://doi.org/10.1016/j.ecolecon.2011.11.011

Costanza R, d'Arge R, De Groot R, Farber S, Grasso M, Hannon B, Limburg K, Naeem S, O'neill RV, Paruelo J, Raskin RG, Sutton P, van den Belt M (1997) The value of the world's ecosystem services and natural capital. Nature 387(6630):253–260. https://doi.org/10.1038/387253a0

Daily GR (1998) Nature's services: societal dependence on natural ecosystems. Environ Values 7(3)

Dariyabayevichc UO (2023) A comprehensive analysis of ecological law in the cyber era. Int J Law Policy 1(4). https://doi.org/10.59022/ijlp.58

de la Vega CM, Gutnisky J, Montilla D, Maid S, Salimbeni S (2023) The impact of Industry 4.0 on sustainable development goals: sustainability, solar energy and IoT on clean water for marginalised communities. In: Proceedings of the IS23 conference, pp 89–94. https://www.iim.ftn.uns.ac.rs/is23/proceedings/papers/T7.1/T7.1-3_08941.pdf

Degato DD (2017) Innovation and paths to social-ecological sustainability. J Innov Sustain RISUS 8(2):3–33. https://doi.org/10.24212/2179-3565.2017v8i2p3-33

Drejeris R, Oželienė D (2019) New approach to the technological aspect of corporate sustainable development. Bus: Theory Pract 20:363–371. https://doi.org/10.20944/preprints201902.0044.v1

Driessen PH, Hillebrand B, Kok RAW, Verhallen TMM (2013) Green new product development: the pivotal role of product greenness. IEEE Trans Eng Manag 60(2):315–326. https://doi.org/10.1109/tem.2013.2246792

Elmqvist T, Folke C, Nyström M, Peterson G, Bengtsson J, Walker B, Norberg J (2003) Response diversity, ecosystem change, and resilience. Front Ecol Environ 1(9):488–494. https://doi.org/10.1890/1540-9295(2003)001[0488:RDECAR]2.0.CO;2

FAO (2020) The state of the world's forests 2020: forests, biodiversity and people. Food and Agriculture Organization of the United Nations. https://doi.org/10.4060/ca8642en

Feng B, Xia P (2014) Study on the method of landscape ecological design for the urban furniture. Appl Mech Mater 522:1645–1648. https://doi.org/10.4028/www.scientific.net/amm.522-524.1645

Fischer J, Gardner TA, Bennett EM, Balvanera P, Biggs R, Carpenter S, Daw T, Folke C, Hill R, Hughes TP, Luthe T, Maass M, Meacham M, Norstrom VA, Peterson G, Queiroz C, Seppelt R, Spierenburg M, Tenhune J (2015) Advancing sustainability through mainstreaming a social–ecological systems perspective. Curr Opin Environ Sustain 14:144–149. https://doi.org/10.1016/j.cosust.2015.06.002

Fried E, Martin A, Esler A, Tran A, Corwin L (2020) Design-based learning for a sustainable future: student outcomes resulting from a biomimicry curriculum in an evolution course. Evol: Educ Outreach 13:1–22. https://doi.org/10.1186/s12052-020-00136-6

Geels FW (2004) From sectoral systems of innovation to socio-technical systems: insights about dynamics and change from sociology and institutional theory. Res Policy 33(6–7):897–920. https://doi.org/10.1016/j.respol.2004.01.015

Gómez-Baggethun E, Ruiz-Pérez M (2011) Economic valuation and the commodification of ecosystem services. Prog Phys Geogr 35(5):613–628. https://doi.org/10.1177/0309133311421708

Harrington R, Anton C, Dawson TP, de Bello F, Feld CK, Haslett JR, Kluvánkova-Oravská T, Kontogianni A, Lavorel S, Luck GW, Rounsevell MD, Samways MJ, Settele J, Skourtos M, Spangenberg JH, Vandewalle M, Zobel M, Harrison PA (2010) Ecosystem services and biodiversity conservation: concepts and a glossary. Biodivers Conserv 19:2773–2790. https://doi.org/10.1007/s10531-010-9834-9

Hellin J, Fisher E, Loboguerrero AM (2021) Reflections on enhancing the impact of climate risk management through transformative adaptation. Front Climate 3:751691. https://doi.org/10.3389/fclim.2021.751691

Hlukhaniuk V, Shymkova I, Tsvilyk S, Marushchak O, Buha O (2023) Content modeling the environmental training of future labor training teachers in universities. In: Society. Integration. Education. Proceedings of the international scientific conference, vol 1, pp 275–288. https://doi.org/10.17770/sie2023vol1.7129

Holling CS (1973) Resilience and stability of ecological systems. Annu Rev Ecol Syst 4(1):1–23. https://doi.org/10.1146/annurev.es.04.110173.000245

Hussain J, Anjum S, Yousuf M, Ahmad F (2023) Greening growth: the environmental implications of technology innovation, green finance, and foreign direct investment in Pakistan. Rev Econ Dev Stud 9(2):133–146. https://doi.org/10.47067/reads.v9i2.490

Ilieva L, Ursano I, Traista L, Hoffmann B, Dahy H (2022) Biomimicry as a sustainable design methodology—introducing the 'Biomimicry for sustainability framework.' Biomimetics 7(2):37. https://doi.org/10.3390/biomimetics7020037

IPCC (2014) Climate change 2014: impacts, adaptation, and vulnerability. Intergovernmental Panel on Climate Change. https://www.ipcc.ch/report/ar5/wg2/

Izah SC, Ogidi OI, Ogwu MC, Salimon SS, Yusuf ZM, Akram M, Raimi MO, Iyingiala A-A (2023b) Historical perspectives and overview of the value of herbal medicine. In: Izah SC, Ogwu MC, Akram M (eds) Herbal medicine phytochemistry. Reference series in phytochemistry. Springer, Cham. https://doi.org/10.1007/978-3-031-21973-3_1-1

Izah SC, Ogwu MC, Hamidifar H (2024) Biomonitoring of pollutants in the global south. Springer, Switzerland. https://doi.org/10.1007/978-981-97-1658-6

Izah SC, Ogwu MC, Akram M (2023a) Herbal medicine phytochemistry: applications and trends. Springer, Cham, Switzerland. https://doi.org/10.1007/978-3-031-21973-3

Izah SC, Ogwu MC, Loukas A, Hamidifar H (2024) Water related challenges and management strategies. Springer, Singapore. https://doi.org/10.1007/978-981-97-4966-9

Izah SC, Richard G, Stanley HO, Ogwu MC, Sawyer WE, Uwaeme OR (2023c) Prospects and application of multivariate and reliability analyses to one health risk assessments of toxic elements. Toxicol Environ Heal Sci. https://doi.org/10.1007/s13530-023-00199-0

Jamei E, Vrcelj Z (2021) Biomimicry and the built environment, learning from nature's solutions. Appl Sci 11(16):7514. https://doi.org/10.3390/app11167514

Jänicke M, Jacob K (2005) Ecological modernisation and the creation of lead markets. In: Towards environmental innovation systems. Springer, Berlin, Heidelberg, pp 175–193. https://doi.org/10.1080/09644010500418772

Jones P, Wynn M, Hillier D, Comfort D (2017) The sustainable development goals and information and communication technologies. Indonesian J Sustain Account Manag 1(1):1–15. https://doi.org/10.28992/ijsam.v1i1.22

Jørgensen SE (2006) Application of ecological engineering principles in lake management. Lakes Reserv Res Manag 11(2):103–109. https://doi.org/10.1111/j.1440-1770.2006.00296.x

Kaiser J, Haase D, Krueger T (2021) Payments for ecosystem services: a review of definitions, the role of spatial scales, and critique. Ecol Soc 26(2):1–13. https://doi.org/10.5751/es-12307-260212

Kapoor RT, Rafatullah M, Qamar M, Qutob M, Alosaimi AM, Alorfi HS, Hussein MA (2022) Review on recent developments in bioinspired materials for sustainable energy and environmental applications. Sustainability 14(24):16931. https://doi.org/10.3390/su142416931

Kowalczyk M, Sudra P (2014) Ecosystem services in spatial planning. Europa XXI 27:5–18. https://doi.org/10.7163/eu21.2014.27.1

La Notte A, Rhodes C (2020) The theoretical frameworks behind integrated environmental, ecosystem, and economic accounting systems and their classifications. Environ Impact Assess Rev 80:106317. https://doi.org/10.1016/j.eiar.2019.106317

Layman CA, Jud ZR, Archer SK, Riera D (2014) Provision of ecosystem services by human-made structures in a highly impacted estuary. Environ Res Lett 9(4):044009. https://doi.org/10.1088/1748-9326/9/4/044009

Levin SA (1998) Ecosystems and the biosphere as complex adaptive systems. Ecosystems 1(5):431–436. https://doi.org/10.1007/s100219900037

Loorbach D, Van Bakel JC, Whiteman G, Rotmans J (2010) Business strategies for transitions towards sustainable systems. Bus Strateg Environ 19(2):133–146. https://doi.org/10.1002/bse.645

Loreau M, Naeem S, Inchausti P (eds) (2002) Biodiversity and ecosystem functioning: synthesis and perspectives. Oxford University Press, Oxford

Malik MN, Khan HH, Chofreh AG, Goni FA, Klemeš JJ, Alotaibi Y (2019) Investigating students' sustainability awareness and the curriculum of technology education in Pakistan. Sustainability 11(9):2651. https://doi.org/10.3390/su11092651

Mariyono J, Dewi HA, Daroini PB, Latifah E, Hakim AL, Luther GC (2020) Farmer field schools for improving economic sustainability performance of Indonesian vegetable production. Int J Product Perform Manag 71(4):1188–1211. https://doi.org/10.1108/ijppm-09-2019-0445

McDonough W, Braungart M (2010) Cradle to cradle: remaking the way we make things. North Point Press

McPhearson T, Raymond CM, Gulsrud N, Albert C, Coles N, Fagerholm N, Nagatsu M, Olafsson AS, Soininen N, Vierikko K (2021) Radical changes are needed for transformations to a good Anthropocene. NPJ Urban Sustain 1(1):5. https://doi.org/10.1038/s42949-021-00017-x

Meadows DH (2008) Thinking in systems: a primer. Sustainability Institute

Morelli F, Møller AP (2015) Concerns about the use of ecosystem services as a tool for nature conservation: from misleading concepts to providing a "price" for nature, but not a "value." Eur J Ecol 1(1):68–70. https://doi.org/10.1515/eje-2015-0009

Morse-Jones S, Luisetti T, Turner RK, Fisher B (2011) Ecosystem valuation: some principles and a partial application. Environmetrics 22(5):675–685. https://doi.org/10.1002/env.1073

Navas-Molina JA, Peralta-Sánchez JM, González A, McMurdie PJ, Vázquez-Baeza Y, Xu Z, Ursell LK, Lauber C, Zhou H, Song SJ, Huntley J, Ackermann GL, Berg-Lyons D, Holmes S, Caporaso JG, Knight R (2013) Advancing our understanding of the human microbiome using QIIME. In: Methods in enzymology, vol 531. Academic, pp 371–444. https://doi.org/10.1016/b978-0-12-407863-5.00019-8

Obahiagbon EG, Ogwu MC (2024) The nexus of business, sustainability, and herbal medicine. In: Izah SC, Ogwu MC, Akram M (eds) Herbal medicine phytochemistry. Reference series in phytochemistry. Springer, Cham. https://doi.org/10.1007/978-3-031-21973-3_67-1

Odum EP (1993) Ecology and our endangered life-support systems, 2nd edn. Sinauer Associates, Sunderland, MA

Ogwu MC (2019) Towards sustainable development in Africa: the challenge of urbanization and climate change adaptation. In: Cobbinah PB, Addaney M (eds) The geography of climate change adaptation in urban Africa. Springer, Switzerland, pp 29–55. https://doi.org/10.1007/978-3-030-04873-0_2

Ogwu MC (2020) Value of *Amaranthus* [L.] species in Nigeria. In: Waisundara V (ed) Nutritional value of Amaranth. IntechOpen, UK, pp 1–21. https://doi.org/10.5772/intechopen.86990

Ogwu MC (2023) Plants as monitors and managers of pollution. In: Srivastav AL, Grewal AS, Tiwari M, Pham TD (eds) Role of green chemistry in ecosystem restoration to achieve environmental

sustainability. Elsevier, Amsterdam, Netherlands, pp 51–60. ISBN: 978-0-443-15291-7. https://doi.org/10.1016/B978-0-443-15291-7.00022-5
Ogwu MC, Izah SC (2023) One health implications of agrochemicals and their sustainable alternatives (First Edition). In: Ramawat KG (ed) Sustainable development and biodiversity series, vol 34. Springer, Singapore, 826p. Hardcover. ISBN: 978-981-99-3438-6. https://doi.org/10.1007/978-981-99-3439-3
Ogwu MC, Izah SC, Dessureault-Rompré J, Gasparatos D (2024) Sustainable soil systems in global south. Springer, Switzerland. https://doi.org/10.1007/978-981-97-5276-8
Ogwu MC, Malikia CN, Stansfield A, Gonzales-Torres AD, Izah SC (2023) Sustainable food processing waste management for environmental protection. In: Srivastav AL, Grewal AS, Tiwari M, Pham TD (eds) Role of green chemistry in ecosystem restoration to achieve environmental sustainability. Elsevier, Amsterdam, Netherlands, pp 291–299. ISBN: 978-0-443-15291-7. https://doi.org/10.1016/B978-0-443-15291-7.00022-5
Ogwu MC, Osawaru ME, Ahana CM (2014) Challenges in conserving and utilizing plant genetic resources (PGR). Int J Genet Mol Biol 6(2):16–22. https://doi.org/10.5897/IJGMB2013.0083
Ostrom E (2009) A general framework for analyzing sustainability of social-ecological systems. Science 325(5939):419–422. https://doi.org/10.1126/science.1172133
Park S, Jun S (2017) Statistical technology analysis for competitive sustainability of three dimensional printing. Sustainability 9(7):1142. https://doi.org/10.3390/su9071142
Payne CL, Van Itterbeeck J (2017) Ecosystem services from edible insects in agricultural systems: a review. Insects 8(1):24. https://doi.org/10.3390/insects8010024
Perricone V, Arcamone S, Monti F (2022) Digital architectural design inspired by and created with nature. DIID—Disegno Industriale Industrial Design 76:18. https://doi.org/10.30682/diid7622p
Pettorelli N, Schulte to Bühne H, Tulloch A, Dubois G, Macinnis-Ng C, Queirós AM, Keith DA, Wegmann M, Schrodt F, Stellmes M, Sonnenschein R, Geller GN, Roy S, Somers B, Murray N, Bland L, Geijzendorffer I, Kerr JT, Broszeit S, Leitao PJ, Duncan C, El Serafy G, He KS, Blanchard JL, Lucas R, Mairota P, Webb TJ, Nicholson (2017) Satellite remote sensing of ecosystem functions: opportunities, challenges and way forward. Remote Sens Ecol Conserv 4(2):71–93. https://doi.org/10.1002/rse2.59
Pitt J (2011) The technology of goodness: philosophical reflections on the nature of technology. Sci Eng Ethics 17(2):277–291. https://doi.org/10.1007/s11948-010-9214-0
Porter ME, Kramer MR (2011) Creating shared value. Harvard Bus Rev 89(1/2):62–77
Pretty J (2008) Agricultural sustainability: concepts, principles and evidence. Philos Trans Roy Soc B: Biol Sci 363(1491):447–465. https://doi.org/10.1098/rstb.2007.2163
Raymond CM, Bryan BA, MacDonald DH, Cast A, Strathearn S, Grandgirard A, Kalivas T (2009) Mapping community values for natural capital and ecosystem services. Ecol Econ 68(5):1301–1315. https://doi.org/10.1016/j.ecolecon.2008.12.006
Regassa Y, Dabasa T, Amare G, Lemu HG (2023) Bio-inspired design trends for sustainable energy structures. In: IOP conference series: materials science and engineering, vol 1294, no 1. IOP Publishing, p 012044. https://doi.org/10.1088/1757-899x/1294/1/012044
Rendon P, Steinhoff-Knopp B, Saggau P, Burkhard B (2020) Assessment of the relationships between agroecosystem condition and the ecosystem service soil erosion regulation in Northern Germany. PLoS ONE 15(12):e0234288. https://doi.org/10.1371/journal.pone.0234288
Rockström J, Steffen W, Noone K, Persson Å, Chapin FS, Lambin EF, Lenton TM, Scheffer M, Folke C, Schellnhuber HJand Nykvist B, de Wit CA, Hughes T, van der Leeuw S, Rodhe H, Sorlin S, Snyder PK, Costanza R, Svedin U, Falkenmark M, Karlberg L, Corell RW, Fabry VJ, Hansen J, Walker B, Liverman D, Richardson K, Crutzen P, Foley JA (2009) A safe operating space for humanity. Nature 461(7263):472–475. https://doi.org/10.1038/461472a
Salim NAA, Hassan AS (2022) Simulation on building envelope design using biomimicry conceptual of palm leaf pattern for self cleaning maintenance. J Adv Res Appl Sci Eng Technol 28(3):376–392. https://doi.org/10.37934/araset.28.3.376392

Sayar G (2023) The nexus between environmental taxes, environmental technologies and ecological footprint: an empirical assessment from selected EU countries. https://doi.org/10.58830/ozgur.pub274

Schulp CJ, Burkhard B, Maes J, Van Vliet J, Verburg PH (2014) Uncertainties in ecosystem service maps: a comparison on the European scale. PLoS ONE 9(10):e109643. https://doi.org/10.1371/journal.pone.0109643

Schumpeter JA (2013) Capitalism, socialism and democracy. Routledge. http://debracollege.dspaces.org/bitstream/123456789/441/1/schumpeter-joseph-a-capitalism-socialism-and-democracy.pdf

Semenets-Orlova I, Teslenko V, Dakal A, Zadorozhnyi V, Marusina O, Klochko A (2021) Distance learning technologies and innovations in education for sustainable development. Stud Appl Econ 39(5). https://doi.org/10.25115/eea.v39i5.5065

Semyachkov K (2021) Smart city as a model for sustainable development of territories. In: SHS web of conferences, vol 110. EDP Sciences, p 05003. https://doi.org/10.1051/shsconf/202111005003

Shrivastava P, Ivanaj S, Ivanaj V (2016) Strategic technological innovation for sustainable development. Int J Technol Manag 70(1):76–107. https://doi.org/10.1504/ijtm.2016.074672

Singer L, Fouda A, Bourauel C (2023) Biomimetic approaches and materials in restorative and regenerative dentistry. BMC Oral Health 23(1):105. https://doi.org/10.1186/s12903-023-02808-3

Smil V (2005) Energy at the crossroads: global perspectives and uncertainties. MIT Press

Smith A, Stirling A (2010) The politics of social-ecological resilience and sustainable socio-technical transitions. Ecol Soc 15(1). https://doi.org/10.5751/ES-03218-150111

Somoza-Norton A, Pabigian L, Drap M (2023) Biomimetic leadership: core beliefs for sustainable organizations. J Ecohuman 2(1):77–91. https://doi.org/10.33182/joe.v2i1.2713

Stahel WR (2016) The circular economy. Nat News 531(7595):435. https://doi.org/10.1038/531435a

Tabrizian S (2019) Technological innovation to achieve sustainable development—renewable energy technologies diffusion in developing countries. Sustain Dev 27(3):537–544. https://doi.org/10.1002/sd.1918

Tilman D (1999) The ecological consequences of changes in biodiversity: a search for general principles. Ecology 80(5):1455–1474. https://doi.org/10.1890/0012-9658(1999)080[1455:TECOCI]2.0.CO;2

Toner J, Desha C, Reis K, Hes D, Hayes S (2023) Integrating ecological knowledge into regenerative design: a rapid practice review. Sustainability 15(17):13271. https://doi.org/10.3390/su151713271

Truu J, Truu M, Espenberg M, Nõlvak H, Juhanson J (2015) Phytoremediation and plant-assisted bioremediation in soil and treatment wetlands: a review. Open Biotechnol J 9(1). https://doi.org/10.2174/1874070701509010085

UNEP (2019) Global environment outlook 6: healthy planet, healthy people. United Nations Environment Programme. https://doi.org/10.1017/9781108627146

Unruh GC (2000) Understanding carbon lock-in. Energy Policy 28(12):817–830. https://doi.org/10.1016/S0301-4215(00)00070-7

Urban RA, Bakshi BR (2013) Techno-ecological synergy as a path toward sustainability of a North American residential system. Environ Sci Technol 47(4):1985–1993. https://doi.org/10.1021/es303025c

van den Oord A, Van Witteloostuijn A (2018) A multi-level model of emerging technology: an empirical study of the evolution of biotechnology from 1976 to 2003. PLoS ONE 13(5):e0197024. https://doi.org/10.1371/journal.pone.0197024

Vincent JF, Bogatyreva OA, Bogatyrev NR, Bowyer A, Pahl AK (2006) Biomimetics: its practice and theory. J R Soc Interface 3(9):471–482. https://doi.org/10.1098/rsif.2006.0127

Volstad NL, Boks C (2012) On the use of biomimicry as a useful tool for the industrial designer. Sustain Dev 20(3):189–199. https://doi.org/10.1002/sd.1535

Vymazal J (2011) Constructed wetlands for wastewater treatment: five decades of experience. Environ Sci Technol 45(1):61–69. https://doi.org/10.1021/es101403q

Walker B, Holling CS, Carpenter SR, Kinzig A (2004) Resilience, adaptability and transformability in social–ecological systems. Ecol Soc 9(2). https://doi.org/10.5751/ES-00650-090205

Wang P, Wong YH, Tan CY, Li S, Chong WT (2022) Vertical greening systems: technological benefits, progresses and prospects. Sustainability 14(20):12997. https://doi.org/10.3390/su142012997

Wang Y, Qamruzzaman M, Kor S (2023) Greening the future: harnessing ICT, innovation, eco-taxes, and clean energy for sustainable ecology—insights from dynamic seemingly unrelated regression, continuously updated fully modified, and continuously updated bias-corrected models. Sustainability 15(23):16417. https://doi.org/10.3390/su152316417

Wei X, Song W, Shao Y, Cai X (2022) Progress of ecological restoration research based on bibliometric analysis. Int J Environ Res Public Health 20(1):520. https://doi.org/10.3390/ijerph20010520

Wells P, Orsato RJ (2005) Redesigning the industrial ecology of the automobile. J Ind Ecol 9(3):15–30. https://doi.org/10.1162/1088198054821645

WWF (2020) Living planet report 2020: bending the curve of biodiversity loss. World Wide Fund for Nature. https://www.worldwildlife.org/publications/living-planet-report-2020

Gu GX, Su I, Sharma S, Voros JL, Qin Z, Buehler MJ (2016) Three-dimensional-printing of bioinspired composites. J Biomech Eng 138(2):021006. https://doi.org/10.1115/1.4032423

Xu Y, Zhang RQ, Xiao-min F, Wang QJ (2022) How does green technology innovation affect urbanization? An empirical study from provinces of China. Environ Sci Pollut Res 29(24):36626–36639. https://doi.org/10.1007/s11356-021-18117-7

Yusuf MF, Ashari H, Razalli MR (2018) Environmental technological innovation and its contribution to sustainable development. Civ Eng 9(8):1569. https://doi.org/10.14716/ijtech.v9i8.2748

Zhang Y, Du J, Xing B, Geller G, Rustige H (2015) Experiences with hybrid constructed wetland and pond system for polytypic wastewater treatment. https://doi.org/10.2166/wpt.2015.095

Zhang W, Gruber P (2020) Wind-resilient civil structures: what can we learn from nature. Botany 98(1):37–48. https://doi.org/10.1139/cjb-2019-0034

Chapter 5
Science and Theory of Pollution: Sources, Pathways, Effects and Pollution Credit

Matthew Chidozie Ogwu

Abstract This chapter delves into the theories and frameworks that form the foundation of pollution science, with a focus on its sources, pathways, effects, and innovative mitigation strategies, including pollution credits. It provides an interdisciplinary analysis, integrating perspectives from economics, sociology, environmental science, and governance to understand pollution dynamics. Pollution, a multifaceted environmental challenge, is examined through its anthropogenic and natural sources, transport and deposition pathways, and diverse impacts on ecosystems and human health. The chapter discussed theoretical frameworks, including economic, sociological, and ecological perspectives, to analyze pollution dynamics and propose actionable solutions for its mitigation. Pollution credits are integral to cap-and-trade systems, which allow entities to trade emissions allowances to achieve cost-effective pollution reduction. Rooted in economic theories of externalities, these credits align environmental goals with economic incentives, fostering innovation in cleaner technologies while reducing compliance costs. Real-world applications, such as the U.S. Acid Rain Program and the European Union Emissions Trading System, are discussed to illustrate the effectiveness of these mechanisms in achieving emission reductions. The chapter also addresses critical challenges, including market volatility, carbon leakage, monitoring inefficiencies, and equity concerns. These obstacles highlight the need for robust governance, technological innovation, and equitable policy frameworks. Emerging trends, such as blockchain integration for transparency, linking regional markets, and expanding pollution credit systems to agriculture and land-use sectors, are explored as avenues for enhancing their effectiveness and scalability. By aligning with global sustainability goals, climate change mitigation and equity in resource distribution, pollution credits, and broader pollution management strategies offer transformative potential in reducing environmental harm and promoting sustainable development. The chapter aligns these discussions with the broader objectives of sustainable development, emphasizing the interplay between economic, social, and environmental dimensions. Proposing actionable pathways and

M. C. Ogwu (✉)
Goodnight Family Department of Sustainable Development, 212 Living Learning Center, Appalachian State University, Boone, NC, USA
e-mail: ogwumc@appstate.edu

M. C. Ogwu and S. Chibueze Izah (eds.), *Evaluating Environmental Processes and Technologies*, Environmental Science and Engineering,
https://doi.org/10.1007/978-3-031-85327-2_5

future directions for research and innovation underscores the potential for interdisciplinary approaches to reduce pollution, enhance resilience, and promote equitable and sustainable outcomes.

Keywords Pollution credits · Cap-and-trade · Emissions trading · Market-based mechanisms · Environmental policy · Climate change · Emissions reduction

5.1 Introduction

Pollution, a multifaceted phenomenon, can be defined as the introduction of harmful substances or products into the environment, leading to adverse effects on ecosystems and human health. Its scope encompasses various forms, including air, water, soil, and noise pollution, each with distinct sources and impacts. The complexity of pollution necessitates an interdisciplinary approach, integrating insights from environmental science, economics, sociology, and policy development to understand its mechanisms and devise effective management strategies (Han et al. 2023; Ma et al. 2020; Xu 2023). As a complex and multifaceted environmental challenge, pollution requires a multidisciplinary approach to understand its sources, pathways, impacts, and mitigation strategies. Theories and frameworks of pollution provide structured ways to analyze and address these issues, offering insights from economics, sociology, environmental science, and governance. These perspectives enable a deeper understanding of pollution dynamics and guide the development of effective interventions.

Pollution has been a persistent issue, evolving alongside human industrialization and urbanization. The Industrial Revolution marked a significant turning point, as increased manufacturing and resource extraction led to unprecedented environmental degradation. For instance, the rise in air pollution during this period has been linked to respiratory health issues, particularly in urban populations (Gauderman et al. 2015; Naik et al. 2017). Furthermore, contemporary studies highlight the ongoing challenges posed by pollution, such as the exacerbation of water scarcity in regions like China, where degraded water quality significantly threatens water availability (Ma et al. 2020). This historical perspective underscores the necessity for robust frameworks to address the legacy of past pollution and the emerging threats posed by new pollutants, including microplastics and nanoplastics (Tamayo-Belda et al. 2022; Alexy et al. 2019; Erhunmwunse et al. 2024).

Theories and frameworks for understanding pollution have evolved significantly. Economic theories often focus on the cost–benefit analysis of pollution control measures, emphasizing the need for policies that balance economic growth with environmental sustainability (Manhart 2010; Becker and Shadbegian 2005; Akudugu and Ogwu 2024). Sociological perspectives examine the social dimensions of pollution, including public health impacts and community responses to environmental degradation (Naik et al. 2017). Environmental science provides empirical data on pollution's effects, while policy development frameworks aim to create actionable pathways for pollution management, such as integrating health impact assessments into urban planning (Xu 2023). These interdisciplinary approaches are crucial for developing comprehensive strategies that mitigate pollution and promote sustainable practices across various sectors (Han et al. 2023; Dong 2023).

Economic theories, such as the concept of externalities and the Polluter Pays Principle, highlight the market failures that often lead to unchecked pollution. They propose mechanisms like taxation, cap-and-trade systems, and market incentives to internalize the environmental costs, promoting sustainable behavior among industries and individuals. Sociological frameworks, such as environmental justice and social constructionism, emphasize the inequities in pollution exposure and its societal framing, advocating for inclusive and equitable environmental policies. From an ecological standpoint, concepts like assimilative capacity and planetary boundaries underscore ecosystems' limits to absorbing pollutants without degradation. These frameworks are critical for understanding the tipping points beyond which ecological damage becomes irreversible. Policy and governance frameworks further provide actionable solutions, aligning economic and ecological goals through regulations, incentives, and international agreements. Emerging technological frameworks, including artificial intelligence and big data analytics, integrate with traditional theories to offer dynamic and predictive capabilities for monitoring and managing pollution. These innovations are reshaping how pollution is addressed, enabling real-time analysis and targeted interventions. Together, these theories and frameworks form the foundation for addressing pollution comprehensively, integrating scientific, economic, and social dimensions to foster sustainable development and environmental resilience. This chapter delves into these perspectives, offering a holistic understanding of pollution and its management.

This chapter delves into the theories and frameworks that form the foundation of pollution science, with a focus on its sources, pathways, effects, and innovative mitigation strategies, including pollution credits. It provides an interdisciplinary analysis, integrating perspectives from economics, sociology, environmental science, and governance to understand pollution dynamics. Economic theories, such as externalities, Pigouvian taxes, and cap-and-trade systems, are critically evaluated as tools for pollution control. At the same time, sociological frameworks, including environmental justice and social constructionism, underscore the equity and societal dimensions of pollution. Ecological concepts, such as assimilative capacity and planetary boundaries, are explored to emphasize environmental thresholds and ecosystem resilience. The chapter also investigates policy and governance mechanisms, such as the Polluter Pays Principle, emissions trading schemes, and international agreements, which aim to align economic incentives with environmental goals. Additionally, it examines the transformative role of emerging technologies like artificial intelligence, blockchain, and predictive modeling in enhancing pollution monitoring, transparency, and management. By addressing the challenges and limitations of existing frameworks—such as market volatility, data accuracy, and equity concerns—this chapter identifies opportunities for integrating innovative solutions with traditional approaches. Ultimately, the chapter aligns these discussions with the broader objectives of sustainable development, emphasizing the interplay between economic, social, and environmental dimensions. Proposing actionable pathways and future directions for research and innovation underscores the potential for interdisciplinary approaches to reduce pollution, enhance resilience, and promote equitable and sustainable outcomes.

5.2 Sources of Pollution

Pollution is a complex phenomenon that arises from various sources, which can be broadly categorized into natural and anthropogenic (human-made) sources. Understanding these sources is crucial for developing effective pollution management strategies. Natural sources of pollution include phenomena that occur without human intervention, significantly impacting the environment. These sources encompass volcanic eruptions, wildfires, dust storms, and natural weathering processes. Volcanic eruptions release significant quantities of ash and gases, such as sulfur dioxide, which can lead to acid rain and respiratory issues in nearby populations (Tchounwou et al. 2012). Wildfires contribute to air pollution by releasing particulate matter and carbon monoxide, affecting air quality over vast areas (Niveditha et al. 2021). Dust storms are another significant natural source of pollution, particularly in arid and semi-arid regions. Studies have shown that dust transported from deserts can elevate particulate matter levels in urban areas, leading to health risks such as exacerbated chronic respiratory diseases (Assi et al. 2022). Additionally, natural weathering processes can release heavy metals into the environment, contributing to soil and water pollution (Tchounwou et al. 2012). These natural sources underscore the importance of considering environmental factors when assessing pollution levels and their impacts.

Human activities primarily drive anthropogenic sources of pollution and are responsible for most of the pollution in urban and industrialized areas. Industrial activities are significant contributors to air, water, and soil pollution. Factories emit various pollutants, including volatile organic compounds, particulate matter, and heavy metals, which can have severe health implications for nearby communities (Salama et al. 2016; Raffee et al. 2022). For example, emissions from shipping and industrial operations in port areas have been linked to elevated levels of air pollutants, exceeding air quality guidelines and posing health risks to local populations (Salama et al. 2016). Agricultural practices also play a crucial role in pollution. Using fertilizers and pesticides can lead to runoff that contaminates nearby water bodies, resulting in eutrophication and harm to aquatic ecosystems (Yu et al. 2021). Additionally, livestock farming contributes to air pollution by releasing ammonia and methane, which can affect air quality and contribute to greenhouse gas emissions (Sarwar et al. 2019). The impact of agricultural pollution is compounded by practices such as burning crop residues, which releases particulate matter and other harmful pollutants into the atmosphere (Dandotiya 2021).

Urbanization further exacerbates pollution, as increased population density increases transportation, construction, and energy consumption emissions. Transportation is often the largest source of urban air pollution, with studies indicating

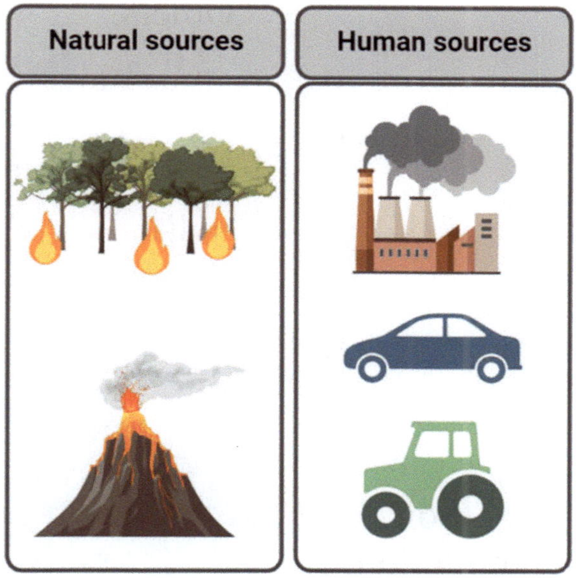

Fig. 5.1 Sources of pollution. *Source* Adapted from Antenozio et al. (2024)

that vehicle emissions significantly contribute to elevated nitrogen oxides and particulate matter levels in urban areas (Samaranayake et al. 2014; Singh 2023). The spatial distribution of pollutants is influenced by urban design, with areas near significant traffic routes experiencing higher concentrations of air pollutants compared to more remote locations ("Spatial variation of air pollutants by using GIS modeling," 2023). Moreover, urban areas face challenges related to indoor air pollution, which can arise from household activities such as cooking, cleaning, and heating. Research has shown that indoor sources, including burning candles and incense, can contribute significantly to indoor air quality issues, particularly in homes with asthmatic children (Habre et al. 2013; Husna et al. 2018) (Fig. 5.1).

5.3 Types of Pollution and Their Related Sciences

Pollution has diverse forms—air, water, soil, noise, and light, and poses significant challenges to environmental health and human well-being. Each type of pollution has distinct sources and impacts, necessitating a multidisciplinary approach to address and mitigate their effects effectively. Figure 5.2 highlights the distribution and severity of different types of pollution in developing countries. Air pollution is the most significant contributor at 73.8%, noise pollution at 63.0%, and water pollution at 55.2%. Soil pollution constitutes a smaller fraction at 6.5%, with other pollution accounting for 2.0%. Air and noise pollution plays a dominant role in

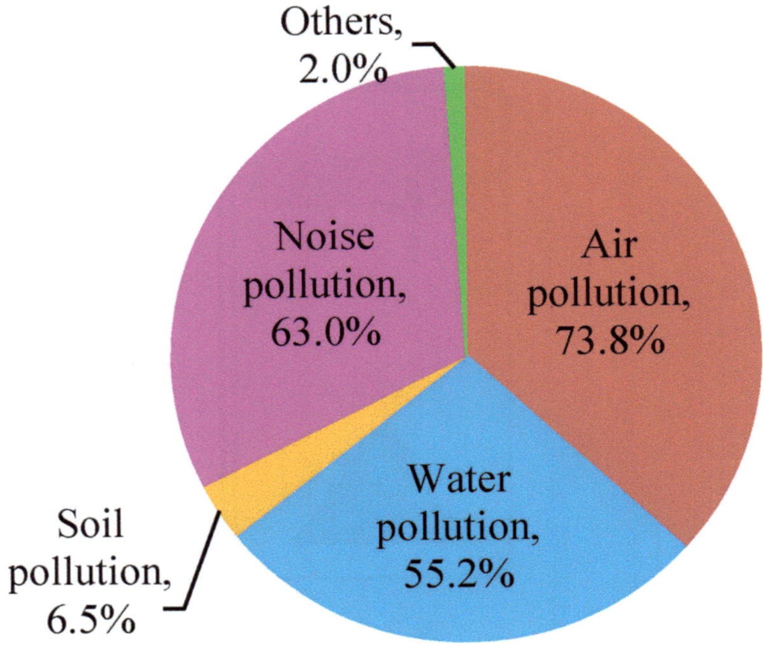

Fig. 5.2 Severity of effects and distribution of the different types of pollution. *Source* Nahar et al. (2021)

environmental degradation, and this points to the need for comprehensive mitigation strategies across all categories. Continued research and policy development are essential to enhance our understanding of these pollutants and implement effective management strategies that protect ecosystems and public health.

Air Pollution: Air pollution is primarily composed of a mixture of harmful substances, including particulate matter (PM), nitrogen oxides (NO_x), sulfur dioxide (SO_2), carbon monoxide (CO), volatile organic compounds (VOCs), and ozone (O_3). These pollutants originate from various anthropogenic activities, such as transportation, industrial processes, and agricultural practices (Yang 2019, 2020; Anderson et al. 2011). For instance, transportation is a significant contributor to urban air pollution, with studies indicating that road traffic emissions are a substantial source of nitrogen oxides and particulate matter, which are linked to respiratory diseases and cardiovascular conditions (Vienneau et al. 2015; Anderson et al. 2011). The World Health Organization (WHO) estimates that particulate matter alone is responsible for approximately 800,000 premature deaths annually, highlighting the urgent need for effective air quality management strategies (Anderson et al. 2011).

Water Pollution: Water pollution involves the introduction of harmful substances into water bodies, leading to the degradation of water quality and adverse effects on aquatic ecosystems and human health. Common contaminants include heavy metals, pesticides, pharmaceuticals, and pathogens, which can enter water systems through agricultural runoff, industrial discharges, and wastewater (Awuchi and Amagwula 2021; Xu et al. 2014; Ogwu et al. 2023). For example, heavy metals from non-ferrous metal smelting have been shown to contaminate soil and water, posing risks to food safety and public health (Xu et al. 2014; Ogwu et al. 2024). Additionally, emerging contaminants such as microplastics and pharmaceuticals are increasingly recognized for their potential ecological and health impacts, necessitating further research and regulatory attention (Awuchi and Amagwula 2021).

Soil Pollution: Soil pollution is primarily caused by accumulating hazardous substances, including heavy metals, organic pollutants, and agricultural chemicals. Factors contributing to soil degradation include industrial activities, improper waste disposal, and excessive use of fertilizers and pesticides (Tang and Jianping 2016; Dai et al. 2014). The infiltration of these pollutants can disrupt soil structure and function, leading to reduced agricultural productivity and contamination of the food chain (Tang and Jianping 2016; Ikhajiagbe and Ogwu 2020). For instance, studies have indicated that heavy metal pollution from industrial processes can lead to bioaccumulation in crops, posing significant health risks to consumers (Xu et al. 2014; Dai et al. 2014). Furthermore, organic pollutants, such as polycyclic aromatic hydrocarbons (PAHs), have been shown to adversely affect soil microbiota, further complicating soil health and ecosystem stability (Dai et al. 2014; Ogwu and Izah 2023).

Noise Pollution: Noise pollution, often overlooked compared to other forms of pollution, has emerged as a significant environmental concern due to its detrimental effects on human health and well-being. Characterized by unwanted or harmful sounds, noise pollution primarily originates from transportation, industrial activities, and urban development (Hemmat 2023; Mousavi et al. 2020). Research indicates that chronic noise exposure can lead to various health issues, including hypertension, sleep disturbances, and cognitive impairments in children (Stansfeld 2015; Elmenhorst et al. 2019). The WHO has identified noise pollution as a critical risk factor for public health, emphasizing the need for effective noise mitigation strategies in urban planning and policy (Hemmat 2023). Moreover, studies have shown that noise pollution can interact with air pollution, exacerbating health risks associated with both environmental stressors (Vienneau et al. 2015; Stansfeld 2015).

Light Pollution: Light pollution, defined as excessive or misdirected artificial light, poses significant threats to human health and ecological systems. It disrupts natural light cycles, affecting the behavior and physiology of various species, particularly nocturnal animals (Peng et al. 2015; Kunc et al. 2016). For instance, artificial lighting can interfere with the migratory patterns of birds and disrupt the reproductive cycles of insects, leading to declines in biodiversity (Kunc et al. 2016). Additionally, light pollution has been linked to adverse health effects in humans, including sleep disturbances and increased risk of chronic conditions (Peng et al. 2015). As urban areas expand, addressing light pollution through sustainable lighting practices and urban design becomes increasingly critical for preserving ecological integrity and promoting public health (Peng et al. 2015; Kunc et al. 2016).

5.4 Pathways of Pollution and Their Effects on Ecosystems and Human Health

5.4.1 *Pathways of Pollution*

5.4.1.1 Atmospheric Transport of Pollutants

Atmospheric transport is a primary pathway through which pollutants such as particulate matter (PM), nitrogen oxides (NO_x), sulfur dioxide (SO_2), and volatile organic compounds (VOCs) are distributed. These pollutants can travel long distances from their source, facilitated by wind and atmospheric currents (Seinfeld and Pandis 2016). Secondary pollutants like ground-level O_3 and particulate matter ($PM_{2.5}$) form through chemical reactions in the atmosphere, exacerbating air quality issues (WHO 2018). Acid rain, resulting from SO_2 and NO_x emissions, exemplifies atmospheric transport's transboundary impacts, leading to soil and water acidification and associated ecological disruptions (Likens and Bormann 1974).

5.4.1.2 Hydrological Pathways: Surface Runoff and Groundwater Movement

Pollutants such as pesticides, fertilizers, heavy metals, and industrial effluents enter water bodies through surface runoff and groundwater infiltration (Posthuma et al. 2019). Urbanization and agricultural intensification exacerbate these processes by increasing impervious surfaces and reducing natural filtration. Pollutants transported via hydrological pathways contribute to water quality degradation, threatening aquatic ecosystems and human water supplies (Carpenter et al. 1998). Groundwater contamination by nitrates and heavy metals, often from agricultural sources, poses significant risks to drinking water safety (WHO 2011).

5.4.1.3 Soil and Sediment Interaction with Pollutants

Soil is a sink and conduit for pollutants, influencing their persistence and bioavailability. Pollutants such as pesticides, heavy metals, and hydrocarbons bind to soil particles or accumulate in sediments, where they may persist for extended periods (Kumar et al. 2019). Soil degradation from pollutants disrupts microbial communities and soil fertility, impairing ecosystem functions. Sediment pollution, particularly in aquatic systems, risks benthic organisms and disrupts food chains (Förstner and Salomons 2010).

5.4.2 Effects on Ecosystems

5.4.2.1 Bioaccumulation and Biomagnification in Ecosystems

Pollutants such as mercury, polychlorinated biphenyls (PCBs), and POPs bioaccumulate in organisms and biomagnify through food webs, producing toxic effects on higher trophic levels (Borga et al. 2012). Apex predators and species with long lifespans are particularly vulnerable, experiencing reproductive and neurological impairments (Tanabe 2004). Biomagnification reduces biodiversity and destabilizes ecosystems by altering species interactions and ecosystem services.

5.4.2.2 Impact on Biodiversity and Ecosystem Services

Pollution-induced habitat degradation and toxicity reduce biodiversity and impair ecosystem services such as pollination, nutrient cycling, and water purification. For instance, pesticide use in agriculture has been linked to pollinator declines, threatening crop productivity and food security. Aquatic ecosystems are similarly affected, with pollutants disrupting species composition and ecosystem resilience.

5.4.2.3 Eutrophication and Aquatic System Disruptions

Excessive nutrient inputs, particularly nitrogen and phosphorus from agricultural runoff, lead to eutrophication in water bodies, characterized by algal blooms, hypoxia, and biodiversity loss (Smith et al. 1999). Hypoxic zones, or "dead zones," impair fish populations and aquatic food webs, with cascading effects on fisheries and livelihoods (Diaz and Rosenberg 2008). Eutrophication also reduces water quality, affecting drinking water supplies and recreational activities.

5.4.2.4 Acidification of Soil and Water Systems

Acid rain and acidifying emissions alter soil pH, leach essential nutrients, and mobilize toxic metals, negatively affecting plant growth and soil biota. In aquatic systems, acidification compromises water quality and biodiversity, particularly for calcium-dependent organisms such as mollusks and corals (Gattuso and Hansson 2011). Acidification exacerbates ecological vulnerability to other stressors, including climate change.

5.4.3 *Effects on Human Health*

5.4.3.1 Short- and Long-Term Health Effects of Airborne Pollutants

Exposure to airborne pollutants, including $PM_{2.5}$, ozone, and heavy metals, poses significant health risks, contributing to respiratory and cardiovascular diseases, cancer, and neurological impairments (Dockery et al. 1993). Long-term exposure to air pollution has been linked to reduced life expectancy, particularly in urban and industrial regions. Children and elderly populations are disproportionately affected, with increased susceptibility to asthma, chronic bronchitis, and other respiratory ailments.

5.4.3.2 Waterborne Diseases and Contaminants

Contaminated water sources are a significant vector for waterborne diseases such as cholera, typhoid, and hepatitis. Pathogens and chemical pollutants in water supplies compromise human health, particularly in regions with inadequate sanitation and water treatment facilities (WHO 2017). Emerging contaminants, including microplastics and pharmaceuticals, present additional risks by entering the food chain through aquatic systems (Eerkes-Medrano et al. 2015).

5.4.3.3 Soil Pollution and Food Safety Concerns

Soil contamination with heavy metals, pesticides, and industrial pollutants affects food safety by introducing toxins into agricultural products. Cadmium and lead, for example, are frequently detected in crops grown in polluted soils, posing risks to human health through dietary exposure. Soil pollution also reduces agricultural productivity, undermining food security in vulnerable regions (Alloway 2013).

Other pollutants, such as POPs, including dioxins, PCBs, and organochlorine pesticides, persist in the environment due to their resistance to degradation. They bioaccumulate in fatty tissues, posing chronic risks to human health and ecosystems. POPs have been linked to endocrine disruption, carcinogenicity, and developmental disorders in humans (UNEP 2009). Ecologically, POPs disrupt reproductive systems in wildlife, contributing to population declines and biodiversity loss (Tanabe 2004). The global transport of POPs underscores the need for international cooperation, as exemplified by the Stockholm Convention on POPs (2001). Addressing the complex effects of pollution requires integrative approaches that combine technological innovation, policy frameworks, and community engagement. Technologies such as bioremediation, constructed wetlands, and precision agriculture reduce pollutant loads in the environment while enhancing ecosystem services (Singh and Ward 2004). Policy instruments, including emissions trading schemes and pollution taxes, incentivize sustainable practices and internalize environmental costs (Stavins 1998). Public awareness and community participation are essential for effective implementation, particularly in vulnerable regions.

5.5 Origins and Theoretical Foundations of Pollution Credits

Pollution credits, often associated with cap-and-trade systems and emissions trading schemes, are market-based mechanisms designed to control and reduce environmental pollution. These credits enable organizations to trade the rights to emit a specific amount of pollutants, incentivizing industries to adopt cleaner technologies and reduce their environmental impact. By establishing a market for emissions allowances, pollution credits aim to align economic growth with environmental sustainability, offering a flexible and cost-effective solution to achieving emission reduction targets (Stavins 1998).

The concept of pollution credits stems from the broader field of environmental economics, particularly the work of Pigou (1920) and Coase (1960). Pigouvian taxation principles suggest that externalities, such as pollution, can be internalized through corrective taxes. Building on this foundation, Coase's theorem emphasizes the role of property rights and private negotiations in achieving efficient pollution outcomes. Cap-and-trade systems operationalize these theories by creating a market for emissions allowances, where pollution credits represent tradable property rights.

The U.S. Acid Rain Program was the first large-scale pollution-credit system application established under the Clean Air Act Amendments of 1990 (Table 5.1). This program targeted sulfur dioxide (SO_2) emissions and demonstrated the effectiveness of market-based approaches in achieving significant emission reductions at lower costs than traditional regulatory methods (Ellerman et al. 2000).

Pollution credit systems are built on a cap-and-trade framework comprising three core components:

1. **Emission Caps**: Regulatory authorities set an upper limit or cap on the total allowable emissions for a specific pollutant within a defined region or sector. This cap decreases over time to align with environmental goals.
2. **Allocation of Credits**: Pollution credits, representing the right to emit a specific quantity of pollutants (e.g., one metric ton of CO_2), are distributed for free (grandfathering) or auctioned to participants.
3. **Trading of Credits**: Entities can buy or sell excess credits. Firms that can reduce emissions at a lower cost often sell their surplus credits to firms with higher abatement costs, promoting cost-effective pollution reduction.

Pollution credit systems have been implemented in various sectors, with notable examples including:

Table 5.1 Examples of pollution credit programs

Program name	Region	Pollutant(s) targeted	Program focus	Environmental impact
U.S. Acid Rain Program (1990)	United States	Sulfur dioxide (SO_2), and nitrogen oxides (NO_x)	The cap-and-trade system with emission allowances and credit banking allowed	Reduced SO_2 emissions by over 50% from 1990 levels, significantly improving air quality
European Union Emissions Trading System (EU ETS)	European Union	Carbon dioxide (CO_2)	Most extensive carbon trading program; phased allocation (free and auctioned credits)	Achieved substantial reductions in GHG emissions while incentivizing renewable energy use
California Cap-and-Trade Program (2013)	California, USA	Carbon dioxide (CO_2), methane, and nitrous oxide	Covers multiple sectors, including power, transportation, and industry	Reduced GHG emissions while generating revenue for clean energy initiatives
China's National Carbon Market (2021)	China	Carbon dioxide (CO_2)	Focused on the power sector; uses intensity-based targets linked to GDP growth	It aims to peak emissions by 2030 and promote decarbonization in the world's largest emitter
Regional Greenhouse Gas Initiative (RGGI)	Northeastern USA	Carbon dioxide (CO_2)	Market-based program for the electricity sector; funds energy efficiency projects	Reduced CO_2 emissions in the power sector by 45% between 2005 and 2020
New Zealand Emissions Trading Scheme (NZ ETS)	New Zealand	Carbon dioxide (CO_2), and methane	Includes forestry sector; allows offsets from carbon sequestration projects	Promoted afforestation and sustainable forest management while reducing GHG emissions
Tokyo Cap-and-Trade Program (2010)	Tokyo, Japan	Carbon dioxide (CO_2)	Targets buildings and facilities; focuses on energy efficiency improvements	Reduced emissions from participating facilities by 25% during the first compliance period
Kyoto Protocol's Clean Development Mechanism (CDM)	Global	Carbon dioxide (CO_2), methane, and nitrous oxide	Allows industrialized nations to earn credits by funding projects in developing countries	Supported renewable energy and sustainable development in over 140 countries
Alberta Emission Offset System	Alberta, Canada	Carbon dioxide (CO_2), and methane	Allows offsets from sectors outside the regulatory framework, such as forestry	Encouraged innovation in agriculture and renewable energy while reducing emissions

(continued)

Table 5.1 (continued)

Program name	Region	Pollutant(s) targeted	Program focus	Environmental impact
South Korea Emissions Trading Scheme (2015)	South Korea	Carbon dioxide (CO_2)	Covers large industrial emitters and incentivizes energy efficiency	Reduced emissions while promoting low-carbon technologies and renewable energy
Australia's Safeguard Mechanism (2016)	Australia	Carbon dioxide (CO_2)	Sets baseline emissions limits for large facilities; allows credit trading	Supported emissions reductions in the industrial sector through market-based mechanisms
Mexico Pilot Emissions Trading System (2020)	Mexico	Carbon dioxide (CO_2)	Covers power and industrial sectors; aligns with national climate goals	Pilot phase promotes capacity building for a full-scale market launch in 2023
Kazakhstan Emissions Trading Scheme (2013)	Kazakhstan	Carbon dioxide (CO_2)	Covers energy and industrial sectors; aligns with national GHG reduction goals	Improved efficiency in the power and industrial sectors while reducing carbon emissions

- **Energy and Power Generation**: The European Union Emissions Trading System (EU ETS), launched in 2005, is the most extensive cap-and-trade program globally, covering power plants, industrial facilities, and aviation. It has significantly reduced greenhouse gas (GHG) emissions while driving investments in renewable energy and energy efficiency.
- **Industrial Manufacturing**: In China, pilot emission trading schemes have targeted industrial sectors, including steel, cement, and petrochemicals, aiming to integrate market mechanisms into the country's broader environmental strategy (Zhang et al. 2017).
- **Aviation and Transportation**: The aviation sector's inclusion in the EU ETS highlights the potential for pollution credits to address emissions from fuel-intensive industries. California's Low Carbon Fuel Standard also incorporates credit trading to incentivize cleaner fuels (CARB 2020).
- **Agriculture**: Emerging programs explore pollution credits for reducing methane and nitrous oxide emissions in agriculture, promoting sustainable farming practices and innovative technologies like biogas capture.

Pollution credit systems offer several advantages, including:

1. **Cost-Effectiveness**: By allowing firms to trade credits, cap-and-trade systems ensure that emissions are reduced where it is cheapest, minimizing the overall cost of compliance (Burtraw et al. 2005).
2. **Incentives for Innovation**: The economic value of credits motivates firms to invest in cleaner technologies and energy efficiency measures, fostering long-term innovation.
3. **Environmental Impact**: Programs like the U.S. Acid Rain Program and EU ETS have significantly reduced pollutants, improving air quality and public health (Ellerman et al. 2000).
4. **Revenue Generation**: Auctions of pollution credits generate government revenue that can be reinvested in climate adaptation, renewable energy, and other sustainability initiatives.

Pollution credits are often affected by fluctuations in credit prices, which can undermine the predictability and stability of these systems, discouraging investments in abatement technologies. Industries may relocate to regions with weaker environmental regulations, offsetting the environmental benefits of pollution credits. The method of credit allocation, whether through grandfathering or auctions, has sparked debates about fairness and effectiveness. Free allocation risks overcompensation, while auctions may burden smaller firms disproportionately. Effective implementation requires robust monitoring and enforcement mechanisms to prevent fraudulent reporting and non-compliance (Ogwu 2019, 2020, 2023). Critics argue that pollution credit systems disproportionately benefit large corporations and wealthier regions while exposing vulnerable communities to localized pollution hotspots (Fowlie et al. 2012). Innovations in pollution credit systems aim to address existing challenges and expand their applicability in blockchain technology is being explored to enhance transparency and traceability in credit trading, reducing the risk of fraud (Andoni et al. 2019). Efforts to link regional cap-and-trade systems, such as California's and Quebec's carbon markets, promote greater market efficiency and environmental impact (Borghesi et al. 2016). Programs incorporating reforestation, wetland restoration, and soil carbon sequestration are emerging as avenues for generating pollution credits while enhancing ecosystem services (Griscom et al. 2017). Advances in satellite imaging, remote sensing, and AI are improving the accuracy and efficiency of emissions monitoring, which is critical for the integrity of credit systems (Chen et al. 2021a, b, c).

Today, pollution credit systems have diverse applications and outcomes (Table 5.1). The U.S. Acid Rain Program reduced SO_2 emissions by over 50% from 1990, achieving environmental targets at a fraction of the projected cost (Ellerman et al. 2000). In a related manner, by covering over 40% of EU emissions, the EU ETS has driven significant GHG reductions and fostered innovation in low-carbon technologies. In China's Pilot Programs, initial results from China's regional cap-and-trade systems indicate the potential for scaling up to a national program despite challenges in enforcement and market liquidity (Zhang et al. 2017). Despite current strides, pollution credit systems must evolve to address global environmental challenges effectively. Priorities include integrating these systems into international climate frameworks, enhancing equity considerations, and leveraging emerging technologies for improved efficiency and scalability. Expanding pollution credits to sectors like agriculture and land-use change offers new opportunities for mitigating emissions while promoting sustainable development.

5.6 Pollution Models, Theories, and Frameworks

Pollution represents a complex environmental challenge that demands a multidisciplinary approach to analyze its sources, pathways, and impacts while designing effective mitigation strategies. Theories and frameworks have evolved across domains such as economics, sociology, environmental science, and policy development, offering robust insights into the mechanisms of pollution and actionable pathways for its management. Theoretical frameworks and pollution models play a crucial role in understanding pollution dynamics and its impacts on human health and the environment. Pollution dispersion models provide valuable insights into the transport and transformation of pollutants, while dose–response relationships help establish safe exposure limits and inform public health policies. Environmental risk assessment methodologies enable comprehensive evaluations of pollution risks, guiding decision-making processes for pollution management. As pollution poses significant global challenges, developing and refining these frameworks and models will be essential for effective pollution mitigation and management strategies.

5.6.1 Some Pollution Models

Pollution Dispersion Models

Pollution dispersion models are essential for understanding how pollutants spread in the environment. These models simulate the transport and transformation of air, water, and soil pollutants, providing insights into their potential impacts on human health and ecosystems (Izah et al. 2024a, b). Various models exist, from simple empirical models to complex numerical simulations incorporating meteorological and environmental data. One widely used approach is the Gaussian dispersion model, which assumes that pollutants disperse in a bell-shaped curve around a point source. This model is beneficial for estimating the concentration of air pollutants emitted from stationary sources, such as factories or power plants (Ascenso et al. 2021). However, Gaussian models have limitations, particularly in complex terrains or urban environments where wind patterns and topography can significantly influence pollutant dispersion (Jaiswal et al. 2022). Advanced models such as computational fluid dynamics (CFD) and mesoscale chemical transport models (CTMs) have been developed to address these limitations. These models provide higher resolution and more accurate predictions by simulating the physical and chemical processes affecting pollutant dispersion (Ascenso et al. 2021). For example, Ascenso et al. demonstrated the effectiveness of CTMs in assessing the impact of green infrastructure on urban air quality, highlighting the importance of spatial resolution in pollution modeling (Ascenso et al. 2021). Moreover, pollution dispersion models can be integrated with GIS to visualize pollution patterns and assess spatial variability. This integration allows for a more comprehensive understanding of pollution sources and their impacts on regions (Chen et al. 2019). For instance, Chen et al. utilized GIS-based models to predict the spatial distribution of airborne pollutants, emphasizing the role of land-use planning in mitigating pollution (Chen et al. 2019).

Dose–Response Relationship Models

The dose–response relationship is a critical concept in pollution studies, as it describes the relationship between the exposure level to a pollutant and the resulting health effects. Research has shown that the dose–response relationship for many pollutants is not linear, meaning that even low levels of exposure can lead to adverse health effects (Eze et al. 2015). For example, Eze et al. conducted a systematic review and meta-analysis on the association between ambient air pollution and diabetes mellitus, highlighting the need for future studies to assess the dose–response relationship more nuancedly (Eze et al. 2015; Izah et al. 2023a, b, c). This approach can help identify critical exposure thresholds where interventions yield significant health benefits. Furthermore, the complexity of the dose–response relationship is influenced by various factors, including individual susceptibility, the presence of multiple pollutants, and the duration of exposure. Sciascia et al. (2022) emphasized the importance of accurately quantifying environmental exposure to pollutants, as this can significantly impact the understanding of the dose–response relationship and its implications for health policy. In addition to epidemiological studies, toxicological research plays a vital role in elucidating dose–response relationships. For instance, Buonanno et al. examined the health effects of daily airborne particle doses in children, demonstrating a direct association between personal exposure levels and respiratory health outcomes (Buonanno et al. 2013). Such studies provide valuable insights into how pollutants affect human health and inform the development of exposure guidelines.

Environmental Risk Assessment Models

Environmental risk assessment (ERA) is a systematic process used to evaluate the potential adverse effects of pollutants on human health and the environment. ERA involves several key steps, including hazard identification, dose–response assessment, exposure assessment, and risk characterization. This framework allows for a comprehensive evaluation of the risks associated with specific pollutants and informs decision-making processes for pollution management. One of the primary challenges in conducting effective ERAs is the need for accurate data on pollutant concentrations and exposure levels. For example, Riebschleager et al. (2012) utilized spatially explicit modeling techniques to estimate potential *E. coli* sources in a watershed, demonstrating the importance of integrating spatial data into risk assessments. This approach allows for a more accurate identification of pollution sources and their potential impacts on water quality. Moreover, integrating advanced modeling techniques into ERA can enhance the accuracy and reliability of risk assessments. For instance, fuzzy logic models have been proposed to assess health risks associated with heavy metal contamination in water bodies, providing a more comprehensive understanding of the complexities involved in pollution assessment (Jaiswal et al. 2022). In addition to traditional risk assessment methodologies, there is a growing emphasis on incorporating sustainability and resilience into environmental risk assessments. Pettit et al. (2011) proposed an integrated approach to urban pollution management that considers pollution's environmental, health, and socio-economic impacts. This holistic perspective is essential for developing sustainable solutions to pollution challenges.

5.6.2 Some Pollution Theories

Economic Theories of Pollution

Economic theories provide foundational insights into the market dynamics of pollution. The concept of pollution as a negative externality, wherein polluters impose environmental costs on society without internalizing those costs, is central to this perspective (Pigou 1920). This market failure necessitates corrective measures such as Pigouvian taxes, which impose levies equal to the social cost of pollution, incentivizing firms to reduce emissions (Baumol and Oates 1988). Similarly, the Coase theorem argues that under well-defined property rights and low transaction costs, private negotiations between affected parties can achieve efficient pollution mitigation without governmental intervention (Coase 1960). These principles underpin market-based mechanisms such as carbon taxes and emissions trading systems (cap-and-trade), which have been widely adopted for managing air pollution and greenhouse gas emissions.

Sociological Theories and Environmental Justice Framework for Pollution

Sociological frameworks emphasize pollution's societal and equity dimensions, particularly the disproportionate burden marginalized community's bear. The environmental justice movement critiques systemic inequities that result in higher pollution exposure for low-income and minority populations due to industrial siting, weak regulations, and inadequate protections (Bullard 1994). Social constructionism further explores how cultural values, and societal priorities shape the framing of pollution issues, influencing public discourse and policymaking (Hannigan 2006). These perspectives advocate for inclusive and equitable environmental policies that address the needs of all stakeholders.

Ecological and Environmental Science Frameworks for Pollution

Ecological theories focus on the environment's capacity to absorb pollutants without adverse effects. The concept of assimilative capacity underscores natural systems' limits to neutralizing pollutants, with exceedances leading to ecosystem degradation and biodiversity loss (Odum 1971). The planetary boundaries framework expands this view by identifying pollution as a critical threshold that, if crossed, risks irreversible ecological damage (Rockström et al. 2009). This framework emphasizes the urgency of sustainable practices and global cooperation to maintain ecological stability and resilience.

Policy and Governance Frameworks for Pollution

Policy and governance frameworks address the mechanisms for institutionalizing pollution control. The Polluter Pays Principle mandates that those responsible for pollution bear mitigation costs, providing an economic incentive to reduce environmental harm (OECD 1975). Cap-and-trade systems, which set emission limits and allow the trading of permits, have proven effective in reducing pollutants like sulfur dioxide under cost-effective conditions (Stavins 1998). These frameworks highlight the critical alignment of economic incentives with environmental protection goals through regulations, international agreements, and collaborative initiatives.

Emerging Frameworks in Technological Contexts

Recent technological advancements have enabled the development of predictive and adaptive models for pollution management. Artificial intelligence (AI) and big data analytics revolutionize environmental monitoring by providing real-time insights into pollution trends, optimizing resource allocation, and forecasting future scenarios (Chen et al. 2021a, b, c). These technologies integrate with traditional frameworks to enhance decision-making and enable dynamic, data-driven approaches to pollution mitigation.

5.7 Geography of Pollution

Pollution is a global challenge transcending national boundary, affecting health, ecosystems, and economies worldwide. Pollution trends vary significantly across regions, influenced by factors such as industrialization, urbanization, and environmental policies. According to Landrigan et al. (2018), pollution is a leading cause of disease and premature death globally, with an estimated 7 million deaths attributed to air pollution alone each year. High-income countries have made strides in pollution control through robust policies and technological advancements, while many low- and middle-income countries continue to grapple with severe pollution challenges due to rapid industrialization and inadequate regulatory frameworks (Landrigan et al. 2018; Le et al. 2022). In China, economic policy uncertainty has been shown to exacerbate pollution emissions from industrial enterprises, as businesses may delay investments in cleaner technologies due to concerns about future regulations (Le et al. 2022; Yang et al. 2022). This uncertainty can lead to increased emissions, highlighting the need for stable and clear environmental policies to encourage sustainable practices. Conversely, regions that have implemented stringent environmental regulations have observed improved air quality and public health outcomes (Landrigan et al. 2018; Chen et al. 2021a, b, c). Seasonal variations also play a critical role in pollution trends. Tefera et al. (2020) conducted a study in Addis Ababa, revealing significant seasonal fluctuations in particulate matter ($PM_{2.5}$) concentrations, which were highest during dry seasons due to increased dust and vehicle emissions. Such findings underscore the importance of considering local environmental conditions when assessing pollution levels and developing mitigation strategies.

Transboundary Pollution Issues

Transboundary pollution refers to pollution originating in one country but affecting the environment and health in another. This phenomenon poses significant challenges to international cooperation and policymaking. For instance, Moran and Kanemoto (2016) highlighted that developed countries often contribute to pollution hotspots in developing nations through their consumption patterns, leading to significant health impacts in those regions. The authors argue that global supply chains often obscure the true environmental costs of production, necessitating greater accountability and transparency in international trade practices. Furthermore, Torre et al. (2019) emphasize the need for coordinated efforts to address transboundary pollution, as emissions from one country can have far-reaching effects on neighboring regions. The complexity of transboundary pollution is compounded by the fact that countries may adopt different environmental policies, leading to a "free-rider" problem where some nations benefit from the pollution control efforts of others without contributing themselves (Torre et al. 2019). This dynamic necessitates international agreements and collaborative frameworks to manage pollution across borders effectively. The strategic behavior of countries in response to pollution control measures can also lead to unintended consequences. For example, Ben-David et al. (2021) discuss how multinational corporations may relocate polluting activities to countries with less stringent regulations, effectively "exporting" pollution rather than reducing it. This practice raises ethical concerns about environmental justice and the responsibilities of developed nations towards developing countries. Transboundary pollution poses significant challenges for environmental management, as pollutants can travel across borders, affecting air and water quality in neighboring regions. Long et al. (2012) examined non-point source pollution in the Lianshui watershed, emphasizing the complexities associated with monitoring and controlling NPSP due to its diffuse nature. The study highlights the need for collaborative efforts among jurisdictions to address transboundary pollution effectively. In urban areas, air pollution complaints can serve as indicators of pollution distribution and sources. Wang and Su (2023 spatialized air pollution complaint data in Beijing, combining it with regional air quality monitoring data to analyze pollution patterns. Their findings demonstrate the utility of GIS technology in visualizing pollution data, facilitating targeted investigations, and improving pollution control measures. Furthermore, the occurrence of antibiotics in aquatic environments has raised concerns about their ecological risks. Zhou et al. (2016) studied the spatiotemporal distribution of antibiotics in Lake Taihu, China, finding that the distribution of different antibiotic classes varies based on environmental processes and pollution sources. This research underscores the need for comprehensive monitoring and management strategies to mitigate the impacts of pharmaceutical pollution in aquatic ecosystems.

Spatial Distribution of Pollution

The spatial distribution of pollutants is influenced by various factors, including the type of pollutant, its sources, and local environmental conditions. For instance, Xu et al. (2021) analyzed vehicle emissions in Zhangdian District, revealing that the emission intensity of vehicle exhaust pollutants is significantly higher during peak periods and varies by road type, with expressways exhibiting the highest emission levels. This finding underscores the importance of considering traffic patterns and road infrastructure when assessing air quality in urban areas. In coastal regions, heavy metals in sediments can pose significant ecological risks. Chen et al. conducted a study in Weitou Bay, China, identifying chromium (Cr), zinc (Zn), and copper (Cu) as common heavy metal pollutants, primarily originating from industrial activities such as electroplating and agriculture (Chen et al. 2021a, b, c). The study highlights the spatial variability of heavy metal contamination in sediments, emphasizing the need for targeted monitoring and remediation efforts in affected areas. Similarly, Zhang et al. (2021) explored soil heavy metal pollution in the river irrigation area of Baoji, China. Their findings indicated that human activities significantly influence the spatial distribution of heavy metals, with variability observed across different sampling points. The study's results suggest that pollution sources are complex and multifaceted, necessitating comprehensive assessments to inform effective management strategies.

Land use and urbanization play critical roles in shaping pollution patterns. Wei et al. (2015) investigated the effects of urban land expansion on regional meteorology and air quality in Eastern China, finding that urbanization leads to increased mean surface temperatures and altered wind patterns, which can exacerbate air pollution. This research highlights the interconnectedness of land use changes and air quality, emphasizing the need for integrated urban planning that considers environmental impacts. The relationship between land use and air quality is further illustrated by the work of Xu et al. (2016), who examined the impacts of land use on air quality in Wuhan, China. Their study revealed a quantitative relationship between different land use types (e.g., built-up land, water bodies, and vegetation) and air quality indicators such as SO_2 and NO_2. The findings suggest that urban green spaces can mitigate air pollution, underscoring the importance of incorporating green infrastructure into urban planning. Moreover, the spatial distribution of non-point source pollution (NPSP) in agricultural watersheds has garnered attention in recent studies. Huang et al. conducted a study on agricultural NPSP in the Zhangxi watershed, utilizing GIS to simulate runoff and pollutant loading (Huang et al. 2010). Their research highlights the significance of landscape structure and agricultural practices in influencing NPSP patterns, providing valuable insights for managing agricultural pollution.

The geography of pollution has significant implications for environmental management and policymaking. Understanding the spatial distribution of pollutants and their sources is essential for developing effective monitoring and remediation strategies. For instance, Wei et al. (2022) assessed heavy metal contamination in river sediments in Huanghua City, utilizing GIS mapping to identify pollution sources and ecological risks. Their findings can inform targeted interventions to address contamination in affected areas. Moreover, the integration of advanced modeling techniques and GIS technology can enhance the accuracy of pollution assessments. Zhao et al. (2021) conducted a spatiotemporal analysis of urban air pollutants throughout China, revealing trends in PMs and gaseous pollutants. Their research highlights the importance of continuous monitoring and data analysis to inform air quality management strategies.

5.8 Technological and Policy Solutions to Pollution

Pollution remains one of the most pressing environmental challenges of our time, necessitating a multifaceted approach that combines technological innovations and robust policy frameworks. This literature review explores advances in pollution control technologies and the role of environmental regulations and international agreements in mitigating pollution. Technological advancements have played a crucial role in addressing pollution across various sectors. One significant area of innovation is in air pollution control technologies. Charles and Bakshi (2021) discuss the evolution of air pollution control technologies, which have advanced from early inventions like stoker systems and scrubbers to modern solutions that integrate spatially explicit techno-ecological synergy. These advancements aim to improve air quality, particularly in urban and industrial settings, by enhancing the efficiency of pollutant removal processes.

Another promising area of research is the development of transition metal catalysts for atmospheric heavy metal removal. Ma (2023) highlights the potential of heterogeneous catalysis with transition metals, which has become a key approach for removing heavy metals from the atmosphere. Innovations in catalyst design, particularly bimetallic systems, are being explored to improve specificity and cost-effectiveness, translating laboratory innovations into scalable technologies for industrial applications. Indoor air quality is also a critical concern, especially given that urban populations spend a significant amount of time indoors. Kumar et al. (2023) conducted a systematic review on the mitigation of common indoor air pollutants using plant-based methods, emphasizing the potential of phytoremediation as an effective solution. Utilizing natural processes to detoxify indoor pollutants offers a sustainable approach to improving indoor air quality.

In the realm of water pollution, Fawzy et al. (2011) explored the biomonitoring and phytoremediation potential of aquatic macrophytes in the River Nile. Their findings suggest that aquatic plant-based treatment systems are low-cost technologies that can effectively recycle and treat wastewater contaminated by heavy metals, providing an alternative to conventional methods. The integration of digital technologies is also transforming pollution monitoring and management. Kaur and Sharma (2023) discuss IoT-enabled air pollution monitoring systems, which leverage connectivity and real-time data analysis to enhance pollution control efforts. These systems can provide valuable insights into pollution sources and trends, enabling more effective responses to air quality issues.

While technological innovations are vital for pollution control, effective environmental regulations and international agreements such as those in Table 5.2 are equally important in driving systemic change. Calvo and Rubio (2013) discuss dynamic models of international environmental agreements, emphasizing the need for cooperative frameworks that encourage countries to invest in pollution reduction technologies. Their research highlights the potential for club membership models, where countries benefit from collective efforts to address transboundary pollution issues. The role of international agreements in addressing global pollution challenges cannot be overstated. For example, the Paris Agreement aims to limit global warming and reduce greenhouse gas emissions, which are significant contributors to air pollution. By fostering international cooperation, such agreements create a framework for countries to share best practices and technologies for pollution reduction. Furthermore, Gopalakrishnan et al. (2019) argue that nature-based solutions can complement technological approaches to mitigate air emissions. Policies that encourage the restoration of ecosystems not only enhance biodiversity but also provide significant societal benefits, including improved air quality. This integrated approach underscores the importance of considering both technological and ecological solutions in environmental policy. Prata et al. (2019) propose integrated strategies for controlling and mitigating plastic and microplastic pollution. Their review emphasizes the need for a comprehensive approach that includes regulation of production and consumption, eco-design, and improved waste management practices. By addressing the entire life-cycle of plastics, these strategies aim to reduce the environmental impacts of plastic pollution.

5.9 Conclusion

Pollution arises from a combination of natural and anthropogenic sources, each contributing to environmental degradation and health risks. Natural sources, while significant, are often overshadowed by the extensive impacts of human activities, particularly in industrial, agricultural, and urban contexts. Pollution pathways—through air, water, soil, and biological systems—interact with ecosystems and human health in profound and interconnected ways. Atmospheric transport, hydrological movement, and soil-sediment interactions facilitate pollutant distribution, while processes such as bioaccumulation and biomagnification exacerbate ecological and health impacts. Addressing these challenges requires interdisciplinary research and collaborative action to reduce pollution sources, restore ecosystems, and protect human health. Future strategies must prioritize sustainable development, technological innovation, and equitable resource management to mitigate pollution's far-reaching effects. Pollution is a complex and evolving issue that requires a multi-faceted understanding and response. Historical insights reveal the persistent nature of pollution, while contemporary theories provide a framework for addressing its challenges. Pollution theories and frameworks provide a comprehensive understanding of its dynamics and offer diverse pathways for mitigation. From economic models that internalize environmental costs to sociological perspectives advocating equity, and from ecological thresholds to cutting-edge technological tools, these frameworks collectively inform sustainable and effective pollution management strategies. By leveraging these interdisciplinary insights, policymakers and stakeholders can address pollution's multifaceted challenges while promoting environmental resilience and sustainability. Pollution credit systems represent a critical tool for balancing economic development with environmental sustainability. By incentivizing cost-effective pollution reduction, fostering innovation, and generating revenue for sustainability initiatives, these systems align with global efforts to combat climate change and improve air quality. However, addressing challenges related to market volatility, equity, and enforcement is essential to maximizing their potential. As the global community continues to refine and expand these mechanisms, pollution credits will remain a cornerstone of modern environmental policy.

Table 5.2 Key pollution-focused environmental regulations and international agreements

Regulation/agreement	Focus	Key features	Impacts
Kyoto Protocol (1997)	Global agreement targeting greenhouse gas (GHG) emissions reduction	Legally binding targets for industrialized nations; introduction of carbon trading mechanisms	Initiated global GHG reduction efforts, promoting carbon markets and emission reduction strategies
Paris Agreement (2015)	Global commitment to limit global warming to below 2 °C (preferably 1.5 °C)	Nationally Determined Contributions (NDCs); climate financing for developing nations	Encouraged global collaboration on climate action and fostered renewable energy transitions
Clean Air Act (1970, USA)	National regulation aimed at reducing air pollution	Establishes air quality standards for pollutants like SO_2, NO_x, and PM; permits for emissions	Improved air quality, reduced acid rain, and mitigated public health risks from air pollution
European Union Water Framework Directive (2000)	EU-wide regulation for protecting water resources	Requires member states to achieve "good" water quality for all water bodies	Enhanced water quality across Europe; reduced pollution from agricultural and industrial sources
Basel Convention (1989)	International treaty controlling the transboundary movement of hazardous waste	Prohibits the export of hazardous waste to countries lacking the capacity for safe disposal	Reduced global environmental risks from hazardous waste mismanagement
Montreal Protocol (1987)	Treaty targeting ozone-depleting substances (ODS)	Phased out the production and use of ODS like CFCs and HCFCs	Successfully reversed ozone layer depletion trends and reduced related environmental damages
Stockholm Convention on Persistent Organic Pollutants (2001)	Global treaty aimed at eliminating or restricting POPs	Bans or limits the use of chemicals like DDT, PCBs, and dioxins	Reduced bioaccumulation of harmful chemicals in ecosystems and food chains
Convention on Biological Diversity (1992)	Global framework for biodiversity conservation	Three objectives: conservation, sustainable use, and fair sharing of genetic resource benefits	Promoted ecosystem preservation and sustainable management of natural resources

(continued)

Table 5.2 (continued)

Regulation/agreement	Focus	Key features	Impacts
ASEAN Agreement on Transboundary Haze Pollution (2002)	Regional agreement to address haze pollution from land and forest fires in Southeast Asia	Encourages member states to adopt preventive measures and respond to transboundary haze issues	Improved air quality and reduced environmental degradation caused by forest fires
Environmental Protection Act (1986, India)	National legislation to protect and improve the environment	Provides authority to regulate industrial pollution and manage hazardous substances	Strengthened pollution control and enforced stricter compliance across industries
International Maritime Organization MARPOL Convention (1973/78)	Regulations to prevent marine pollution from ships	Prohibits discharges of oil, chemicals, and waste from vessels	Significantly reduced marine pollution and safeguarded marine biodiversity
UN Framework Convention on Climate Change (UNFCCC 1992)	Global agreement addressing climate change	Facilitated negotiations for the Kyoto Protocol and Paris Agreement	Established a foundation for global climate governance and action
Environmental Impact Assessment (EIA) Regulations	National and international requirements for assessing environmental impacts of proposed projects	Mandates evaluation of environmental consequences before project approval	Ensured mitigation of adverse environmental impacts through proactive planning

References

Akudugu MA, Ogwu MC (2024) Sustainable development policies and interventions: a bibliometric analysis of the contributions of the academic community. J Clean Prod. Article # 139919. https://doi.org/10.1016/j.jclepro.2023.139919

Alexy P, Anklam E, Emans T, Furfari A, Galgani F, Hanke G, Kluettgen B et al (2019) Managing the analytical challenges related to micro- and nanoplastics in the environment and food: filling the knowledge gaps. Food Addit Contam Part A 37(1):1–10. https://doi.org/10.1080/19440049.2019.1673905

Alloway BJ (2013) Heavy metals in soils: trace metals and metalloids in soils and their bioavailability. Springer

Anderson J, Thundiyil J, Stolbach A (2011) Clearing the air: a review of the effects of particulate matter air pollution on human health. J Med Toxicol 8(2):166–175. https://doi.org/10.1007/s13181-011-0203-1

Andoni M, Robu V, Flynn D, Abram S, Geach D, Jenkins D, Peacock A (2019) Blockchain technology in the energy sector: a systematic review of challenges and opportunities. Renew Sustain Energy Rev 100:143–174. https://doi.org/10.1016/j.rser.2018.10.014

Antenozio ML, Caissutti C, Caporusso FM, Marzi D, Brunetti P (2024) Urban air pollution and plant tolerance: omics responses to ozone, nitrogen oxides, and particulate matter. Plants 13(15):2027. https://doi.org/10.3390/plants13152027

Ascenso A, Augusto B, Silveira C, Rafael S, Coelho S, Monteiro A, Miranda A et al (2021) Impacts of nature-based solutions on the urban atmospheric environment: a case study for eindhoven, the Netherlands. Urban for Urban Green 57:126870. https://doi.org/10.1016/j.ufug.2020.126870
Assi H, Meouchy P, Mahmoud A, Massouh A, Zerdan M, Alameh I, Noureddine S et al (2022) A survey on the knowledge, attitudes, and practices of Lebanese physicians regarding air pollution. Int J Environ Res Public Health 19(13):7907. https://doi.org/10.3390/ijerph19137907
Awuchi C, Amagwula I (2021) Environmental pollutants and contaminants of emerging concern: an African perspective. J La Lifesci 2(3):39–50. https://doi.org/10.37899/journallalifesci.v2i3.418
Baumol WJ, Oates WE (1988) The theory of environmental policy. Cambridge University Press
Becker R, Shadbegian R (2005) A change of pace: comparing the 1994 and 1999 pollution abatement costs and expenditures surveys. J Econ Soc Meas 30(1):63–95. https://doi.org/10.3233/jem-2005-0249
Ben-David I, Jang Y, Kleimeier S, Viehs M (2021) Exporting pollution: where do multinational firms emit CO_2? Econ Policy. https://doi.org/10.1093/epolic/eiab009
Borga K, Gabrielsen GW, Skaare JU (2012) Biomagnification of organochlorines along a Barents Sea food chain. Environ Pollut 113(2):187–198. https://doi.org/10.1016/S0269-7491(00)00118-6
Borghesi S, Montini M, Barreca A (2016) Linking emission trading schemes: a short review. FEEM working paper series, 2016.57
Bullard RD (1994) Dumping in dixie: race, class, and environmental quality. Westview Press
Buonanno G, Marks G, Morawska L (2013) Health effects of daily airborne particle dose in children: direct association between personal dose and respiratory health effects. Environ Pollut 180:246–250. https://doi.org/10.1016/j.envpol.2013.05.039
Burtraw D, Palmer K, Kahn D (2005) Allocation of CO_2 emission allowances in the Regional Greenhouse Gas Cap-and-Trade Program. RFF discussion paper
California Air Resources Board (CARB) (2020) Low carbon fuel standard: credit trading activity report
Calvo E, Rubio S (2013) Dynamic models of international environmental agreements: a differential game approach. Int Rev Environ Resour Econ 6(4):289–339. https://doi.org/10.1561/101.00000053
Carpenter SR, Caraco NF, Correll DL, Howarth RW, Sharpley AN, Smith VH (1998) Nonpoint pollution of surface waters with phosphorus and nitrogen. Ecol Appl 8(3):559–568. https://doi.org/10.1890/1051-0761(1998)008[0559:NPOSWW]2.0.CO;2
Charles M, Bakshi B (2021) Designing industrial landscapes for mitigating air pollution with spatially-explicit techno-ecological synergy. Aiche J 67(10). https://doi.org/10.1002/aic.17347
Chen L, Li L, Yang X, Chen L, Ma X (2019) Assessing the impact of land-use planning on the atmospheric environment through predicting the spatial variability of airborne pollutants. Int J Environ Res Public Health 16(2):172. https://doi.org/10.3390/ijerph16020172
Chen J, Wang W, Zhang H (2021a) Big data analytics for environmental pollution monitoring and management: a systematic review. Environ Res 197:111060. https://doi.org/10.1016/j.envres.2021.111060
Chen Q, Huang F, Cai A (2021b) Spatiotemporal trends, sources and ecological risks of heavy metals in the surface sediments of Weitou Bay, China. Int J Environ Res Public Health 18(18):9562. https://doi.org/10.3390/ijerph18189562
Chen Y, Shen X, Wang L (2021c) The heterogeneity research of the impact of EPU on environmental pollution: empirical evidence based on 15 countries. Sustainability 13(8):4166. https://doi.org/10.3390/su13084166
Coase RH (1960) The problem of social cost. J Law Econ 3(1):1–44
Dai D, Li C, Tian L (2014) Thermal desorption experiment of polycyclic aromatic hydrocarbons (PAHs) contaminated soil used as cement raw meal. Appl Mech Mater 587–589:860–864. https://doi.org/10.4028/www.scientific.net/amm.587-589.860
Dandotiya B (2021) Air pollution, health and perception. https://doi.org/10.5772/intechopen.96622

Diaz RJ, Rosenberg R (2008) Spreading dead zones and consequences for marine ecosystems. Science 321(5891):926–929. https://doi.org/10.1126/science.1156401

Dockery DW, Pope CA, Xu X, Spengler JD, Ware JH, Fay ME, Ferris BG Jr Speizer FE (1993) An association between air pollution and mortality in six U.S. cities. N Engl J Med 329(24):1753–1759. https://doi.org/10.1056/NEJM199312093292401

Dong K (2023) Assessing the role of green investment in energy efficiency: does digital economy matter? Energy Explor Exploit 42(4):1450–1471. https://doi.org/10.1177/01445987231216763

Eerkes-Medrano D, Thompson RC, Aldridge DC (2015) Microplastics in freshwater systems: a review of the emerging threats, identification, and quantification. Environ Sci Technol 49(19):11459–11480. https://doi.org/10.1021/acs.est.5b01556

Ellerman AD, Joskow PL, Schmalensee R, Montero JP, Bailey EM (2000) Markets for clean air: the U.S. acid rain program. Cambridge University Press

Elmenhorst E, Griefahn B, Rolny V, Basner M (2019) Comparing the effects of road, railway, and aircraft noise on sleep: exposure–response relationships from pooled data of three laboratory studies. Int J Environ Res Public Health 16(6):1073. https://doi.org/10.3390/ijerph16061073

Erhunmwunse NO, Pajiah TJ, Ogwu MC (2024) Microplastics as water pollutants and sustainable management strategies. In: Izah SC, Ogwu MC, Loukas A, Hamidifar H (eds) Water crises and sustainable management in the global south. Springer, Singapore. https://doi.org/10.1007/978-981-97-4966-9_8

Eze I, Hemkens L, Bucher H, Hoffmann B, Schindler C, Künzli N, Probst-Hensch N et al (2015) Association between ambient air pollution and diabetes mellitus in Europe and North America: systematic review and meta-analysis. Environ Health Perspect 123(5):381–389. https://doi.org/10.1289/ehp.1307823

Fawzy M, Badr N, El-Khatib A, Aboelkassem A (2011) Heavy metal biomonitoring and phytoremediation potentialities of aquatic macrophytes in river nile. Environ Monit Assess 184(3):1753–1771. https://doi.org/10.1007/s10661-011-2076-9

Förstner U, Salomons W (2010) Sediment research, management, and policy. J Soils Sediments 10(2):138–149. https://doi.org/10.1007/s11368-010-0199-9

Fowlie M, Holland SP, Mansur ET (2012) What do emissions markets deliver and to whom? Evidence from Southern California's NO_x trading program. Am Econ Rev 102(2):965–993. https://doi.org/10.1257/aer.102.2.965

Gattuso J-P, Hansson L (2011) Ocean acidification. Oxford University Press

Gauderman W, Urman R, Avol E, Berhane K, McConnell R, Rappaport E, Gilliland F et al (2015) Association of improved air quality with lung development in children. N Engl J Med 372(10):905–913. https://doi.org/10.1056/nejmoa1414123

Gopalakrishnan V, Ziv G, Hirabayashi S, Bakshi B (2019) Nature-based solutions can compete with technology for mitigating air emissions across the united states. Environ Sci Technol 53(22):13228–13237. https://doi.org/10.1021/acs.est.9b01445

Griscom BW et al (2017) Natural climate solutions. PNAS 114(44):11645–11650

Habre R, Coull B, Moshier E, Godbold J, Grunin A, Nath A, Koutrakis P et al (2013) Sources of indoor air pollution in New York City residences of asthmatic children. J Eposure Sci Environ Epidemiol 24(3):269–278. https://doi.org/10.1038/jes.2013.74

Han J, Li M, Sheng Z (2023) Research on environmental pollution control strategies based on grey-scale models. Acad J Sci Technol 5(1):159–161. https://doi.org/10.54097/ajst.v5i1.5541

Hannigan J (2006) Environmental sociology. Routledge

Hemmat W (2023) Exploring noise pollution, causes, effects, and mitigation strategies: a review paper. Eur J Theoret Appl Sci 1(5):995–1005. https://doi.org/10.59324/ejtas.2023.1(5).86

Huang B, Yang H, Wang S, Yang L (2010) Spatial and temporal heterogeneity study of agricultural non-point source pollutant loading of Zhangxi Watershed, pp 1–4. https://doi.org/10.1109/icbbe.2010.5516909

Husna IDY, Karyanto Y, Sunoko H (2018) Indoor air pollution in non AC passenger bus. E3s Web Conf 31:06012. https://doi.org/10.1051/e3sconf/20183106012

Ikhajiagbe B, Ogwu MC (2020) Hazard quotient, microbial diversity and plant composition of spent crude oil polluted-soil. Beni-Suef Univ J Basic Appl Sci 9(26). https://doi.org/10.1186/s43088-020-00052-0

Izah SC, Ogwu MC, Akram M (2023a) Herbal medicine phytochemistry: applications and trends. Springer, Cham. https://doi.org/10.1007/978-3-031-21973-3

Izah SC, Ogidi OI, Ogwu MC, Salimon SS, Yusuf ZM, Akram M, Raimi MO, Iyingiala A-A (2023b) Historical perspectives and overview of the value of herbal medicine. In: Izah SC, Ogwu MC, Akram M (eds) Herbal medicine phytochemistry. Reference series in phytochemistry. Springer, Cham. https://doi.org/10.1007/978-3-031-21973-3_1-1

Izah SC, Richard G, Stanley HO, Ogwu MC, Sawyer WE, Uwaeme OR (2023c) Prospects and application of multivariate and reliability analyses to one health risk assessments of toxic elements. Toxicol Environ Heal Sci. https://doi.org/10.1007/s13530-023-00199-0

Jaiswal M, Gupta S, Chabukdhara M, Nasr M, Nema A, Hussain J, Malik T et al (2022) Heavy metal contamination in the complete stretch of yamuna river: a fuzzy logic approach for comprehensive health risk assessment. PLoS ONE 17(8):e0272562. https://doi.org/10.1371/journal.pone.0272562

Izah SC, Ogwu MC, Loukas A, Hamidifar H (2024a) Water related challenges and management strategies. Springer Nature, Singapore. https://doi.org/10.1007/978-981-97-4966-9

Izah SC, Ogwu MC, Hamidifar H (2024b) Biomonitoring of pollutants in the global south. Springer Nature, Switzerland. https://doi.org/10.1007/978-981-97-1658-6

Kaur S, Sharma S (2023) IoT-enabled air pollution monitoring systems: technologies, solutions, and challenges. https://doi.org/10.3233/atde221280

Kumar M, Saurabh S, Lee DW (2019) Heavy metal pollution: mechanisms of detoxification and sustainable removal. Ecotoxicol Environ Saf 181:201–213. https://doi.org/10.1016/j.ecoenv.2019.05.065

Kumar R, Verma V, Thakur M, Singh G, Bhargava B (2023) A systematic review on mitigation of common indoor air pollutants using plant-based methods: a phytoremediation approach. Air Qual Atmos Health 16(8):1501–1527. https://doi.org/10.1007/s11869-023-01326-z

Kunc H, McLaughlin K, Schmidt R (2016) Aquatic noise pollution: implications for individuals, populations, and ecosystems. Proc R Soc B Biol Sci 283(1836):20160839. https://doi.org/10.1098/rspb.2016.0839

Landrigan P, Fuller R, Hu H, Caravanos J, Cropper M, Hanrahan D, Suk W et al (2018) Pollution and global health—an agenda for prevention. Environ Health Perspect 126(8). https://doi.org/10.1289/ehp3141

Le D, Ren F, Tang Y, Zhu Y (2022) The effect of environmental policy uncertainty on enterprises' pollution emissions: evidence from chinese industrial enterprise. Int J Environ Res Public Health 19(16):9849. https://doi.org/10.3390/ijerph19169849

Likens GE, Bormann FH (1974) Acid rain: a serious regional environmental problem. Science 184(4142):1176–1179. https://doi.org/10.1126/science.184.4142.1176

Long Y, Jiang C, Shi H, Bei C (2012) The simulation research on non-point source pollution in lianshui watershed. Appl Mech Mater 209–211:2018–2022. https://doi.org/10.4028/www.scientific.net/amm.209-211.2018

Ma Q (2023) Transition metal catalysts for atmospheric heavy metal removal: a review of current innovations and advances. Molecules 28(22):7620. https://doi.org/10.3390/molecules28227620

Ma T, Sun S, Fu G, Hall J, Ni Y, He L, Zhou C et al (2020) Pollution exacerbates China's water scarcity and its regional inequality. Nat Commun 11(1). https://doi.org/10.1038/s41467-020-14532-5

Manhart A (2010) International cooperation for metal recycling from waste electrical and electronic equipment. J Ind Ecol 15(1):13–30. https://doi.org/10.1111/j.1530-9290.2010.00307.x

Moran D, Kanemoto K (2016) Tracing global supply chains to air pollution hotspots. Environ Res Lett 11(9):094017. https://doi.org/10.1088/1748-9326/11/9/094017

Mousavi S, Amiri Z, Darvishi P, Mahmoudi A, Salari N, Nayeri D (2020) Investigation of knowledge, attitude, and practice of people in Kermanshah, Iran, toward the effects of traffic noise pollution on human health. Arch Hyg Sci 9(4):246–255. https://doi.org/10.52547/archhygsci.9.4.246
Nahar N, Mahiuddin S, Hossain Z (2021) The severity of environmental pollution in the developing countries and its remedial measures. Earth 2(1):124–139. https://doi.org/10.3390/earth2010008
Naik Y, Jones S, Christmas H, Roderick P, Cooper D, McGready K, Gent M et al (2017) Collaborative health impact assessment and policy development to improve air quality in west yorkshire—a case study and critical reflection. Climate 5(3):62. https://doi.org/10.3390/cli5030062
Niveditha S, Balaji Ganesh S, Arivarasu L (2021) Covid-19 lockdown and its impact on air pollution. J Contemp Issues Bus Gov 26(02). https://doi.org/10.47750/cibg.2020.26.02.169
Obahiagbon EG, Ogwu MC (2024) Soil business: exploring business opportunities in soil systems for sustainable growth in the global south. In: Ogwu MC, Izah SC, Dessureault-Rompré J, Gasparatos D (eds) Sustainable soil systems in global south. Springer, Singapore. https://doi.org/10.1007/978-981-97-5276-8_11
Odum EP (1971) Fundamentals of ecology. Saunders
Ogwu MC (2019) Towards sustainable development in Africa: the challenge of urbanization and climate change adaptation. In: Cobbinah PB, Addaney M (eds) The geography of climate change adaptation in Urban Africa. Springer Nature, Switzerland, pp 29–55. https://doi.org/10.1007/978-3-030-04873-0_2
Ogwu MC (2020) Value of Amaranthus [L.] species in Nigeria. In: Waisundara V (ed) Nutritional value of Amaranth. IntechOpen, UK, pp 1–21. https://doi.org/10.5772/intechopen.86990
Ogwu MC (2023) Plants as monitors and managers of pollution. In: Srivastav AL, Grewal AS, Tiwari M, Pham TD (eds) Role of green chemistry in ecosystem restoration to achieve environmental sustainability. Elsevier, Amsterdam, Netherlands, pp 51–60. ISBN: 978-0-443-15291-7. https://doi.org/10.1016/B978-0-443-15291-7.00022-5
Ogwu MC, Izah SC (2023) One health implications of agrochemicals and their sustainable alternatives. In: Ramawat KG (ed) Sustainable development and biodiversity series, vol 34, 1st edn. Springer Nature, Singapore, p 826. Hardcover ISBN 978-981-99-3438-6. https://doi.org/10.1007/978-981-99-3439-3
Ogwu MC, Osawaru ME, Chime AO (2014) Comparative assessment of plant diversity and utilization patterns of tropical home gardens in Edo State, Nigeria. Sci Africana 13(2):146–162
Ogwu MC, Malikia CN, Stansfield A, Gonzales-Torres AD, Izah SC (2023) Sustainable food processing waste management for environmental protection. In: Srivastav AL, Grewal AS, Tiwari M, Pham TD (eds) Role of green chemistry in ecosystem restoration to achieve environmental sustainability. Elsevier, Amsterdam, Netherlands, pp 291–299. ISBN: 978-0-443-15291-7. https://doi.org/10.1016/B978-0-443-15291-7.00022-5
Ogwu MC, Izah SC, Dessureault-Rompré J, Gasparatos D (2024) Sustainable soil systems in global south. Springer Nature, Switzerland. https://doi.org/10.1007/978-981-97-5276-8
Organisation for Economic Co-operation and Development (OECD) (1975) The polluter pays principle: OECD analyses and recommendations. OECD Publishing
Osawaru ME, Ogwu MC (2020) Survey of plant and plant products in local markets within Benin City and environs. In: Filho LW, Ogugu N, Ayal D, Adelake L, da Silva I (eds) African handbook of climate change adaptation. Springer Nature, Switzerland. pp 1–24. https://doi.org/10.1007/978-3-030-42091-8_159-1
Peng C, Zhao X, Liu G (2015) Noise in the sea and its impacts on marine organisms. Int J Environ Res Public Health 12(10):12304–12323. https://doi.org/10.3390/ijerph121012304
Pettit C, Chung W, Sharifi V, Chalabi Z, Fletcher T, Cleall P, Azapagic A et al (2011) Sustainable management of urban pollution: an integrated approach. Build Serv Eng Res Technol 32(1):21–34. https://doi.org/10.1177/0143624410394528
Pigou AC (1920) The economics of welfare. Macmillan
Posthuma L, Alink FM, Verweij RA (2019) Water quality and its influence on biodiversity. Hydrobiologia 828(1):23–35. https://doi.org/10.1007/s10750-018-3805-8

Prata J, Silva A, Costa J, Mouneyrac C, Walker T, Rocha-Santos T (2019) Solutions and integrated strategies for the control and mitigation of plastic and microplastic pollution. Int J Environ Res Public Health 16(13):2411. https://doi.org/10.3390/ijerph16132411

Raffee A, Hamid H, Rahmat S, Jaffar M (2022) The cause-and-effect analysis of ground level ozone (O_3), air pollutants and meteorological parameters using the causal relationship approach. J Eng Res. https://doi.org/10.36909/jer.17233

Riebschleager K, Karthikeyan R, Srinivasan R, McKee K (2012) Estimating potential *E. coli* sources in a watershed using spatially explicit modeling techniques. JAWRA J Am Water Resourc Assoc 48(4):745–761. https://doi.org/10.1111/j.1752-1688.2012.00649.x

Rockström J, Steffen W, Noone K, Persson Å, Chapin FS III et al (2009) A safe operating space for humanity. Nature 461(7263):472–475. https://doi.org/10.1038/461472a

Salama K, Alrashed Y, Alghamdi F, Alrashed R (2016) Assessment of the air quality levels in the King Abdul Aziz Port in Dammam. Int J Med Sci Publ Health 5(2):282. https://doi.org/10.5455/ijmsph.2016.14032015104

Samaranayake S, Glaser S, Holstius D, Monteil J, Tracton K, Seto E, Bayen A et al (2014) Real-time estimation of pollution emissions and dispersion from highway traffic. Comput Aid Civ Infrastruct Eng 29(7):546–558. https://doi.org/10.1111/mice.12078

Sarwar M, Hui Z, Maqbool A (2019) Causes and control measures of urban air pollution in China. Environ Ecosyst Sci 3(1):35–36. https://doi.org/10.26480/ees.01.2019.35.36

Sciascia S, Gilcrease G, Roccatello L, Padovan D, Peano C, Ricceri F (2022) Air pollution from global health to individual risk factor—is it time for enviropathies in everyday clinical practice? Int J Environ Res Public Health 19(15):9595. https://doi.org/10.3390/ijerph19159595

Seinfeld JH, Pandis SN (2016) Atmospheric chemistry and physics: from air pollution to climate change, 3rd edn. Wiley

Singh A (2023) Design and development of a air pollution monitoring using smartphone. Int J Eng Technol Manag Sci 7(3):489–493. https://doi.org/10.46647/ijetms.2023.v07i03.68

Singh S, Ward OP (2004) Bioremediation and environmental sustainability. Environ Bioremed Technol 1:1–16. https://doi.org/10.1007/978-3-662-05794-3_1

Smith VH, Tilman GD, Nekola JC (1999) Eutrophication: Impacts of excess nutrient inputs on freshwater, marine, and terrestrial ecosystems. Environ Pollut 100(1–3):179–196. https://doi.org/10.1016/S0269-7491(99)00091-3

Stansfeld S (2015) Noise effects on health in the context of air pollution exposure. Int J Environ Res Public Health 12(10):12735–12760. https://doi.org/10.3390/ijerph121012735

Stavins RN (1998) What can we learn from the grand policy experiment? Lessons from SO_2 allowance trading. J Econ Perspect 12(3):69–88. https://doi.org/10.1257/jep.12.3.69

Tamayo-Belda M, Pulido-Reyes G, Rosal R, Fernández-Piñas F (2022) Nanoplastic toxicity towards freshwater organisms. Water Emerg Contam Nanoplast 1(4):19. https://doi.org/10.20517/wecn.2022.17

Tanabe S (2004) PCB problems in the future: foresight from current knowledge. Environ Pollut 137(2):245–258. https://doi.org/10.1016/j.envpol.2004.01.031

Tang M, Jianping H (2016) Statistical analysis and evaluation of heavy metal ions in soil environment. Oalib 03(05):1–8. https://doi.org/10.4236/oalib.1102690

Tchounwou P, Yedjou C, Patlolla A, Sutton D (2012) Heavy metal toxicity and the environment, pp 133–164. https://doi.org/10.1007/978-3-7643-8340-4_6

Tefera W, Kumie A, Berhane K, Gilliland F, Lai A, Sricharoenvech P, Schauer J et al (2020) Chemical characterization and seasonality of ambient particles ($PM_{2.5}$) in the city centre of addis Ababa. Int J Environ Res Publ Health 17(19):6998. https://doi.org/10.3390/ijerph17196998

Torre D, Liuzzi D, Marsiglio S (2019) Transboundary pollution externalities: think globally, act locally?. https://doi.org/10.48550/arxiv.1910.04469

UNEP (2009) Stockholm convention on persistent organic pollutants. United Nations Environment Programme

Vienneau D, Pérez L, Schindler C, Lieb C, Sommer H, Probst-Hensch N, Röösli M et al (2015) Years of life lost and morbidity cases attributable to transportation noise and air pollution: a comparative health risk assessment for Switzerland in 2010. Int J Hyg Environ Health 218(6):514–521. https://doi.org/10.1016/j.ijheh.2015.05.003

Wang L, Su Y (2023) Economic loss and financial risk assessment of ecological environment caused by environmental pollution under big data. Sustainability 15(4):3834. https://doi.org/10.3390/su15043834

Wei T, Liu J, Ban-Weiss G, Hauglustaine D, Zhang L, Zhang Q, Tao S et al (2015) Effects of urban land expansion on the regional meteorology and air quality of Eastern China. https://doi.org/10.5194/acpd-15-10299-2015

Wei H, Wang Y, Liu J, Zeng R (2022) Heavy metal in river sediments of Huanghua City in water diversion area from yellow river, China: contamination, ecological risks, and sources. Water 15(1):58. https://doi.org/10.3390/w15010058

World Health Organization (WHO) (2011) Guidelines for drinking-water quality. WHO Press

World Health Organization (WHO) (2017) Water quality and health: review of impacts and mitigation measures. WHO Press

World Health Organization (WHO) (2018) Ambient (outdoor) air pollution. https://www.who.int/news-room/fact-sheets/detail/ambient-(outdoor)-air-quality-and-health

Xu J (2023) Managing the risks of new pollutants in China: the perspective of policy integration. Environ Health 1(6):360–366. https://doi.org/10.1021/envhealth.3c00054

Xu J, Feng C, Xi G, Niu L, Ju C, Zhou S (2014) Pollution analysis and control of non-ferrous metal smelting production. Adv Mater Res 955–959:2182–2186. https://doi.org/10.4028/www.scientific.net/amr.955-959.2182

Xu G, Jiao L, Zhao S, Yuan M, Li X, Han Y, Dong T et al (2016) Examining the impacts of land use on air quality from a spatio-temporal perspective in Wuhan, China. Atmosphere 7(5):62. https://doi.org/10.3390/atmos7050062

Xu Y, Hao Y, Zhang T, Yan W, Wang Y, Guo D, Li J et al (2021) Emission reduction efficiency analysis based on characteristics of vehicle emissions. Emiss Control Sci Technol 7(4):359–373. https://doi.org/10.1007/s40825-021-00200-7

Yang T (2019) Impacts of perceived environmental pollution on urban and rural residents' health—China national study. https://doi.org/10.21203/rs.2.12990/v1

Yang T (2020) Association between perceived environmental pollution and health among urban and rural residents—a Chinese national study. BMC Publ Health 20(1)

Yang W, Zhang Y, Hu Y (2022) Heterogeneous impact of economic policy uncertainty on provincial environmental pollution emissions in China. Sustainability 14(9):4923. https://doi.org/10.3390/su14094923

Yu D, Wang J, Wang Y, Du X, Li G, Li B (2021) Identifying the source of heavy metal pollution and apportionment in agricultural soils impacted by different smelters in china by the positive matrix factorization model and the PB isotope ratio method. Sustainability 13(12):6526. https://doi.org/10.3390/su13126526

Zhang X, Karplus VJ, Cassisa C, Zhang D (2017) Emissions trading in China: progress and prospects. Energy Policy 75:9–16. https://doi.org/10.1016/j.enpol.2017.01.002

Zhang J, Gao Y, Yang N, Dai E, Yang M, Wang Z, Yani G et al (2021) Ecological risk and source analysis of soil heavy metals pollution in the river irrigation area from Baoji, China. PLoS ONE 16(8):e0253294. https://doi.org/10.1371/journal.pone.0253294

Zhao C, Sun Y, Zhong Y, Xu S, Liang Y, Liu S, He M et al (2021) Spatio-temporal analysis of urban air pollutants throughout China during 2014–2019. Air Qual Atmos Health 14(10):1619–1632. https://doi.org/10.1007/s11869-021-01043-5

Zhou L, Wu Q, Zhang B, Zhao Y, Zhao B (2016) Occurrence, spatiotemporal distribution, mass balance and ecological risks of antibiotics in subtropical Shallow Lake Taihu, China. Environ Sci Process Impacts 18(4):500–513. https://doi.org/10.1039/c6em00062b

Chapter 6
Green Technologies: Theories, Systems, Principles and Applications

Prasann Kumar and Debjani Choudhury

Abstract Green technologies, often called environmentally sustainable technologies, are transformative tools that significantly mitigate environmental impacts by optimising the use of natural resources, reducing pollution, and minimising waste. These technologies encompass many innovations, from biotechnological advancements to diverse industry applications. This chapter offers a comprehensive synthesis of green technologies, integrating theoretical underpinnings with practical usage and developmental insights in key sectors, including agriculture, energy, waste management and recycling, transportation, and construction. The discussion highlights the integration of renewable energy systems, carbon capture and storage mechanisms, circular economy principles, and sustainable material management practices. Moreover, the chapter emphasises the role of emerging digital technologies such as the Internet of Things and Artificial Intelligence in optimising green solutions, enhancing efficiency, and accelerating their adoption across various domains. Real-world case studies illustrate successful implementations of green technologies, offering practical insights into their applications and benefits. These examples are complemented by analysing existing policy frameworks that support or hinder the adoption of such technologies, providing a balanced perspective on challenges and opportunities. The chapter also delves into how collaborative approaches between governments, industries, and communities can drive innovation and scalability in green technology solutions. By bridging theory with practice and aligning technological innovations with sustainable development goals, this chapter paints a holistic picture of the potential of green technologies to enable a sustainable future. It underscores the urgency of leveraging these innovations to address global environmental challenges, fostering resilience and sustainability across ecosystems and industries. This multidimensional

P. Kumar (✉)
Department of Agronomy, School of Agriculture, Lovely Professional University, Phagwara, Punjab, India
e-mail: prasann0659@gmail.com

D. Choudhury
Department of Plant Pathology, School of Agriculture, Lovely Professional University, Phagwara, Punjab, India

M. C. Ogwu and S. Chibueze Izah (eds.), *Evaluating Environmental Processes and Technologies*, Environmental Science and Engineering,
https://doi.org/10.1007/978-3-031-85327-2_6

exploration is a valuable resource for researchers, policymakers, and practitioners seeking to understand and advance the transformative power of green technologies.

Keywords Agriculture · Green · Technology · Carbon capture · Energy security · Biotechnological interventions · Waste management

6.1 Introduction

Often called sustainable or eco-friendly technologies, green technologies are ideas and methods meant to reduce environmental impact while advancing resource economy and sustainability (Zeng et al. 2024). These technologies seek to maximise renewable energy sources, lower greenhouse gas emissions, cut waste, and design socially fair, ecologically benign, and financially feasible systems. Green technologies are applied in many fields, including waste recycling, transportation, water management, building, energy generation, and agriculture (Abbas et al. 2024; Dhumal et al. 2024; Fegade 2024). Emphasising sustainable practices, renewable resource use, pollution prevention, and long-term environmental stewardship, they combine Prime examples of how green technologies transform sectors to fit global sustainability objectives, such as solar panels, wind turbines, biodegradable materials, and water purification systems. Furthermore, not limited to physical innovations are green technologies, which include digital tools like the Internet of Things (IoT), artificial intelligence (AI), and blockchain, which significantly help monitor environmental effects and maximise resource use. Green technologies' all-encompassing approach guarantees they solve the linked problems of resource constraint, ecological damage, and climate change (Krasnodębski 2024).

Ecological issues caused by humans for centuries have necessitated green technologies. The Industrial Revolution was a turning point because rapid industrialisation caused unprecedented environmental damage and economic growth. Water, air, habitat pollution, habitat destruction, and deforestation resulted from coal and oil dependence. Urbanisation and industry exacerbated health issues, biodiversity loss, and climate change (Oyefusi et al. 2024). In the mid-twentieth century, scientists and activists stressed the adverse effects of human activity on the environment, igniting the environmental movement. The Cuyahoga River fire and Rachel Carson's Silent Spring highlighted the need for sustainable solutions. In response, governments, businesses, and researchers developed environmental protection technologies. Early projects included industrial scrubbers and car catalytic converters. Solar panels and wind generators became fossil fuel substitutes in the late twentieth century. Environmentalism, including the founding of the EPA and the Kyoto Protocol, provided a legislative framework for these advances. The rising global population and changing consumption patterns caused problems even with these changes. Integrating social, economic, and technical factors to create sustainable systems is a twenty-first century trend. Green technologies are leading efforts to ensure a sustainable future, preserve biodiversity, and slow global warming (Imamoglu 2024).

Green technologies for sustainable solutions are essential for human and environmental survival. Climate change threatens ecosystems, businesses, and populations worldwide with rising global temperatures, severe storms, and sea-level rise. These challenges make polluting methods unfeasible and require non-renewable resources (Sridhar et al. 2024). Green technologies help decarbonise economies, reduce resource use, and build environmental resilience. Renewable energy sources like solar and wind power are reliable, affordable, and reduce greenhouse gas emissions. Organic and precision farming improves soil quality and reduces chemical inputs, ensuring food security for future generations.

For social, economic, and environmental reasons, green technologies are essential. Green technologies boost innovation, employment, and economic growth in new industries. The global sustainability movement has created a multibillion-dollar green economy in sustainable transportation, green building, and renewable energy. Green technologies improve public health by reducing pollution and improving living conditions (Arfelis et al. 2024). They also address equity issues by providing safe drinking water and clean energy, especially in underdeveloped areas. Green technologies align with global goals like the UN Sustainable Development Goals (SDGs), emphasising social, environmental, and financial sustainability. Thus, transitioning to green technologies is necessary to address environmental issues and ensure global stability and prosperity.

This chapter comprehensively explores green technologies, focusing on their principles, applications, and potential to address pressing environmental and societal challenges. Defining green technologies and distinguishing them from conventional approaches establishes a framework grounded in sustainability, resource efficiency, and pollution prevention. Tracing their evolution from early pollution control initiatives to cutting-edge innovations in renewable energy, agriculture, and waste management, the chapter highlights key milestones and insights for future advancements. It emphasises the critical need to transition to environmentally friendly solutions, demonstrating the social, economic, and environmental benefits of adoption while addressing the risks of inaction. Through practical examples and sector-specific applications—from energy and transportation to agriculture and construction—the chapter showcases the transformative power of green technologies in promoting sustainable development. Tackling barriers to adoption, such as technological, financial, and policy challenges, equips readers with actionable strategies to embrace and advocate for these innovations. This chapter is a valuable resource for professionals, policymakers, and students, inspiring informed action and fostering a collective commitment to building a sustainable future.

6.2 Principles of Green Technologies

Green technologies are essential to creating, implementing, and advancing sustainable environmental, economic, and social solutions. These ideas emphasise sustainability, which balances economic growth, social well-being, and environmental protection. This approach recognises that economic growth should not sacrifice social justice or the environment. Green technologies create financial value while preserving natural resources and promoting social inclusion (Afonso et al. 2024). Renewable energy systems like solar and wind power reduce greenhouse gas emissions, create jobs, and power underprivileged areas, addressing all three sustainability dimensions. Sustainable agricultural methods like precision farming and organic production increase crop yields, soil health, and food security, reflecting the triple bottom line.

Another key concept is resource efficiency, which optimises energy, water, and raw material use to reduce waste and boost production. Industrial development and population growth are causing resource scarcity, making resource efficiency essential. Green technologies use sophisticated tools and systems to monitor, analyse, and maximise resource use. Intelligent irrigation systems with sensors and IoT technologies deliver water precisely to crops, reducing waste and increasing output. Green technologies like material recycling and energy-efficient machinery reduce manufacturing energy consumption and dependence on virgin raw materials (Huang and Li 2024). These methods save money and resources. Thus, resource efficiency benefits companies and the environment. Green building technologies use natural lighting, insulation, and renewable energy sources to maximise energy consumption in the built environment. This increases urban development's sustainability and lowers operating costs.

Pollution prevention reduces emissions and waste at its source, unlike end-of-pipe solutions. Green technologies, which design processes and products to naturally emit less pollution, rely on this proactive approach. Electric vehicles (EVs) reduce tailpipe emissions, improving urban air quality. Green chemistry methods minimise side effects to protect ecosystems and human health. Pollution prevention is crucial to waste management, and waste-to-energy and composting recycle organic waste instead of dumping it. Sustainable consumption and manufacturing reduce pollution and environmental damage and meet public health and resource economy goals (Irianto et al. 2025).Green technologies prioritise renewable resources over scarce ones. This idea is essential to creating infinitely renewable systems (Dhumal et al. 2024; Ogwu et al. 2023). Renewable energy sources like solar, wind, and hydropower provide low-impact, inexhaustible fossil fuel substitutes. Biomass for biofuels and biodegradable materials from agricultural waste support the transition to a renewable economy. Wastewater recycling and rainwater collection enable sustainable water management. Green technologies reduce resource use and use natural regeneration to ensure ecosystem and economic survival (Standal and Ahlquist 2025).

The circular economy, another green technology, promotes reuse, recycling, and low-waste production and consumption. A circular economy uses materials as long as possible to maximise value before recovering and rejuvenating them. Green technologies help enable this change by developing disassembly, repair, and recycling products. Modular electronics reduce electronic waste by allowing simple part replacement and extending device lifespan (Fiscal-Ladino et al. 2024). Recycling technologies can recover valuable materials from complex waste streams like plastic and electronic waste and turn them into new products. Crop rotation, composting, and biofertilisation create closed-loop systems that improve soil quality and reduce agricultural inputs. The circular economy optimises resource efficiency and waste minimisation to minimise environmental impact and boost industry resilience and innovation.

Decarbonisation—reducing greenhouse gas emissions and fighting climate change—is central to green technologies. Decarbonisation involves switching from fossil fuel-based energy, transportation, and industrial systems to low- or zero-carbon alternatives. Energy systems can integrate battery storage and grid management, but renewable energy technologies like solar photovoltaics and wind turbines are essential for decarbonising electricity generation. Moving to electric vehicles, hydrogen fuel cells, and biofuels in transportation reduces emissions from a primary industry. Carbon capture and use (CCU) technologies collect factory emissions and turn them into synthetic fuels and building materials. Beyond technical fixes, decarbonisation requires systemic and behavioural changes, such as energy efficiency, sustainable living, and low-carbon laws. Decarbonisation ensures that green technologies help limit temperature rise and protect ecosystems by addressing climate change 's root causes (Addo-Bankas et al. 2024). Sustainability, resource efficiency, pollution prevention, renewability, circular economy, and decarbonisation guide green technology solutions that balance environmental, economic, and social goals. These ideas solve ecological issues and build solid systems that can adapt. Green technologies incorporate these ideas into energy, agriculture, manufacturing, and urban development for a holistic approach to sustainable development. These values are increasingly essential to create a sustainable and fair future as global issues like pollution, resource constraints, and climate change worsen (Table 6.1).

6.3 Core Applications of Green Technologies

6.3.1 Agriculture

Green technologies can transform agriculture by addressing resource inefficiency, soil degradation, and environmental pollution while ensuring sustainable food production. Precision agriculture improves farming using GPS, remote sensing, and the IoT. Precision agriculture uses data-driven insights to apply water, fertilisers, and pesticides precisely when and where needed (Lin et al. 2024). IoT-based sensors measure soil moisture, nutrient levels, and crop health in real-time, enabling real-time adjustments to reduce resource waste and boost productivity. Intelligent irrigation systems using IoT sensors deliver water directly to plant roots in precise amounts, reducing water consumption. Multispectral drones can assess crop health and identify problem areas, enabling targeted yield increases and environmental protection. These technologies boost farmers' productivity and profitability while reducing agriculture's environmental impact, supporting global sustainability goals. Bio fertilisers and biodegradable mulching films improve soil management and crop

Table 6.1 Different principles of green chemistry and its application

Principle	Definition	Applications
Energy efficiency	Optimizing energy use while maintaining functionality	Energy-efficient building designs, LED lighting, high-efficiency HVAC systems
Renewable energy integration	Harnessing renewable energy sources to reduce dependence on fossil fuels	Solar panels, wind turbines, geothermal heating and cooling
Sustainable materials	Using eco-friendly, recyclable, and renewable materials in construction	Bamboo, recycled metals, low-carbon concrete
Water conservation	Reducing water usage and promoting water reuse	Low-flow plumbing fixtures, rainwater harvesting, water-efficient landscaping
Waste reduction and recycling	Minimizing construction waste and encouraging recycling	Modular construction, on-site waste segregation, recycled materials
Green building design	Creating buildings that maximize energy and resource efficiency while minimizing environmental impact	LEED-certified designs, passive houses, net-zero energy homes
Urban green spaces	Incorporating green areas into urban settings to improve ecological balance	Urban parks, green rooftops, tree-lined streets
Smart cities technology	Employing IoT, AI, and data analytics to enhance urban efficiency and sustainability	Smart traffic systems, IoT-enabled waste management, energy-efficient lighting
Climate resilience	Designing infrastructure to withstand extreme weather events	Flood-resistant buildings, stormwater systems, resilient transportation infrastructure
Carbon footprint reduction	Minimizing emissions from construction and urban development	Carbon–neutral designs, low-carbon materials, electric vehicles
Local sourcing of materials	Reducing transportation emissions by sourcing materials locally	Locally sourced timber, regional stone, nearby construction aggregates
Vertical farming	Urban food production to enhance food security and reduce supply chain emissions	Rooftop farms, hydroponic systems, indoor vertical farming
Passive solar design	Using solar energy for heating and lighting buildings naturally	Solar thermal heating, south-facing windows, daylighting techniques
Urban heat island mitigation	Reducing urban temperature by incorporating cooling solutions	Cool roofs, urban tree canopies, reflective pavements

(continued)

Table 6.1 (continued)

Principle	Definition	Applications
Sustainable land use	Promoting land utilization that conserves biodiversity and minimizes disruption	Compact urban developments, greenfield conservation, mixed-use neighborhoods
Environmental awareness	Educating communities to adopt sustainable practices	Workshops, eco-building tours, educational campaigns
Energy storage systems	Storing renewable energy for consistent supply	Solar battery systems, grid-scale energy storage, pumped hydro systems
Green transportation	Developing low-emission transport infrastructure	EV charging stations, bike-sharing networks, pedestrian-friendly zones
Modular construction	Using pre-fabricated components for efficient and low-waste building	Prefabricated units, modular walls and roofing, off-site assembly
Biodiversity preservation	Protecting ecosystems in urban areas	Wildlife corridors, pollinator-friendly landscaping, green roofs
Sustainable drainage systems	Managing stormwater sustainably to reduce flooding	Permeable paving, rain gardens, bio-retention swales
Circular economy	Reusing and recycling resources to minimize waste generation	Design for disassembly, reclamation of materials, modular product designs
Zero-waste construction	Striving for construction with minimal landfill contributions	Recycled materials, demolition reuse, waste sorting systems
Healthy buildings	Ensuring buildings promote health and well-being of occupants	Improved indoor air systems, natural lighting, non-toxic materials
Urban agriculture	Growing food locally within cities to improve sustainability	Community gardens, aquaponic systems, urban farms
Carbon capture and sequestration	Capturing and storing carbon emissions from urban infrastructure	Urban carbon sinks, industrial carbon capture, green walls
Regenerative design	Creating systems that restore and enhance the environment	Ecosystem restoration, closed-loop water recycling, renewable energy systems

Sources Antenucci et al. (2022), Outili et al. (2023), Dabas et al. (2024), Fegade (2024), Lei (2024), Simi (2024) and Zhang et al. (2024)

yield in agriculture (de Souza Mesquita et al. 2024). Biofertilisers from natural microorganisms fix nitrogen, solubilise phosphates, and promote plant growth. Bio fertilisers improve soil structure and fertility without polluting waterways like synthetic fertilisers. Fertilisers often contain nitrogen-fixing bacteria like Rhizobium and phosphate-solubilizing fungi like Aspergillus to improve crop nutrient availability. Biodegradable mulching films from starch or cellulose control weeds and retain soil moisture. These films biodegrade naturally after use, eliminating the need for laborious disposal and reducing plastic pollution from traditional mulching films. Biodegradable mulching films conserve water, reduce soil erosion, and regulate soil temperature, improving crop yields and sustainable land use (Gupta et al. 2024). Using fertilisers and biodegradable mulching films ensures regenerative agriculture, replenishing natural resources. Phytoremediation and rejuvenation systems use green technologies to solve soil contamination and degradation. Phytoremediation uses plants that extract, degrade, or stabilise heavy metals, pesticides, and hydrocarbons. Hyperaccumulator plants like Brassica juncea (Indian mustard) and Helianthus annuus (sunflower) can remove lead, cadmium, and arsenic from the soil (Zhu et al. 2024). These plants clean the soil and recover biomass for bioenergy. Advanced phytoremediation systems use genetically modified plants that absorb metal better, improving efficiency. To restore degraded land, soil rejuvenation systems use organic amendments, microbial inoculants, and sustainable land management. Soil health and productivity depend on composting, vermiculture, and biochar, which increase soil organic matter, water retention, and microbial diversity. Phytoremediation and soil rejuvenation technologies clean up contamination and promote arable land restoration for future agricultural use while reducing environmental risks. Precision agriculture, biofertilisers, biodegradable mulching films, and phytoremediation systems are examples of green technologies that can solve agricultural problems while protecting the environment. These technologies enhance productivity and resource efficiency while promoting ecosystem restoration and climate resilience. With the rising global demand for food and sustainable land use, green technologies in agriculture are essential for a future where farming practices harmonise with nature (Kumar and Chand 2024).

6.3.2 Energy

Green technologies have transformed energy generation, storage, and use, emphasising sustainability, efficiency, and greenhouse gas reduction. These initiatives rely on renewable energy sources like solar, wind, hydro, and geothermal. Photovoltaic (PV) cells directly convert sunlight into electricity, making solar power a clean and abundant energy source. High-efficiency panels, solar thermal systems, and floating solar farms have made solar power more feasible and scalable in various geographic and economic contexts (Feng et al. 2024). Wind energy is a renewable resource that uses turbines to generate electricity. Due to more substantial and more consistent winds, large-scale renewable energy generation is promising in offshore wind farms. Hydropower, the most common renewable energy source, uses dams or run-of-river systems to generate electricity. Earth's internal heat provides reliable, low-emission geothermal energy for electricity generation and direct heating. These renewable energy technologies form the basis of a sustainable energy future, reducing fossil fuel use and energy production's environmental impact (Karthick Raja Namasivayam et al. 2023). Renewable energy integration requires energy storage and intelligent grid solutions to address intermittency and demand variability. By storing surplus power from peak generation and discharging it during high demand or low generation, energy storage technologies, especially lithium-ion batteries, stabilise renewable energy supply. Solid-state and flow batteries have improved energy density, longevity, and safety, making large-scale storage systems possible. Hydrogen fuel cells and thermal storage systems can extend energy storage beyond batteries (Yue et al. 2023; Ogwu 2019). Smart grids use advanced communication and automation technologies with storage solutions to improve energy distribution networks' efficiency, reliability, and resilience. Smart grids enable real-time energy flow monitoring and management, integrating distributed energy resources like rooftop solar panels and EVs while reducing gearbox losses. Demand-response programs allow consumers to adjust energy usage based on price signals or grid conditions, optimising energy consumption and reducing peak loads. Energy storage and smart grids help overcome renewable energy limitations and provide stable, efficient energy to modern societies.

CCU technologies help meet the urgent need to reduce atmospheric CO_2 levels and combat climate change. Carbon capture involves extracting CO_2 from industrial emissions or the atmosphere using solvents, membranes, or adsorption. Either sequester CO_2 in geological formations or use it to create value-added products (Feder-Kubis et al. 2024). Captured CO_2 can be converted into synthetic fuels to reduce fossil fuel dependence or used as a feedstock in concrete production to strengthen materials. Research is exploring the use of CO_2 in chemical manufacturing, polymer feedstock, and other industrial applications. These utilisation pathways reduce CO_2 emissions and encourage industries to adopt carbon capture technologies. Integrating carbon capture and utilisation (CCU) with renewable energy systems improves sustainability by powering energy-intensive CO_2 conversion processes, creating closed-loop systems that follow circular economy principles. CCU technologies are crucial to net-zero emissions strategies because they reduce emissions and valorise resources. Integrating renewable energy, energy storage, smart grids, and carbon capture and use solutions shows how green technologies can transform the energy sector. Renewable energy is clean and sustainable; storage and intelligent grids ensure reliable distribution. CCU technologies reduce residual emissions from hard-to-decarbonize sectors, easing the low-carbon transition. These innovations minimise energy production's environmental impact while boosting economic growth and energy security, creating a sustainable and resilient global energy framework (Assen et al. 2024).

6.3.3 Water and Waste Management

Green water and waste management technologies are changing how we handle water scarcity, waste disposal, and resource recovery, promoting sustainability and environmental health. The latest water purification systems remove contaminants from various sources to provide clean, safe water. These systems effectively treat heavy metals, pathogens, and organic compounds using reverse osmosis, ultrafiltration, and advanced oxidation processes (AOPs). Membrane filtration improves water quality and cuts energy and operational costs. Graphene oxide membranes and nanotechnology-enhanced filters improve water purification speed and accuracy. In urban and industrial settings, integrated water recycling systems reuse treated wastewater for irrigation, industrial processes, and potable use. International initiatives to promote sustainable water resource management use these technologies to reduce water scarcity and environmental impact (Akinsemolu 2023). Developing waste-to-energy (WTE) technologies has transformed waste from a burden to an asset. Waste-to-energy (WTE) processes convert organic and inorganic waste into electricity, heat, or fuel while reducing landfill waste. Incineration, a popular waste-to-energy (WTE) method, uses advanced combustion systems to generate electricity and filtration mechanisms to capture emissions, meeting environmental standards. Anaerobic digestion creates biogas from organic waste using biochemical properties. You can use biogas for cooking or heating or convert it to biomethane for natural gas injection. Pyrolysis and gasification can convert waste into syngas and biochar, increasing the variety of energy outputs from waste streams. These technologies reduce municipal solid waste and improve circular economy frameworks by maximising material value (Nahata et al. 2024). Recycling and composting advances help these initiatives recover resources and reintegrate organic matter into the soil. Intelligent sorting technologies like AI-driven robotics and optical sensors have improved recycling efficiency and precision, allowing for the clean recovery of plastics, metals, and glass. Chemical recycling, which breaks down plastics into their molecular components, may solve the problem of non-recyclable plastics. Additionally, aerobic digesters and in-vessel systems have accelerated the decomposition of organic waste into nutrient-dense compost. Urban areas with limited space and waste require efficient and scalable systems. These processes produce compost, which improves soil quality and reduces synthetic fertiliser use, completing the organic waste management cycle. Modern green water and waste management technologies include advanced water purification systems, waste-to-energy technologies, and recycling and composting innovations. These solutions solve urgent environmental issues and recover and use resources sustainably. These technologies can significantly reduce ecological impact and promote resource conservation in urban, industrial, and agricultural systems, creating a cleaner, more sustainable future (Wang and Zhang 2024; Ogwu 2019).

6.3.4 Transportation

Green transport and logistics technologies are leading carbon emission reduction, energy efficiency, and sustainable mobility efforts. Electric vehicles (EVs) and alternative fuels are replacing fossil-fuel-based transportation systems. With advanced lithium-ion or solid-state batteries, EVs are clean and efficient and eliminate tailpipe emissions. Governments and businesses worldwide are investing heavily in charging networks and battery recycling facilities to promote EV adoption. Long-range EVs and faster charging have reduced EV adoption barriers. Beyond battery-powered EVs, hydrogen fuel cell vehicles (FCVs) are becoming a sustainable option for heavy-duty transport, emitting only water vapour instead of emissions (Szekely 2024). Biodiesel, ethanol, and CNG also help green transport systems, especially in areas where electrification is impossible. These renewable fuels from agricultural waste and algae reduce lifecycle emissions compared to conventional fuels.

Innovative and sustainable urban transport systems complement these advancements to maximise mobility and minimise environmental impact. Electric and hybrid buses, light rail, e-bikes and e-scooters are changing public transportation networks. These systems reduce urban congestion and emissions by reducing private vehicle use. New technologies like IoT and AI enable real-time tracking, route planning, and adaptive traffic management, making urban transport systems responsive and efficient. Mobility as a Service (MaaS) platforms, combining public transit, ride-sharing, and micro-mobility into one digital interface, simplify urban mobility. Sustainable transport planning includes pedestrian-friendly infrastructure and green corridors to promote walking and cycling and reduce vehicle emissions. Transit-oriented development (TOD) models prioritise high-density urban development near transport hubs to reduce commuting distances and encourage sustainable living (Amoneit et al. 2024).

The logistics and supply chain sectors are also greening by adopting low-emission logistics practices. Electric last-mile delivery vans and trucks powered by renewable energy reduce emissions. Drone deliveries and autonomous vehicles boost logistics efficiency and reduce carbon emissions. Transporting perishable goods via cold chain logistics increasingly uses solar-powered and energy-efficient refrigeration. Intelligent supply chain systems use AI, blockchain, and IoT to optimise inventory management, waste reduction, and supply chain transparency. Intermodal transport solutions combine rail, road, and sea freight to reduce fuel-intensive trucking. Energy-efficient buildings, renewable energy installations, and automated systems reduce storage facilities' environmental impact. Using electric vehicles, alternative fuels, intelligent urban transport systems, and low-emission logistics for green transport is comprehensive. These methods reduce environmental impact, boost energy efficiency, and sustain economic growth. These innovations can help societies address climate change and urbanisation by creating cleaner, more resilient mobility systems (Table 6.2).

6.3.5 Construction and Urban Development

Green construction and urban development technologies improve space design, construction, and management by emphasising sustainability, energy efficiency, and liveability. This transformation relies on environmentally friendly building materials and methods from construction to deconstruction. Bamboo, recycled concrete aggregates, fly ash, and cross-laminated timber supplement cement and steel. These materials reduce construction's carbon footprint while improving thermal insulation,

Table 6.2 The core application of green chemistry with examples

Application	Description	Examples
Biodegradable pesticides	Pesticides that degrade quickly, reducing environmental impact	Biopesticides made from plant extracts or microorganisms
Natural fertilizers	Use of organic and naturally occurring substances as fertilizers	Compost, manure, and algae-based fertilizers
Crop protection with botanical extracts	Plant-derived substances used to protect crops from pests and diseases	Neem oil for pest control; garlic and chili-based sprays for pest repulsion
Precision agriculture	Technologies to optimize the application of fertilizers, pesticides, and water	GPS and drone-guided fertilization; variable rate pesticide application
Sustainable herbicides	Natural herbicides that are safer for the environment	Acetic acid and vinegar-based herbicides for weed control
Bio-based plastics for mulching	Use of renewable biopolymers for agricultural mulching	Biodegradable mulch films made from starch or PLA (polylactic acid)
Green solvents for pesticide formulation	Non-toxic solvents used in pesticide formulations	Vegetable oil-based solvents; ionic liquids
Phytoremediation	Using plants to remove contaminants from soil and water	Sunflowers for heavy metal absorption; grass species for soil nitrogen regulation
Nutrient release control	Fertilizers that release nutrients slowly and in a controlled manner	Controlled-release urea; bio-based slow-release phosphorus fertilizers
Microbial inoculants for soil health	Beneficial microbes introduced to enhance soil fertility and plant health	Rhizobium inoculants for legumes; mycorrhizal fungi for soil improvement
Enzyme-based fertilizers	Fertilizers enhanced with enzymes to improve nutrient uptake efficiency	Enzyme-enhanced composts; phosphatase enzymes for phosphorus release in soils
Bio-based growth regulators	Natural plant growth regulators used to enhance crop yields	Auxins and cytokinins from seaweed extracts; gibberellic acid from fungi
Agroecology	Application of ecological principles to minimize agricultural harm	Agroforestry practices; polyculture systems promoting biodiversity
Organic pesticides	Natural or minimally processed compounds for pest control	Diatomaceous earth; insecticidal soaps
Bio-based fungicides	Natural substances used to prevent fungal infections in crops	Bacillus thuringiensis; clove oil for fungal protection
Sustainable irrigation techniques	Water-efficient irrigation methods to reduce waste	Drip irrigation systems; soil moisture sensors for optimized water usage

(continued)

Table 6.2 (continued)

Application	Description	Examples
Recycling of agricultural waste	Converting by-products into valuable resources	Composting crop residues; rice husks into biochar
Bio-based pest traps	Traps using natural materials or substances for pest control	Fermented molasses traps; biodegradable pheromone traps
Plant-extract-based antioxidants	Using plant extracts to prevent oxidative stress in crops	Green tea or turmeric extracts for oxidative damage protection
Natural fungus resistance	Enhancing crops' resistance to fungal diseases with natural compounds	Chitosan for plant defense; essential oils for fungal inhibition
Microbial biofertilizers	Microbes supplying essential nutrients to plants	Azospirillum for nitrogen fixation; phosphate-solubilizing bacteria
Green chemistry in seed coatings	Eco-friendly coatings for seeds to protect and enhance germination	Natural wax-based coatings; essential oil-based pest repellents
Soil remediation with green chemicals	Natural substances used to restore soil health	Humic acid for soil rejuvenation; algal-based amendments
Eco-friendly aquaculture	Green approaches for sustainable aquaculture	Plant-based fish feed; eco-friendly pond water treatments
Biocontrol agents for IPM	Using natural predators or pathogens for pest control	Ladybugs for aphid control; nematodes for soil pests
Natural herbicide production	Herbicides developed from naturally occurring substances	Citric acid; essential oils like cinnamon or clove oil
Recyclable plant-based packaging	Plant-derived materials for reducing plastic waste	PLA-based food packaging; cellulose biodegradable films
Green chemistry in crop genetic engineering	Genetically engineered crops requiring fewer chemicals	Bt cotton for pest resistance; drought-resistant crops
Sustainable livestock feed	Eco-friendly feed developed using green chemistry	Algae-based feed; insect-derived protein sources
Microbial biodegradation	Microbes breaking down harmful chemicals in the environment	Bioremediation of pesticide-contaminated soils; microbial degradation of herbicides
Waste-to-energy conversion	Transforming agricultural waste into energy	Biomass gasification; anaerobic digestion to produce biogas
Green chemistry in composting	Optimizing composting with bio-based additives	Accelerators for composting; microbial inoculants
Sustainable crop protection from invasive species	Eco-friendly methods for controlling invasive pests	Mulching with natural fibers; biological control agents

(continued)

Table 6.2 (continued)

Application	Description	Examples
Organic crop certification standards	Standards emphasizing natural inputs in agriculture	Organic pesticides; natural produce preservatives
Green chemical fertilizer synthesis	Producing fertilizers sustainably with reduced energy use	Green ammonia; phosphorus from renewable sources
Nutrient cycling	Recycling nutrients in soil for sustainability	Cover crops for nitrogen fixation; crop rotation
Low-toxicity insecticides	Safer insecticides made using green chemistry principles	Pyrethrins; insecticidal soaps
Green chemistry in agricultural equipment	Energy-efficient farming tools with lower emissions	Solar-powered irrigation; electric tractors
Eco-friendly soil amendments	Natural materials improving soil properties	Vermicompost; lime for pH regulation
Sustainable water management	Green chemistry solutions for efficient water use in agriculture	Water-saving polymers; bio-based water-retaining gels
Green crop protection products	Non-toxic and biodegradable crop protection solutions	Plant-derived pest repellents; eco-friendly insecticides
Agro-chemical recycling	Recycling fertilizers and pesticides to reduce waste	Pesticide reuse; fertilizer recovery from runoff
Sustainable herbicide resistance management	Strategies to manage herbicide resistance sustainably	Crop rotation; non-chemical weed control
Low-impact soil fertilization	Techniques minimizing environmental impact from fertilization	Green manure crops; precision agriculture to reduce chemical use
Biological pest control agents	Natural predators and parasitoids for pest control	Predatory beetles; Trichogramma wasps
Eco-friendly crop storage	Green methods for preserving crops during storage	Essential oils for fungal prevention; natural wax coatings for fruits
Crop pest monitoring with eco-friendly devices	Non-toxic devices for monitoring pests	Natural attractant traps; non-toxic sticky boards
Green chemistry in greenhouse production	Green chemistry for optimizing greenhouse conditions	Biodegradable pest control; natural climate control systems
Composting with bio-based additives	Bio-based solutions enhancing composting efficiency	Accelerators; microbial inoculants
Green crop seed treatment	Sustainable chemicals used to treat seeds pre-planting	Neem oil; bio-based protectants
Green agriculture supply chains	Green practices across agriculture supply chains	Eco-friendly packaging; sustainable logistics

Sources Hussain et al. (2024), Ibrahim et al. (2024), Anitha and Beda (2024), Verma et al. (2024), Wang et al. (2024), Zain et al. (2024) and Zheng et al. (2024)

durability, and aesthetics. 3D printing in construction reduces waste and allows for local, low-impact materials. Modular construction and prefabrication optimise material use and construction time. Green construction improves indoor air quality and energy efficiency with advanced insulation, low-VOC finishes, and reflective roofing (Rani et al. 2024).

Meanwhile, net-zero energy buildings (NZEBs) are essential to sustainable urban development. These structures use energy-efficient systems and renewable energy to generate energy equal to their consumption over their lifespan. High-performance insulation, energy-efficient lighting, nd advanced HVAC systems reduce energy use. Instead, rooftop solar panels, wind turbines, and geothermal heating systems generate clean energy. Innovative building technologies use IoT sensors and devices to monitor and optimise energy use in real-time. Rainwater harvesting, greywater recycling, and low-flow fixtures help NZEBs conserve water. By reducing greenhouse gas emissions and operational costs, these technologies help NZEBs achieve climate goals. These buildings prioritise occupant health and well-being by optimising natural light, ventilation, biophilic features, and energy efficiency. City greening and innovative designs change the relationship between cities, their environments, and their residents. Green roofs, vertical gardens, and urban forests improve air quality, reduce the urban heat island effect, and increase biodiversity. Porous pavements, rain gardens, and bioswales reduce runoff and recharge groundwater, enabling sustainable water management. Innovative city initiatives use technology to create more efficient, resilient, and inclusive cities. IoT-enabled systems manage energy, water, and transportation networks to maximise resource use and minimise waste. Advanced traffic management systems reduce congestion and emissions, while smart grids and energy-sharing platforms let neighbourhoods share surplus renewable energy. Smart cities create multifunctional public spaces with recreational, ecological, and social functions to improve community well-being. A comprehensive sustainable urban development strategy includes green building materials, net zero energy buildings (NZEBs), urban greening, and smart city technologies. These innovations reduce environmental impact and address social and economic issues by creating healthier, more efficient, and resilient cities. These principles and technologies are essential for a sustainable future as the global population urbanises (Table 6.3).

6.4 Integration of Digital Technologies in Green Solutions

Integrating digital technologies into green solutions has catalysed a transformative shift in addressing global environmental challenges, paving the way for more innovative, efficient, and sustainable practices. Central to this transformation is the **role of IoT (Internet of Things), AI (Artificial Intelligence), and Blockchain**, which collectively create an ecosystem where data-driven decisions optimise resource usage, reduce waste, and enhance system efficiencies. IoT is the backbone of interconnected systems by enabling real-time data collection from sensors embedded in diverse applications such as energy grids, water management systems, agricultural

Table 6.3 Use of green chemistry for construction and urban development

Principle	Definition	Applications
Energy efficiency	Optimization of energy use to minimize waste and reduce consumption	Smart energy management systems in buildings, LED lighting systems, energy-efficient HVAC systems
Renewable energy Integration	Utilization of renewable energy sources like solar, wind, and geothermal	Rooftop solar panels, urban wind turbines, geothermal heating and cooling systems
Sustainable materials	Use of eco-friendly, recyclable, or renewable materials in construction	Bamboo, recycled steel, low-impact concrete, eco-friendly insulation, and cladding materials
Water conservation	Reducing water use and reusing water in urban settings	Rainwater harvesting systems, low-flow faucets and toilets, water-efficient landscaping
Waste management and recycling	Managing construction waste through reduction, reuse, and recycling	On-site recycling facilities, modular construction for reuse, recycled building materials
Green building design	Building designs that optimize energy, water, and material use	LEED-certified buildings, passive house designs, net-zero energy buildings
Urban green spaces	Incorporating green spaces within urban environments for ecological benefits	Green rooftops, urban parks, community gardens, green streetscapes, and tree-lined avenues
Low-impact development	Minimizing environmental impact through sustainable land-use planning	Permeable pavements, bioretention swales, green infrastructure for stormwater management
Smart cities technology	Integration of IoT, AI, and data analytics to enhance urban sustainability	Smart traffic management systems, intelligent building controls, and real-time energy usage monitoring
Climate resilience	Designing buildings and cities to withstand extreme weather events	Flood-resistant infrastructure, stormwater management systems, and heat-resistant materials
Carbon footprint reduction	Reducing carbon emissions associated with urban activities	Carbon–neutral buildings, electrified public transport, and low-carbon building materials
Local sourcing of materials	Using locally sourced materials to reduce transport emissions	Locally harvested timber, local stone and sand, and regionally produced construction materials
Vertical farming	Growing food in urban settings to improve food security and reduce transport	Rooftop gardens, indoor vertical farming systems, and hydroponic farming in commercial spaces

(continued)

Table 6.3 (continued)

Principle	Definition	Applications
Integrated transportation systems	Sustainable and connected transport networks reduce emissions and congestion	Electric bus fleets, bicycle-sharing programs, car-free city zones, walkable neighbourhoods
Passive solar design	Designing buildings to utilise sunlight for heating, lighting, and cooling	South-facing windows, solar heat gain control, and thermal mass for energy retention
Urban heat Island mitigation	Reducing urban heat retention through reflective or green technologies	Cool roofs, green walls, tree canopy coverage, and shade provision
Sustainable land use	Efficient land use promotes biodiversity and minimises habitat destruction	Mixed-use developments, compact communities, protected green spaces, and wildlife corridors
Environmental education	Promoting awareness of sustainability within urban communities	Educational programs on sustainability, community workshops, and eco-certifications for buildings
Energy storage systems	Technologies storing renewable energy for high-demand periods	Lithium-ion battery storage, pumped hydro storage, flywheel energy systems
Green transportation infrastructure	Infrastructure supporting sustainable transport systems	Electric vehicle charging stations, bicycle lanes, public transport hubs and networks
Modular construction	Prefabricated construction, reducing waste and speeding up processes	Modular housing units, prefabricated walls and roofs, off-site assembly
Low-carbon concrete	Alternative materials to replace traditional cement and reduce emissions	Geopolymer concrete, recycled aggregates, and low-carbon concrete additives
Energy-efficient urban lighting	Energy-saving lighting technologies in urban areas	Solar-powered streetlights, LED lighting in public spaces, motion-sensor street lighting
Biophilic design	Incorporating nature into building and urban design	Indoor plants, living walls, natural ventilation, daylighting, green terraces, and balconies
Biodiversity preservation	Protecting or enhancing local biodiversity in urban settings	Urban wildlife corridors, bird and bat houses, and pollinator-friendly plantings
Sustainable drainage systems	Managing stormwater sustainably	Green roofs, rain gardens, porous pavements
Eco-friendly transport networks	Promoting sustainable urban mobility	Carpool lanes, EV public transport, pedestrian zones, and walking trails

(continued)

Table 6.3 (continued)

Principle	Definition	Applications
Sustainable urbanization	Urban planning strategies create environmentally friendly cities	Transit-oriented development, zoning policies for sustainable buildings, energy-efficient public buildings
Urban agriculture	Incorporating agriculture in urban areas for food production	Community gardens, urban farms, and agricultural education programs
Green roofing	Installing vegetation on building roofs for energy efficiency and biodiversity	Green rooftops on residential and commercial buildings, as well as living roofs
Circular economy	Designing buildings for repairability, reuse, and recycling	Design for disassembly, building material reuse programs, and circular supply chains in construction
Zero-waste construction	Building without generating landfill waste	On-site sorting of construction waste, reclaimed materials, and demolition waste reuse
Healthy buildings	Creating spaces that promote occupant health and comfort	Natural ventilation, non-toxic materials, and optimised indoor air quality systems
Carbon capture and sequestration	Technologies capturing and storing carbon emissions	Urban industrial carbon capture, bio-based sequestration in green spaces
Urban green infrastructure	Ecosystem services within urban areas	Urban wetlands, green urban corridors, and water filtration systems
Building integrated photovoltaics	Integration of solar panels into building structures	Solar windows, building-integrated solar roofs and facades
Low-carbon footprint Construction	Construction minimising greenhouse gas emissions	Electrified construction machinery, prefabricated low-carbon materials, carbon-conscious planning
Smart waste management	Technology-driven urban waste management systems	Smart bins with IoT sensors, automated recycling systems, and waste-to-energy plants
Water-efficient landscaping	Landscaping designs reduce water usage	Xeriscaping, smart irrigation systems, recycled water landscaping
Climate adaptive design	Adapting infrastructure to climate change	Elevated structures, thermal insulation, weather-resistant materials
Low-impact urban mobility	Sustainable urban transport systems minimizing environmental impact	Car-sharing programs, electric scooters, mobility-as-a-service platforms

Sources Abbas et al. (2024), Huang and Li (2024), Imamoglu (2024), Krasnodębski (2024) and Zhang et al. (2024)

fields, and waste recycling facilities (Ahmad et al. 2024; de Souza Mesquita et al. 2024; Feder-Kubis et al. 2024; Lei 2024; Standal and Ahlquist 2025). This interconnected network facilitates seamless communication between devices, allowing for efficient monitoring and control of traditionally manual and resource-intensive processes. For instance, IoT-powered smart meters in energy systems track electricity usage in real-time, helping consumers and utilities make informed decisions to conserve energy and reduce costs. Similarly, IoT sensors monitor water flow, pressure, and quality in water management, detecting leaks or contaminants early and enabling timely corrective actions.

Conversely, AI adds a layer of intelligence to the data collected by IoT systems. By leveraging machine learning algorithms and predictive analytics, AI can analyse vast amounts of data to identify patterns, predict outcomes, and recommend optimal solutions. In agriculture, AI-driven models use data from IoT devices, satellite imagery, and weather forecasts to optimise irrigation schedules, reduce water wastage, and enhance crop yields. AI also plays a pivotal role in energy management, where it predicts energy demand patterns and aligns renewable energy supply accordingly, ensuring grid stability and minimising reliance on fossil fuels. Furthermore, blockchain technology complements these advancements by providing a secure, transparent, immutable ledger for tracking and verifying data. In green supply chains, blockchain ensures the traceability of sustainable practices, allowing consumers to verify the environmental footprint of products. It also facilitates carbon credit trading by providing an auditable record of emissions reductions, incentivising industries to adopt cleaner technologies (Dabas et al. 2024; Dhumal et al. 2024; Oyefusi et al. 2024).

A critical application of these technologies is in **real-time monitoring and predictive analytics for resource optimisation**, essential for achieving sustainability goals. Real-time monitoring systems provide instantaneous feedback on energy, water, and material consumption, enabling users to identify inefficiencies and implement corrective measures. Powered by AI, predictive analytics further enhances this capability by forecasting future resource requirements based on historical data and trends. For example, in smart cities, predictive models analyse traffic patterns to optimise traffic light timings, reducing congestion and emissions. In industrial settings, predictive maintenance systems monitor machinery performance and anticipate failures, minimising downtime and reducing energy consumption associated with inefficient operations. Moreover, advanced analytics help forecast weather and environmental conditions, allowing renewable energy systems like solar and wind farms to adjust operations dynamically and maximise energy outpu (Outili et al. 2023; Yue et al. 2023; Zhu et al. 2024)t. These technologies collectively ensure that resources are used judiciously, waste is minimised, and sustainability targets are met.

Intelligent energy and water management systems are among the most impactful applications of digital technologies in green solutions. Smart or intelligent energy grids leverage IoT and AI to create a two-way communication network between energy producers and consumers. These grids integrate renewable energy sources such as solar and wind, manage energy storage systems, and optimise energy distribution to minimise losses. For instance, intelligent grids can store excess energy in batteries or redirect it to higher-demand areas during high renewable energy generation periods. They also allow for dynamic pricing models, encouraging consumers to shift their energy usage to off-peak times, thereby reducing strain on the grid and promoting efficient energy use. Intelligent energy management systems in residential and commercial buildings use IoT-enabled devices to control lighting, HVAC systems, and appliances, ensuring they operate only when needed and at optimal settings. These systems significantly reduce energy consumption and contribute to the broader decarbonisation goal.

Similarly, intelligent water management systems transform how water resources are utilised and conserved. IoT sensors installed in water distribution networks monitor flow rates, detect leaks and measure water quality in real time, ensuring efficient water delivery and minimising losses. In agriculture, precision irrigation systems use soil moisture sensors and weather data to determine the water needed for crops, reducing water wastage and enhancing agricultural productivity. AI-powered water treatment plants optimise purification processes by adjusting parameters like chemical dosing and filtration rates based on real-time water quality data. Smart water meters in households and industries also enable users to track their water consumption, identify patterns, and adopt conservation measures. These systems address the challenge of water scarcity and ensure equitable distribution and sustainable usage (Amoneit et al. 2024; Anitha and Beda 2024; Nahata et al. 2024).

The convergence of IoT, AI, and Blockchain has also enabled innovative applications such as microgrids and decentralised water management systems, which promote resilience and sustainability at the community level. Microgrids are localised energy networks that operate independently or in conjunction with the primary grid, integrating renewable energy sources, energy storage, and intelligent controls. These grids enhance energy security, reduce transmission losses, and empower communities to adopt clean energy solutions. Similarly, decentralised water systems, such as rainwater harvesting and greywater recycling, utilise IoT and AI to optimise collection, storage, and treatment processes, reducing reliance on centralised infrastructure and promoting sustainable water use.

Integrating digital technologies into green solutions represents a paradigm shift in managing energy, water, and other critical resources. IoT, AI, and Blockchain form a powerful trio that enables real-time monitoring, predictive analytics, and transparent data management, driving efficiency and sustainability across sectors. Intelligent energy and water management systems exemplify these technologies' transformative potential, offering scalable and adaptable solutions to some of the most pressing environmental challenges. By embracing these advancements, industries, governments, and individuals can contribute to a sustainable future where resource optimisation and ecological preservation go hand in hand. As digital and green technologies evolve, their synergy will play an increasingly vital role in shaping a cleaner, more innovative, and more resilient world (Antenucci et al. 2022; Fiscal-Ladino et al. 2024; Lin et al. 2024).

6.5 Policy and Frameworks Supporting Green Technologies

6.5.1 Global Agreements and Initiatives (E.g., Paris Agreement, SDGs)

Urgent environmental issues and global sustainable development require international agreements and initiatives. The Paris Agreement and SDGs unite nations to fight climate change, protect ecosystems, and advance society and business. The 2015 Paris Agreement under the UNFCCC pledges to limit global warming to below 2 °C above pre-industrial levels to maintain the rise below 1.5 °C. Nearly 200 nations have agreed to ambitious greenhouse gas emission reduction goals in this legally binding agreement. Based on the "common but differentiated responsibilities principle," the framework recognises nations' diverse capacities and historical climate change contributions. To increase ambition, countries submit nationally determined contributions (NDCs) detailing their climate action strategies and should update them every five years. The Paris Agreement emphasises emission reductions, climate adaptation, financial aid for developing countries, and climate resilience.

The UN's 2015 Sustainable Development Goals (SDGs) complement the Paris Agreement by providing a comprehensive 2030 agenda to address global issues. The 17 Sustainable Development Goals and 169 targets aim to end poverty and inequality, promote peace, and protect the environment. SDG 13 (Climate Action), SDG 14 (Life Below Water), and SDG 15 (Life on Land) require nations to fight climate change, protect marine ecosystems, and preserve biodiversity. Assimilation of environmental goals with economic growth, social inclusion, and governance is a component of the SDGs. Renewable energy (SDG 7) reduces carbon emissions, creates jobs (SDG 8), and improves health by reducing air pollution (SDG 3). This comprehensive approach emphasises cohesive solutions to ensure everyone participates in the sustainability shift. In addition to these main initiatives, many international agreements and frameworks support the global environmental effort. The Kyoto Protocol preceded the Paris Agreement, set emission reduction targets for developed nations and laid the groundwork for international climate negotiations. The Montreal Protocol, which eliminated ozone-depleting substances, shows how global collaboration can solve environmental problems. The CBD and UNCCD address biodiversity loss, land degradation, and ecosystem restoration through sustainable land management. The Green Climate Fund (GCF) and Global Environment Facility (GEF) are crucial for funding environmental projects in developing nations. These funds promote renewable energy, climate-resilient agriculture, and ecosystem conservation to benefit at-risk communities from global sustainability initiatives. For carbon sequestration and biodiversity preservation, the Coalition of Rainforest Nations and REDD + prioritise forest conservation. Despite their potential, these agreements face political opposition, insufficient funding, and differences in country commitment. Nations like the US leaving and rejoining the Paris Agreement show the fragility of international climate action consensus. In addition, financial and technical differences often prevent developing countries from meeting their obligations, worsening inequality.

Resolving these issues requires improving capacity building, technology transfer, and equitable financing to foster collective responsibility and reciprocal advantage. Non-state actors like municipalities, corporations, and civil society organisations have boosted global environmental and development efforts. After the provisional U.S. exit from the Paris Agreement, the We Are Still In coalition shows how subnational and private-sector leadership can promote climate action. Businesses are adopting SDGs through sustainability strategies, green innovations, and carbon–neutral commitments, demonstrating the importance of public–private partnerships. The Paris Agreement and Sustainable Development Goals (SDGs) demonstrate humanity's commitment to environmental protection and sustainable development. These frameworks work best when implemented, collaborate globally, and involve everyone. The international community can overcome challenges and create a sustainable future by increasing commitments, mobilising resources, and encouraging innovation. These agreements are legal and policy tools and symbols of hope and unity in facing global existential threats.

6.5.2 National and Regional Policies Promoting Green Technology Adoption

By creating regulatory frameworks and financial incentives and encouraging innovation to promote sustainable practices, national and regional policies help green technology adoption. Global governments recognise the need to adopt sustainable technologies to combat climate change, pollution, and energy security. The Paris Agreement and SDGs often match international commitments. They consider each country's socio-economic and environmental conditions (Fegade 2024; Oyefusi et al. 2024; Standal and Ahlquist 2025). These policies focus on energy, agriculture, transportation, and industry, promoting green technologies and ensuring sustainability benefits all communities.Renewable energy—solar, wind, hydro, and geothermal—has been a priority due to national policies. Governments encourage clean energy infrastructure investments with feed-in tariffs, renewable energy certificates, and tax credits. EU member states must meet renewable energy targets, encouraging regional adoption. The Jawaharlal Nehru National Solar Mission in India aims to make solar energy a significant part of the country's energy mix. Production and investment tax credits for renewable energy projects have boosted US wind and solar power generation. These policies reduce greenhouse gas emissions and stimulate economic growth by creating renewable energy jobs.

Energy efficiency is key to green technology policies. Governments promote energy-efficient appliances, industrial processes, and architectural designs through standards, labelling, and financial incentives. The Energy Efficiency Directive requires EU member states to save energy by retrofitting buildings and improving industrial energy use. Japan's Top Runner Program sets performance standards based on cutting-edge technology to encourage manufacturers to make energy-efficient products. These measures reduce energy use, consumer costs, and fossil fuel use, advancing sustainability goals. National and regional agriculture policies encourage precision farming, fertilisers, and sustainable irrigation systems to boost productivity and reduce environmental impact. Drip irrigation subsidies and IoT-based monitoring have transformed water management in arid regions. Organic farming policies and biodegradable mulching films reduce chemical inputs and soil degradation, improving agricultural sustainability. Government-supported initiatives in developing nations emphasise capacity building and technology transfer to help smallholder farmers adopt sustainable practices.

Green technologies benefit significantly from transport policies. Global governments are setting ambitious EV adoption goals, supported by subsidies for EV purchases, charging infrastructure investments, and long-term bans on internal combustion engine vehicles. Due to tax exemptions and incentives, Norway leads the world in electric vehicle adoption. China leads the world in electric vehicle production and sales thanks to its New Energy Vehicle initiative. Regional policies like the EU's vehicle CO_2 emission standards force manufacturers to innovate and create cleaner, more efficient models. Waste management and circular economy initiatives are growing in national and regional policies. Mandates for waste segregation, landfill taxes, and extended producer responsibility (EPR) frameworks promote recycling, composting, and waste-to-energy technologies. Germany has high recycling rates due to strict policies and public participation, while Japan excels in waste separation and resource reclamation. Regional initiatives, like the EU's Circular Economy Action Plan, promote product reuse and recycling to reduce supply chain waste.

Governments fund R&D to accelerate green technology adoption and innovation ecosystems. Grants, loans, and public–private partnerships help green tech startups and businesses—the US Department of Energy funds research for renewable energy, advanced battery, and carbon capture. The European Innovation Council and Japan's Green Innovation Fund prioritise green technology innovation funding. National policies prioritise education and skill development to train workers for a green economy. National green technology strategies also require financial instruments like green bonds and climate funds. India, China, and Brazil have used the Green Climate Fund to finance renewable energy and sustainable urban development. Domestic policies often benefit green technology companies by providing concessional loans, guarantees, and subsidies.

Green technology adoption and sustainable development depend on national and regional policies. Governments encourage adopting eco-friendly technologies by aligning economic incentives, regulatory frameworks and innovation support. These policies address urgent environmental issues while fostering economic growth, social equity, and resilience for a sustainable future. National and regional policies will promote a greener, more sustainable world as nations cooperate and innovate.

6.5.3 Role of Public–Private Partnerships in Driving Innovation

Public–private partnerships (PPPs) are essential for fostering innovation, particularly in tackling intricate global issues such as climate change, resource scarcity, and sustainable development. These collaborations merge the public sector's regulatory and financial frameworks with the private sector's agility, technical proficiency, and investment capacity, generating a synergy that promotes innovative solutions. Public–private partnerships facilitate the distribution of risks, resources, and responsibilities, thereby fostering the advancement and implementation of innovative technologies that may otherwise be impractical due to elevated costs or market volatility. Public–private partnerships (PPPs) have become essential instruments in green technologies and sustainable practices, effectively addressing financing, research, and implementation deficiencies while fostering inclusive growth and enduring resilience.

Public–private partnerships play a crucial role in advancing clean energy technologies. Governments frequently need technical proficiency and market insight to commercialise renewable energy initiatives. Simultaneously, private enterprises may be reluctant to invest in high-risk endeavours absent policy backing or assured returns. Collaborative efforts between public and private sectors can expedite the implementation of solar, wind, and other renewable energy sources (Assen et al. 2024; Anitha and Beda 2024; Wang and Zhang 2024). Partnerships such as the U.S. Department of Energy's SunShot Initiative have engaged with private enterprises to decrease the cost of solar energy and enhance its accessibility for consumers. Likewise, the Clean Energy Finance Corporation in Australia utilises public funds to mitigate risks associated with private investments in renewable energy, facilitating consistent advancement towards national decarbonisation objectives. These collaborations foster innovation throughout the value chain, encompassing research and development (R&D) to extensive implementation.

Public–private partnerships (PPPs) significantly contribute to urban infrastructure and the advancement of sustainable city development. Global cities implement innovative urban concepts that utilise digital technologies to improve resource efficiency and sustainability. Initiatives like the European Union's Smart Cities Marketplace unite governments, businesses, and research organisations to collaboratively devise solutions for urban challenges. Initiatives related to intelligent traffic management, waste-to-energy systems, and sustainable building technologies frequently depend on public–private partnerships to distribute financial responsibilities and expertise. The Smart Cities Mission in India illustrates how public–private partnerships can enhance urban infrastructure by combining private-sector innovation with public financing to provide energy-efficient transport, water conservation systems, and digital governance solutions (Fiscal-Ladino et al. 2024; Huang and Li 2024; Oyefusi et al. 2024).

A pivotal area where PPPs thrive is sustainable agriculture. In areas dependent on conventional farming practices, collaborations among governments, private agrotech companies, and research institutions have implemented precision agriculture, IoT monitoring systems, and bio-fertilisers. Organisations such as the Alliance for a Green Revolution in Africa (AGRA) collaborate with public entities and private enterprises to grant farmers access to contemporary technologies and financial resources, enhancing productivity while maintaining environmental sustainability. These collaborations improve food security and diminish the ecological impact of agriculture.

Public–private partnerships have been pivotal in promoting innovative water and waste management solutions. Public entities establish the foundational infrastructure, whereas private companies implement wastewater recycling, desalination, and waste-to-energy conversion technologies. In partnership with private enterprises through the Public Utilities Board, the Singaporean government has developed innovative water recycling systems such as NEWater, effectively tackling water scarcity while maintaining high-efficiency standards. Waste management systems in Sweden and the Netherlands illustrate public–private partnerships, wherein private enterprises function under governmental regulations to attain elevated recycling rates and transform waste into renewable energy.

The transport sector derives substantial advantages from public–private partnerships. As governments establish ambitious objectives for decarbonising transport systems, private enterprises are intervening to innovate electric vehicles (EVs), charging infrastructures, and low-emission logistics solutions. The EV100 initiative unites businesses and policymakers to facilitate the global transition to electric cars. Private innovation in vehicle design and battery technology enhances public investment in infrastructure, including charging stations, promoting the swift adoption of cleaner transportation solutions.

Public–private partnerships (PPPs) promote research and development (R&D) in advanced green technologies. Public funding underpins foundational research, whereas private sector participation enhances commercialisation and scalability. Joint initiatives in carbon capture and storage (CCS), advanced battery technologies, and sustainable materials underscore the efficacy of collective endeavours in attaining significant innovations. For example, initiatives such as Mission Innovation—a global alliance of governments and enterprises—seek to double investments in clean energy research and development, illustrating the efficacy of collaboration in addressing climate change (Huang and Li 2024; Kaya et al. 2024; Sridhar et al. 2024; Zhang et al. 2024).

Moreover, PPPs facilitate capacity building and knowledge transfer, guaranteeing that innovations are disseminated to underserved communities and developing countries. Public–private partnerships (PPPs) utilise the resources and expertise of both sectors to offer training programs, financing access, and tailored solutions that meet specific requirements. Partnerships between multinational corporations and African governments have implemented off-grid solar systems in rural regions, supplying electricity and diminishing dependence on fossil fuels.

Notwithstanding their potential, PPPs encounter challenges, including the alignment of priorities, the assurance of equitable benefit-sharing, and the maintenance of transparency. The efficacy of such collaborations relies on explicit agreements, robust governance frameworks, and reciprocal accountability. Governments must establish conducive environments via supportive policies and incentives, while private entities must emphasise long-term sustainability rather than short-term profits. When effectively organised, PPPs stimulate innovation and guarantee the equitable distribution of its advantages, promoting social, economic, and environmental advancement.

Public–private partnerships (PPPs) are essential for fostering innovation in green technologies and sustainable solutions. They merge the public sector's dedication to social welfare with the private sector's efficiency and technological proficiency, effectively addressing intricate challenges neither sector can resolve independently. Collaborations in renewable energy, urban development, agriculture, and waste management facilitate a sustainable future, illustrating that shared responsibility and cooperation are essential for transformative progress (Table 6.4).

6.6 Challenges and Barriers

6.6.1 Economic Constraints and Investment Requirements

Economic constraints and investment requirements are critical in adopting and scaling green technologies. Transitioning to sustainable practices often requires significant upfront capital, posing challenges for governments, businesses, and individuals. The high costs associated with research and development (R&D), infrastructure deployment, and technology adoption can create barriers, particularly in developing nations with limited financial resources (Rani et al. 2024; Verma et al. 2024). Furthermore, the economic landscape is shaped by market volatility, competing budgetary priorities, and inadequate access to financing mechanisms, making it challenging to meet the investment demands of green innovation. Addressing these economic constraints necessitates a multi-faceted approach involving policy reforms, innovative financial instruments, and international cooperation.

One of the primary economic constraints is the initial cost of implementation for green technologies. Solar panels, wind turbines, electric vehicles (EVs), and energy-efficient infrastructure often involve higher upfront expenditures than conventional alternatives. For example, renewable energy sources offer long-term cost savings, but installation and setup costs can be prohibitive for small businesses, households, or governments in resource-constrained regions. Similarly, transitioning to sustainable agriculture through technologies like precision farming and bio fertilisers requires farmers to invest in new equipment and inputs, which may be unaffordable without subsidies or loans (Antenucci et al. 2022; Abbas et al. 2024; Zhang et al. 2024; Standal and Ahlquist 2025). These economic hurdles are exacerbated in countries with limited access to affordable credit or financial incentives for green investments.

In addition to initial costs, operational and maintenance expenses can be challenging for many stakeholders. Advanced technologies such as carbon capture and storage (CCS), smart grids, and water purification systems require skilled personnel,

Table 6.4 Different policies and respective framework for green chemistry and its key details

Policy/framework	Focus	Example/key details
Paris agreement (UNFCCC)	International treaty aiming to limit global warming and promote green technologies for climate action	Countries committed to net-zero carbon emissions by mid-century, enhancing clean energy and green tech
UN sustainable development goals (SDGs)	Global framework promoting sustainability across all sectors, including green technologies	SDG 7 (Affordable and Clean Energy), SDG 9 (Industry, Innovation, and Infrastructure), SDG 13 (Climate Action)
EU green deal	Roadmap for making the EU's economy sustainable by 2050, promoting green tech and circular economy	We aim for carbon neutrality by 2050, boosting renewable energy and investing in clean technologies
Clean energy standard (US)	U.S. policy aimed to increase the share of clean energy sources and reduce carbon emissions	They introduced federal and state-level renewable energy targets, tax credits, and funding for green tech innovation
Green new deal (US)	Proposal to create green jobs and tackle climate change with green technology innovation	Focus on clean energy, infrastructure improvements, and sustainable agriculture
International energy agency (IEA) framework	Promotes global energy policies and clean energy technology innovations	Annual reports on renewable energy technologies and energy efficiency strategies
Clean development mechanism (CDM)	Allows industrialised countries to invest in emission-reducing projects in developing countries	Supports low-carbon technology transfer and capacity building for green innovations in developing nations
Global green growth institute (AGI)	Multilateral organisation promoting green growth policies and technologies	Supports policies that enhance green technologies across energy, water, agriculture, and waste management
Circular economy action plan (EU)	Promotes a circular economy focusing on resource efficiency and reducing waste	Encourages the use of renewable materials, product lifespan extension, and eco-design
Renewable energy directive (EU)	An EU law to promote the use of renewable energy in the EU by setting binding targets	Targets for renewable energy sources, including wind, solar, and bioenergy, with funding for clean technology
Carbon pricing mechanisms	Tools for pricing carbon to encourage the adoption of low-carbon technologies	Cap-and-trade programs, carbon taxes, and carbon credits to incentivise green innovation

(continued)

Table 6.4 (continued)

Policy/framework	Focus	Example/key details
National bioenergy policy (India)	Policy framework supporting the production and use of bioenergy technologies	Promotes using biomass, biogas, and biofuels to meet energy demands sustainably
Global climate change act (US)	U.S. framework to address climate change through policy actions, including green technology support	Sets greenhouse gas emission targets, supporting green tech R&D and clean energy development
Energy efficiency directive (EU)	EU directive aiming to improve energy efficiency across sectors with support for green technologies	Mandatory energy-saving measures for companies and public bodies promote energy-efficient technologies
Environmental protection agency (EPA) regulations	U.S. agency regulations that enforce environmental standards and promote sustainable technologies	Enforces air and water quality policies, chemical safety, and waste management, promoting green tech solutions
Green investment bank (UK)	Establishes a bank to invest in green technologies and sustainable infrastructure	Provides funding for renewable energy, energy efficiency, and green infrastructure projects
Green technology fund (Singapore)	National fund to support research, development, and commercialisation of green technologies	Provides financial support for clean energy, waste management, and water technologies
Ecodesign directive (EU)	EU framework for improving the environmental performance of products throughout their life cycle	Focuses on energy efficiency, material use reduction, and recycling during product design
Global environment facility (GEF)	International organisations support environmental projects through funding and green technology development	Funds projects focusing on biodiversity, climate change, and sustainable resource management
World bank's green growth initiative	Promotes sustainable growth by supporting green technologies in developing countries	Provides loans and grants for clean energy, sustainable agriculture, and climate resilience projects
Sustainable energy for all (SEforALL)	UN initiative focuses on ensuring universal access to sustainable energy	Focus on renewable energy technologies, energy efficiency, and innovative green tech solutions
The Kyoto protocol	International agreement to reduce greenhouse gas emissions, supporting green technologies	Mechanisms like carbon trading and emissions reduction projects encourage the adoption of sustainable technologies

(continued)

Table 6.4 (continued)

Policy/framework	Focus	Example/key details
Japan's green innovation fund	A Japanese government fund to foster green tech development in clean energy and environment	Funds R&D for carbon–neutral technologies and supports innovation in green industries
China's 14th five-year plan for ecological civilization	National framework promoting green technologies for sustainable development	Focuses on clean energy, environmental protection, and green innovation in critical sectors like transportation
Clean energy ministerial	International initiative promoting the adoption of clean energy technologies worldwide	A collaborative effort between governments, businesses, and non-profits to support green energy solutions
Green growth and development framework (OECD)	Promotes green growth policies and technology in member countries for sustainable economic development	Focuses on innovation, investment in green tech, and creating sustainable markets
Net-zero emissions Act (Australia)	Australian policy targeting net-zero emissions through green technology adoption	Incentivises low-carbon technologies, clean energy, and industrial decarbonisation
Zero waste international alliance (ZWIA)	Advocates for zero-waste policies and technologies to minimise waste generation	Encourages recycling technologies, composting, and sustainable waste management systems
Low carbon economy transition act (Canada)	Canadian policy supporting the transition to a low-carbon economy through green tech adoption	Targets emission reductions, clean energy investments, and green tech innovation funding
Sustainable transport policy (EU)	Policy framework for reducing emissions from the transport sector with support for green tech	Promotes electric vehicles, alternative fuels, and green transport technologies
Circular economy promotion law (China)	The law promotes circular economy practices and the adoption of green technologies in the industry	Focus on reducing waste, increasing recycling, and supporting green innovation in manufacturing
EU emissions trading system (EU ETS)	A cap-and-trade system to reduce industrial emissions and promote clean technologies	Companies trade carbon allowances, promoting investments in emission-reducing green technologies
Carbon clean tech fund (India)	Investment fund supporting clean technology innovation in the Indian energy sector	Focuses on funding projects for carbon capture, renewable energy, and energy efficiency

(continued)

Table 6.4 (continued)

Policy/framework	Focus	Example/key details
International renewable energy Agency (IRENA)	International organisation supporting the deployment and development of renewable energy technologies	Focuses on wind, solar, hydropower, and bioenergy technologies globally
Net zero by 2050 strategy (UK)	UK government strategy for achieving net-zero emissions by 2050 through green technologies	Sets sector-specific goals for decarbonisation, encouraging innovation in clean tech and sustainable solutions
Federal renewable energy incentives (US)	US policy framework to support the transition to renewable energy through tax credits and grants	Includes incentives for solar, wind, and bioenergy technologies, supporting green tech adoption
Green public procurement (EU)	EU policy promotes the purchase of environmentally friendly goods and services	Encourages the use of green technologies in public contracts, promoting sustainability in procurement
India's national action plan on climate Change	The Indian government plans to address climate change, including deploying green technology	Focus on renewable energy, energy efficiency, and green technologies in agriculture and infrastructure
Renewable energy expansion law (Germany)	German law promotes the use of renewable energy sources through incentives and support for green technology	Guarantees feed-in tariffs for renewable energy producers, promoting clean tech innovation
Green growth strategy (Republic of Korea)	National policy framework supporting eco-friendly industrialisation and green technology	Focuses on clean energy, green industries, and promoting green technologies for economic growth
Brazil's national policy on climate change	Brazilian framework targeting reductions in emissions through sustainable development practices	Focus on clean energy, deforestation reduction, and green technologies for industrial applications
World trade organization (WTO) environmental trade policy	WTO framework promoting environmentally friendly trade policies that support green technologies	Focuses on removing barriers to green tech trade and encouraging innovation through international collaboration
European investment bank (EIB) green bond	The EU uses financial instruments to fund green projects, including clean technologies	Financing for renewable energy, sustainable agriculture, and green infrastructure
Green bonds (Global)	Financial products to raise funds for environmental projects and green technology adoption	Supports financing for energy efficiency, renewable energy, and other sustainable infrastructure

(continued)

Table 6.4 (continued)

Policy/framework	Focus	Example/key details
Corporate social responsibility (CSR) guidelines	Business policies promoting sustainability, including green tech investment and eco-friendly practices	Companies are encouraged to integrate green technologies in their operations and production
E-waste management rules (India)	Regulations for managing electronic waste, promoting the recycling and safe disposal of electronics	Encourages the use of green technologies for recycling e-waste and reducing hazardous materials
National clean energy fund (India)	Funding body for clean energy research, development, and commercialisation of green technologies	Supports innovative energy solutions, like solar power, wind, and energy storage technologies
Bioenergy policy (Brazil)	Policy framework encouraging bioenergy sources to reduce dependence on fossil fuels	It focuses on biofuels and biomass for energy generation and supports green technologies in agriculture and energy
Green hydrogen strategy (Germany)	German government strategy to develop green hydrogen technologies for industrial and transport sectors	Focus on hydrogen production through renewable energy and its use in various sectors
Sustainable development act (UK)	Legislation supporting sustainable development across sectors, including green technology	Encourages investment in low-carbon technologies and the reduction of carbon emissions across industries
US department of energy's office of energy efficiency and renewable energy (EERE)	U.S. government office supporting green energy innovation through grants and funding	Provides financial support for energy efficiency, renewable energy, and green technology development projects
Carbon neutral cities alliance (CNCA)	Alliance of Global Cities working to accelerate the adoption of green technologies for carbon neutrality	Focuses on urban sustainability, clean energy, and green infrastructure technologies in cities
Green manufacturing policy (South Korea)	South Korea's policy supports clean manufacturing technologies and sustainable industrial practices	Promotes energy efficiency, waste reduction, and green technologies in industrial processes
Renewable energy deployment act (EU)	EU law to enhance the deployment of renewable energy systems and technologies	Encourages renewable energy projects and technology development across member states
Sustainable urban development policy (China)	Policy supporting sustainable urban planning and development with green technologies	Focuses on energy-efficient buildings, green transportation, and eco-friendly urban planning solutions

(continued)

Table 6.4 (continued)

Policy/framework	Focus	Example/key details
Carbon-free transportation act (US)	U.S. policy to promote carbon-free transportation technologies such as electric vehicles	Incentives for electric vehicle adoption, charging infrastructure, and low-emission vehicles
Environmental sustainability framework (Australia)	National framework promoting sustainable environmental practices, including green technologies	Encourages eco-friendly practices in agriculture, energy, and urban development
Zero emissions vehicle regulation (California, US)	U.S. policy promotes the adoption of zero-emissions vehicles through regulations and incentives	Focus on electric vehicle adoption, infrastructure development, and reducing transportation emissions
Carbon capture and storage (CCS) regulation (EU)	EU regulations promote the development and deployment of carbon capture technologies	Focus on funding CCS projects to reduce industrial emissions through green technologies
World bank green technology strategy	Strategy to promote green technology solutions for climate mitigation and sustainable development	Focus on renewable energy, energy efficiency, and low-carbon technologies globally

Sources Assen et al. (2024), de Oliveira et al. (2024), Wang and Zhang (2024) and Zheng et al. (2024)

regular maintenance, and sophisticated monitoring, which can strain budgets over time. While green technologies are designed to reduce long-term costs through energy savings, waste reduction, and improved efficiency, the lack of immediate returns may discourage investments, particularly for small and medium-sized enterprises (SMEs) and low-income communities.

Market dynamics and global inequalities also influence economic constraints. In many developing nations, subsidies for fossil fuels and conventional energy sources continue to distort market incentives, making green alternatives less competitive. Additionally, the global distribution of investment flows often favours developed countries, leaving emerging economies with inadequate funding to build their green infrastructure. For instance, while renewable energy investments have surged in regions like Europe and North America, many sub-Saharan African and Southeast Asia countries still need help to secure financing for similar projects. These disparities underscore the need for more equitable resource allocation and international support to address global sustainability challenges.

Substantial investment requirements must be met through innovative financing mechanisms and supportive policies to overcome these economic barriers. Green bonds, climate funds, and public–private partnerships (PPPs) are emerging as practical tools for mobilising capital. For example, the Green Climate Fund provides financial assistance to developing nations for climate adaptation and mitigation projects. At the same time, PPPs bring together public resources and private expertise to share costs and risks (Ahmad et al. 2024; Fiscal-Ladino et al. 2024; Oyefusi et al. 2024). Additionally, carbon pricing mechanisms, such as carbon taxes and cap-and-trade systems, create economic incentives for businesses to adopt greener practices by internalising the environmental costs of pollution.

Governments are crucial in mitigating economic constraints through policy interventions and fiscal measures. Subsidies, tax breaks, and grants for renewable energy, electric vehicles, and sustainable agriculture can lower the cost of adoption, making green technologies more accessible. For instance, the U.S. Inflation Reduction Act of 2022 includes significant investments in clean energy, EV incentives, and carbon capture technologies, providing a blueprint for other nations. Similarly, targeted financial support for SMEs, rural communities, and underrepresented groups ensures that the benefits of green investments are distributed equitably, fostering inclusive economic growth (Nahata et al. 2024; Verma et al. 2024).

International cooperation is also essential to address global investment gaps and economic disparities. Multilateral agreements, such as the Paris Agreement, encourage developed nations to provide financial and technical assistance to less developed countries, ensuring a collective effort toward sustainability. Global financial institutions, including the World Bank and International Monetary Fund, can facilitate low-interest loans and green project grants while fostering knowledge-sharing and capacity-building initiatives.

Economic constraints and investment requirements present significant challenges to adopting green technologies. However, policy support, financial innovation, and international collaboration can address these barriers effectively. Governments, businesses, and global institutions must collaborate to mobilise the necessary resources, reduce upfront costs, and create enabling environments that encourage green investments. By overcoming these economic hurdles, societies can unlock the transformative potential of green technologies, paving the way for a sustainable and resilient future (Amoneit et al. 2024; Rani et al. 2024).

6.6.2 Technological Limitations and Scalability Issues

Technological constraints and scalability challenges present substantial obstacles in multiple fields, impeding the implementation and widespread acceptance of innovative solutions. A principal area for improvement is the discrepancy between prototype performance and practical application. Although numerous technologies operate efficiently in controlled settings, their scalability frequently encounters obstacles from infrastructural deficiencies, resource limitations, or difficulties in integrating with current systems. In the renewable energy sector, advanced battery technologies for energy storage may excel in laboratory conditions but encounter challenges in scalability due to elevated manufacturing costs, restricted raw material availability, and the environmental repercussions of large-scale production (Assen et al. 2024; Fegade 2024; Imamoglu 2024). Likewise, precision agriculture technologies, including IoT-enabled sensors and automated machinery, present significant opportunities for enhancing crop yields and resource efficiency. Nevertheless, their extensive implementation could be improved by the substantial installation costs, insufficient technical proficiency among farmers, and connectivity challenges in isolated regions. Scalability faces challenges in nascent domains like artificial intelligence and machine learning, where the computational power and data storage needed for training intricate models frequently surpass the capacities of current systems. The expenses associated with operating extensive data centres and apprehensions regarding data privacy and energy usage intensify these challenges. Moreover, scalability challenges are intrinsically linked to interoperability, as systems developed in isolation frequently need help to integrate smoothly with others, resulting in deployment bottlenecks. Although blockchain technology is groundbreaking for secure transactions, it faces challenges in scalability due to the substantial computational overhead necessary for processing a large volume of transactions. This constraint is especially apparent in finance or supply chain management sectors, where instantaneous processing is essential. Technological constraints are evident in accessibility, as innovations frequently cater to developed markets, disregarding the specific needs of developing regions (Assen et al. 2024; de Oliveira et al. 2024; Ibrahim et al. 2024). This disparity is apparent in healthcare, where sophisticated diagnostic tools or telemedicine solutions are unavailable to rural or economically disadvantaged populations due to insufficient infrastructure or excessive costs. Moreover, scalability challenges frequently entail trade-offs among efficiency, cost, and environmental consequences. Attaining scalability often necessitates optimising processes or implementing mass production methods, potentially undermining quality or sustainability. Increasing production to satisfy global demand for electric vehicles requires extracting rare earth materials such as lithium and cobalt, which raises issues regarding resource depletion and ethical sourcing practices. The swift progression of technological development results in obsolescence, wherein previous systems fail to accommodate newer technologies, exacerbating scalability challenges. Cybersecurity represents a crucial domain where technological constraints hinder scalability. As digital systems increase, the attack surface enlarges, necessitating robust security measures that frequently trail behind emerging threats. These constraints require a multidisciplinary strategy for scalability, highlighting the need for collaboration among stakeholders, governmental involvement, and advancing adaptive technologies. Confronting these challenges necessitates significant investment in research and development, promoting innovation while guaranteeing inclusivity and sustainability. Emphasising modularity, resource efficiency, and user-centred design enables the surmounting of technological constraints and the development of scalable solutions that address varied requirements in global markets.

6.6.3 Social and Cultural Resistance to Adoption

Social and cultural opposition to adoption constitutes a substantial obstacle to the assimilation and broad acceptance of new technologies, practices, and innovations. This resistance arises from entrenched beliefs, customs, and societal norms that influence individual and collective perspectives. Change frequently disrupts established lifestyles, resulting in scepticism and resistance, particularly when novel practices confront cultural values or existing power dynamics. In agricultural communities, implementing genetically modified crops or advanced farming technologies may encounter scepticism due to concerns regarding environmental damage, erosion of traditional farming practices, or economic reliance on large corporations. In healthcare, opposition to telemedicine or AI-based diagnostic tools frequently stems from a preference for in-person consultations with trusted medical practitioners, exacerbated by apprehensions regarding data privacy and confidence in digital platforms (Feng et al. 2024; Fiscal-Ladino et al. 2024; Lin et al. 2024).

Cultural resistance is apparent in the workplace, where digital transformation initiatives may encounter opposition from employees who perceive automation as a threat or feel inundated by the necessity to acclimatise to new technologies. This resistance frequently stems from apprehension regarding job displacement or a perceived decline in human interaction and creativity. Moreover, older generations may need help adopting digital tools due to insufficient familiarity or comfort with technology, resulting in a generational divide that exacerbates implementation challenges (Assen et al. 2024; de Oliveira et al. 2024; Wang and Zhang 2024). In education, online learning platforms and digital classrooms frequently face opposition from parents, teachers, and students who prioritise traditional in-person learning methods, viewing technology as a replacement rather than an enhancement to human instruction.

Religious and cultural convictions significantly influence resistance. Innovations like reproductive technologies, vaccines, and genetically modified foods are occasionally contested due to ethical considerations or conflicts with religious beliefs. Vaccination campaigns in certain regions have encountered opposition due to misinformation or misconceptions regarding their compatibility with spiritual practices. Renewable energy initiatives, such as wind farms or solar installations, may face resistance from Indigenous communities if viewed as infringing upon sacred territories or local ecosystems (Outili et al. 2023; Zhu et al. 2024).

The digital divide intensifies social resistance, as unequal access to technology fosters exclusion and mistrust within marginalised communities. Individuals who perceive themselves as marginalised by technological progress may regard innovations as instruments of domination or exploitation by more affluent factions. This perception is widespread in developing countries, where technology advantages are inequitably allocated, exacerbating existing disparities. Moreover, the velocity of technological advancement frequently surpasses society's capacity to adapt, resulting in a need for more comprehension and acceptance (Hussain et al. 2024; Roveda et al. 2024).

To mitigate social and cultural resistance, prioritising community engagement and inclusive dialogue is imperative. Educational initiatives elucidating technology and emphasising its advantages in culturally pertinent manners can cultivate trust and acceptance. Policymakers and innovators must consider local contexts by engaging stakeholders in the design and implementation process to ensure that solutions are congruent with cultural values and societal requirements. Addressing the digital divide through equitable access and capacity-building initiatives can alleviate resistance, enabling communities to accept change. Ultimately, surmounting social and cultural resistance necessitates empathy, collaboration, and a dedication to confronting the fundamental fears and apprehensions that fuel opposition (Afonso et al. 2024; Fiscal-Ladino et al. 2024; Oyefusi et al. 2024).

6.6.4 Regulatory Hurdles and Gaps in Enforcement

Regulatory obstacles and deficiencies in enforcement pose considerable challenges to developing, deploying, and adopting innovative technologies, practices, and systems across diverse sectors. These challenges frequently stem from inflexible or antiquated regulatory systems that need help to keep pace with swift advancements, posing obstacles for innovators and enterprises aiming to implement groundbreaking solutions. In biotechnology, renewable energy, or artificial intelligence sectors, the regulatory framework must often address emerging risks sufficiently or offer clear guidelines, resulting in uncertainty and potential delays. Innovators frequently encounter protracted approval procedures, varying standards, or contradictory regulations across different regions, which can hinder creativity and deter investment.

The absence of cohesive global regulatory standards is a significant challenge in our current environmental landscape. In sectors such as pharmaceuticals or electronics, varying regulations among nations can create challenges in the production and distribution of products. A company might need to adhere to a specific set of rules in its domestic environment while navigating a different framework in international markets, which can elevate expenses and complicate logistical operations (Abbas et al. 2024; Dhumal et al. 2024; Zhang et al. 2024). This poses significant challenges for small and medium-sized enterprises (SMEs), which often need more resources to manage intricate regulatory environments. Similarly, the landscape of data privacy is marked by diverse legal frameworks, including the robust General Data Protection Regulation (GDPR) of the European Union and the comparatively lenient regulations found in other areas, posing significant compliance challenges for multinational corporations.

Insufficient enforcement exacerbates these challenges, as even thoughtfully crafted regulations may only fulfil their intended purposes with adequate implementation and supervision. For instance, inadequate enforcement measures can compromise environmental regulations to reduce emissions, enabling industries to continue polluting without substantial consequences. This situation frequently arises from insufficient financial resources, a shortage of skilled professionals, or political meddling, which can undermine public confidence in regulatory bodies. In developing nations, enforcement challenges are notably significant, as inadequate infrastructure and resources hinder the ability to monitor compliance and tackle violations effectively. This fosters a detrimental cycle, where businesses might use weak regulations to reduce expenses, jeopardising equitable competition and long-term ecological objectives (Krasnodębski 2024; Zhang et al. 2024; Irianto et al. 2025).

Regulatory challenges arise from the careful stance that authorities take regarding new technologies. While this caution is reasonable—considering the potential risks linked to innovations such as autonomous vehicles, artificial intelligence, or gene editing—it may result in excessively restrictive policies that impede advancement. Regulatory uncertainty surrounding the use of drones has hindered their integration into agriculture, logistics, and disaster management, even though their advantages are well-documented. Furthermore, regulatory frameworks often struggle to keep pace with rapid technological progress, resulting in ambiguous situations where innovations function beyond the scope of current legal structures (Amoneit et al. 2024; Assen et al. 2024; Feng et al. 2024). This situation may result in unmanaged risks or innovators' need to pause their activities until regulatory frameworks are established.

Confronting these challenges necessitates a multifaceted strategy. Regulators must actively collaborate with industry stakeholders, researchers, and the public to create adaptive and forward-looking policies. Necessary actions include optimising approval procedures, promoting global partnerships for unified standards, and enhancing enforcement frameworks. Regulatory sandboxes provide a unique opportunity to explore innovations within a controlled setting, allowing for experimentation while maintaining a degree of oversight. Clear communication and responsibility are crucial for fostering public confidence and guaranteeing that regulations benefit the community while allowing innovation to thrive (Dash et al. 2024; Ibrahim et al. 2024; Roveda et al. 2024).

6.7 Case Studies and Success Stories

6.7.1 Green Technology Initiatives in Agriculture (E.g., Sustainable Farming in India)

Initiatives focused on environmentally friendly technology in agriculture, especially in a nation such as India with its extensive rural economy, are crucial for attaining sustainable development, improving productivity, and tackling the challenges presented by climate change. These initiatives involve various practices and innovations designed to minimise the ecological footprint of agriculture while sustaining or enhancing productivity to satisfy the needs of an expanding population. A crucial aspect to consider is the implementation of precision farming methods that utilise cutting-edge technologies like GPS, IoT sensors, and drones to enhance resource efficiency. By offering immediate insights into soil vitality, moisture content, and crop status, these innovations empower farmers to make educated choices, thereby minimising the excessive application of fertilisers, pesticides, and water resources. Comparably, India has witnessed the advancement of organic farming as a sustainable option, bolstered by initiatives such as the Paramparagat Krishi Vikas Yojana (PKVY), which fosters traditional agricultural methods devoid of synthetic inputs, thereby improving soil health and minimising chemical runoff. A vital aspect of sustainable agriculture practices involves the integration of renewable energy sources, such as solar-powered irrigation systems (Addo-Bankas et al. 2024; de Souza Mesquita et al. 2024; Kumar and Chand 2024; Lei 2024). This is especially pertinent in regions like India, where rural areas often face challenges in accessing reliable electricity. Solar pumps offer an eco-friendly water solution, decreasing dependence on fossil fuels, minimising carbon emissions, and cutting energy expenses for agricultural practitioners. Furthermore, advancements in biofertilisers and biopesticides are reshaping nutrient management, providing sustainable alternatives to traditional agrochemicals. Microbial formulations, including those based on Trichoderma or neem, are increasingly recognised for enhancing crop resilience against pests and diseases, all while fostering the vitality of soil ecosystems (Erinle et al. 2021; Izah et al. 2022, 2023). Incorporating waste management practices into agricultural systems, mainly via bioenergy and composting technologies, is a prime example of sustainable initiatives. Initiatives that transform agricultural byproducts into biogas or biochar minimise waste and offer farmers an extra source of energy or soil enhancements, thereby fostering a more sustainable farming ecosystem. India has adopted climate-smart agricultural practices to tackle the challenges of unpredictable weather patterns, water scarcity, and land degradation (Amabie et al. 2024). Practices like conservation agriculture, characterised by minimal soil disturbance, crop rotation, and cover cropping, play a crucial role in maintaining soil structure, enhancing biodiversity, and boosting water retention (Amoneit et al. 2024; Nahata et al. 2024; Roveda et al. 2024; Verma et al. 2024). Agroforestry represents a sustainable approach incorporating trees within agricultural landscapes, providing advantages like carbon capture, enhanced soil health, and supplementary revenue from fruits or timber. Technology-driven platforms and mobile applications significantly improve the ability of farmers to access vital information such as weather forecasts, market prices, and sustainable farming practices. Initiatives such as eNAM (National Agriculture Market) and applications that advocate for drip irrigation systems are effectively closing the knowledge gap and fostering the uptake of sustainable technologies. Despite these advancements, challenges persist, such as the necessity for enhanced awareness, increased financial backing, and improved infrastructure to effectively scale these initiatives. Collaborative efforts among governmental entities, private sector companies, and non-profit organisations are essential to tackle these deficiencies. Innovative agriculture practices resonate with international objectives like the Sustainable Development Goals (SDGs) and have the potential to rejuvenate India's farming landscape by enhancing resilience, productivity, and a commitment to environmental care.

6.7.2 Innovative Renewable Energy Projects (E.g., Tesla's Powerwall, Wind Farms in Denmark)

Cutting-edge renewable energy initiatives are reshaping the worldwide energy paradigm, demonstrating pioneering technologies that diminish reliance on fossil fuels while tackling the pressing necessity for climate change alleviation. Among these, Tesla's Powerwall and Denmark's wind farms exemplify the profound impact of innovation in advancing the adoption of sustainable energy solutions. Tesla's Powerwall, a home battery system, signifies a groundbreaking advancement in energy storage. This system enables households and enterprises to accumulate electricity from renewable sources, like solar panels, for utilisation during peak demand or power outages. By facilitating decentralised energy management, the Powerwall significantly bolsters energy autonomy for users while simultaneously aiding in the stabilisation of the grid by mitigating the variances in renewable energy supply. The elegant design, coupled with cutting-edge lithium-ion battery technology, has rendered it both appealing and attainable for an increasing demographic of consumers globally, catalysing a transition towards autonomous energy systems.

Denmark's wind energy initiatives are a prominent illustration of exemplary leadership in the national adoption of renewable energy. Denmark is a global leader in wind power, where extensive onshore and offshore wind farms are integral to its energy strategy. The nation's dedication to wind energy is exemplified by initiatives such as the Horns Rev and Anholt offshore wind farms, which have established standards for efficiency and innovation. The sophisticated advancements in turbine technology, along with careful and strategic planning, have allowed Denmark to attain remarkable levels of wind energy integration on a global scale, frequently generating more electricity than it utilises. The excess energy generated is exported, enhancing regional energy security and solidifying Denmark's status as a frontrunner in renewable energy technology (Akinsemolu 2023; Yue et al. 2023; Amoneit et al. 2024; Feder-Kubis et al. 2024; Raza et al. 2024).

An exemplary instance of pioneering renewable energy initiatives is the emergence of floating solar farms. The installations, constructed upon reservoirs or lakes, effectively navigate land-use limitations while optimising the potential for solar energy generation. Nations such as Japan and Singapore have significantly invested in floating solar technology, showcasing its efficacy in areas with high population density. These systems mitigate water evaporation while enhancing solar panels' efficiency, owing to water's cooling properties, thus providing a dual advantage for energy generation and water preservation (Fegade 2024; Fiscal-Ladino et al. 2024; Huang and Li 2024; Lei 2024).

Innovations in hydropower, including applying kinetic turbines to river and ocean currents, underscore the possibilities for diversifying renewable energy sources. Initiatives such as Scotland's MeyGen tidal energy farm use the consistent nature of tidal flows to produce electricity, presenting a dependable alternative to more erratic renewable sources such as wind and solar. The recent developments underscore the significance of utilising various natural resources to achieve a balanced and robust energy portfolio.

Green hydrogen initiatives exemplify the advanced possibilities inherent in renewable energy. The generation of hydrogen through renewable electricity presents an opportunity for a sustainable fuel source applicable across various sectors, including industry, transportation, and energy storage solutions. For example, Germany's HyEnergy program and Australia's hydrogen hubs illustrate the potential of innovative collaborations and investments to facilitate the integration of this multifaceted energy carrier.

Incorporating digital technologies into renewable energy frameworks, including AI-driven grid management and blockchain-facilitated peer-to-peer energy trading, fundamentally alters energy production, distribution, and consumption processes. These initiatives refine operational effectiveness, enable consumer agency, and augment the scalability of renewable resources.

These groundbreaking renewable energy initiatives collectively exemplify the vast possibilities of integrating technology, policy, and creativity to forge a sustainable energy future (Kosoe et al. 2023; Obahiagbon and Ogwu 2023). They lay the groundwork for a more sustainable and environmentally friendly future, motivating worldwide initiatives to shift towards renewable energy sources.

6.7.3 Waste Management Solutions in Urban Areas (E.g., Singapore's Zero Waste Strategy)

Managing waste in urban environments represents a significant challenge that contemporary cities face, influenced by swift urbanisation, a growing populace, and escalating consumption behaviours. Effective strategies are crucial for reducing environmental consequences, enhancing public health, and attaining sustainability objectives. Singapore's zero-waste strategy is a paradigm of progressive thinking in tackling these challenges, utilising innovative waste management practices, technological advancements, and active public engagement. The city-state has embraced a circular economy framework, with the objective of minimising waste production, optimising resource recovery, and fostering recycling initiatives throughout all sectors of society (Kaya et al. 2024; Oyefusi et al. 2024; Zeng et al. 2024; Irianto et al. 2025).

The foundation of Singapore's strategy lies in its comprehensive waste segregation and recycling infrastructure. The National Environment Agency (NEA) spearheads initiatives to promote categorising waste into recyclable and non-recyclable segments among citizens and businesses. Advanced recycling bins, integrated with sensors and IoT technology, have been strategically implemented throughout the city, offering real-time insights into bin utilisation and enabling optimised collection pathways. Initiatives such as the "Recycle Right" campaign enlighten the populace regarding appropriate recycling methodologies, thereby reducing contamination levels and enhancing the calibre of recyclable substances. Singapore's initiatives to address food waste, including implementing on-site food waste digesters in commercial establishments and establishing a centralised food waste recycling facility, exemplify the city's dedication to minimising organic waste while generating compost or biogas as beneficial byproducts.

The Tuas Nexus is a crucial component of Singapore's zero-waste strategy, serving as an integrated facility that merges solid waste treatment with water reclamation processes. This advanced facility enhances resource recovery by transforming waste into energy and water recycling, exemplifying the advantages of integrating complementary processes nearby. In Singapore, incineration facilities significantly diminish the volume of solid waste by as much as 90%, facilitate the recovery of metals, and produce electricity in the process. Semakau Landfill, the city's waste disposal facility, exemplifies a sustainable model aimed at reducing environmental repercussions while fostering biodiversity within the surrounding marine ecosystems.In addition to Singapore, various urban centres across the globe are implementing cutting-edge waste management strategies designed to address their specific challenges. Cities like San Francisco have adopted ambitious zero-waste objectives, attaining elevated recycling and composting rates through rigorous legislation, financial incentives, and active community engagement. Kamikatsu, located in Japan, has established a remarkable benchmark by attaining near-zero waste through an intricate waste sorting system encompassing 45 distinct categories of recyclables. Comparably, Sweden's waste-to-energy initiatives have achieved such remarkable efficiency that it now imports waste from other nations to satisfy its energy requirements. Digital solutions are fundamentally reshaping the landscape of waste management within urban environments. Advanced urban technologies, including AI-driven waste management systems and blockchain-enabled recycling incentive platforms, significantly improve efficiency and transparency (Akinsemolu 2023; Karthick Raja Namasivayam et al. 2023; Dash et al. 2024; Gupta et al. 2024; Zhu et al. 2024). In India, the Swachh Bharat Abhiyan (Clean India Mission) has implemented waste-to-compost and waste-to-energy initiatives, complemented by public awareness campaigns, to tackle the complexities of urban waste management.

In light of recent developments, urban waste management continues grappling with persistent challenges, such as insufficient infrastructure in developing countries, limited public engagement, and the complexities of handling non-recyclable and hazardous materials. Confronting these challenges necessitates amalgamating innovative policy frameworks, strategic technological investments, and a transformation in behavioural paradigms. Cooperation among governmental entities, commercial enterprises, and local communities is essential for developing frameworks that emphasise waste minimisation, resource reclamation, and ecological sustainability. Strategies such as Singapore's zero-waste initiative exemplify a model for urban centres globally, demonstrating how cohesive and progressive methodologies can elevate waste management to a fundamental element of urban resilience and sustainability.

6.8 Future Trends and Innovations

Technological advances will change how industries, economies, and societies address sustainability issues. Nanotechnology, biotechnology, and eco-friendly solutions in Industry 4.0 are growing. Bio-based materials and hydrogen fuel cells are promising innovations that are changing our view of resource use and energy generation. Renewable materials like agricultural waste, algae, and bioplastics can replace petroleum-based products. The packaging, construction, and automotive industries use these biodegradable, recyclable materials more because they reduce carbon footprints (Evivie et al. 2020). In the meantime, hydrogen fuel cells are becoming a clean energy option in transportation and heavy industry. Hydrogen and oxygen react to generate electricity, producing only water, making them essential to decarbonising sectors that have resisted electrification.

Nano and biotechnology are advancing sustainability by creating innovative ways to reduce waste, increase energy efficiency, and protect the environment. Nanotechnology manipulates atomic and molecular structures to improve solar panels, water purification systems, and lightweight, durable transportation and construction materials. Nanomaterials create self-cleaning surfaces and coatings that reduce maintenance and resource use. Biotechnology improves these advances by using living organisms and biological systems to make environmentally friendly products. Genetically engineered microbes for bioremediation, algae-based biofuels, and CRISPR-based precision agriculture are changing industries and promoting environmental stewardship. Biotechnological tools boost productivity, reduce agricultural chemical use, and enable sustainable manufacturing and waste management.

Industry 4.0, the current era of automation and data exchange in manufacturing and beyond, will incorporate green technologies faster, creating a more sustainable industrial ecosystem. Industry 4.0 uses IoT, AI, and blockchain to improve operational efficiency, waste reduction, and resource use. Future predictions call for advanced manufacturing facilities with AI-powered systems that monitor and control energy consumption in real-time, reducing waste and emissions. Internet of Things devices will enable predictive maintenance, reduce equipment failures' environmental impact, and improve supply chain logistics to reduce transportation emissions. Blockchain can generate transparent and immutable records, improving traceability and accountability in sustainable practices like bio-based material certification and ethical sourcing in global supply chains.

The integration of these technologies will also accelerate the development of circular economy frameworks, where waste is reimagined as an asset and products are carefully designed for reuse, repair, or recycling. Modular product designs, material passports, and advanced recycling technologies will help industries achieve sustainability and economic growth. Emerging technologies, nanotechnology, biotechnology, and Industry 4.0 will create a future where green innovation is the norm as governments and businesses prioritise climate action and environmental goals. Current trends can mitigate climate change while promoting economic growth, job creation, and international cooperation.

6.9 Conclusion

Green technologies are a revolutionary way to address climate change, resource depletion, and environmental degradation. Sustainability, efficiency, and low ecological impact underpin the technologies, which reduce carbon footprints, advance renewable energy, and optimise resource use across sectors. Green technologies include renewable energy sources like wind, solar, and hydroelectric power, energy efficiency improvements, and waste management systems that reduce, reuse, and recycle materials. Green technologies emphasise bio-based materials, low-emission manufacturing, and sustainable agriculture to reduce chemical inputs and improve soil health. Energy generation, transportation, construction, and agriculture use these technologies. Solar panels, wind turbines, and biofuel technologies make renewable energy more affordable and accessible. Electric vehicles, hydrogen fuel cells, and alternative fuels promote cleaner, more efficient transportation. Sustainable buildings with advanced technologies use less energy, water, and waste. The effects of green technologies are vast and complex. They significantly reduce greenhouse gas emissions, air pollution, and land degradation. They promote biodiversity conservation through sustainable land-use methods and water conservation through water recycling and advanced irrigation systems. Green technologies can boost innovation, create jobs in new industries, and open new markets for sustainable goods and services. They improve public health, community resilience, and air, water, and energy quality. Infrastructure investment, new regulatory frameworks, and industry resistance to change are all challenges to the shift to sustainable technologies.

Governments, industries, and communities must adopt green technologies for long-term sustainability. Governments must create green innovation policies like renewable energy subsidies, energy-efficient technology tax incentives, and stricter environmental regulations that promote corporate accountability. They must invest in education and infrastructure to promote green initiatives and a supportive regulatory environment. Industries must adopt green technologies to stay competitive, cut costs, and satisfy consumer demand for eco-friendly products. Sustainable practices reduce risks of climate change and resource depletion and boost energy efficiency and brand loyalty. Communities benefit from the adoption of green technology in terms of health, employment, and quality of life. Public awareness and participation in environmentally sustainable initiatives like waste minimisation, conservation, and community-based renewable energy projects are essential for grassroots change.

Promoting sustainable technologies is essential. Governments, businesses, and communities must work together to accelerate sustainability. Policymakers should prioritise green technologies in national frameworks and provide incentives for innovation. Companies must invest in R&D to advance technology and implement sustainable supply chains. Communities should promote sustainability, green initiatives, and environmental policies. Through green technology adoption, every stakeholder can help create a sustainable, resilient, and equitable future for future generations. Time is running out to address climate change and protect our natural resources.

References

Abbas Z, Usmani T, Shaikh SF, Parveen A, Zuhair Abbas Shah S, Al-Enizi AM (2024) A systematic first-principles quantum analysis of the physical properties of Lead-Free bismuth titanate structures $ABi_4Ti_4O_{15}$ (A = Ca, Ba) for low-cost green energy applications. Mater Sci Eng B 310:117750. https://doi.org/10.1016/j.mseb.2024.117750

Addo-Bankas O, Wei T, Zhao Y, Bai X, Núñez AE, Stefanakis A (2024) Revisiting the concept, urban practices, current advances, and future prospects of green infrastructure. Sci Total Environ 954:176473. https://doi.org/10.1016/j.scitotenv.2024.176473

Afonso IS, Cardoso B, Nobrega G, Minas G, Ribeiro JE, Lima RA (2024) Green synthesis of nanoparticles from olive oil waste for environmental and health applications: a review. J Environ Chem Eng 12(5):114022. https://doi.org/10.1016/j.jece.2024.114022

Ahmad S, Jaiswal R, Yadav R, Verma S (2024) Recent advances in green chemistry approaches for pharmaceutical synthesis. Sustain Chem One World 4:100029. https://doi.org/10.1016/j.scowo.2024.100029

Akinsemolu AA (2023) Principles of green microbiology: the microbial blueprint for sustainable development. Environ Adv 14:100440. https://doi.org/10.1016/j.envadv.2023.100440

Amabie T, Izah SC, Ogwu MC, Hait M (2024) Harmonizing tradition and technology: the synergy of artificial intelligence in traditional medicine. In: Izah SC, Ogwu MC, Akram M (eds) Herbal medicine phytochemistry. Reference series in phytochemistry. Springer, Cham. https://doi.org/10.1007/978-3-031-21973-3_70-1

Amoneit M, Weckowska D, Spahr S, Wagner O, Adeli M, Mai I, Haag R (2024) Green chemistry and responsible research and innovation: moving beyond the 12 principles. J Clean Prod:144011. https://doi.org/10.1016/j.jclepro.2024.144011

Anitha L, Beda DP (2024) Implementation of quality by design and green chemistry principles to validate the stability indicating method on UHPLC for the simultaneous quantification of glycopyrronium, fluticasone, vilanterol in inhaled formulation. Green Anal Chem 11:100160. https://doi.org/10.1016/j.greeac.2024.100160

Antenucci A, Nejrotti S, Plata MJM, Mariotti N, Barbero N (2022) Unconventional and sustainable synthesis of polymethine dyes: critical overview and perspectives within the framework of the twelve principles of green chemistry. Eur J Org Chem 2022(44):29–48. https://doi.org/10.1002/ejoc.202200943

Arfelis S, Martín-Perales AI, Nguyen R, Pérez A, Cherubin I, Len C, Malpartida I, Bala A, Fullana-i-Palmer P (2024) Linking mechanochemistry with the green chemistry principles: review article. Heliyon 10(14):e34655. https://doi.org/10.1016/j.heliyon.2024.e34655

Assen AH, Adil K, Belmabkhout Y (2024) Lab-scale insights into green metal–organic framework sorbents for gas separation or purification. Curr Opin Green Sustain Chem 49:100948. https://doi.org/10.1016/j.cogsc.2024.100948

Dabas S, Kumar M, Veer Singh D, Kumar P (2024) Cutting-edge technologies and recent modernization on multifunctional perovskite materials for green energy conversion and spintronic applications. Mater Lett 377:137389. https://doi.org/10.1016/j.matlet.2024.137389

Dash S, Arjun Singh K, Jose S, Vincent Herald Wilson D, Elangovan D, Surapraraju SK, Natarajan SK (2024) Advances in green hydrogen production through alkaline water electrolysis: A comprehensive review. Int J Hydrogen Energy 83:614–629. https://doi.org/10.1016/j.ijhydene.2024.08.157

de Oliveira ML, Rodrigues LM, Silva da Veiga MAM, Souza LRR (2024) Advancing nanomaterial synthesis: harnessing green chemistry for sustainable innovation. Green Anal Chem 11:100148. https://doi.org/10.1016/j.greeac.2024.100148

de Souza Mesquita LM, Contieri LS, e Silva FA, Bagini RH, Bragagnolo FS, Strieder MM, Sosa FHB, Schaeffer N, Freire MG, Ventura SPM, Coutinho JAP, Rostagno MA (2024) Path2Green: introducing 12 green extraction principles and a novel metric for assessing sustainability in biomass valorization†† Green Chem 26(19):10087–10106. Electronic supplementary information (ESI) available. See https://doi.org/10.1039/d4gc02512a

Dhumal P, Chakraborty S, Ibrahim B, Kaur M, Valsami-Jones E (2024) Green-synthesised carbon nanodots: A SWOT analysis for their safe and sustainable innovation. J Clean Prod 480:144115. https://doi.org/10.1016/j.jclepro.2024.144115

Erinle KO, Ogwu MC, Evivie SE, Zaheer MS, Ogunyemi SO, Adeniran SO (2021) Impacts of COVID-19 on agriculture and food security in developing countries: potential mitigation strategies. CAB Rev 16(16):1–16. 10.1079PAVSNNR202116016

Evivie SE, Ogwu MC, Ebabhamiegbebho PA, Abel ES, Imaren JO, Igene JO (2020) Packaging and the Nigerian food industry: challenges and opportunities. In: Ogunlade CA, Adeleke KM, Oladejo MT (eds) Food technology and culture in Africa. Reamsworth Publishing, Ibadan, pp 28–99

Feder-Kubis J, Wirwis A, Policht M, Singh J, Kim K-H (2024) Principles and practice of greener ionic liquid–nanoparticles biosystem†† Green Chem 26(6):3072–3124. Electronic supplementary information (ESI) available. See https://doi.org/10.1039/d3gc04387h

Fegade SL (2024) Assessment of greenness of catalytic deoxygenation of crop oil for green diesel production. Clean Circ Bioeconomy 8:100091. https://doi.org/10.1016/j.clcb.2024.100091

Feng S, Liu G, Shan T, Li K, Lai S (2024) Predicting green technology innovation in the construction field from a technology convergence perspective: a two-stage predictive approach based on interpretable machine learning. J Environ Manage 372:123203. https://doi.org/10.1016/j.jenvman.2024.123203

Fiscal-Ladino J, Lozada-Castro JJ, Rios-Acevedo JJ, Montaño-Montoya D, Carasek E, Richter P, Pulzara-Mora A, Martínez LM, Rosero-Moreano M (2024) New solid phase microextraction fibers with green clay coating via radio frequency magnetron sputtering for detecting low-polar compounds in water samples. Anal Chim Acta 1332:343375. https://doi.org/10.1016/j.aca.2024.343375

Gupta AK, Boruah T, Ghosh P, Ikram A, Rana SS, Shanker MA, Bachetti A, Jha AK, Naik B, Kumar V, Rustagi S (2024) Green chemistry revolutionizing sustainability in the food industry: a comprehensive review and call to action. Sustain Chem Pharm 42:101774. https://doi.org/10.1016/j.scp.2024.101774

Huang X, Li W (2024) Application of low energy building thermal energy optimization based on image recognition and light sensors in urban green environment planning. Therm Sci Eng Prog 55:102954. https://doi.org/10.1016/j.tsep.2024.102954

Hussain CG, Keçili R, Hussain CMBT-CAC (2024) Lab on chip with green miniaturized technologies. Elsevier

Ibrahim MA, Algohary AM, Babaker MA, Ibrahim AM (2024) Design of experiment and green analytical chemistry principles for developing a robust and sustainable stability-indicating HPLC method for bempedoic acid impurity profiling. Microchem J 207:112133. https://doi.org/10.1016/j.microc.2024.112133

Imamoglu E (2024) Green extraction processes from renewable biomass to sustainable bioproducts. Bioresour Technol Reports 27:101952. https://doi.org/10.1016/j.biteb.2024.101952

Irianto I, Putra NR, Yustisia Y, Abdullah S, Syafruddin S, Paesal P, Irmadamayanti A, Herawati H, Raharjo B, Agustini S, Asmaliyah A, Krismawati A, Airlangga B (2025) Green technologies in food colorant extraction: a comprehensive review. S Afr J Chem Eng 51:22–34. https://doi.org/10.1016/j.sajce.2024.10.013

Izah SC, Ovuru KF, Ogwu MC (2022) Lassa fever in Nigeria: social and ecological risk factors exacerbating transmission and sustainable management strategies. Int J Trop Dis 5(2):065. https://doi.org/10.23937/2643-461x/1710065

Izah SC, Ogwu MC, Akram M (2023) Herbal medicine phytochemistry: applications and trends. Springer, Cham. https://doi.org/10.1007/978-3-031-21973-3

Karthick Raja Namasivayam S, Srinivasan S, Samrat K, Priyalakshmi B, Dinesh Kumar R, Bharani A, Ganesh Kumar R, Kavisri M, Moovendhan M (2023) Sustainable approach to manage the vulnerable rodents using eco-friendly green rodenticides formulation through nanotechnology principles—a review. Process Saf Environ Prot 171:591–606. https://doi.org/10.1016/j.psep.2023.01.050

Kaya SI, Ozcelikay-Akyildiz G, Ozkan SA (2024) Green metrics and green analytical applications: a comprehensive outlook from developing countries to advanced applications. Green Anal Chem 11:100159. https://doi.org/10.1016/j.greeac.2024.100159

Kosoe EA, Achana GTW, Ogwu MC (2023) Regulations and policies for herbal medicine and practitioners. In: Izah SC, Ogwu MC, Akram M (eds) Herbal medicine phytochemistry. Reference series in phytochemistry. Springer, Cham. https://doi.org/10.1007/978-3-031-21973-3_33-1

Krasnodębski M (2024) The bumpy road to sustainability: reassessing the history of the twelve principles of green chemistry. Stud Hist Philos Sci 103:85–94. https://doi.org/10.1016/j.shpsa.2023.12.001

Kumar D, Chand P (2024) First principle study of structural, electronic, optical, and thermoelectric properties of Zr_{1-x} M_xNiSn (M = Ta, Nb and x = 0.25, 0.5, 0.75) half Heusler compounds: Green energy applications. Phys B Condens Matter 678:415783. https://doi.org/10.1016/j.physb.2024.415783

Lei M (2024) Application of energy sustainability model based on optical sensing technology in intelligent warehousing performance management in green manufacturing industry. Therm Sci Eng Prog 54:102789. https://doi.org/10.1016/j.tsep.2024.102789

Lin Y-C, Wyżga P, Macyk J, Macyk W, Guzik MN (2024) Solar-driven (photo)electrochemical devices for green hydrogen production and storage: working principles and design. J Energy Storage 82:110484. https://doi.org/10.1016/j.est.2024.110484

Nahata A, Patel M, Krishna Muchakayala S (2024) Strategic approaches to elevate quality and sustainability in drug development: comprehensive pretomanid (PA-824) chemical stability study using QbD and green chemistry principles. Microchem J 200:110413. https://doi.org/10.1016/j.microc.2024.110413

Obahiagbon EG, Ogwu MC (2023) Consumer perception and demand for sustainable herbal medicine products and market. In: Izah SC, Ogwu MC, Akram M (eds) Herbal medicine phytochemistry. Reference series in phytochemistry. Springer, Cham. https://doi.org/10.1007/978-3-031-21973-3_65-1

Ogwu MC (2019) Towards sustainable development in Africa: the challenge of urbanization and climate change adaptation. In: Cobbinah PB, Addaney M (eds) The geography of climate change adaptation in Urban Africa. Springer Nature, Switzerland, pp 29–55. https://doi.org/10.1007/978-3-030-04873-0_2

Ogwu MC, Malikia CN, Stansfield A, Gonzales-Torres AD, Izah SC (2023) Sustainable food processing waste management for environmental protection. In: Srivastav AL, Grewal AS, Tiwari M, Pham TD (eds) Role of green chemistry in ecosystem restoration to achieve environmental sustainability. Elsevier, Amsterdam, pp 291–299. ISBN: 978-0-443-15291-7. https://doi.org/10.1016/B978-0-443-15291-7.00022-5

Outili N, Kerras H, Meniai AH (2023) Recent conventional and non-conventional WCO pretreatment methods: implementation of green chemistry principles and metrics. Curr Opin Green Sustain Chem 41:100794. https://doi.org/10.1016/j.cogsc.2023.100794

Oyefusi ON, Enegbuma WI, Brown A, Zari MP (2024) From green to regenerative supply chain management in construction: towards a conceptual framework. Environ Dev 52:101097. https://doi.org/10.1016/j.envdev.2024.101097

Rani A, Negi S, Fan C, Lam SS, Kim H, Pan S-Y (2024) Revitalizing plastic wastes employing bio-circular-green economy principles for carbon neutrality. J Hazard Mater 472:134394. https://doi.org/10.1016/j.jhazmat.2024.134394

Raza S, Hayat A, Bashir T, Ghasali E, Abdel Hafez AA, Chen C, Shen L, Orooji Y, Lin H (2024) Recent progress in green thin film membrane based materials for desalination: design, properties and applications. Desalination 591:117973. https://doi.org/10.1016/j.desal.2024.117973

Roveda LM, Ottoni VF, de Carvalho CT, Rodrigues R, Corazza MZ, Trindade MAG (2024) Merging 3D-printing technology and disposable materials for electrochemical purposes: a sustainable alternative to ensure greener electroanalysis. Talanta 272:125814. https://doi.org/10.1016/j.talanta.2024.125814

Simi A (2024) The principles of green architecture and sustainable development to preserve the environment. In: Second E (ed) Abraham MABT-E of ST. Elsevier, Oxford, pp 174–189

Sridhar K, Akram A, Banat F (2024) Green extraction technologies for valorization of date seed waste to achieve sustainable development goals: biofunctional and innovative food applications. Food Res Int:115392. https://doi.org/10.1016/j.foodres.2024.115392

Standal D, Ahlquist IH (2025) Transforming coastal fisheries in Norway: institutional implications of green technology implementation. Mar Policy 171:106471. https://doi.org/10.1016/j.marpol.2024.106471

Szekely G (2024) The 12 principles of green membrane materials and processes for realizing the United Nations' sustainable development goals†† RSC Sustain 2(4):871–880. Electronic supplementary information (ESI) available. See https://doi.org/10.1039/d4su00027g

Verma C, Chauhan DS, Aslam R, Banerjee P, Aslam J, Quadri TW, Zehra S, Verma DK, Quraishi MA, Dubey S, AlFantazi A, Rasheed T (2024) Principles and theories of green chemistry for corrosion science and engineering: design and application. Green Chem 26(8):4270–4357. https://doi.org/10.1039/d3gc05207a

Wang S, Zhang H (2024) Green entrepreneurship success in the age of generative artificial intelligence: the interplay of technology adoption, knowledge management, and government support. Technol Soc 79:102744. https://doi.org/10.1016/j.techsoc.2024.102744

Wang X, Wang S, Wu K, Zhai C, Li Y (2024) Environmental protection tax and enterprises' green technology innovation: evidence from China. Int Rev Econ Financ 96:103617. https://doi.org/10.1016/j.iref.2024.103617

Yue C-J, Gu L-P, Ma S-S, Sun X, Dai N, Zhang Z-W, Liu B-L (2023) Improving undergraduate graduation chemical design using green techniques based on green engineering principles. Educ Chem Eng 44:96–103. https://doi.org/10.1016/j.ece.2023.05.003

Zain NNM, Yahaya N, Madurani KA, Shahabuddin S, Sambasevam KP, Baharin SNA, Jeerapan I (2024) Green techniques: Revolutionizing deep eutectic solvents-based modified electrodes for electrochemical sensing of natural antioxidant. Microchem J 206:111491. https://doi.org/10.1016/j.microc.2024.111491

Zeng XQ, Jiang ZN, Duan JM, Li YR, Peng SY, Dong CF, Zhang GA (2024) Developing a novel amino acid derivative as high-efficient green corrosion inhibitor for mild steel in acidic solution: experiments and first-principles calculations. Ind Crops Prod 210:118032. https://doi.org/10.1016/j.indcrop.2024.118032

Zhang J, Xie S, Li X, Xia X (2024) Adoption of green production technologies by farmers through traditional and digital agro-technology promotion—an example of physical versus biological control technologies. J Environ Manage 370:122813. https://doi.org/10.1016/j.jenvman.2024.122813

Zheng S, Zhang J, Jian L (2024) Green technology diffusion mechanism in China's aviation industry cluster based on complex network game model. Energy 313:133634. https://doi.org/10.1016/j.energy.2024.133634

Zhu X, Zhou S, Chen Z, Xing F-B, Chen W-B, Zhang L (2024) Application of indocyanine green in thoracic surgery: a review article. Asian J Surg. https://doi.org/10.1016/j.asjsur.2024.08.194

Chapter 7
Sustainable Water Management and Treatment: Systems, Processes and Technologies

Bharat Mishra and Archita Tiwari

Abstract Global water resources are rapidly diminishing, driven by population growth, climate change, and expanding industrialization. Experts estimate that by 2050, 52% of the projected 9.7 billion people worldwide will reside in areas experiencing water stress or scarcity. The global challenge of accessing clean, potable water will persist as sustainable solutions remain elusive. Water sustainability involves meeting the current generation's water needs without jeopardizing future generations' ability to meet their own. Water is the cornerstone of sustainable development, serving as a common thread linking global challenges such as energy, food security, health, peace, security, and poverty eradication. Our survival and well-being depend heavily on effective water resource systems. However, with growing development pressures on land in watersheds and increasing demands for water in streams, rivers, lakes, and aquifers, it is unrealistic to expect these water systems to return to or maintain their pristine, most productive states. Sustainable water management (SWM) is crucial for addressing these pressures and achieving sustainable development goals. SWM ensures that current water needs are met for all users without compromising the ability of future generations to meet their own needs. This concept aligns with broader sustainability principles, addressing both present and future water challenges. Enhancing the efficiency of conventional membrane technologies for water treatment is now crucial to minimizing their environmental impact. Wastewater treatment removes pollutants, coarse particles, and toxic substances while killing pathogens and producing bio-methane and manure for agriculture. It is crucial in reducing water waste, easing pressure on natural water sources, and supporting clean energy, forming the foundation for sustainable waste management. Membrane technologies are increasingly favored for wastewater treatment due to their sustainability advantages, including cost-effectiveness, operational ease, and safety. Sustainable water treatment technologies utilize innovative methods such as membrane filtration,

B. Mishra (✉)
Shakuntala Misra National Rehabilitation University, Lucknow, India
e-mail: bmishra_pharmacy@dsmnru.ac.in

A. Tiwari
Khwaja Moinuddin Chishti Language University, Lucknow, India

M. C. Ogwu and S. Chibueze Izah (eds.), *Evaluating Environmental Processes and Technologies*, Environmental Science and Engineering,
https://doi.org/10.1007/978-3-031-85327-2_7

advanced oxidation processes, and nanotechnology. Techniques like reverse osmosis and ultrafiltration are highly effective in removing contaminants, microorganisms, and nanoparticles from water. Sustainable water technologies include wastewater treatment plants, intelligent irrigation systems, fog catchers, rainwater harvesting, tap aerators, seawater desalination, portable filters, and solar-powered desalination units.

Keywords Water scarcity · Membrane technology · Electro deionization · Waterborne diseases · Water management · Water Stress Index · Water pollution

7.1 Introduction

Water is essential for all living organisms, serving key roles as a solvent, metabolite, temperature regulator, and structural support for cells. However, rapid population growth and unsustainable modernization have significantly increased freshwater demand over recent years (Khan et al. 2023). Water is a vital natural resource that meets our essential needs and is a precious asset. It is crucial to take action to conserve and make the most of every drop. Water resources can be evaluated based on surface and groundwater bodies (Mishra 2023). Water is essential for life, playing a crucial role in supporting vital functions such as respiration, nutrition, circulation, excretion, and reproduction. It is a fundamental resource for sustaining human life and maintaining bodily processes (Issaoui et al. 2022). Meeting the water needs of a rapidly growing population has become a significant challenge, particularly as rising food production must rely on the same limited water resources. The global water crisis is further aggravated by population growth and climate change. Over 1.1 billion people lack access to clean drinking water, and around 2.6 billion lack basic sanitation. Water stress escalates, especially in developing nations (Mishra et al. 2021). Water is essential, but climate extremes and human activities increasingly threaten its availability (Scanlon et al. 2023). Industrialization, agriculture, urbanization, and deforestation contribute to the overuse and pollution of water resources, while climate change disrupts water cycles by causing droughts, floods, and glacial melt. Large river basins, key areas for human settlement, are particularly vulnerable, facing water scarcity that impacts health, agriculture, and ecosystems(Li et al. 2022; Ogwu et al. 2023).

Water, in its various forms, such as rain, snow, clouds, or ice, significantly influences climate processes. It cools the surface through evaporation, while clouds and snow reflect sunlight, and water vapor helps trap heat, ensuring a delicate balance that sustains Earth's temperature. This intricate interplay of water and energy underscores water's complex role in regulating the planet's climate system (Mishra 2023). The chemical structure of a water molecule consists of covalently bonded one oxygen atom to two hydrogen atoms, forming a polar molecule. The electronegative nature of the oxygen atom pulls electrons from the hydrogen atoms, resulting a partial negative charge on the oxygen atom and a partial positive charge on the hydrogen atoms. This asymmetrical charge distribution leads to strong intermolecular attraction, as positively charged hydrogen atoms are drawn to negatively charged oxygen atoms in nearby water molecules. The high energy of these O–H bonds (5.5 kcal/mol) contributes to water's high boiling point and specific heat, allowing it to remain liquid over a broad temperature range suitable for physiological processes. The H–O–H bond angle of about 104.5° further reinforces water's polarity, supporting its role in biological systems (Riveros-Perez and Riveros 2018). In its abundant presence, water is fundamental to life on Earth and serves as a vital link between land, oceans, and the atmosphere, forming an interconnected system. At the heart of this system is the hydrological cycle, a continuous process of water movement that includes precipitation, evaporation, freezing, melting, and condensation. This cycle drives water transfer from clouds to land and back again, playing a crucial role in Earth's climate dynamics.

The hydrological cycle is intrinsically connected to energy exchanges among the atmosphere, ocean, and land, shaping both global climate patterns and natural variability (Mishra 2023). An environmental system is rooted in the natural environment and consists of both living (biotic) and non-living (abiotic) components that interact with one another. The primary environmental systems include the atmosphere (air), biosphere (all living organisms), hydrosphere (water bodies), and lithosphere (earth's solid rock layer) (The Environmental System 2024). Water bodies [hydrosphere], a natural resource that has its own economic value. Water is continuously renewed through the movement of water from the earth's surface to the atmosphere and back to the surface. Ecosystems play a vital role in supplying, purifying, and safeguarding freshwater resources. These ecosystems regulate water quality and mitigate the effects of floods and droughts; coastal mangroves and riverbank vegetation shield land from erosion, while wetlands, glaciers, and aquifers balance water flow and storage, particularly during dry periods (*Water and Ecosystems*).

Water resources of sufficient quality and quantity are essential for human survival and sustainable development. To effectively protect water ecosystems and conserve these vital resources, efficient management, purification, and transport systems are crucial. Protecting and restoring freshwater ecosystems is essential to reversing biodiversity loss and balancing the needs of communities, businesses, and the environment. Governments are urged to enforce policies for sustainable management at national and river basin levels. Nature-based solutions, like restoring wetlands and riverbank vegetation, are crucial for climate resilience, while wetlands naturally filter wastewater, improving water quality. In a circular economy, treated wastewater can further support ecosystem health by recycling nutrients (*Water and Ecosystems*). Natural water systems are closely monitored to support better resource management and pollution control through pollutant tracking and toxicity assessments. In engineered systems, raw water is purified in drinking water treatment plants (DWTPs) using processes like coagulation, sedimentation, filtration, and disinfection to remove contaminants before distribution to users (Huang et al. 2021a, b; Ogwu 2023).

7.2 Natural Water Resources

Life on Earth relies on healthy ecosystems, with freshwater systems like wetlands, rivers, mangroves, and aquifers playing an essential role in the global water cycle by supplying, purifying, and safeguarding freshwater resources. Natural water resources are necessary for domestic, agricultural, industrial, recreational, and environmental needs. As all living beings depend on water for survival, it is often said, "water is life, and life is water." (Hossain 2019). However, poor management practices and climate change severely damage many of these ecosystems, weakening their capacity to deliver critical freshwater services. This degradation poses significant risks to both human health and natural environments. Although about 70% of our planet is covered in water, only a tiny fraction of it is fresh water. Most of this freshwater, approximately 2.5% is locked in ice or permanent snow cover, leaving only about one-third of that amount accessible for human use. Some forecasts indicate that by 2025, approximately 1.8 billion people may lack access to safe drinking water. With the global population exceeding 7.5 billion and projected to surpass 10 billion by 2050, the water demand will become even more critical (Water and Natural Resources 2019).

Water resources encompass all forms of natural water found on Earth, whether in vapor, liquid, or solid states, that humans can use. In the broadest sense, this includes all water within the hydrosphere, such as oceans, seas, rivers, lakes, groundwater, and glaciers (Khilchevskyi and Karamushka 2021). Over the past 20 years, the demand of freshwater in India has surged due to population growth, urbanization, industrialization, and inefficient agricultural practices. Consequently, per capita water availability has declined from 1820 cubic meters in 2001 to 1508 cubic meters in 2015. There is also significant regional variation, with per capita water availability in major river basins ranging from 300 to 2000 cubic meters annually (Water Resources 2021).

7.2.1 Types of Natural Water Resources

Water resources are categorized into four primary types: drinking water, groundwater, surface water (lakes, rivers, rain, and springs), and seawater (oceans and seas) (Hojjati-Najafabadi et al. 2022). Ecological processes influence the quality of water and the way it moves through a system, and soil formation, erosion, sediment transport and deposition can exert significant influences on hydrology (Water Resources 2018). Groundwater is one of Earth's most vital natural resources, offering numerous benefits to humanity. It provides a key water source for extraction and use while also serving essential in situ roles, such as regulating climate and supporting wet ecosystems (van der Gun 2021) Groundwater serves as a vast natural reservoir beneath the Earth's surface. The deterioration of ecosystems is a major driver of escalating challenges in water resource management. Although around 30% of global land is still forested, at least two-thirds of these forests are degraded. Most of the world's soil resources, particularly those on farmland, are in fair to deplorable condition, with projections indicating further decline. This worsening soil quality significantly disrupts the water cycle, leading to increased evaporation, reduced soil water retention, and more significant surface runoff, accelerating erosion (Water Resources 2018).

Arid regions frequently depend on groundwater for drinking water, irrigation, and industrial use. It provides nearly half of drinking water supplies, with about 43% used for irrigation (Ahmed et al. 2024) The Central Groundwater Board (CGWB) monitors groundwater levels nationwide four times a year in April/May, August, November, and January. Groundwater level observations 2017 reveal that monsoon rainfall is the primary source of groundwater replenishment, with over 55% of the country showing groundwater depths of 0–10 m during this season. However, states like Haryana, Rajasthan, and Delhi consistently display deeper groundwater levels the entire year, highlighting significant water scarcity in these regions (Water Resources 2021).

Anusha et al. (2022) reported that from 2000 to 2020, surface water bodies increased by 0.36%, primarily attributed to rising rainfall levels and intensity. India possesses 4% of the world's water resources, making it a water-rich country. The India-WRIS web portal gathers, stores, and manages surface water data from multiple sources. Snow Glacial Lakes report provides data on glacial lakes and water bodies in the Himalayan regions from 2011 to 2021, covering the Indus, Ganga, and Brahmaputra River basins. Glacial lakes and water bodies more significant than 50 hectares were mapped using AWiFS satellite imagery. It includes water-spread areas (WSA) from June to October each year and changes since the 2009 baseline, as per a joint study by CWC and NRSC (*Water Data*). Lakes are vital water resources they account for only about 0.3% of the Earth's surface water. However, the condition of lakes has been deteriorating due to increasing human activities around them. The most significant impacts of human activity on the ecosystem are sedimentation and habitat modification, which alter the biota's habitat and physiology. Nutrient enrichment from the phosphorous loading into the water overproduces the algae and cyanobacteria, decreasing water transparency and inducing high oxygen fluctuation. High organic matter load into the water causes oxygen depletion; however, the supply of organic matter is often ignored when monitoring lake waters (Lukman 2023). Assessing lake water quality involves evaluating various physico-chemical and biological parameters based on the lake's Designated Best Use (DBU) for different purposes. A critical tool for determining lake water quality is the Water Quality Index (WQI), which helps gauge overall water health and suitability for use (Vasistha and Ganguly 2020).

Oceans are inherently interconnected, forming a unified system that influences energy flows, climate patterns, marine ecosystems, and human activities(Brodie Rudolph et al. 2020) The ocean plays a crucial role in producing oxygen, storing carbon and heat, supplying food, and supporting economic activities such as international trade and transportation (Ogwu 2019a, b; Obahiagbon and Ogwu 2024). Beyond its monetary contributions, the ocean also offers intangible benefits, including the advancement of scientific knowledge, opportunities for collaboration, a sense of place, feelings of awe and reverence, and a free space for recreation and gathering with loved ones. various human activities and climate changes, including carbon dioxide emissions impact (Österblom et al. 2023) Oceans. Burning fossil fuels increases greenhouse gases, leading to warming, acidification, and reduced oxygen levels, as warmer water holds less oxygen than cooler water ("Climate Change Impacts on the Ocean and Marine Resources"). Rivers support both ecosystems and human societies, and maintaining their health is vital for sustainable development on Earth (Chan 2012). Rivers are intricate systems with diverse functions and characteristics, deeply influenced by human actions and climate change. Their flow, development, and usage serve as pathways for natural substances and energy interactions, creating vital links among the atmosphere, lithosphere, biosphere, and human domains (Wang and He 2022).

7.2.2 The Ecological Significance of Maintaining These Resources for Water Security

Water security is defined as the sustainable management and protection of water resources, emphasizing the need for reliable access to essential water services that support human populations and ecological systems. Achieving water security requires maintaining water availability and implementing measures to protect against water-related hazards, including floods and droughts, which pose significant risks to communities and natural environments. This comprehensive approach to water security highlights its dual role in fulfilling human and environmental needs while mitigating potential threats (Wheater and Gober 2015; Iyiola et al. 2023a, b, 2024a, b, c). Water security ensures access to safe, sufficient, and affordable water for a healthy and productive life now and for future generations. Water security involves the sustainable management and protection of water systems, safeguarding against water-related hazards, responsible development of water resources, and preserving water services essential for people and the environment (Mishra et al. 2021). Water scarcity is a pressing global issue, impacting over 40% of the world's population. Projections indicate that by 2025, approximately 3 billion people may lack access to fresh water, facing severe water stress (Manetu and Karanja 2021) Nearly two-thirds of the world's population—around four billion people—experience severe water scarcity for at least one month each year, and over two billion people live in countries with inadequate water supplies. By 2025, half of the global population could live in areas facing water scarcity. The situation is projected to worsen, with intense water scarcity potentially displacing 700 million people by 2030. By 2040, approximately one in four children worldwide may be living in regions under extremely high water stress (*Water scarcity*). Pakistan is the 17th most water-scarce country, with 79% of its drinking water contaminated. Its primary water sources include precipitation, snowfall, glaciers, and an extensive network of rivers, lakes, and coastal basins. In Pakistan, Water pollution affects 89% of rural and 96% of urban residents, while only

1% of industrial wastewater is treated before being discharged into rivers (Ahmad et al. 2023).

7.2.3 Global Availability of Water Resources

Water is an essential resource, yet climate extremes and human activities increasingly threaten its availability. Approximately 70% of the Earth's surface is covered by water, with 97.5% being saltwater and only 2.5% classified as freshwater. Of this 2.5% of freshwater, less than 1% is readily accessible for use (Mishra 2023). Over the past century, regional trends in total water storage have shown considerable variation. GRACE satellite data from the last two decades indicate areas with declining, stable, and rising global water storage. Long-term groundwater monitoring provides deeper insights, revealing increasing water storage in regions like northwest India, central Pakistan, and the northwest United States, contrasted with decreases in the U.S. High Plains and California's Central Valley. Although climate variability affects these changes, human activities, particularly irrigation are significant contributors. It is essential to adopt a more diverse array of management strategies to enhance the resilience of water resources (Scanlon et al. 2023). At the continental level, America holds the largest share of the world's freshwater resources, accounting for 45%, followed by Asia at 28%, Europe at 15.5%, and Africa at 9% ("World Water Resources By Country"). Virtual water trade (VWT) has recently gained significant attention worldwide. It is increasingly viewed as both a policy tool and a practical approach to balancing water resources locally, nationally, and globally. The concept of Virtual Water (VW) has steadily gained recognition, viewing water as an embedded natural resource within various goods and services. VWT typically involves sourcing water-intensive products from regions with abundant water supplies to areas facing scarcity, reducing local production needs. This practice helps support water and food security in the importing regions to some extent (Xu et al. 2020).

7.2.4 An Examination of the Uneven Global Distribution of Water Resources

7.2.4.1 Regions Facing Water Scarcity and Abundance and the Implications for Public Health

At the country level, there is significant variability in Total Renewable Water Resources (TRWR) per capita, ranging from just ten m^3 per person in Kuwait to over 100,000 m^3 per person in Canada, Iceland, Gabon, and Suriname. In 19 countries or territories, the TRWR per capita is less than 500 m^3, while 29 countries or regions have < 1000 m^3 per person. The ten countries with the lowest water resources per capita are Bahrain, Jordan, Kuwait, Libya, the Maldives, Malta, Qatar, Saudi Arabia, the United Arab Emirates, and Yemen (World Water Resources By Country). Global water scarcity poses a significant challenge to sustainable human development and achieving the Sustainable Development Goals (Dolan et al. 2021). Water scarcity poses a significant challenge to sustainable development, influenced by the balance between available freshwater resources and human water withdrawal, both impacted by climate change and human activity. However, pinpointing the primary drivers of global water scarcity remains challenging due to substantial uncertainties in estimating how changes in water withdrawal and availability each contribute to this issue. Most current water scarcity assessments rely on simulations from global hydrological models (GHMs) that account for the effects of climate change and human activities.

The Water Stress Index (WSI), the ratio of water demand to water availability, is commonly employed to evaluate water scarcity conditions. Water demand is typically estimated by aggregating water withdrawal and consumption across various sectors, while water availability is determined based on river discharge measurements (Huang et al. 2021a, b). The Falkenmark Indicator, or Water Stress Index, compares a country's water footprint to its total renewable water resources, including both surface and groundwater (blue water) and soil moisture (green water). A higher index indicates more tremendous pressure on a country's water resources. A region is under water stress if its per capita water availability falls below 1700 m^3 yearly (Ding and Ghosh 2017).

7.3 Types of Water Pollution and Management Options: An Overview

Water is essential to life, sustaining both our existence and well-being. However, due to natural and human activities, water can become contaminated with various pollutants (Morin-Crini et al. 2022). Water pollution refers to the contamination or degradation of water bodies due to the entry of harmful substances or excessive quantities of natural or synthetic pollutants from industrial discharge, agricultural runoff, and household discharge, as shown in Fig. 7.1. Principal types include point source pollution (from specific, identifiable sources), nonpoint source pollution (diffuse sources such as runoff), groundwater pollution, and surface water pollution. Point source pollution refers to contamination from a single location, such as a factory or sewage treatment plant. One of the most significant examples is the 2010 Deepwater Horizon oil spill in the Gulf of Mexico, which released approximately 134 million gallons of oil, making it the most significant point source of oil pollution in U.S. history. Runoff from urban and suburban areas significantly contributes to non-point source pollution, with discarded trash often becoming part of this runoff. For the past decade, NOAA's Marine Debris Program has addressed marine debris as a form of non-point pollution in Tijuana River, California and Shuyak Island, Alaska (*Office of Response and Restoration*).

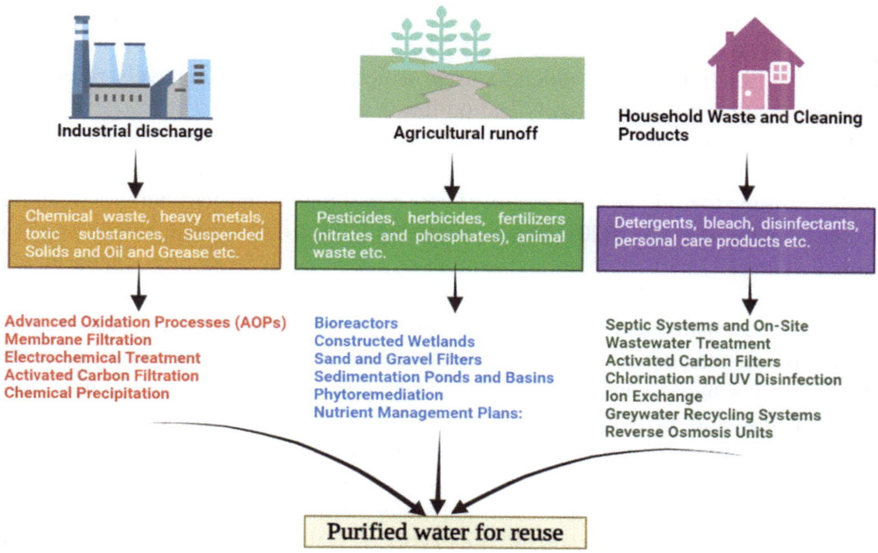

Fig. 7.1 Water pollutants their sources, and treatment technologies

Human activities like industrial processes, agricultural runoff, mining, waste disposal, and oil spills are leading contributors, introducing toxic chemicals, heavy metals, organic pollutants, and oils into water systems, which harms aquatic ecosystems and poses significant health risks (Singh 2024a, b).

Removing these contaminants is crucial to safeguard public health and prevent diseases associated with water pollution (Singh et al. 2020). Water pollution is a growing and critical issue in India. Approximately 71% of surface water and increasing groundwater resources are contaminated with organic and inorganic pollutants (Kumar et al. 2021). Contaminants in water bodies present severe risks to human health and aquatic ecosystems. Water pollutants can be broadly categorized into several types: inorganic, organic, pathogens, thermal, and radioactive. Each category contributes uniquely to environmental and health challenges, affecting water quality and disrupting ecosystems (Madhav et al. 2020).

According to The Pacific Institute for Studies in Development, Environment, and Security, each day, 2 million tons of sewage, industrial, and agricultural waste are discharged into water worldwide, equivalent to the weight of 6.8 billion people (Water Pollution Statistics 2020).

Water quality is determined by assessing physical, chemical, and biological parameters, which help monitor its suitability for specific uses. Exceeding acceptable limits for these parameters can pose health risks. Water Pollution Index (WPI) was applied to evaluate the water quality, which simplifies water quality assessment by comparing observed concentrations (Ci) to standard permissible levels (Si) across selected variables. The index categorizes water quality into four levels: excellent (WPI < 0.5), good ($0.5 \leq$ WPI < 0.75), moderately polluted ($0.75 \leq$ WPI < 1), and highly polluted (WPI > 1), providing a flexible and accessible tool for water quality monitoring (Hossain and Patra 2020). Another standard method is the Heavy Metal Pollution Index (HPI), which focuses on metal contamination levels, providing valuable insights into water safety and pollution management.

Chemical pollution has become an increasingly critical environmental concern. Various regulations have been established to monitor, manage, and improve water quality, aiming to mitigate the harmful effects of pollutants on ecosystems and public health (Posthuma et al. 2020). Water pollutants fall into three main categories: (1) those that directly harm humans and animals by causing disease or physical harm; (2) those that reduce water's oxygen levels, leading to anaerobic conditions and the death of aquatic organisms; and (3) those that indirectly impact water quality by making it less enjoyable or safe to use and by damaging the natural beauty and ecological health of lakes, rivers, and oceans (Moeller 2012). Chemical pollutants are among the most widespread sources of water contamination, often originating from various industries. Industrial processes can release metals and solvents, while agriculture contributes fertilizers and pesticides, and pest control activities add their chemicals to the mix. When these substances infiltrate soil and water systems, they pose severe risks to human health, livestock, and aquatic life (Types of Water Pollution 2022). Industries such as paper, textiles, leather, food, cosmetics, and pharmaceuticals heavily rely on dyes and colorants. When untreated wastewater from these industries is discharged, it introduces toxic colors, chemicals, and organic pollutants into water bodies, harming aquatic life and ecosystems. For example, paper and textile manufacturing produce dye-rich wastewater, which lowers water quality and poses health risks to organisms (Svensson and Berntsson 2014).

Microbiological pollution is a natural form of water contamination to which humans are particularly vulnerable. As a result, effective drinking water systems are essential for removing this type of pollution and ensuring water safety (Types of Water Pollution 2022). When contaminants such as bacteria, viruses, and protozoa infiltrate water sources, they can pollute groundwater, making drinking and household use unsafe. Consuming or using this contaminated water can lead to various health issues, including waterborne diseases like cholera. Protecting groundwater from pollution is essential for ensuring safe and healthy water for communities (Arif et al. 2020).

7.3.1 Various Management Strategies for Reducing Pollution and Protecting Water Quality

Water pollution worsens as traditional treatment methods struggle to meet increasingly stringent environmental standards (Alkhadra et al. 2022). Water pollution control methods are generally categorized into three main stages:

- Primary treatment includes screening, sedimentation, homogenization, neutralization, mechanical flocculation, and chemical coagulation.

- Secondary treatment, involves biological methods such as aerobic and anaerobic treatment, activated sludge process, aerated lagoons, trickling filters, oxidation ditches, and ponds.
- Tertiary treatment utilizes advanced techniques like membrane technologies, oxidation, adsorption, coagulation and flocculation, electrochemical processes, ion exchange, evaporation and crystallization (Senthil Kumar and Saravanan 2018).

7.3.2 Statistics on Water Pollution, Diseases, and Management

Water pollution occurs when harmful substances contaminate water, making it unsuitable for sustaining life. The release of hazardous pollutants into water bodies is a major global issue, posing serious risks to human health, aquatic life, and ecosystems. Exposure to polluted water can harm humans, marine organisms, plants, animals, and the environments that rely on these vital resources (Chowdhary et al. 2020; Hojjati-Najafabadi et al. 2022). According to WHO data from 2023, microbiologically contaminated drinking water can transmit diseases like diarrhoea, cholera, and typhoid, causing an estimated 505,000 diarrhoeal deaths annually (Drinking Water 2023). Pesticides and fertilizers can significantly affect aquatic ecosystems and water quality. When pesticide residues reach ponds, rivers, and lakes through surface runoff or spray drift, they can lead to both acute toxicity (such as fish kills) and chronic poisoning, where non-lethal levels accumulate in wildlife over time. Similarly, fertilizers particularly nitrogen (N) and phosphorous (P) used to boost agricultural productivity, may seep into waterways and groundwater, leading to degraded water quality and eutrophication. Eutrophication promotes excessive algae growth, disturbing aquatic ecosystems by triggering phytoplankton blooms and increasing organic matter production. This can result in oxygen depletion, harming benthic organisms that live on or near the seabed (Wato et al. 2020). Over one-third of the Earth's accessible renewable freshwater is utilized for agricultural, industrial, and domestic purposes. These essential activities, however, often lead to water contamination by synthetic chemicals and naturally occurring geogenic substances. As a result, the pollution of natural water bodies has emerged as a pressing global concern, affecting ecosystems and human health. The widespread contamination has highlighted the need for sustainable water management practices and stricter controls on pollutants entering freshwater systems (Schwarzenbach et al. 2010). The textile industry is one of the most water-intensive and polluting sectors worldwide. Many wet textile processes require substantial fresh water and generate large volumes of wastewater, typically containing highly concentrated pollutants. This wastewater is often characterized by vibrant colors, high levels of organic compounds, and a wide range of contaminants such as inorganic finishing agents, surfactants, chlorine compounds, salts, phosphates, polymers, and other organic substances (Rather et al. 2019).

More than 80% of sewage produced by human activities is released into rivers and oceans without treatment, causing widespread environmental pollution and contributing to over 50 different diseases. Poor water quality accounts for 80% of diseases globally and 50% of child deaths worldwide. The most prevalent illness from water pollution is diarrhoea, primarily spread through enteroviruses in contaminated water (Lin et al. 2022). Diarrhoea is the most common illness from contaminated food and water, but in 2021, over 251.4 million people required treatment for schistosomiasis, a parasitic infection from contaminated water (Drinking Water 2023). A significant issue linked to water pollution is the potential harm to the heart and kidneys when contaminated water is consumed regularly. Other health risks include poor blood circulation, vomiting, skin lesions, cholera, gastroenteritis, and damage to the nervous system (Afroz et al. 2014).

Water contamination from both natural and human activities presents a significant global public health risk. Natural sources include microbial activity, geological formations, and water supply pollutants. In contrast, anthropogenic sources stem from human actions such as industrial processes, agricultural practices, improper waste disposal, and inadequate sewage systems. These activities compromise the safety and purity of water supplies, posing significant health risks to communities . The health impacts of exposure to polluted water can vary greatly, influenced by factors such as the chemical makeup of the contaminants, the duration of exposure, and the concentration of pollutants present (Babuji et al. 2023).

Water pollution poses serious health risks, causing diseases like cholera, typhoid, and dysentery, as well as long-term issues such as organ damage, cancer, and developmental disorders. Addressing water pollution is crucial to protecting public health (Singh 2024a, b). Inadequate water treatment is a significant global issue, with over 80% of sewage from human activities being released into rivers and oceans untreated which leads to severe environmental pollution and contributes to the spread of over 50 diseases. Poor water quality is linked to approximately 80% of diseases worldwide and is responsible for 50% of child deaths globally. Waterborne and water-related diseases result from direct and indirect contact with water, whether through drinking, bathing, or recreational use. Such diseases may occur due to ingesting contaminated water or exposure to pathogens and pollutants present in the water (Nwabor et al. 2016). Waterborne diseases arise primarily from consuming water contaminated with human or animal waste that contains harmful microorganisms. Worldwide, diarrheal illnesses from contaminated water are a leading cause of mortality and morbidity, resulting in 1.8 million deaths and around 4 billion cases of illness each year (Manetu and Karanja 2021). Lin et al. reviewed approximately 85 studies on water pollution and associated diseases, concluding that diarrhea is the most prevalent illness linked to water contamination, primarily spread by enteroviruses in water sources (Marmot et al. 2007). WHO statistics reveal that 19.3 million people were diagnosed with cancer in 2020, and cancer-related deaths reached 10 million. Currently, one in five people worldwide is expected to develop cancer over their lifetime. The types and concentrations of carcinogens in drinking water can vary depending on the contamination point, whether from the water source, during treatment, or in the distribution process. In water sources, contaminants such as arsenic, nitrate, and chromium are closely linked to cancer. Consuming arsenic-contaminated water, in particular, is associated with risks of skin, kidney, and bladder cancers (Lin et al. 2022).

Viruses are the smallest microorganisms among all parasites, measuring approximately 0.03–0.1 μm in size. Although they can be found in drinking water sources, their effects on human health are not as widely recognized or understood. Nevertheless, ingesting these viruses can lead to significant health issues (Adelodun et al. 2021). Over 100 different enteric viruses from humans and animals have been identified as waterborne. Viral infections such as rotavirus, enterovirus, norovirus, and hepatitis A and E can be transmitted through contaminated water (Annan et al. 2018).

Research has shown that various water treatment methods are employed in developing countries, reducing waterborne diseases by 50–90%. Among these methods, ceramic-based water purification systems are considered one of the most modern and cost-effective solutions, known for their high efficiency and local manufacturability. As a result, ceramic water filtration systems are popular due to their affordability and ease of maintenance (Shayo et al. 2023).

7.4 Global and Regional Efforts in Water Resource Management: A Current Scenario

Water is an essential natural resource and a critical environmental factor, playing an irreplaceable role in supporting sustainable socio-economic development. Its cyclical nature makes it renewable, ensuring its continued availability for both ecosystems and human needs. Water resource management focuses on understanding water's formation, distribution, and movement to meet human needs while addressing environmental concerns. Global warming and shifts in atmospheric systems have intensified the timing and distribution of water cycle changes, worsening both global and regional water scarcity (Yang et al. 2021). India's Ministry of Water Resources developed the National Water Policy to guide water resource planning and development. One major initiative, the National River Linking Project (NRLP), aims to connect rivers across the country to address water scarcity and improve resource distribution. India has also launched several sustainable water management programs, including the Jal Shakti Abhiyan, Pradhan Mantri Krishi Sinchayee Yojana, National Water Mission, National Rural Drinking Water Programme, Jal Jeevan Mission, and Atal Bhujal Yojana, to promote effective water management (Singh 2024a, b).

The water cycle is vital for sustainable development, especially climate change. It links the atmosphere, lithosphere, biosphere, and human activities, which are increasingly altering water patterns. Recent climate and land use changes have heightened water variability, creating challenges for global water security.Water scarcity, both acute and chronic, impacts 4 billion people globally (Mekonnen and Hoekstra 2016). A monthly monitoring system, using hydrologic data from the Famine Early Warning System Network (FEWS NET), the Land Data Assimilation System (FLDAS), and World population data, calculates per capita water availability and generates monthly maps of water scarcity. These maps, based on Falkenmark's classifications and deviations from historical averages, identify acute water scarcity events and provide decision-makers with timely information to support water conservation efforts (McNally et al. 2019; Tzanakakis et al. 2020). Feeding 10 billion people by 2050 will demand a 50% increase in agricultural production (which currently uses 70% of water resources) and a 15% rise in water withdrawals (Water Resources Management).

7.5 Sustainable Water Management Technologies for Water Purification, Conservation, and Management

Sustainability is a dynamic concept, shaped by the continuous and varied changes happening across different timeframes and geographic areas. Our reliance on water resource systems is vital for both our survival and well-being. However, expecting these systems to revert to or maintain their untouched, most productive natural states is unrealistic, given the increasing demands for land and water due to development in their surrounding watersheds and water bodies such as streams, rivers, lakes, and aquifers (Loucks 2000). Sustainable water management is crucial for optimizing the combined and efficient use of various water resources and developing allocation strategies that maximize economic and social benefits while enhancing water productivity. Additionally, it is essential to prioritize equitable access to water and consider the social impacts of water allocation policies (Mishra et al. 2021).

Sustainable water resource management involves effectively meeting the growing demand for water today while ensuring that future generations have the same ability to meet their own water needs. It emphasizes the need for careful planning, responsible consumption, and innovative solutions to address the challenges posed by climate change, population growth, and industrial development (Issaoui et al. 2022). Industrial effluents release various pollutants into natural water sources, including hazardous chemicals used in the production of sportswear, fashion, and luxury clothing. These pollutants, such as heavy metals, pesticides, ammonia, and infectious microorganisms, are harmful to aquatic life and humans. Therefore, it is crucial to remove toxic substances from industrial wastewater to protect both the environment and public health (Senthil Kumar and Saravanan 2018). Water treatment involves removing biological, chemical, and physical contaminants that could harm water supplies for human and domestic use. This process produces water that is safe, pleasant-tasting, clear, colourless, and odour-free. Treating water before distribution is essential to enhance its quality, supporting a sustainable and healthy life (Pakharuddin et al. 2021).

Water from natural sources often contains various physical, chemical, and microbiological contaminants, rendering it unsafe for human consumption. The purification process, which transforms contaminated, suspect, or unclean water into drinkable water, is complex and varies based on the type and concentration of contaminants

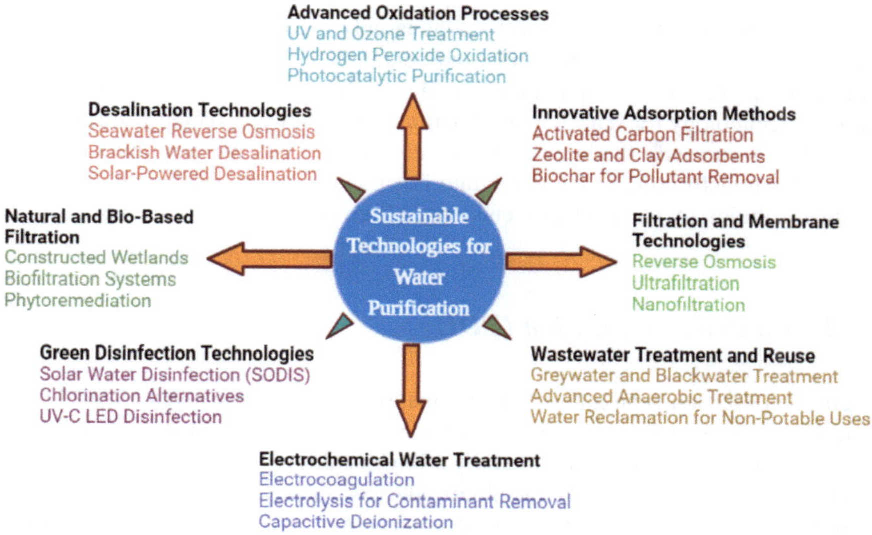

Fig. 7.2 Advance water purification technologies

present. Some impurities pose significant health risks, while others affect sensory properties such as taste, odor, and appearance. Organizations like the EPA and WHO have established guidelines for permissible contaminant levels in drinking water. To comply with these standards, a variety of purification technologies are employed, including reverse osmosis, electro-deionization (EDI), sedimentation, ultrafiltration, ion exchange, biosorption, and ultraviolet (UV) treatment. Adsorption using activated carbon and biochar traps pollutants, and IoT-enabled purification systems enable real-time monitoring and treatment adjustments (Fig. 7.2). These methods efficiently remove harmful substances and improve the water's safety, clarity, and overall quality (Ahmad and Azam 2019). Rapid industrialization, agricultural expansion, and population growth have led to economic and physical water scarcity. High-efficiency desalination and advanced treatment technologies with strong ion removal capabilities are essential to address these challenges. Electrochemical separation offers promising features, including chemical selectivity, compact design, low secondary waste, and versatility for ion separation and purification.

7.5.1 *Reverse Osmosis (RO)*

Using a semi-permeable membrane under pressure, is widely used for advanced drinking water purification. When applied to freshwater, RO's energy consumption is minimized due to the low osmotic pressure difference, reducing the impact of concentration polarization. Recently, interest in RO for purifying freshwater resources has grown significantly. This increased focus is largely because RO offers an effective barrier against emerging micro- and nano-contaminants, which are challenging to remove through conventional treatment methods (Zhai et al. 2022). RO stands as the foremost technology for seawater desalination and wastewater reclamation (Li et al. 2021). RO is currently the leading desalination technology and continues to grow rapidly. By transforming saline water into fresh, clean water, desalination provides a critical solution to water scarcity challenges (Lee et al. 2011).

7.5.2 *Electro Deionization (EDI)*

Electro deionization is an integrated desalination process that merges ion exchange with electrodialysis to enable continuous deionization (Liu et al. 2019). Electrochemical separation techniques are particularly promising due to their chemical selectivity, adaptability, compact design, reduced secondary waste production, and broad applicability in ion separation and water purification (Fig. 7.3). To address water pollution and shortages, there is a pressing need for advanced water treatment and desalination technologies that offer high ion removal efficiency and energy savings (Khan et al. 2023). EDI, or continuous deionization (CDI), is a hybrid separation technology that combines ion-exchange resins and ion-exchange membranes. EDI is increasingly recognized for its effectiveness in removing and recovering ions from water. In a continuous electrodeionization (CEDI) system, the ion-depleting compartments and sometimes the ion-concentrating compartments are packed with ion exchange resins. These resins facilitate ion transport and can also act as a medium for electrochemical reactions, like the dissociation of water into H^+ and OH^- ions. Various media configurations are used in EDI, such as a mixture of anion and cation exchange resins or distinct sections of resins with consistent polarity, containing either anion or cation exchange resins in each section (Arar et al. 2014).

7.5.3 *Sedimentation*

Sedimentation is a physical water treatment method that relies on gravity to settle colloidal and suspended solids in water. Commonly used in conventional water treatment, sedimentation (or settling) takes place in a sedimentation basin, also known as a settling tank or clarifier. This process, known as clarification, involves removing turbidity caused by fine suspended solids, colloidal impurities, and organic color through coagulation, flocculation, and sedimentation steps (Prasad 2019). Sedimentation is crucial for removing solid particles, as sedimentation tanks allow for the separation of contaminants from liquid at a relatively low cost through straightforward design. Simpler water purification systems, like sedimentation tanks, are being developed to expand their use in urban areas and enhance purification efficiency while minimizing installation space. Swirl sedimentation tanks, available

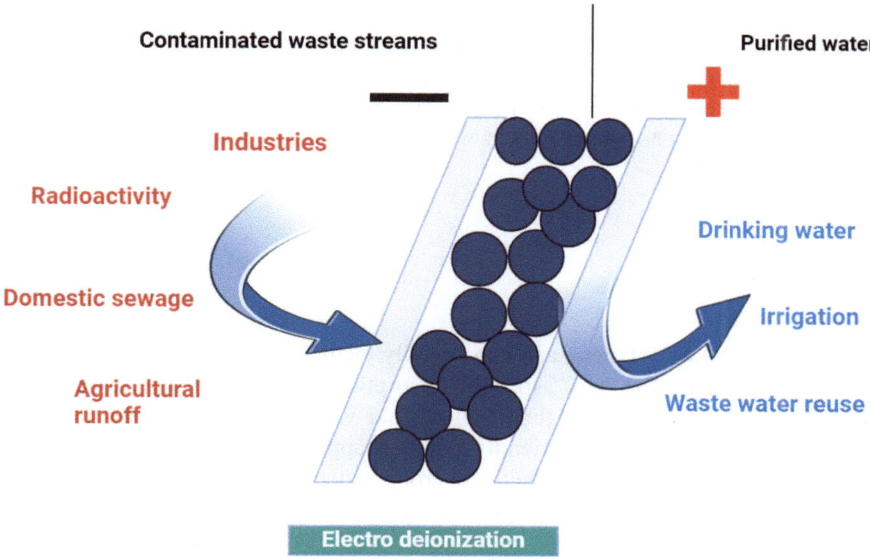

Fig. 7.3 Mechanism of electro deionization technique for waste water treatment

from various providers, are designed to capture solid contaminants, sludge, and slurries from rainwater, snowmelt, and wastewater moving through sewers (Ochowiak et al. 2017).

7.5.4 Membrane Technology

In wastewater treatment, micro- and ultra-filtration membranes are employed to separate activated sludge (biomass) from treated water. This method provides the benefits of thoroughly removing solids, bacteria, and most viruses, particularly those bound to suspended solids (Cornel and Krause 2006).

Membrane technology has gained significant attention in recent years due to its efficiency in both removing contaminants and recovering resources or energy from water. Membrane materials are generally categorized into polymeric, ceramic, and metal membranes, among others (Wang et al. 2022). White water from paper mills using recycled fibers contains various suspended particles, ranging from hundreds of micrometers to sub-micrometer sizes, due to fillers, colloidal materials from latex binders, dispersants from coated grades, and other papermaking additives. Since white water is reused repeatedly in modern paper machines, purifying it is essential to prevent contaminants from affecting paper quality or production. Ultrafiltration is commonly used for this purpose, as it effectively removes fine solid and colloidal particles at the sub-micrometer level (Hassan et al. 2020a, b). In recent years, ceramic membranes have attracted growing interest for their effectiveness in removing contaminants from water. Among these, alumina membranes are particularly notable for their versatile roles, supporting separation processes like microfiltration, ultrafiltration, and nanofiltration, as well as enhanced separation applications involving catalysis and adsorption in water purification and wastewater treatment (Wang et al. 2022).

7.5.5 *Biosorption*

Biosorption is a fast, efficient, and eco-friendly method for detecting and removing water contaminants. The "bio" aspect of biosorption encompasses the use of natural materials, including bacteria, fungi, algae, and various agricultural or industrial wastes, positioning it as an eco-friendly treatment approach (Muhammad et al. 2020). These biosorbents have diverse functional groups, such as carboxyl, sulfonate, amine, phosphonate, and hydroxyl groups, on their cell wall surfaces. These groups play a critical role in facilitating various biosorption mechanisms, including complexation, chelation, ion exchange, and physical adsorption (Brazesh et al. 2021). Biosorption has gained attention as a sustainable approach for purifying water contaminated by industrial activities and natural processes. Biomaterials used in biosorption offer unique reactive sites where contaminants can be effectively adsorbed, making this process particularly suited for thermodynamic analysis. By applying surface complexation theory, researchers can examine the energy and dynamics involved in the interaction between biomaterials and various contaminants. This theoretical framework enables the development of predictive models that simulate the behaviour of biosorption under diverse environmental conditions, allowing for better handling of complex mixtures commonly found in both natural settings and industrial effluents (Ngwenya et al. 2009). Biosorption using fungal pellets is widely acknowledged for effectively removing hazardous pollutants from water sources. This effectiveness is largely attributed to the unique properties of fungal cell walls, which are rich in polysaccharides and proteins with diverse functional groups such as phosphate, carboxyl, carbonyl, amine, and amide groups. These groups enable interactions with both inorganic and organic pollutants through various chemical forces and mechanisms (Legorreta-Castañeda et al. 2020).

7.5.6 *Ultraviolet (UV) Disinfection*

UV disinfection is a highly effective method for inactivating pathogens in water and is gaining popularity in industrial applications. UV radiation offers several advantages over conventional chemical disinfection methods like chlorination or ozonation. It requires no chemical additives, does not produce harmful disinfection by-products

(DBPs), and avoids promoting disinfectant-resistant bacteria. As a result, UV disinfection is increasingly recommended as an alternative to chemical treatments for surface water purification (Song et al. 2016).

7.5.7 Advanced Oxidation Processes (AOPs)

AOPs are promising water treatment methods for breaking down persistent or toxic organic pollutants, especially those resistant to traditional treatments. AOPs utilize hydroxyl radicals (OH) to oxidize contaminants in wastewater and industrial effluents. They can be activated by UV or visible light, oxidants like O_2, H_2O_2, or O_3, and catalysts like TiO_2. AOPs are generally classified into photochemical (light-dependent) and non-photochemical methods. Non-photochemical methods, which generate OH radicals without light, include ozonation (O_3), ozonation with hydrogen peroxide (O_3/H_2O_2), catalytic ozonation (O_3/catalysts), the Fenton system (Fe^{2+}/H_2O_2), and ultrasound (Cardoso et al. 2021; Schwätter et al. 2007).

7.5.8 Innovative Techniques for Water Conservation and Purification

Innovative techniques for water conservation and purification include rainwater harvesting systems, greywater recycling, desalination technologies, smart irrigation systems, and advanced water purification methods, which help address global challenges.

In a world where water is a finite and vital resource, innovative water management is essential for sustainable development. Water scarcity, pollution, and inefficient usage pose serious global threats, impacting the environment, human health, agriculture, and industry ("Innovations In Water Conservation And Purification"). There are number of examples of low-energy, cost-effective, and environmentally friendly methods as Ceramic membranes (Dong et al. 2022), photocatalysis (Niknejad et al. 2023), bioremediation (Haripriyan et al. 2022), biosorbents, magnetic sorbents (Osman et al. 2023), and carbon-based materials such as activated carbon, carbon nanotubes, graphene, and graphene oxide (Kotia et al. 2020) are promising, cost-effective, and energy-efficient options for eco-friendly water treatment. The chemical and thermal stability, low fouling propensity, and long lifespan of ceramic membranes make this technology highly attractive. Additionally, ceramic membranes can be integrated with advanced oxidation processes, such as in-situ ozonation, which cannot be used with polymeric membranes due to the risk of degradation over long-term exposure (Asif and Zhang 2021).

Photocatalysis is an emerging wastewater treatment method with great potential, particularly for effluents containing toxic and refractory organics that resist conventional treatment. It is an effective technology for mineralizing complex contaminants. The process uses photocatalysts to convert photonic energy (from sunlight, UV, or visible light) into chemical energy, generating electron–hole (e^--h^+) pairs. These energized pairs reduce and oxidize contaminants on the photocatalyst surface, enabling the degradation of pollutants (Mishra and Sundaram 2023).

Bioremediation is an innovative, cost-effective technology for removing heavy metals from polluted water and soil. Using microorganisms, it mineralizes organic contaminants into end products like water, CO_2, and metabolites, which support cell growth, making it a promising solution for future environmental cleanup (Saeed et al. 2022).

Magnetic biosorbents, including metal or metal nanoparticles on biosorbents, are efficient and eco-friendly materials for remediating heavy metals and organic contaminants in water treatment. Their increasing popularity stems from their adjustable active sites, large surface area, easy separation, and reusability. The performance of magnetic biosorbents depends on factors such as raw biomass, metal nanoparticle properties, modification methods, and process parameters, which affect their contaminant removal efficiency (Hassan 2020a, b). Carbon nanomaterials, including activated carbon, carbon nanotubes, graphene, graphene oxide, and carbon/graphene quantum dots, have shown great promise as adsorbents for wastewater treatment and purification (Sheoran et al. 2022).

7.6 Advancements in Water Treatment Technologies

ECs like endocrine-disrupting compounds, personal care products, and pharmaceuticals pose serious environmental and health threats. These contaminants originate from everyday cosmetics, medicines, and pesticides. As global water resources decline, effective treatment methods for safely removing ECs from wastewater are increasingly critical, even at low concentrations.

Advanced treatment technologies, including NF, UF, RO (Yusuf et al. 2020), MBRs, and catalytic oxidation, are effective for EC removal. Still, their high operational costs present significant challenges. To address the limitations of both conventional and advanced systems, integrated treatment processes are being developed and studied. These processes combine physical, chemical, and biological treatments to remove ECs from water effectively.

Physical treatment methods include adsorption and filtration, while biological treatments involve techniques such as activated sludge, trickling filters, biological activated carbon, constructed wetlands, and MBRs. Chemical processes essential for EC removal include ozonation, coagulation, advanced oxidation processes, Sono chemical methods, Fenton and photo-Fenton processes, and photocatalysis (Shahid et al. 2021).

Conventional treatment methods, which combine physical, chemical, and biological processes, are proving to be less efficient. Recently, carbon nanotubes (CNTs) have emerged as a promising and adaptable solution for effective water treatment due to their outstanding material properties. These include a large surface area, impressive porosity, hollow and layered structures, and the ability to be easily chemically activated and functionalized, making them ideal for adsorption applications (Goyal et al. 2023).

In some instances, multiple treatment methods are used in conjunction to enhance water purification. Among the various modern technologies, adsorption has proven to be one of the most effective and economical methods for removing pollutants from wastewater. This technique is favored for its low operational costs, straightforward implementation, and simple design (Chang et al. 2018; Younas et al. 2021).

Heavy metals such as zinc (Zn), copper (Cu), lead (Pb), nickel (Ni), cadmium (Cd), and mercury (Hg) contribute significantly to environmental issues due to their toxicity levels (Table 7.1) (Carolin et al. 2017). Primarily resulting from industrial activities like electrolysis, electroplating, metal smelting, and chemical manufacturing. These metals pose serious health risks due to their teratogenic and carcinogenic nature. According to Saleh et al. (2022), heavy metals in open waters can destroy aquatic life, cause oxygen depletion, and cause the growth of harmful algal blooms. When these metals are released into rivers, they are converted into hydrated ions, which are even more toxic than the original metal atoms. Ionic forms of metals like Cd^{2+}, Pb^{2+}, Hg^{2+}, Ag^{+}, and As^{3+} bind with biological particles to form toxic compounds, highlighting the importance of isolating them due to health risks. These ions interfere with enzymatic processes and are more readily absorbed by organisms, exacerbating the threat. As a result, removing heavy metals from water is essential to mitigate public health risks. Organizations such as the WHO and the EPA have established safe discharge limits for heavy metals to control water pollution (Carolin et al. 2017). Conventional removal techniques, including ion exchange, membrane filtration, and pH-sensitive methods, have limitations: ion exchange is less effective in concentrated metal solutions, membrane filtration has high initial costs, and pH-sensitive methods require continuous monitoring, which can be challenging. Beyond these traditional approaches, newer methods like photocatalysis, flotation, remediation, electrochemical treatment, adsorption, and coagulation/flocculation offer promising solutions. Their industrial scalability and practical feasibility make them more adaptable and effective for treating heavy metal contamination in water sources (Carolin et al. 2017; Saleh et al. 2022).

7.7 Recent Innovations in Water Treatment

Water treatment using advanced filtration membranes, such as MF and UF, provides an energy-efficient solution for removing emulsified contaminants, oils, and suspended solids without needing heat or chemicals. These systems are versatile and suitable for high-pressure environments and on-site treatments, including off-shore

Table 7.1 Heavy metals present in polluted water and their methods of removal

S. No.	Heavy metals	Effect on human health	Removal techniques	References
1.	Chromium	Cancer, vomiting, pulmonary congestion, liver, skin inflammation, and kidney damage, and ulcer	Electrocoagulation, membrane filtration, adsorption, ion exchange, bioremediation, chemical reduction	Assefa et al. (2024), Wan et al. (2024)
2.	Cadmium	"Itai-Itai" softening of the bones and fractures to human beings, lung cancer, hepatic toxicity, harmful impact on respiratory system, liver, kidney, and reproductive organs	Adsorption, ion exchange, fluidized bed reactor, nanotechnology, hybrid system method, advanced oxidation process, chelation, foam fractionation, cryogens, solvent extraction, carbon nanotubes, graphene, activated carbon, microbial fuel cells, zeolites, and phytoremediation	Ahmad et al. (2024a, b), Nguyen et al. (2023), Türkmen et al.
3.	Nickel	Dry cough, creates breathing problems, chest pain, nausea, pulmonary fibrosis, diarrhea, skin eruption, gastrointestinal ache, renal edema	Adsorption, Ion exchange, and electrocoagulation	Nizeyimana et al. (2024), Wang et al. (2023)
4.	Copper	Anemia, hair loss, kidney damage, and headache	Membrane separation, ion exchange, electrocoagulation, electrochemistry, chemical precipitation, adsorption, and biotechnology	Hansen et al. (2023)
5.	Lead	Mental retardation, kidney and nervous system damage, and cancer	Biochar-based adsorption Adsorption through adsorptive membrane made of the Zeolitic Imidazolate Framework-8 (ZIF-8)	Bayar et al. (2024), Chee et al. (2023)
6.	Mercury	Damage to the kidney, brain, respiratory and reproductive system	Biochar-based adsorption, bioremediation	Ahmed et al. (2024), Bayar et al. (2024)
7.	Zinc	Vomiting, pain, skin inflammation, fever, anaemia	Filtration, ion exchange, precipitation, and electrolysis	Berdimurodov et al. (2023), Mehenktaş and Arar (2023)

oil rigs, marine vessels, and industrial water processes. For strict water quality standards, MF and UF filters can serve as pre-treatment for RO systems. While RO effectively removes contaminants, its membranes are prone to fouling from oils, leading to costly maintenance and frequent replacements (Walters 2024). Modern water treatment membranes, made of polymers with varied porosities, filter solutes from angstrom to micron scales. This technology effectively removes dissolved contaminants and suspended particles, making it essential in desalination and wastewater treatment (Issaoui et al. 2022).

Desalination technologies are evolving to meet changing needs and sustainability goals. However, desalination can be complex, costly, and energy-intensive without careful consideration. According to the International Desalination Association (IDA), 150 countries provide desalinated water to over 3 billion people. Between 2016 and 2019, desalination plants increased by 12.4%, and daily water output grew by 41.2%, highlighting the technology's rapid advancement ("Innovations in Desalination Process: Efforts in making the Desalination Process Efficient and Sustainable"). The global installed capacity of desalination plants is approximately 86,572 MLD, with 44% located in the Middle East and North Africa. Currently, 1% of the world's population relies on desalinated water daily, a figure expected to rise to 14% by 2025 (Water Resources 2021). Nanotechnology is crucial in integrated water management by improving treatment efficiency and expanding water supplies from unconventional sources. It offers advanced nanomaterials for treating contaminated surface water, groundwater, and wastewater, effectively removing harmful metals, solutes, and microorganisms. Current applications in wastewater treatment include nano adsorbents, nano zero-valent iron, nano biocides, nanofiltration, magnetic nanoparticles, and hybrid systems like catalytic wet air oxidation. These methods offer efficient, cost-effective solutions for managing emerging hazardous wastes with reduced energy usage (Nishu and Kumar 2023) (Fig. 7.4).

Water treatment technologies are transforming access to clean water and significantly enhancing public health worldwide. Advanced techniques such as membrane filtration, reverse osmosis, UV purification, and advanced oxidation processes are now more widely implemented, making water safer by effectively removing

Fig. 7.4 Innovations in water conservation and purification for sustainable development

pathogens, chemicals, and heavy metals. Additionally, innovations like photocatalysis, nanotechnology, and electrochemical water treatment enable the removal of complex pollutants, such as pharmaceuticals and microplastics, that are challenging for traditional methods to eliminate. These advancements in water treatment are crucial for reducing diseases caused by waterborne pathogens, like cholera and dysentery, and for protecting against chronic health issues linked to chemical contaminants, such as arsenic and lead. Modern water treatment technologies are laying the foundation for improved health outcomes, particularly in vulnerable communities, by ensuring more reliable and safer water supplies.

7.8 Conclusion

This chapter emphasizes the urgent need for sustainable water management as global water resources face rapid depletion due to population growth, climate change, and industrialization. Human activities, such as industrial processes and agricultural runoff, introduce harmful pollutants that threaten ecosystems and public health. According to WHO data (2023), microbiologically contaminated drinking water causes diseases like diarrhea, cholera, and typhoid, leading to approximately 505,000 diarrheal deaths annually. To combat water pollution, various control methods, including sedimentation, neutralization, flocculation, membrane technologies, and advanced oxidation processes, are employed. However, these methods often fall short of addressing all types of pollutants, particularly ECs from cosmetics, medicines, and pesticides advanced technologies like NF, UF, RO, and MBRs effectively remove ECs but face challenges due to high operational costs. Integrated treatment processes combining physical, chemical, and biological methods are being developed to enhance pollutant removal efficiency. To address all these issues Sustainable water management requires understanding the principles of water resource formation, distribution, and use while addressing environmental challenges. In India, initiatives like the National Water Policy, NRLP, and programs such as Jal Shakti Abhiyan and Jal Jeevan Mission aim to promote effective water management and equitable distribution. Innovative techniques like rainwater harvesting, greywater recycling, and smart irrigation systems are also employed to conserve and purify water efficiently. In conclusion, water is an essential resource, and its sustainable management is crucial to ensure clean water access for current and future generations. By adopting advanced, cost-effective, environment friendly technologies, integrated treatment approaches, and robust policies, we can safeguard water resources, mitigate pollution, and significantly enhance global public health.

Abbreviations

SWM	Sustainable water management
ML	Machine learning
DWTPs	Drinking water treatment plants
CGWB	Central Groundwater Board
WRIS	Water Resource Information System
CWC	Central Water Commission
NRSC	National Remote Sensing Centre
DBU	Designated best use
WQI	Water Quality Index
GRACE	Gravity recovery and climate experiment
VWT	Virtual water trade
VW	Virtual water
TRWR	Total renewable water resources
GHMs	Global hydrological models
WSI	Water Stress Index
NOAA	National Oceanic and Atmospheric Administration
WPI	Water Pollution Index
HPI	Heavy Metal Pollution Index
N	Nitrogen
P	Phosphorous
NRLP	National River Linking Project
FEWS NET	Famine Early Warning System Network
FLDAS	Famine Land Data Assimilation System
EDI	Electro-deionization
UV	Ultraviolet
RO	Reverse osmosis
CDI	Continuous deionization
CEDI	Continuous electro deionization
DBPs	Disinfection by-products
AOPs	Advanced Oxidation Processes
OH	Hydroxyl radicals
O_2	Ozone
O_2/H_2O_2	Ozonation with hydrogen peroxide
O_2/catalysts	Catalytic ozonation
Fe^{2+}/H_2O_2	The Fenton system
CO_2	Carbon dioxide
ECs	Emerging contaminants
EDCs	Endocrine disrupting compounds
PCPs	Personal care products
NF	Nanofiltration
UF	Ultrafiltration
RO	Reverse osmosis

MBRs	Membrane bioreactors
CNTs	Carbon nanotubes
Zn	Zinc
Cu	Copper
Pb	Lead
Ni	Nickel
Cd	Cadmium
Hg	Mercury
WHO	World Health Organization
EPA	Environmental Protection Agency
MF	Microfiltration
IDA	International Desalination Association
MLD	Minimal liquid discharge

References

Adelodun B, Ajibade FO, Ighalo JO, Odey G, Ibrahim RG, Kareem KY et al (2021) Assessment of socioeconomic inequality based on virus-contaminated water usage in developing countries: a review. Environ Res 192:110309

Afroz R, Masud MM, Akhtar R, Duasa JB (2014) Water pollution: Challenges and future direction for water resource management policies in Malaysia. Environ Urban Asia 5(1):63–81

Ahmad A, Azam T (2019) 4—Water purification technologies. In: Grumezescu AM, Holban AM (eds) Bottled and packaged water. Woodhead Publishing, pp 83–120

Ahmad S, Jia H, Ashraf A, Yin D, Chen Z, Xu C et al (2023) Water resources and their management in Pakistan: a critical analysis on challenges and implications. Water-Energy Nexus 6:137–150. https://doi.org/10.1016/j.wen.2023.10.001

Ahmad I, Asad RU, Maryam L, Masood M (2024) Treatment methods for cadmium removal from wastewater. In: Jha AK, Kumar N (eds) Cadmium toxicity in water: challenges and solutions. Springer, Cham, pp 139–174

Ahmad I, Fatima N, Naz E, Farooqi ZUR, Bulgariu L (2024) Treatment methods for mercury removal from soil and wastewater. In: Mercury toxicity mitigation: sustainable nexus approach. Springer, Cham, pp 257–281

Ahmed AA, Sayed S, Abdoulhalik A, Moutari S, Oyedele L (2024) Applications of machine learning to water resources management: a review of present status and future opportunities. J Clean Prod 441:140715. https://doi.org/10.1016/j.jclepro.2024.140715

Alkhadra MA, Su X, Suss ME, Tian H, Guyes EN, Shocron AN et al (2022) Electrochemical methods for water purification, ion separations, and energy conversion. Chem Rev 122(16):13547–13635

Annan E, Agyei-Tuffour B, Bensah YD, Konadu DS, Yaya A, Onwona-Agyeman B, Nyankson E (2018) Application of clay ceramics and nanotechnology in water treatment: a review. Cogent Eng 5(1):1476017

Anusha BN, Babu KR, Kumar BP, Kumar PR, Rajasekhar M (2022) Geospatial approaches for monitoring and mapping of water resources in semi-arid regions of Southern India. Environ Challenges 8:100569. https://doi.org/10.1016/j.envc.2022.100569

Arar Ö, Yüksel Ü, Kabay N, Yüksel M (2014) Various applications of electrodeionization (EDI) method for water treatment—a short review. Desalination 342:16–22. https://doi.org/10.1016/j.desal.2014.01.028

Arif A, Malik MF, Liaqat S, Aslam A, Mumtaz K, Afzal A et al (2020) Water pollution and industries. Pure Appl Biol 9(4):2214–2224

Asif MB, Zhang Z (2021) Ceramic membrane technology for water and wastewater treatment: a critical review of performance, full-scale applications, membrane fouling and prospects. Chem Eng J 418:129481. https://doi.org/10.1016/j.cej.2021.129481

Assefa H, Singh S, Olu FE, Dhanjal DS, Mani D, Khan NA et al (2024) Advances in adsorption technologies for hexavalent chromium removal: mechanisms, materials, and optimization strategies. Desalin Water Treat 319:100576. https://doi.org/10.1016/j.dwt.2024.100576

Babuji P, Thirumalaisamy S, Duraisamy K, Periyasamy GJW (2023) Human health risks due to exposure to water pollution: a review. Water 15(14):2532

Bayar J, Ali N, Dong Y, Ahmad U, Anjum MM, Khan GR et al (2024) Biochar-based adsorption for heavy metal removal in water: a sustainable and cost-effective approach. Environ Geochem Health 46(11):428. https://doi.org/10.1007/s10653-024-02214-w

Berdimurodov E, Berdimuradov K, Eliboev I, Azimov L, Rajabov Y, Mamatov J et al (2023) Chemical methods of heavy metal management—filtration, ion exchange, and electrolysis. In: Heavy metals in the environment: management strategies for global pollution. ACS Publications, pp 229–245

Brazesh B, Mousavi SM, Zarei M, Ghaedi M, Bahrani S, Hashemi SA (2021) Chapter 9—Biosorption. In: Ghaedi M (ed) Interface Science and Technology, vol 33. Elsevier, Amsterdam, pp 587–628

Brodie Rudolph T, Ruckelshaus M, Swilling M, Allison EH, Österblom H, Gelcich S, Mbatha P (2020) A transition to sustainable ocean governance. Nat Commun 11(1):3600. https://doi.org/10.1038/s41467-020-17410-2

Cardoso IM, Cardoso RM, da Silva JCE (2021) Advanced oxidation processes coupled with nanomaterials for water treatment. Nanomaterials 11(8):2045

Carolin CF, Kumar PS, Saravanan A, Joshiba GJ, Naushad M (2017) Efficient techniques for the removal of toxic heavy metals from aquatic environment: a review. J Environ Chem Eng 5(3):2782–2799. https://doi.org/10.1016/j.jece.2017.05.029

Chan NW (2012) Managing urban rivers and water quality in malaysia for sustainable water resources. Int J Water Resour Dev 28(2):343–354. https://doi.org/10.1080/07900627.2012.668643

Chang H, Quan X, Zhong N, Zhang Z, Lu C, Li G et al (2018) High-efficiency nutrients reclamation from landfill leachate by microalgae *Chlorella vulgaris* in membrane photobioreactor for biolipid production. Bioresour Technol 266:374–381

Chee DNA, Aziz F, Ismail AF, Kueh ABH, Amin MAM, Amran M (2023) Adsorptive zeolitic imidazolate framework-8 membrane on ceramic support for the removal of lead(II) ions. Chem Eng Sci 276:118775. https://doi.org/10.1016/j.ces.2023.118775

Chowdhary P, Bharagava RN, Mishra S, Khan N (2020) Role of industries in water scarcity and its adverse effects on environment and human health. In: Shukla V, Kumar N (eds) Environmental concerns and sustainable development, vol 1. Air, water and energy resources. Springer, Singapore, pp 235–256

Climate Change Impacts on the Ocean and Marine Resources. Climate change impacts

Cornel P, Krause S (2006) Membrane bioreactors in industrial wastewater treatment—European experiences, examples and trends. Water Sci Technol 53(3):37–44. https://doi.org/10.2166/wst.2006.074

Ding GKC, Ghosh S (2017) Sustainable water management—a strategy for maintaining future water resources. In: Abraham MA (ed) Encyclopedia of sustainable technologies. Elsevier, Oxford, pp 91–103

Dolan F, Lamontagne J, Link R, Hejazi M, Reed P, Edmonds J (2021) Evaluating the economic impact of water scarcity in a changing world. Nat Commun 12(1):1915. https://doi.org/10.1038/s41467-021-22194-0

Dong Y, Wu H, Yang F, Gray S (2022) Cost and efficiency perspectives of ceramic membranes for water treatment. Water Res 220:118629. https://doi.org/10.1016/j.watres.2022.118629

Drinking Water (2023) https://www.who.int/news-room/fact-sheets/detail/drinking-water
Goyal N, Nawaz A, Chandel KS, Devnarayan D, Gupta L, Singh S et al (2023) A cohesive effort to assess the suitability and disparity of carbon nanotubes for water treatment. Environ Sci Pollut Res Int 30(60):124832–124853. https://doi.org/10.1007/s11356-022-23137-y
Hansen HK, Gutiérrez C, Leiva Gonzalez J, Lazo A, Hansen ME, Lazo P et al (2023) Combined electrodialysis and electrocoagulation as treatment for industrial wastewater containing arsenic and copper. Membranes 13(3):264
Haripriyan U, Gopinath KP, Arun J, Govarthanan M (2022) Bioremediation of organic pollutants: a mini review on current and critical strategies for wastewater treatment. Arch Microbiol 204(5):286. https://doi.org/10.1007/s00203-022-02907-9
Hassan M, Naidu R, Du J, Liu Y, Qi F (2020a) Critical review of magnetic biosorbents: their preparation, application, and regeneration for wastewater treatment. Sci Total Environ 702:134893. https://doi.org/10.1016/j.scitotenv.2019.134893
Hassan ML, Fadel SM, Abouzeid RE, Abou Elseoud WS, Hassan EA, Berglund L, Oksman K (2020b) Water purification ultrafiltration membranes using nanofibers from unbleached and bleached rice straw. Sci Rep 10(1):11278. https://doi.org/10.1038/s41598-020-67909-3
Hojjati-Najafabadi A, Mansoorianfar M, Liang T, Shahin K, Karimi-Maleh H (2022) A review on magnetic sensors for monitoring of hazardous pollutants in water resources. Sci Total Environ 824:153844. https://doi.org/10.1016/j.scitotenv.2022.153844
Hossain MF (2019) Chapter Six—Water. In: Hossain MF (ed) Sustainable design and build. Butterworth-Heinemann, pp 301–418
Hossain M, Patra PK (2020) Water pollution index—a new integrated approach to rank water quality. Ecol Ind 117:106668. https://doi.org/10.1016/j.ecolind.2020.106668
Huang R, Ma C, Ma J, Huangfu X, He Q (2021a) Machine learning in natural and engineered water systems. Water Res 205:117666. https://doi.org/10.1016/j.watres.2021.117666
Huang Z, Yuan X, Liu X (2021b) The key drivers for the changes in global water scarcity: water withdrawal versus water availability. J Hydrol 601:126658. https://doi.org/10.1016/j.jhydrol.2021.126658
Innovations in Desalination Process: Efforts in making the Desalination Process Efficient and Sustainable. https://epcmholdings.com/innovations-in-desalination-process-efforts-in-making-the-desalination-process-efficient-and-sustainable/
Innovations in water conservation and purification. https://fastercapital.com/topics/innovations-in-water-conservation-and-purification.html
Issaoui M, Jellali S, Zorpas AA, Dutournie P (2022) Membrane technology for sustainable water resources management: challenges and future projections. Sustain Chem Pharm 25:100590. https://doi.org/10.1016/j.scp.2021.100590
Iyiola AO, Kolawole AS, Ajayi FO, Ogidi OI, Ogwu MC (2023a) Sustainable water use and management for agricultural transformation in Africa. In: Madhav S, Srivastav AL, Izah SC, van Hullebusch E (eds) Water resources management for rural development. Elsevier, Amsterdam, pp 287–299. https://doi.org/10.1016/B978-0-443-18778-0.00014-3
Iyiola AO, Izah SC, Morya S, Akinsorotan A, Ogwu MC (2023b) Owalla reservoir in South-western Nigeria: assessment of fish distribution, biological diversity, and Water Quality Index. Indones J Agric Res 6(02):106–125. https://doi.org/10.32734/injar.v6i2.12207
Iyiola AO, Afolabi OA, Alimi SK, Akingba OO, Izah SC, Ogwu MC (2024a) Climate change and water crisis in the Global South. In: Izah SC, Ogwu MC, Loukas A, Hamidifar H (eds) Water crises and sustainable management in the Global South. Springer, Singapore. https://doi.org/10.1007/978-981-97-4966-9_4
Iyiola AO, Kolawole AS, Setufe SB, Bilikoni J, Ofori E, Ogwu MC (2024b) Fish as a sustainable biomonitoring tool in aquatic environments. In: Izah SC, Ogwu MC, Hamidifar H (eds) Biomonitoring of pollutants in the Global South. Springer, Singapore. https://doi.org/10.1007/978-981-97-1658-6_12
Iyiola AO, Setufe SB, Ofori E, Bilikoni J, Ogwu MC (2024c) Biomarkers for the detection of pollutants from the water environment. In: Izah SC, Ogwu MC, Hamidifar H (eds) Biomonitoring of

pollutants in the Global South. Springer, Singapore. https://doi.org/10.1007/978-981-97-1658-6_16
Khan ZU, Moronshing M, Shestakova M, Al-Othman A, Sillanpää M, Zhan Z et al (2023) Electro-deionization (EDI) technology for enhanced water treatment and desalination: a review. Desalination 548:116254. https://doi.org/10.1016/j.desal.2022.116254
Khilchevskyi V, Karamushka V (2021) Global water resources: distribution and demand. In: Clean Water and Sanitation, Springer, Cham, pp 1–11
Kotia A, Yadav A, Rohit Raj T, Gertrud Keischgens M, Rathore H, Sarris IE (2020) Carbon nanoparticles as sources for a cost-effective water purification method: a comprehensive review. Fluids 5(4):230
Kumar N, Kumar A, Marwein BM, Verma DK, Jayabalan I, Kumar A, Kumar N (2021) Agricultural activities causing water pollution and its mitigation—a review. Int J Modern Agric 10(1):590–609
Lee KP, Arnot TC, Mattia D (2011) A review of reverse osmosis membrane materials for desalination—development to date and future potential. J Membr Sci 370(1):1–22. https://doi.org/10.1016/j.memsci.2010.12.036
Legorreta-Castañeda AJ, Lucho-Constantino CA, Beltrán-Hernández RI, Coronel-Olivares C, Vázquez-Rodríguez GA (2020) Biosorption of water pollutants by fungal pellets. Water 12(4):1155
Li W-X, Yang Z, Liu W-L, Huang Z-H, Zhang H, Li M-P et al (2021) Polyamide reverse osmosis membranes containing 1D nanochannels for enhanced water purification. J Membr Sci 618:118681. https://doi.org/10.1016/j.memsci.2020.118681
Li P, Wang D, Li W, Liu L (2022) Sustainable water resources development and management in large river basins: an introduction. Environ Earth Sci 81(6):179. https://doi.org/10.1007/s12665-022-10298-9
Lin L, Yang H, Xu X (2022) Effects of water pollution on human health and disease heterogeneity: a review. Front Environ Sci 10:880246
Liu Y, Wang J, Xu Y, Wu B (2019) A deep desalination and anti-scaling electrodeionization (EDI) process for high purity water preparation. Desalination 468:114075. https://doi.org/10.1016/j.desal.2019.114075
Liu Y, Wang H, Cui Y, Chen N (2023) Removal of copper ions from wastewater: a review. Int J Environ Res Public Health 20(5):3885
Loucks DP (2000) Sustainable water resources management. Water International 25(1):3–10. https://doi.org/10.1080/02508060008686793
Lukman L (2023) Anthropogenic Impact on Lake Ecosystem. In Ali AA (ed) Science of lakes, Chap 5. IntechOpen, Rijeka
Madhav S, Ahamad A, Singh AK, Kushawaha J, Chauhan JS, Sharma S, Singh P (2020) Water pollutants: sources and impact on the environment and human health. In: Pooja D, Kumar P, Singh P, Patil S (eds) Sensors in water pollutants monitoring: role of material. Springer, Singapore, pp 43–62
Manetu WM, Karanja AM (2021) Waterborne disease risk factors and intervention practices: a review. Open Access Library Journal 8(5):1–11
Marmot M, Atinmo T, Byers T, Chen J, Hirohata T, Jackson A et al (2007) Food, nutrition, physical activity, and the prevention of cancer: a global perspective
McNally A, Verdin K, Harrison L, Getirana A, Jacob J, Shukla S et al (2019) Acute water-scarcity monitoring for Africa. Water 11(10):1968
Mehenktaş C, Arar Ö (2023) Removal of zinc (Zn^{2+}) through biopolymer-enhanced ultrafiltration. J Polym Environ 31(4):1373–1382
Mekonnen MM, Hoekstra AY (2016) Four billion people facing severe water scarcity. Sci Adv 2(2):e1500323
Ministry of Jal Shakti. https://nwic.gov.in/data
Mishra RK (2023) Fresh water availability and its global challenge. Br J Multidiscip Adv Stud 4(3):1–78

Mishra S, Sundaram B (2023) A review of the photocatalysis process used for wastewater treatment. Mater Today Proc. https://doi.org/10.1016/j.matpr.2023.07.147
Mishra BK, Kumar P, Saraswat C, Chakraborty S, Gautam A (2021) Water security in a changing environment: concept, challenges and solutions. Water 13(4):490
Moeller T (2012) Chemistry: with inorganic qualitative analysis. Elsevier, Amsterdam
Morin-Crini N, Lichtfouse E, Liu G, Balaram V, Ribeiro ARL, Lu Z et al (2022) Worldwide cases of water pollution by emerging contaminants: a review. Environ Chem Lett 20(4):2311–2338. https://doi.org/10.1007/s10311-022-01447-4
Muhammad G, Mehmood A, Shahid M, Ashraf RS, Altaf M, Hussain MA, Raza MA (2020) Biochemical methods for water purification. In: Inamuddin M, Ahamed I, Lichtfouse E, Asiri AM (eds) Methods for bioremediation of water and wastewater pollution. Springer, Cham, pp 181–212
Nguyen TT, Thi TMLP, Thi TN, Le TT, Thi CTN, Nguyen NH (2023) Adsorption optimization for the removal of cadmium in water by aluminum (Hydr) oxide on cation exchange resin. Curr Appl Sci Technol 10:55003
Ngwenya BT, Tourney J, Magennis M, Kapetas L, Olive V (2009) A surface complexation framework for predicting water purification through metal biosorption. Desalination 248(1):344–351. https://doi.org/10.1016/j.desal.2008.05.074
Niknejad N, Nazari B, Foroutani S, Hussin AR (2023) A bibliometric analysis of green technologies applied to water and wastewater treatment. Environ Sci Pollut Res 30(28):71849–71863. https://doi.org/10.1007/s11356-022-18705-1
Nishu, Kumar S (2023) Smart and innovative nanotechnology applications for water purification. Hybrid Adv 3:100044. https://doi.org/10.1016/j.hybadv.2023.100044
Nizeyimana JC, Ndagijimana P, Khan J, Xiangru L, Twagirayezu G, Manzi HP et al (2024) A hybrid system for Nickel ions removal from synthesized wastewater using adsorption assisted with electrocoagulation. Environ Sci Pollut Res 31(19):28321–28340
Nwabor OF, Nnamonu EI, Martins PE, Ani OC (2016) Water and waterborne diseases: a review. Int J Trop Dis Health 12(4):1–14
Obahiagbon EG, Ogwu MC (2024) The nexus of business, sustainability, and herbal medicine. In: Izah SC, Ogwu MC, Akram M (eds) Herbal medicine phytochemistry. Reference series in phytochemistry. Springer, Cham. https://doi.org/10.1007/978-3-031-21973-3_67-1
Ochowiak M, Matuszak M, Włodarczak S, Ancukiewicz M, Krupińska A (2017) The modified swirl sedimentation tanks for water purification. J Environ Manage 189:22–28. https://doi.org/10.1016/j.jenvman.2016.12.023
Office of Response and Restoration. NOAA. https://response.restoration.noaa.gov/point-vs-non-point-water-pollution-what-s-difference
Ogwu MC, Malikia CN, Stansfield A, Gonzales-Torres AD, Izah SC (2023) Sustainable food processing waste management for environmental protection. In: Srivastav AL, Grewal AS, Tiwari M, Pham TD (eds) Role of green chemistry in ecosystem restoration to achieve environmental sustainability. Elsevier, Amsterdam, pp 291–299. https://doi.org/10.1016/B978-0-443-15291-7.00022-5
Ogwu MC (2019a) Towards sustainable development in Africa: the challenge of urbanization and climate change adaptation. In: Cobbinah PB, Addaney M (eds) The geography of climate change adaptation in Urban Africa. Springer Nature, Switzerland, pp 29–55. https://doi.org/10.1007/978-3-030-04873-0_2
Ogwu MC (2019b) Understanding the composition of food waste: an "-Omics" approach to food waste management. In: Gunjal AP, Waghmode MS, Patil NN, Bhatt P (eds) Global initiatives for waste reduction and cutting food loss. IGI Global, Pennsylvania, pp 212–236. https://doi.org/10.4018/978-1-5225-7706-5.ch011
Ogwu MC (2023) Plants as monitors and managers of pollution. In: Srivastav AL, Grewal AS, Tiwari M, Pham TD (eds) Role of green chemistry in ecosystem restoration to achieve environmental sustainability. Elsevier, Amsterdam, Netherlands. pp 51–60. https://doi.org/10.1016/B978-0-443-15291-7.00022-5

Osman AI, El-Monaem EMA, Elgarahy AM, Aniagor CO, Hosny M, Farghali M et al (2023) Methods to prepare biosorbents and magnetic sorbents for water treatment: a review. Environ Chem Lett 21(4):2337–2398. https://doi.org/10.1007/s10311-023-01603-4
Österblom H, Wabnitz CC, Tladi D, Allison EH, Arnaud-Haond S, Bebbington J et al (2023) Towards ocean equity. In: The blue compendium: from knowledge to action for a sustainable ocean economy. Springer, pp 485–521
Pakharuddin N, Fazly M, Sukari SA, Tho K, Zamri W (2021) Water treatment process using conventional and advanced methods: a comparative study of Malaysia and selected countries. Paper presented at the IOP Conference Series: Earth and Environmental Science.
Posthuma L, Zijp MC, De Zwart D, Van de Meent D, Globevnik L, Koprivsek M et al (2020) Chemical pollution imposes limitations to the ecological status of European surface waters. Sci Rep 10(1):14825. https://doi.org/10.1038/s41598-020-71537-2
Prasad KY (2019) Sedimentation in water and used water purification. Springer, Cham, pp 1–27
Rather LJ, Jameel S, Dar OA, Ganie SA, Bhat KA, Mohammad F (2019) 10—Advances in the sustainable technologies for water conservation in textile industries. In: Muthu SS (ed) Water in textiles and fashion. Woodhead Publishing, pp 175–194
Riveros-Perez E, Riveros R (2018) Water in the human body: an anesthesiologist's perspective on the connection between physicochemical properties of water and physiologic relevance. Ann Med Surg 26:1–8. https://doi.org/10.1016/j.amsu.2017.12.007
Saeed MU, Hussain N, Sumrin A, Shahbaz A, Noor S, Bilal M et al (2022) Microbial bioremediation strategies with wastewater treatment potentialities—a review. Sci Total Environ 818:151754. https://doi.org/10.1016/j.scitotenv.2021.151754
Saleh TA, Mustaqeem M, Khaled M (2022) Water treatment technologies in removing heavy metal ions from wastewater: a review. Environ Nanotechnol Monitor Manage 17:100617. https://doi.org/10.1016/j.enmm.2021.100617
Scanlon BR, Fakhreddine S, Rateb A, de Graaf I, Famiglietti J, Gleeson T et al (2023) Global water resources and the role of groundwater in a resilient water future. Nat Rev Earth Environ 4(2):87–101. https://doi.org/10.1038/s43017-022-00378-6
Schwarzenbach RP, Egli T, Hofstetter TB, Von Gunten U, Wehrli B (2010) Global water pollution and human health. Annu Rev Environ Resour 35(1):109–136
Schwätter F, Hannich CB, Nöthe T, Oehlmann J, Fahlenkamp H (2007) Risk assessment for organic trace compounds in wastewater: comparison of conventional and advanced treatment. Water Sci Technol 56(5):9–13. https://doi.org/10.2166/wst.2007.551
Senthil Kumar P, Saravanan A (2018) Sustainable waste water treatment technologies. In Muthu SS (ed) Detox fashion: waste water treatment. Springer, Singapore, pp 1–25
Shahid MK, Kashif A, Fuwad A, Choi Y (2021) Current advances in treatment technologies for removal of emerging contaminants from water—a critical review. Coord Chem Rev 442:213993. https://doi.org/10.1016/j.ccr.2021.213993
Shayo GM, Elimbinzi E, Shao GN, Fabian C (2023) Severity of waterborne diseases in developing countries and the effectiveness of ceramic filters for improving water quality. Bull Natl Res Centre 47(1):113. https://doi.org/10.1186/s42269-023-01088-9
Sheoran K, Kaur H, Siwal SS, Saini AK, Vo D-VN, Thakur VK (2022) Recent advances of carbon-based nanomaterials (CBNMs) for wastewater treatment: synthesis and application. Chemosphere 299:134364. https://doi.org/10.1016/j.chemosphere.2022.134364
Singh J, Yadav P, Pal AK, Mishra V (2020) Water pollutants: origin and status. In: Pooja D, Kumar P, Singh P, Patil S (eds) Sensors in water pollutants monitoring: role of material. Springer, Singapore, pp 5–20
Singh CR (2024) Water resources management in India: current state and challenges. Paper presented at the National Conference on Science Technology Engineering and Management, Departments of Mechanical and Civil Engineering, Amritsar Group of Colleges, Amritsar
Singh V (2024) Water pollution. In: Textbook of Environment and Ecology. Springer, Singapore, pp 253–266

Song K, Mohseni M, Taghipour F (2016) Application of ultraviolet light-emitting diodes (UV-LEDs) for water disinfection: a review. Water Res 94:341–349. https://doi.org/10.1016/j.watres.2016.03.003
Svensson E, Berntsson T (2014) The effect of long lead times for planning of energy efficiency and biorefinery technologies at a pulp mill. Renew Energy 61:12–16
The Environmental System (2024) The geography of transport systems. https://doi.org/10.4324/9781003343196. https://transportgeography.org/contents/chapter4/transportation-and-environment/environmental-system/
Türkmen A, Öztaş F, Öztaş H, Türkmen M. Effects of environmental heavy metal contamination. In: Human Health and Environmental Toxicology
Types of Water Pollution (2022) https://atlas-scientific.com/blog/types-of-water-pollution/#:~:text=The%20main%20types%20of%20water,contaminants%20to%20enter%20our%20waterways
Tzanakakis VA, Paranychianakis NV, Angelakis AN (2020) Water supply and water scarcity. Water 12(9):2347
van der Gun J (2021) Chapter 24—Groundwater resources sustainability. In: Mukherjee A, Scanlon BR, Aureli A, Langan S, Guo H, McKenzie AA (eds) Global groundwater. Elsevier, pp 331–345
Vasistha P, Ganguly R (2020) Water quality assessment of natural lakes and its importance: an overview. Mater Today Proc 32:544–552. https://doi.org/10.1016/j.matpr.2020.02.092
Walters L (2024) The future of water reuse: how innovations in membrane filtration are helping conserve water resources. https://www.wwdmag.com/water-reuse-recycling/article/55131204/the-future-of-water-reuse-how-innovations-in-membrane-filtration-are-helping-conserve-water-resources
Wan J, Yu M, Bi W, Sun Y, Wang W, Hou Y et al (2024) Research on chromium removal mechanism by electrochemical method from wastewater. J Water Process Eng 68:106447. https://doi.org/10.1016/j.jwpe.2024.106447
Wang H, He G (2022) Rivers: linking nature, life, and civilization. River 1(1):25–36
Wang Y, Ma B, Ulbricht M, Dong Y, Zhao X (2022) Progress in alumina ceramic membranes for water purification: status and prospects. Water Res 226:119173. https://doi.org/10.1016/j.watres.2022.119173
Wang M, Zeng M, Wang P, Liu Y (2023) Comparative investigation on Ni (II) removal from electroplating wastewater by mineral adsorbent (CSAM) and ion-exchange resins. Sep Sci Technol 58(11):1959–1971
Water and Ecosystems. https://www.unwater.org/water-facts/water-and-ecosystems
Water Data. Department of Water Resources, River Development and Ganga Rejuvenation
Water and Natural Resources (2019) The concious club. https://www.theconsciouschallenge.org/ecologicalfootprintbibleoverview/water-natural-resources
Water Pollution Statistics (2020) Sourgum waste. https://www.sourgum.com/the-sour-scoop/water-pollution-statistics/
Water Resources Management. World Bank Group. https://www.worldbank.org/en/topic/waterresourcesmanagement
Water Resources (2018) Improving the management of natural water resources through so-called nature-based solutions. https://www.greenfacts.org/en/water-resources-nature-solutions/index.htm#1
Water Resources (2021) Development Monitoring and evaluation office. https://dmeo.gov.in/sites/default/files/2021-08/9a_Sector_Report_Water_Resources.pdf
Water Scarcity. https://www.unicef.org/wash/water-scarcity#:~:text=Four%20billion%20people%20%E2%80%94%20almost%20two,by%20as%20early%20as%202025.
Wato T, Amare M, Bonga E, Demand BBO, Coalition BBR (2020) The agricultural water pollution and its minimization strategies—a review. J Resour Dev Manag 64:10–22
Wheater HS, Gober P (2015) Water security and the science agenda. Water Resour Res 51(7):5406–5424
World Water Resources By Country. https://www.fao.org/4/y4473e/y4473e08.htm#:~:text=At%20the%20continental%20level%2C%20America,and%20Africa%20with%209%20percent

Xu Z, Yao L, Zhang Q, Dowaki K, Long Y (2020) Inequality of water allocation and policy response considering virtual water trade: a case study of Lanzhou city, China. J Clean Prod 269:122326. https://doi.org/10.1016/j.jclepro.2020.122326

Yang D, Yang Y, Xia J (2021) Hydrological cycle and water resources in a changing world: a review. Geogr Sustain 2(2):115–122. https://doi.org/10.1016/j.geosus.2021.05.003

Younas F, Mustafa A, Farooqi ZUR, Wang X, Younas S, Mohy-Ud-Din W et al (2021) Current and emerging adsorbent technologies for wastewater treatment: trends, limitations, and environmental implications. Water 13(2):215

Yusuf A, Sodiq A, Giwa A, Eke J, Pikuda O, De Luca G et al (2020) A review of emerging trends in membrane science and technology for sustainable water treatment. J Clean Prod 266:121867. https://doi.org/10.1016/j.jclepro.2020.121867

Zhai Y, Liu G, van der Meer WGJ (2022) One-step reverse osmosis based on riverbank filtration for future drinking water purification. Engineering 9:27–34. https://doi.org/10.1016/j.eng.2021.02.015

Chapter 8
Air Pollution Control Technologies and Strategies

Odoligie Imarhiagbe, Arthur C. Okafor, and Matthew Chidozie Ogwu

Abstract Air pollution remains a critical global issue, posing significant threats to public health, ecosystems, and climate stability. The chapter aims to comprehensively analyze the sources, dynamics, and mitigation approaches for air pollution. It explores key sources of pollutants, including industrial emissions, vehicular exhaust, and agricultural activities, alongside the role of meteorological factors like temperature and wind in influencing pollutant dispersion. The chapter emphasizes the adverse health impacts of pollutants such as particulate matter, nitrogen oxides, sulfur dioxide, and volatile organic compounds, linking them to respiratory and cardiovascular diseases and ecosystem degradation. Advanced monitoring technologies, such as IoT-enabled air quality sensors, remote sensing systems, and Fourier-transform infrared spectroscopy, are highlighted for their precision and real-time data capabilities. These innovations enable comprehensive pollutant assessment, inform regulatory compliance, and support public awareness. Strategies for emission reduction across industrial, transportation, and agricultural sectors form a core focus. Examples include carbon capture and storage in industries, promoting electric vehicles and public transport in urban areas, and precision agriculture practices like optimized fertilizer application to reduce greenhouse gas emissions in farming. The integration of green technologies is presented as a pathway to sustainable air pollution control. Renewable energy systems, such as solar and wind power, reduce dependency on

O. Imarhiagbe (✉)
Department of Health and Social Science, London School of Science and Technology, Birmingham, UK
e-mail: Odoligie.imarhiagbe@lsst.ac

A. C. Okafor
Institute of Medical Microbiology and Hygiene, Austrian Agency for Health and Food Safety, Vienna, Austria
e-mail: okaforac1@gmail.com

M. C. Ogwu
Goodnight Family Department of Sustainable Development, Appalachian State University, Boone, NC, USA
e-mail: ogwumc@appstate.edu

M. C. Ogwu and S. Chibueze Izah (eds.), *Evaluating Environmental Processes and Technologies*, Environmental Science and Engineering,
https://doi.org/10.1007/978-3-031-85327-2_8

fossil fuels, while nanotechnology-based air purifiers enhance pollutant removal efficiency. Policy and regulatory frameworks, such as the Paris Agreement and national emission standards, are discussed for their critical role in fostering global and local air quality improvements. Public–private partnerships and community engagement initiatives are underscored for their ability to drive innovation, promote education, and enhance inclusivity in decision-making. This chapter offers a plan for sustainable air quality management by addressing air pollution's sociological, technological, and regulatory aspects. It closes the gap between science and practice by providing creative and flexible methods to reduce air pollution, safeguard public health, and support international sustainability objectives.

Keywords Air pollution · Monitoring technologies · Green technologies · Emission control · Sustainable development · Community engagement · Carbon capture · Nanotechnology · Precision agriculture

8.1 Introduction

Air pollution is a significant global issue characterized by harmful substances in the atmosphere, which can adversely affect human health and the environment. Air pollutants can be classified into primary pollutants, which are emitted directly from sources such as vehicles and factories, and secondary pollutants, which forms in the ecosystem through chemical reactions, such as ozone and particulate matter (PM) (Zhao et al. 2022; Liu et al. 2020). The primary sources of air pollution include industrial and agricultural activities and natural phenomena like wildfires and volcanic eruptions (Istiqomah 2023; Munsif et al. 2021). The health impacts of air pollution are profound, contributing to respiratory diseases, cardiovascular conditions, and even premature mortality (Xia et al. 2017; Zhang et al. 2019; Izah et al. 2023). Furthermore, air pollution poses significant environmental risks to ecosystems, climate change, and reduced visibility (Zhao et al. 2022; Saadeh et al. 2022; Ogwu et al. 2023). The relationship between air pollution and meteorological conditions is complex and varies by region. Temperature, humidity, and wind speed can significantly influence pollutant concentrations. For instance, increased wind speed can disperse pollutants, reducing their concentration, while higher temperatures can exacerbate the formation of ground-level ozone (Liu et al. 2020; Rahman et al. 2022). Studies have shown that specific meteorological conditions can mitigate or exacerbate air pollution levels, highlighting the importance of considering these factors to mitigate air pollution (Liang et al. 2017a, b).

Various control technologies and strategies have been implemented globally to address air pollution. These include regulatory measures, technological innovations, and public awareness campaigns. For example, introducing stricter emissions standards for vehicles and industries has significantly reduced pollutant levels (Reddy 2024; Niveditha et al. 2021). Innovative approaches, such as using hydrogen fuel and intelligent transportation systems, are being explored further to mitigate air pollution

(Reddy 2024). Public awareness is also crucial because it can educate communities about the risks associated with air pollution, and promoting behavioral changes can enhance the effectiveness of pollution control measures (Saadeh et al. 2022; Johnson 2011).

The application of smart technologies, such as IoT-enabled air quality monitoring systems, is highlighted for its role in real-time data collection and informed decision-making (Memić et al. 2022). The chapter further examines sector-specific approaches, such as emission reduction in transportation and industry, while emphasizing renewable energy adoption as a sustainable pathway to reduce fossil fuel dependency (Li et al. 2022). Policy and regulatory frameworks, like those outlined in the Paris Agreement, underscore the need for coordinated international efforts to address transboundary air pollution issues (Roelfsema et al. 2020). Through the integration of science, technology, and governance, the chapter provides a holistic perspective on innovative solutions and sustainable practices that mitigate environmental harm and promote global sustainability goals (Steffen et al. 2015).

This chapter aims to provide a comprehensive understanding of air pollution dynamics, including its causes, types, sources, and impacts on environmental and human health, while exploring existing and emerging technologies for effective mitigation. It delves into sector-specific strategies for emission control across industries, transportation, and agriculture, emphasizing the role of renewable energy and smart technologies in fostering cleaner air and sustainable development. By examining international, national, and local regulatory frameworks, the chapter highlights policy implementation and compliance mechanisms while stressing the importance of public participation, awareness, and stakeholder collaboration in achieving air quality goals. It identifies barriers to effective air pollution control, such as technological and policy gaps, and outlines opportunities for innovation, investment, and future advancements. It also contributes to sustainable development discourse by showcasing cutting-edge technologies, emphasizing holistic approaches, and advocating for inclusive practices that align with global sustainability goals.

8.2 Air Quality Monitoring and Assessment

Air quality monitoring and assessment are critical to managing air pollution and protecting public health. The Air Quality Index (AQI) serves as a primary tool for communicating air quality levels to the public, providing a standardized method for assessing and reporting the concentration of various pollutants (Table 8.1). The AQI is calculated based on the levels of key pollutants, including particulate matter (PM_{10} and $PM_{2.5}$), nitrogen dioxide (NO_2), sulfur dioxide (SO_2), carbon monoxide (CO), and ozone (O_3) (Li et al. 2021). Each pollutant is assigned a value corresponding to its potential health effects, allowing the public to understand the implications of current air quality conditions. The AQI is categorized into different levels, ranging from "Good" to "Hazardous," which helps individuals make informed decisions about outdoor activities, especially vulnerable populations such as children and the elderly (Cisneros et al. 2017; Ikhajiagbe and Ogwu 2020). The application of the AQI extends beyond public awareness; it is also utilized by policymakers and environmental agencies to evaluate compliance with air quality standards and to implement necessary regulations. For instance, studies have shown that regions with effective AQI communication strategies can improve public health outcomes and increase community engagement in air quality issues (Cisneros et al. 2017).

Moreover, the AQI can serve as a basis for developing targeted interventions, such as issuing health advisories during high pollution events, thereby enhancing the overall effectiveness of air quality management strategies (Xiao et al. 2018). Monitoring technologies and methods have evolved significantly, enabling more accurate and comprehensive air quality assessments. Traditional monitoring stations, which utilize high-precision instruments to measure pollutant concentrations, remain

Table 8.1 Air Quality Index (AQI) outlining AQI ranges, categories, corresponding pollutant concentrations, and health implications

QI range	Category	Pollutant concentration ($PM_{2.5}$)	Health implications	Precautionary measures
0–50	Good	0–12 μg/m^3	Air quality is satisfactory, and air pollution poses little or no risk	No special measures are required
51–100	Moderate	12.1–35.4 μg/m^3	Acceptable air quality: however, sensitive groups may be concerned	Sensitive individuals should limit outdoor activity
101–150	Unhealthy for sensitive groups	35.5–55.4 μg/m^3	Members of sensitive groups may experience health effects; the public is less likely to be affected	Sensitive groups should reduce prolonged outdoor exertion
151–200	Unhealthy	55.5–150.4 μg/m^3	Everyone may begin to experience health effects; members of sensitive groups may experience more serious effects	Limit outdoor activities; wear masks if necessary
201–300	Very unhealthy	150.5–250.4 μg/m^3	Health alert: everyone may experience more serious health effects	Avoid outdoor activities; use air purifiers indoors
301–500	Hazardous	> 250.4 μg/m^3	Health warning of emergency conditions: the entire population is more likely to be affected	Stay indoors; keep windows closed; use N95 masks

$PM_{2.5}$ refers to particulate matter with a diameter of 2.5 micrometers or smaller
Similar indices exist for other pollutants, including PM_{10}, ozone (O_3), sulfur dioxide (SO_2), nitrogen dioxide (NO_2), and carbon monoxide (CO), with varying concentration thresholds
AQI categories and thresholds may vary slightly by country or region based on local standards (e.g., the U.S. EPA, EU standards, or WHO guidelines)

a cornerstone of air quality assessment. However, the deployment of low-cost sensors has gained traction in recent years, allowing for broader spatial coverage and real-time data collection (Wesseling et al. 2019). These sensors can measure various parameters, including temperature, humidity, and specific pollutant levels, providing valuable insights into local air quality conditions. While low-cost sensors offer advantages in terms of accessibility and community engagement, they also present challenges related to data quality and calibration (Watne et al. 2021). Therefore, it is essential to integrate data from both traditional monitoring stations and low-cost sensors to obtain a more comprehensive understanding of air quality dynamics. Data collection and analysis are fundamental to effective air quality management.

The integration of advanced data analytics, including machine learning and artificial intelligence, has revolutionized the way air quality data is processed and interpreted (Rybarczyk and Žalakevičiūtė 2018). These technologies enable the identification of patterns and trends in air pollution, facilitating predictive modeling and forecasting of air quality levels. For example, machine learning algorithms can analyze historical air quality data alongside meteorological variables to predict future pollution events, allowing for implementing proactive measures (Al Momani and Örmeci 2014). Furthermore, using geographic information systems (GIS) enhances the spatial analysis of air quality data, identifying pollution hotspots, and assessing the effectiveness of control measures (Guo et al. 2020). Citizen science initiatives have also emerged as a valuable approach to air quality monitoring, encouraging community participation in data collection and analysis (Wesseling et al. 2019). These initiatives enhance data availability and raise public awareness about air pollution issues by empowering individuals to contribute to air quality assessments. Studies have indicated that citizen-generated data can complement traditional monitoring efforts, providing localized insights that may not be captured by fixed monitoring stations (Watne et al. 2021). However, it is crucial to ensure the reliability and accuracy of citizen-collected data through proper training and calibration of monitoring equipment (Watne et al. 2021).

Monitoring technologies and methods for air quality play a critical role in understanding and mitigating pollution's impact on human health and the environment. Advanced systems like real-time sensors, satellite-based remote sensing, and air quality monitoring networks enable the detection and measurement of pollutants such as particulate matter ($PM_{2.5}$ and PM_{10}), nitrogen oxides (NO_x), and volatile organic compounds (VOCs). Laser scattering, gas chromatography, and spectroscopy provide precise pollutant analysis, while IoT-enabled devices offer continuous data collection and accessibility. Air quality monitoring employs diverse technologies to measure and analyze pollutants, enabling effective regulatory compliance and public health strategies (Table 8.2). Gravimetric sampling offers precise PM_{10} and $PM_{2.5}$ measurement but is time-intensive, while Beta Attenuation Monitors (BAM) and Optical Particle Counters (OPC) provide automated, real-time particulate monitoring with some sensitivity limitations. Advanced methods like Gas Chromatography (GC) and Fourier-Transform Infrared (FTIR) spectroscopy deliver high-precision gas pollutant analysis, though they require skilled operation and high costs. Portable electrochemical sensors and air quality monitors offer cost-effective, real-time options but need regular calibration. Remote sensing technologies, such as satellites and LIDAR, provide extensive spatial coverage and high-resolution pollutant mapping but face ground-level accuracy and cost challenges. Continuous Emission Monitoring Systems (CEMS) ensure compliance for fixed sources, while passive diffusion tubes and portable monitors offer simple, low-cost solutions for time-integrated or real-time measurements. Together, these methods advance air quality management, supporting environmental protection and public health. These technologies support regulatory compliance, inform policy decisions, and raise public awareness, facilitating proactive measures to improve air quality and ecological sustainability.

8.3 Technological Approaches to Air Pollution Control

Technological approaches to air pollution control are essential for mitigating the adverse effects of air pollutants on human health and the environment. Various systems and methods have been developed to filter, absorb, and reduce airborne contaminants. Air filtration systems such as the one illustrated in Fig. 8.1 are widely employed to remove particulate matter (PM) and other airborne pollutants from the air. This air sampling and particle counting system is used for air quality analysis. The system features an air compressor that supplies a steady airflow, regulated by a flowmeter to ensure precise sampling. Air passes through an air-filtered bottle to remove larger contaminants before reaching a composite membrane that captures fine particulates and pollutants for further analysis. A valve controls airflow direction, while a particle counter provides real-time data on particulate concentration, enabling accurate monitoring of air quality parameters. These systems utilize various materials, including fibrous media, to capture particles through interception, inertial impaction, and diffusion (Zhu et al. 2021). High-efficiency particulate Air (HEPA) filters are a typical example, capable of trapping at least 99.97% of particles that are 0.3 μm in diameter (Maestas et al. 2018). Recent advancements in nanofiber technology have led to the development of highly efficient air filters that can capture smaller particles while maintaining low resistance to airflow (Liu et al. 2015; Hao and Wu 2022). For instance, the use of nanofibers in air filtration has shown promise in enhancing the capture efficiency of $PM_{2.5}$, which is critical given its association with severe health issues, including respiratory and cardiovascular diseases (Wang et al. 2019; Ogwu et al. 2023).

On the other hand, absorption systems utilize chemical processes to remove gaseous pollutants from the air. These systems often employ activated carbon or other adsorbent materials to capture volatile organic compounds (VOCs), odors, and other harmful gases (Capulli 2024). The effectiveness of these systems can be influenced by factors such as humidity, temperature, and the concentration of pollutants. For example, the adsorption capacity of activated carbon can be significantly enhanced by modifying its surface properties through chemical treatments (Souzandeh et al. 2016). Figure 8.2 illustrates a CO_2 absorption system with a Raschig ring absorption column. The process begins with CO_2 gas entering the system, regulated by a flow

Table 8.2 Monitoring technologies and methods for air quality

Technology/method	Principle/technique	Pollutants measured	Advantages	Limitations
Gravimetric sampling	Collects particulate matter (PM) on filters	PM_{10}, $PM_{2.5}$	Accurate and widely accepted	Time-consuming, requires lab analysis
Beta attenuation monitors (BAM)	Measures PM by detecting beta radiation attenuation	PM_{10}, $PM_{2.5}$	Continuous monitoring, automated	Sensitive to high humidity levels
Optical particle counters (OPC)	Detects light scattering by particles	PM_{10}, $PM_{2.5}$	Real-time data, portable	Limited accuracy in high humidity
Gas chromatography (GC)	Separates and measures gaseous pollutants	VOCs, hydrocarbons	High precision and sensitivity	Expensive, requires skilled operators
Fourier-transform infrared (FTIR) spectroscopy	Measures infrared absorption spectra	CO, CO_2, VOCs, SO_2	Simultaneous multi-pollutant detection	High initial cost, complex maintenance
Electrochemical sensors	Measures chemical reactions producing electrical signals	CO, NO_2, O_3, SO_2	Portable, cost-effective	Limited lifespan, calibration needed
Remote sensing (Satellite)	It uses satellite sensors to detect pollutants	PM, O_3, CO, NO_2, SO_2	Large spatial coverage	Low spatial resolution for ground-level measurements
LIDAR (Light detection and ranging)	Uses lasers to measure particulate and gas concentrations	PM, aerosols	High spatial and temporal resolution	High cost, requires expertise
Continuous emission monitoring systems (CEMS)	Measures pollutant emissions from fixed sources	CO_2, SO_2, NO_x	Reliable for compliance monitoring	Not applicable for ambient air monitoring
Photoacoustic spectroscopy	Measures sound waves generated by light absorption	CO, CH_4, NO_2, VOCs	High sensitivity, low interference	Limited field deployment

(continued)

Table 8.2 (continued)

Technology/ method	Principle/ technique	Pollutants measured	Advantages	Limitations
Diffusion tubes	Passive samplers for time-integrated measurements	NO_2, SO_2, VOCs	Low cost, simple to deploy	Limited temporal resolution
Portable air quality Monitors	Combines sensors and digital interfaces	PM, CO, NO_2, O_3	User-friendly, real-time data	Limited accuracy, prone to drift

PM_{10}, Particulate matter $\leq$ 10 μm diameter; $PM_{2.5}$, Particulate matter $\leq$ 2.5 μm diameter; VOCsVolatile organic compounds; NO_2, Nitrogen dioxide; O_3, Ozone; COCarbon monoxideSO_2, Sulfur dioxide

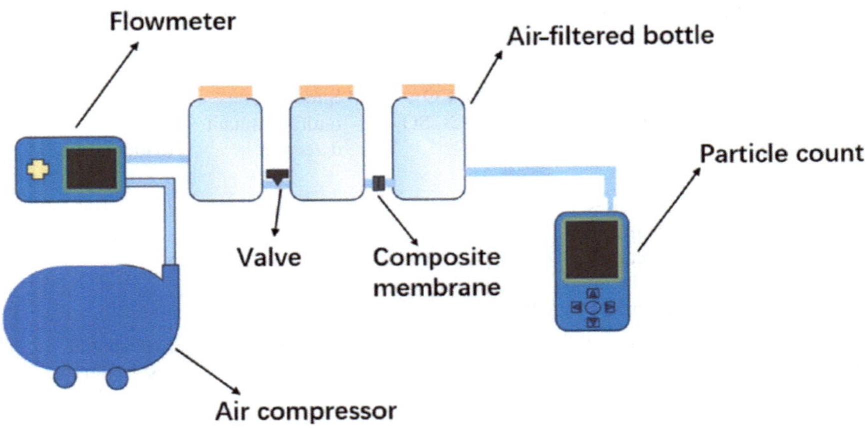

Fig. 8.1 Air sampling and particle counting system with filtration units. *Source* He et al. (2024)

meter and powered by an air compressor. The CO_2 is absorbed in the Raschig ring column, where a NaOH absorbent solution facilitates chemical reactions to capture the CO_2. A CO_2 analyzer monitors the gas concentration post-absorption to evaluate the system's efficiency. The treated air or liquid is collected in a reservoir at the column's base. This system is widely applied in gas treatment, emission control, and industrial processes. Additionally, integrating filtration and absorption technologies can provide a more comprehensive approach to air quality management, simultaneously addressing particulate and gaseous pollutants (Liu et al. 2017a, b). Implementing filtration and absorption systems has yielded significant improvements in indoor air quality. These systems underscore the integration of traditional physicochemical methods and natural biological systems for indoor air purification (Fig. 8.3). It highlights a holistic approach to mitigating indoor air pollutants while enhancing air quality sustainability. Studies have demonstrated that using air filtration units in residential settings can lead to substantial reductions in $PM_{2.5}$ concentrations, lowering the associated health risks (Fig. 8.3; Ward et al. 2015; Allen and Barn 2020). For instance, a randomized controlled trial in Detroit revealed that inexpensive air filtration systems could effectively reduce personal $PM_{2.5}$ exposure, highlighting the potential for such interventions to improve public health outcomes (Maestas et al. 2018). Furthermore, applying filtration technologies in industrial settings can mitigate emissions from processes that generate significant amounts of particulate and gaseous pollutants, contributing to overall air quality improvement (Yu et al. 2020). Technological advancements have led to nanotechnologys application in air purification through three primary mechanisms: filtration, adsorption, and catalysis (Fig. 8.4). The framework addresses volatile organic compounds (VOCs) and nitrogen oxides (NO_x) as key pollutants and explores innovative methods to transform polluted air into fresh air.

Electrostatic Precipitators and Scrubbers Electrostatic precipitators (ESPs) are advanced air pollution control devices that utilize electrical charges to remove particulate matter from industrial exhaust streams. These systems charge particles as they pass through an electric field, causing them to adhere to oppositely charged collection plates (Leung and Kwok 2012). ESPs are particularly effective for capturing fine particulate matter, including $PM_{2.5}$, and have been widely adopted in power generation and cement manufacturing (Mohraz et al. 2018). The efficiency of ESPs can be influenced by factors such as particle size, resistivity, and moisture in the flue gas (Bai et al. 2018). Despite their effectiveness, ESPs can generate ozone as a byproduct,

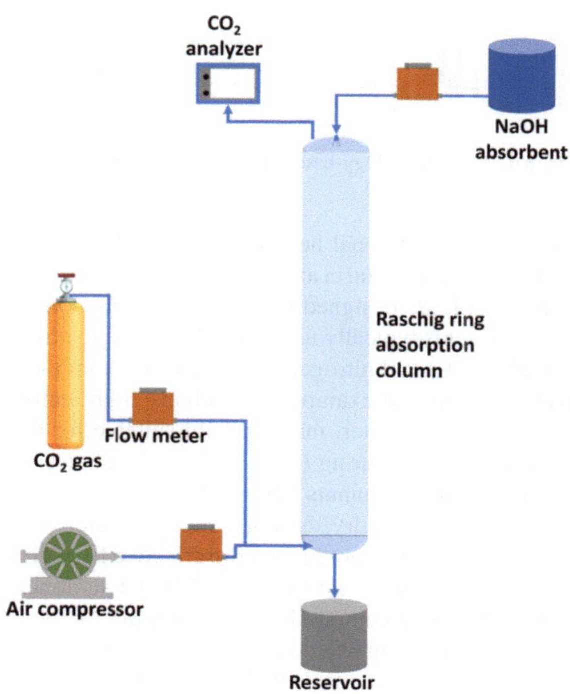

Fig. 8.2 CO_2 absorption system using Raschig Ring column. *Source* Jakfar et al. (2023)

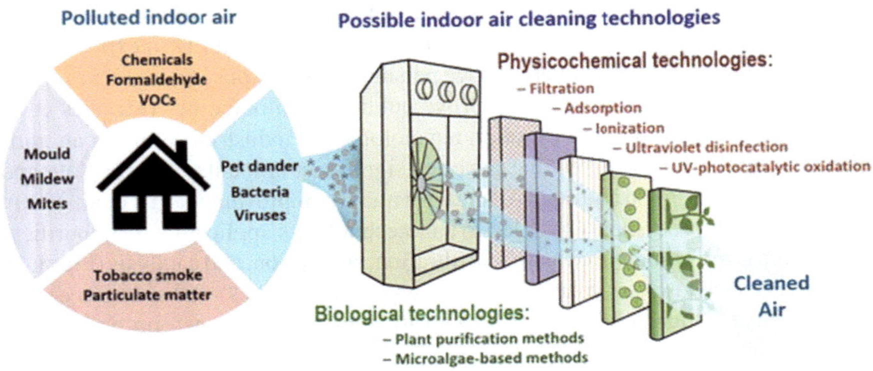

Fig. 8.3 Indoor air pollution sources and cleaning technologies: physicochemical and biological approaches. *Source* Mata et al. (2022)

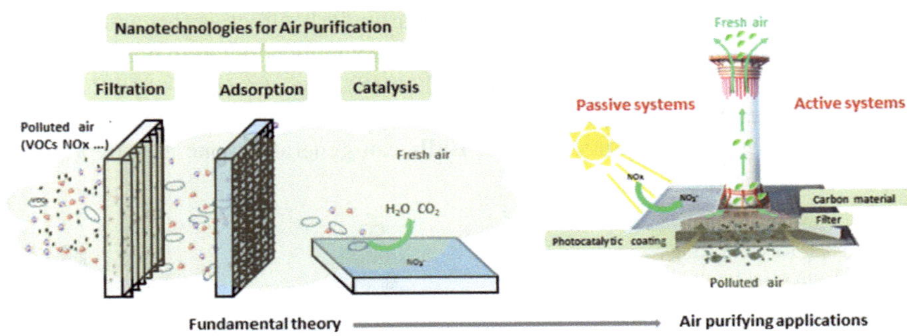

Fig. 8.4 Nanotechnology-based air purifying systems. *Source* Cao et al. (2021)

which poses additional health risks; thus, careful management of operating conditions is necessary (Bai et al. 2018). Scrubbers, another critical technology in air pollution control, are designed to remove gaseous pollutants from industrial emissions. These systems typically use a liquid solution to absorb contaminants, such as sulfur dioxide (SO_2) and nitrogen oxides (NO_x), from the gas stream (Wang et al. 2015). Wet scrubbers, for example, can achieve high removal efficiencies for acidic gases and particulate matter, making them suitable for applications in power plants and chemical manufacturing (Shao 2023). The choice of scrubber technology depends on the specific pollutants being targeted and economic and operational considerations. For instance, dry scrubbers may be preferred when water usage is a concern, while wet scrubbers are often more effective for high concentrations of soluble gases (Huang et al. 2018). Integrating ESPs and scrubbers into industrial processes can significantly reduce emissions, contributing to compliance with air quality regulations and improving overall environmental performance. For example, implementing these technologies in coal-fired power plants has resulted in substantial decreases in

SO_2 and particulate emissions, thereby reducing the environmental impact of energy production (Yu et al. 2020). Moreover, ongoing research into enhancing the efficiency of these systems, such as optimizing the design of collection electrodes in ESPs or developing novel absorbent materials for scrubbers, holds promise for further advancements in air pollution control technologies (Mohraz et al. 2018; Wang et al. 2019).

Catalytic reduction systems convert harmful gaseous pollutants into less toxic substances through chemical reactions facilitated by catalysts. Selective Catalytic Reduction (SCR) is a widely used technology for reducing nitrogen oxide (NO_x) emissions from diesel engines and industrial processes. In SCR systems, a reductant, typically ammonia or urea, is injected into the exhaust stream, where it reacts with NO_x in the presence of a catalyst, converting it into nitrogen and water vapor (Yu et al. 2020). The effectiveness of SCR systems can be influenced by factors such as temperature, catalyst type, and the concentration of NO_x in the exhaust (Becchio et al. 2019). These systems have been shown to achieve removal efficiencies exceeding 90%, making them a critical component of modern emissions control strategies (Yu et al. 2020). Non-catalytic reduction systems, such as Non-Selective Catalytic Reduction (NSCR) and thermal reduction, also play a role in air pollution control. NSCR systems utilize hydrocarbon fuels to reduce NO_x emissions without the need for a catalyst, relying on high temperatures to facilitate the reaction (Wang et al. 2015). While these systems can be effective, they often require careful management of operating conditions to ensure optimal performance.

On the other hand, thermal reduction systems employ high temperatures to decompose pollutants directly, making them suitable for applications with high concentrations of NO_x (Huang et al. 2018). Implementing catalytic and non-catalytic reduction systems has significantly reduced NO_x emissions, particularly in urban areas with stringent air quality standards. For example, introducing SCR technology in heavy-duty diesel vehicles has led to substantial decreases in NO_x emissions, contributing to improved air quality in cities (Yu et al. 2020). Furthermore, ongoing research into novel catalysts and reaction pathways aims to enhance the efficiency and effectiveness of these systems, addressing the challenges posed by emerging pollutants and evolving regulatory requirements (Becchio et al. 2019; Huang et al. 2018).

8.4 Strategies for Reducing Emissions from Key Sectors

Industrial emissions are a significant contributor to air pollution and greenhouse gas emissions. Various regulations and control mechanisms have been established globally to mitigate these emissions. Regulatory frameworks often include emissions standards, permitting processes, and monitoring requirements that compel industries to adopt cleaner technologies and practices (Waisman et al. 2019; Chen et al. 2021). For example, the European Union has implemented stringent regulations under the Industrial Emissions Directive (IED), which aims to minimize pollution from industrial activities by setting limits on emissions of specific pollutants, including sulfur dioxide

(SO_2), nitrogen oxides (NO_x), and particulate matter (Chen et al. 2021). In addition to regulatory measures, industries are increasingly adopting voluntary initiatives and corporate social responsibility (CSR) practices to reduce their environmental impact. These initiatives often involve implementing cleaner production techniques, energy efficiency improvements, and waste minimization strategies (Johansson et al. 2012). For instance, the petroleum refining industry has explored innovative approaches such as on-site biomass gasification to produce hydrogen, which can significantly reduce CO_2 emissions compared to traditional methods (Johansson et al. 2012). Furthermore, the adoption of advanced technologies, such as carbon capture and storage (CCS), is gaining traction to capture CO_2 emissions from industrial processes before they are released into the atmosphere (Table 8.3; Imansyah et al. 2017). Collaboration between government, industry, and research institutions is essential for developing and implementing effective emission reduction strategies. Public–private partnerships can facilitate sharing of knowledge, resources, and technology, leading to more sustainable industrial practices (Jiang et al. 2022). Moreover, integrating economic instruments, such as carbon pricing and emissions trading schemes, can incentivize industries to reduce their emissions and invest in cleaner technologies (Song et al. 2022).

The transportation sector is a significant source of greenhouse gas emissions, mainly CO_2, due to the reliance on fossil fuels for vehicles. Combining innovations and policies is necessary to address this issue and promote cleaner transportation options. One of the most effective strategies is the promotion of electric and hybrid vehicles, which produce zero or significantly lower emissions compared to conventional gasoline or diesel vehicles ("Betweenness-Based Method to Identify Critical Provincial Carbon Transport Sectors in the Yellow River Basin and

Table 8.3 Strategies for reducing emissions from key sectors

Sector	Strategies	Examples/technologies
Industrial emissions	Regulations and emission control standards, cleaner production technologies, energy efficiency improvements, and use of low-emission fuels and raw materials. Monitoring frameworks ensure compliance	Carbon capture and storage (CCS), flue gas desulfurization, scrubbers, renewable energy adoption like solar and wind
Transportation emissions	Promotion of electric and hybrid vehicles, stricter fuel efficiency standards, expansion of public transport networks, and use of alternative fuels. Smart traffic management systems help reduce congestion	Battery-electric vehicles (BEVs), plug-in hybrids, biofuels, hydrogen fuel cells, urban mass transit systems, and traffic optimization tools
Agricultural emissions	Precision agriculture, methane reduction in livestock, sustainable farming practices, agroforestry, and biogas production from agricultural waste	GPS-enabled machinery, feed additives, anaerobic digesters, organic fertilizers, crop rotation, and biogas plants

Explore Spatiotemporal Evolution Driving Factors" 2024). Governments worldwide are implementing incentives, such as tax credits and rebates, to encourage consumers to adopt electric vehicles (EVs) (Jiang et al. 2022). In addition to vehicle electrification, improvements in the public transportation system can significantly reduce transportation sector emissions. Investments in rail and bus infrastructure and the promotion of active transportation modes such as cycling and walking can decrease reliance on personal vehicles (Yuan et al. 2020). For example, cities that have expanded their public transit networks have observed reduced traffic congestion and associated emissions (Table 8.3; Liu et al. 2021). Furthermore, developing smart transportation systems, which utilize technology to optimize traffic flow and reduce idle times, can contribute to lower emissions (Sebos 2023). Policy measures, such as stricter emissions standards for vehicles and fuel efficiency regulations, are also critical in driving emissions reductions in the transportation sector. The U.S. Environmental Protection Agency (EPA) and the Department of Transportation have established ambitious fuel economy standards to reduce greenhouse gas emissions from light-duty vehicles (Jiang et al. 2022). Additionally, promoting alternative fuels, such as biofuels and hydrogen, can further reduce the transportation sector's carbon footprint (Luttrell et al. 2017).

Agriculture is another significant source of greenhouse gas emissions, mainly methane (CH_4) and N_2O, potent greenhouse gases. Various techniques and practices can be implemented to mitigate emissions from this sector. One effective strategy is adopting sustainable agricultural practices, such as precision agriculture, which utilizes technology to optimize resource use and minimize waste (Zhou et al. 2023). For example, precision nutrient management can reduce the over-application of fertilizers, thereby decreasing nitrous oxide emissions (Ali et al. 2020; Ogwu and Izah 2023). Livestock management practices also play a crucial role in reducing agricultural emissions. Strategies such as improving feed efficiency, implementing rotational grazing, and utilizing anaerobic digesters for manure management can significantly lower methane emissions from livestock (Ahsan 2024). Additionally, agroforestry practices integrating trees into agricultural landscapes can enhance carbon sequestration while providing additional benefits such as improved soil health and biodiversity (Lefèvre et al. 2020; Ikhajiagbe et al. 2022; Ogwu et al. 2024). Government policies and incentives are essential for promoting the adoption of emission-reduction techniques in agriculture. Programs that support farmers financially to implement sustainable practices, such as conservation tillage or cover cropping, can facilitate the transition to more environmentally friendly agricultural systems (Chan et al. 2016). Furthermore, education and outreach initiatives to raise awareness among farmers about the benefits of emission-reduction practices can enhance participation in these programs (Bai et al. 2021).

8.5 Advancements in Green Technologies for Air Pollution Control

The increasing concern over air pollution has catalyzed the development and integration of green technologies to mitigate emissions and enhance urban air quality. Transitioning to renewable energy sources is a pivotal strategy for reducing air pollution. Traditional fossil fuel-based energy generation contributes to air pollutants, including SO_2, NO_x, and PM (Gómez et al. 2019). By integrating renewable energy sources such as solar, wind, and hydroelectric power, cities can significantly decrease their reliance on fossil fuels, reducing emissions and improving air quality (Srinivasan 2024). For instance, deploying solar photovoltaic systems has reduced air pollution by displacing the need for coal and natural gas power plants, major air pollutants sources (Table 8.4; Dube 2023). A study conducted in India highlighted the effectiveness of smart energy management systems in industrial clusters, which optimized energy consumption and facilitated the integration of renewable energy sources, leading to substantial reductions in air pollution (Dube 2023). Furthermore, using wind energy has been associated with lower emissions of greenhouse gases and other pollutants, contributing to cleaner air in urban environments (Ali 2023). Implementing policies that incentivize the adoption of renewable energy technologies is crucial. Governments can promote renewable energy through subsidies, tax credits, and feed-in tariffs, encouraging residential and commercial investments in clean energy solutions (Kencono 2023). Additionally, community-based renewable energy projects can empower local populations to participate in the transition to cleaner energy, fostering a sense of ownership and responsibility toward air quality improvement (Wai and Yu 2023).

Smart technologies transform urban air quality management by leveraging data and connectivity to monitor and manage air pollution effectively. Integrating the Internet of Things (IoT) and sensor networks enables real-time air quality monitoring, providing valuable data for decision-makers (Ucar et al. 2020). For example, IoT-based air quality monitoring systems can track pollutants such as particulate matter (PM) and nitrogen dioxide (NO_2) across urban areas, allowing for timely interventions and public health campaigns (Srinivasan 2024; Ababneh 2023). Moreover, smart technologies facilitate predictive modeling and optimization, enhancing the efficiency of urban air quality management strategies. By analyzing historical data and current trends, cities can forecast pollution levels and implement proactive measures to mitigate emissions (Garg et al. 2017). For instance, machine learning algorithms can be employed to predict air quality fluctuations based on traffic patterns, weather conditions, and industrial activities, enabling cities to optimize traffic flow and reduce congestion, which is a significant source of urban air pollution (Wai and Yu 2023; Hashem et al. 2023). The development of smart urban environments also emphasizes the importance of green spaces. Urban forestry and vegetation can significantly improve air quality by absorbing pollutants and providing shade, reducing urban heat islands (Ucar et al. 2020). Integrating smart technologies with

Table 8.4 Advancements in green technologies for air pollution control

Technology	Principle/Mechanism	Applications	Advantages
Electrostatic precipitators (ESP)	It uses electric charge to capture fine particulate matter from industrial emissions	Power plants, cement factories, and steel mills	High efficiency in removing fine particles
Biofilters	It uses microorganisms to degrade air pollutants like VOCs and odors	Industrial odor control, wastewater treatment plants, chemical industries	Eco-friendly and cost-effective
Photocatalytic oxidation	UV light activates a catalyst (e.g., TiO_2) to oxidize pollutants into harmless compounds	Air purifiers, coatings on buildings, tunnels	Breaks down VOCs, NO_x, and biological pollutants
Carbon adsorption	It uses activated carbon to trap gases and odors through adsorption	Industrial exhaust treatment air purification systems	Effective for VOCs and gas-phase contaminants
Green roofs and walls	Vegetation absorbs pollutants and reduces heat islands	Urban areas, buildings, public infrastructure	Improves air quality and provides insulation
Biogas scrubbers	Removes harmful gases (e.g., H_2S, CO_2) from biogas, allowing its use as a cleaner fuel	Biogas plants, waste-to-energy facilities	Converts waste gases into renewable energy
Hybrid filters (ESP + Fabric)	Combines ESP and fabric filters for enhanced particulate matter removal	Industrial facilities requiring stringent air quality control	High capture efficiency with reduced emissions

urban green spaces can enhance their management and effectiveness, ensuring they contribute optimally to air quality improvement (Ababneh 2023).

Numerous case studies illustrate the successful application of green technologies in reducing air pollution and enhancing urban air quality. One notable example is implementing a comprehensive air quality monitoring system in Semarang, Indonesia, which utilizes CCTV and IoT technologies for real-time environmental monitoring (Kencono 2023). This initiative enabled local authorities to track pollution sources and implement targeted interventions, improving air quality and public health outcomes. Another successful case study is the deployment of green roofs in Seoul, South Korea. Green roofs mitigate stormwater runoff and improve air quality by filtering pollutants and providing insulation (Shafique et al. 2018). Integrating green roofs into urban planning has been shown to reduce urban heat and enhance biodiversity, contributing to a more sustainable urban environment. Smart cities, Barcelona has implemented various smart technologies to manage air quality effectively (Jacob et al. 2024). By utilizing data analytics and IoT sensors, the city has optimized traffic flow, reduced evehicle emissions and improved public transportation systems (Croitoru and Năstase 2018). These efforts have led to a measurable decrease in air pollution levels, demonstrating the potential of smart technologies in urban air quality management.

8.6 Policy Frameworks and Regulatory Strategies

International treaties and agreements play a crucial role in addressing transboundary air pollution and setting global standards for air quality. One of the most significant agreements is the Convention on Long-range Transboundary Air Pollution (CLRTAP), established in 1979, which aims to limit and reduce air pollution across borders. The convention has led to several protocols, including the Gothenburg Protocol, which sets legally binding emission reduction targets for key pollutants such as SO_2, NO_x, and PM (Liao et al. 2016; Fann et al. 2011). These international frameworks facilitate cooperation among countries to tackle air pollution collectively, recognizing that air quality issues often transcend national borders. Moreover, the Paris Agreement under the United Nations Framework Convention on Climate Change (UNFCCC) indirectly influences air quality by promoting the reduction of greenhouse gas emissions, which are often correlated with air pollutants. By committing to limit global warming, countries are encouraged to adopt cleaner technologies and practices that also improve air quality (Kumar et al. 2016a, b). The integration of air quality considerations into climate change policies is essential for achieving comprehensive environmental benefits. The effectiveness of these international agreements is contingent upon the commitment of signatory nations to implement and enforce the stipulated measures. Regular reporting and assessment mechanisms are vital for tracking progress and ensuring accountability (Matte et al. 2013). For instance, the Global Environment Facility (GEF) supports projects aimed at reducing air pollution in developing countries, highlighting the importance of financial and technical assistance in achieving international air quality goals (Borghi et al. 2017).

At the national level, regulations and standards for emission control are critical for managing air quality. Countries typically establish air quality standards based on scientific research and public health considerations. In the United States, the Clean Air Act mandates the Environmental Protection Agency (EPA) to set National Ambient Air Quality Standards (NAAQS) for six criteria pollutants: CO, Pb, NO_2, O_3, PM_{10} and $PM_{2.5}$, and SO_2 (Fattahi 2015). These standards are designed to protect public health and the environment and serve as benchmarks for state and local air quality management efforts. National regulations often require states to develop State Implementation Plans (SIPs) that outline how they will achieve and maintain compliance with NAAQS. SIPs must include strategies for reducing emissions from various sources, including transportation, industrial processes, and power generation (Kumar et al. 2016a, b). The effectiveness of these regulations depends on adequate enforcement mechanisms and the allocation of resources for monitoring and compliance (Schumacher and Shandas 2019; Izah et al. 2024a, b).

In addition to regulatory measures, economic instruments such as emissions trading systems and carbon pricing can incentivize industries to reduce their emissions. For example, the European Union Emissions Trading System is a cap-and-trade system that limits greenhouse gas emissions from power plants and industrial facilities, promoting cost-effective emissions reductions while addressing air quality (Mueller et al. 2020). Public–private partnerships (PPPs) are increasingly recognized as effective mechanisms for implementing air quality management strategies. These collaborations leverage the strengths of both sectors to achieve common goals in reducing air pollution. For instance, partnerships between government agencies and private companies can facilitate developing and deploying innovative technologies for air quality monitoring and pollution control (Tang et al. 2017). One successful example of a PPP is the collaboration between local governments and technology firms to deploy air quality monitoring networks using low-cost sensors. These networks enhance data collection and provide real-time information on air pollution levels, enabling more informed decision-making and public awareness (Moreoane et al. 2021). Additionally, private sector involvement in financing and implementing green infrastructure projects, such as urban green spaces and green roofs, can contribute to improved air quality while also providing economic benefits (Barnes et al. 2011). Moreover, public–private partnerships can foster community engagement and education on air quality issues. By involving local communities in monitoring efforts and decision-making processes, these partnerships can enhance public awareness and support for air quality initiatives (Tutmez 2021). For example, initiatives that empower citizens to participate in air quality monitoring can lead to increased advocacy for cleaner air policies and practices.

8.7 Community and Stakeholder Engagement

Effective air quality management requires the active participation of communities and stakeholders. Engaging the public in awareness campaigns, leveraging the role of non-governmental organizations (NGOs), and adopting inclusive approaches to policy implementation are crucial for fostering a collective response to air pollution challenges. Public awareness campaigns are essential for educating communities about the sources, effects, and mitigation strategies related to air pollution. These campaigns aim to inform the public about the health risks associated with poor air quality and encourage behavioral changes that can reduce exposure to pollutants. Research indicates that awareness of air pollution issues can increase public concern and advocacy for cleaner air policies (Ward et al. 2022; Saadeh et al. 2022). For instance, campaigns that utilize innovative communication strategies, such as storytelling and visual arts, have been shown to effectively engage the public and raise awareness about air quality issues (Pope 2024; Robert et al. 2021). One successful approach to public awareness is the use of AQI to communicate pollution levels in an accessible manner. Studies have demonstrated that enhancing public understanding of AQI reports can empower individuals to make informed decisions about outdoor activities, particularly during high-pollution events (Borbet et al. 2018). Additionally, integrating local air quality data into community engagement initiatives allows residents to connect more personally with the issue, fostering a sense of responsibility and action (Masri et al. 2022; Lu et al. 2022). Furthermore, targeted campaigns aimed at vulnerable populations, such as children and the elderly, can be particularly impactful. Research has shown that parents are often concerned about the effects of air pollution on their children's health, making them a key audience for educational initiatives (Krishna et al. 2021; Ngobeni 2023). Public awareness campaigns can successfully inspire community action against air pollution by using a variety of communication channels and customizing messages for populations.

Non-governmental organizations (NGOs) play a vital role in advocating for air quality improvements and engaging communities in environmental issues. NGOs often serve as intermediaries between the public and policymakers, raising awareness about air pollution and advocating for stronger regulations and policies (Jordan et al. 2018; Noël et al. 2021). Their efforts can include organizing community workshops, conducting research, and mobilizing grassroots campaigns to influence decision-makers. For example, NGOs have been instrumental in promoting the use of low-cost air quality sensors to empower communities to monitor their own environments. Such initiatives not only provide valuable data on local air quality but also foster community engagement and advocacy for cleaner air policies (Masri et al. 2022; Lu et al. 2022). By involving community members in the data collection process, NGOs can enhance public understanding of air pollution issues and encourage collective action. Moreover, NGOs often collaborate with academic institutions and government agencies to conduct research and disseminate findings related to air quality. This collaboration can lead to more informed policy decisions and greater public awareness of the health impacts of air pollution (Noël et al. 2021; Wong et al. 2020; Kosoe et al. 2023). Advocacy efforts by NGOs can also highlight the disproportionate impacts of air pollution on marginalized communities, emphasizing the need for equitable solutions (Wong et al. 2020; Mora 2023).

Inclusive approaches to policy implementation are essential for ensuring that all stakeholders, particularly marginalized communities, have a voice in air quality management. Engaging communities in the decision-making process can lead to more effective and equitable policies. Research has shown that participatory approaches, where community members are involved in the development and implementation of air quality initiatives, can enhance trust and cooperation between communities and government agencies (Wong et al. 2020; Sîrbu et al. 2015). One effective strategy for fostering inclusivity is the establishment of community advisory boards or committees that include representatives from diverse stakeholder groups. These boards can provide valuable insights into local air quality issues and help shape policies that reflect the needs and concerns of the community (Masri et al. 2022; Wong et al. 2020). Additionally, utilizing citizen science initiatives, where community members actively participate in monitoring air quality, can empower residents and promote a sense of ownership over local environmental issues (Lu et al. 2022; Mora 2023). Furthermore, public engagement efforts should prioritize transparency and accessibility. Providing clear information about air quality policies, monitoring results, and health impacts can help build public trust and encourage community participation (Saadeh et al. 2022; Borbet et al. 2018). Engaging with local leaders and organizations can also facilitate outreach to underrepresented populations, ensuring that their voices are heard in the policy-making process (Wong et al. 2020; Mora 2023). Community and stakeholder engagement are critical components of effective air quality management. Public awareness campaigns, the active role of NGOs, and inclusive policy implementation strategies can empower communities to act against air pollution. By fostering collaboration and ensuring that all voices are heard, stakeholders can work together to develop and implement effective solutions for improving air quality and protecting public health.

8.8 Challenges and Opportunities in Air Pollution Control

Economic and technological barriers significantly hinder the widespread adoption of air pollution control technologies. One of the primary economic challenges is the high initial investment required for implementing advanced pollution control systems. Industries, particularly in developing countries, often face budget constraints that limit their ability to invest in cleaner technologies (Matte et al. 2013; Dehghani et al. 2022). For instance, the transition to renewable energy sources or the installation of advanced filtration systems can be prohibitively expensive for small and medium-sized enterprises, which may lack the financial resources to undertake such investments (Tsao et al. 2012). Additionally, the economic viability of certain technologies can be influenced by fluctuating energy prices and regulatory frameworks. For example, the cost-effectiveness of biofuels can be compromised by land-use changes and associated emissions, which may negate some of the environmental benefits (Raaschou-Nielsen et al. 2013). Furthermore, the lack of financial incentives or subsidies for adopting cleaner technologies can deter industries from making the necessary investments (Liu et al. 2017a, b). Technological barriers also play a critical role in limiting the effectiveness of air pollution control measures. Many existing technologies require significant maintenance and operational expertise, which may not be readily available in all regions (Brantley et al. 2014). Moreover, the rapid pace of technological advancement can lead to obsolescence, where industries hesitate to invest in current technologies that may soon be surpassed by more efficient solutions (Wang et al. 2016). This uncertainty can stifle innovation and delay the implementation of effective air quality management strategies.

Despite these challenges, there are substantial opportunities for innovation and investment in air pollution control. The growing awareness of air quality issues has spurred interest in developing new technologies and solutions that can effectively reduce emissions. For instance, advancements in low-cost air quality sensors have democratized air monitoring, allowing communities to engage in data collection and advocacy for cleaner air (Liang et al. 2017a, b; Liang et al. 2017a, b). These sensors enable real-time monitoring of air pollutants, empowering citizens to act and hold industries accountable for their emissions. Investment in green technologies presents another significant opportunity. Governments and private investors are increasingly recognizing the potential for clean energy solutions to drive economic growth while addressing environmental concerns. For example, the transition to electric vehicles (EVs) is gaining momentum, supported by government incentives and advancements in battery technology (Benmarhnia et al. 2014). The development of charging infrastructure and the integration of renewable energy sources into transportation systems can further enhance the environmental benefits of EVs (Meier et al. 2015). Moreover, public–private partnerships (PPPs) can facilitate innovation by combining resources and expertise from both sectors. Collaborative initiatives can lead to the development of new air quality management strategies that leverage cutting-edge technologies and community engagement (Pignocchino et al. 2022). For instance, partnerships between local governments and technology firms can enhance air quality monitoring efforts and promote the adoption of cleaner technologies in urban areas (Wang et al. 2022).

The future of air pollution management will likely involve a multifaceted approach that integrates technological advancements, regulatory frameworks, and community engagement. One promising direction is the incorporation of artificial intelligence (AI) and machine learning into air quality modeling and prediction. These technologies can enhance the accuracy of air quality forecasts, enabling cities to implement proactive measures to mitigate pollution (Chin et al. 2019; Amabie et al. 2024). For example, AI can analyze historical data and real-time monitoring information to predict pollution spikes, allowing for timely interventions (Kuo et al. 2018). Another important direction is the emphasis on equity in air quality management. As air pollution disproportionately affects marginalized communities, future policies must prioritize equitable access to clean air and environmental justice (Zheng et al. 2021). This can involve targeted interventions in high-pollution areas and the inclusion of community voices in decision-making processes (Kiesewetter et al. 2014). Engaging communities in air quality monitoring and policy advocacy can empower residents to take ownership of their environment and advocate for necessary changes (Chuwah et al. 2013). Furthermore, the integration of air quality considerations into broader urban planning and public health initiatives is essential. By recognizing the interconnectedness of air quality, health, and urban development, policymakers can create comprehensive strategies that address the root causes of pollution while promoting sustainable growth (Quarmby et al. 2019). For example, urban greening initiatives, such as the development of parks and green roofs, can improve air quality while enhancing community well-being (Ahmed et al. 2019).

8.9 Conclusion

Air quality monitoring and assessment are vital for understanding and managing air pollution. The AQI serves as a key communication tool, while advancements in monitoring technologies and data analysis techniques enhance the accuracy and comprehensiveness of air quality assessments. The integration of citizen science initiatives further enriches the data landscape, fostering community engagement and awareness. As air quality continues to be a pressing global issue, the ongoing development and implementation of effective monitoring strategies will be essential for protecting public health and the environment. Technological approaches to air pollution control encompass a diverse array of systems and methods designed to mitigate the impacts of airborne pollutants. Filtration and absorption systems provide effective means of removing particulate and gaseous contaminants, while electrostatic precipitators and scrubbers play critical roles in industrial emissions control. Catalytic and non-catalytic reduction systems further enhance the ability to reduce harmful emissions, particularly nitrogen oxides. As air pollution continues to pose significant health and environmental challenges, ongoing advancements in these technologies will be essential for achieving cleaner air and protecting public health. Reducing emissions from key sectors—industrial, transportation, and agricultural—is vital for achieving climate goals and improving air quality. Effective strategies involve a combination of regulations, technological innovations, and sustainable practices tailored to the unique challenges of each sector. Collaborative efforts among governments, industries, and communities are essential for implementing these strategies and fostering a transition to a low-carbon economy.

Advancements in green technologies for air pollution control are essential for creating sustainable urban environments. The integration of renewable energy sources, the application of smart technologies, and successful case studies highlight the potential for innovative solutions to mitigate air pollution. As cities continue to grow and face increasing environmental challenges, the adoption of these technologies will be crucial for improving air quality and protecting public health. Effective air quality management requires a multifaceted approach that includes international treaties, national regulations, and public–private partnerships. International agreements provide a framework for global cooperation, while national regulations establish standards and enforcement mechanisms to protect public health. Public–private partnerships offer innovative solutions and community engagement opportunities that enhance the effectiveness of air quality strategies. As air pollution continues to pose significant challenges, the integration of these policy frameworks and regulatory strategies will be essential for achieving cleaner air and promoting public health.

References

Ababneh A (2023) Smart urban management of green space. J Des Resilience Architect Plan 4(3):339–353. https://doi.org/10.47818/drarch.2023.v4i3101

Ahmed F, Hossain S, Hossain S, Fakhruddin A, Abdullah A, Chowdhury M et al (2019) Impact of household air pollution on human health: source identification and systematic management approach. SN Appl Sci. https://doi.org/10.1007/s42452-019-0405-8

Ahsan M (2024) Optimizing environmental impact: MCDM-based approaches for petrochemical industry emission cuts. IEEE Access 12:35309–35324. https://doi.org/10.1109/access.2024.3370164

Ali K, Ahmad M, Yusup Y (2020) Issues, impacts, and mitigations of carbon dioxide emissions in the building sector. Sustainability 12(18):7427. https://doi.org/10.3390/su12187427

Ali S (2023) Enabling technologies for next-generation smart cities: a comprehensive review and research directions. Fut Internet 15(12):398. https://doi.org/10.3390/fi15120398

Allen R, Barn P (2020) Individual- and household-level interventions to reduce air pollution exposures and health risks: a review of the recent literature. Curr Environ Health Rep 7(4):424–440. https://doi.org/10.1007/s40572-020-00296-z

Al Momani FA, Örmeci B (2014) Optimization of polymer dose based on residual polymer concentration in dewatering supernatant. Water Air Soil Pollut 225:2154 https://doi.org/10.1007/s11270-014-2154-z

Amabie T, Izah SC, Ogwu MC, Hait M (2024) Harmonizing tradition and technology: the synergy of artificial intelligence in traditional medicine. In: Izah SC, Ogwu MC, Akram M (eds) Herbal medicine phytochemistry. Reference series in phytochemistry. Springer, Cham. https://doi.org/10.1007/978-3-031-21973-3_70-1

Bai S, Zhang B, Ning Y, Ying W (2021) Comprehensive analysis of carbon emissions, economic growth, and employment from the perspective of industrial restructuring: a case study of China. Environ Sci Pollut Res 28(36):50767–50789. https://doi.org/10.1007/s11356-021-14040-z

Bai Y, Han C, He C, Gu G, Nie J, Shao J et al. (2018) Washable multilayer triboelectric air filter for efficient particulate matter $PM_{2.5}$ removal. Adv Funct Mater. https://doi.org/10.1002/adfm.201706680

Barnes J, Chatterton T, Hayes E, Longhurst J, Olowoporoku A (2011) Assessing the potential for local action to achieve EU limit values. https://doi.org/10.2495/air110261

Becchio C, Bottero M, Corgnati S, Dell'Anna F, Fabi V, Lingua C, et al. (2019) Effects on energy savings and occupant health of an antibacterial filter. E3s Web Conf 111:02056. https://doi.org/10.1051/e3sconf/201911102056

Benmarhnia T, Rey L, Cartier Y, Clary C, Deguen S, Brousselle A (2014) Addressing equity in interventions to reduce air pollution in urban areas: a systematic review. Int J Public Health 59(6):933–944. https://doi.org/10.1007/s00038-014-0608-0

Borbet T, Gladson L, Cromar K (2018) Assessing air quality index awareness and use in mexico city. BMC Public Health. https://doi.org/10.1186/s12889-018-5418-5

Borghi F, Spinazzè A, Rovelli S, Campagnolo D, Buono L, Cattaneo A et al (2017) Miniaturized monitors for assessment of exposure to air pollutants: a review. Int J Environ Res Public Health 14(8):909. https://doi.org/10.3390/ijerph14080909

Brantley H, Hagler G, Deshmukh P, Baldauf R (2014) Field assessment of the effects of roadside vegetation on near-road black carbon and particulate matter. Sci Total Environ 468–469:120–129. https://doi.org/10.1016/j.scitotenv.2013.08.001

Cao J-J, Huang Y, Zhang Q (2021) Ambient air purification by nanotechnologies: from theory to application. Catalysts 11(11):1276. https://doi.org/10.3390/catal11111276

Capulli, D. (2024). Liquid air filtration and continuous monitoring: customized indoor air quality. E3s Web Conf 523:06003. https://doi.org/10.1051/e3sconf/202452306003

Chan W, Sugiyama M, Borges G (2016) Assessment of CO_2 emission mitigation for a Brazilian oil refinery. Braz J Chem Eng 33(4):835–850. https://doi.org/10.1590/0104-6632.20160334s20140149

Chen W, Wu X, Ngabo D (2021) Benefit analysis of low-carbon policy mix innovation based on consumer perspective in smart city. Sci Program 2021:1–10. https://doi.org/10.1155/2021/3282398

Chin Y, Pretto L, Thuppil V, Ashfold M (2019) Public awareness and support for environmental protection—a focus on air pollution in peninsular Malaysia. PLoS ONE 14(3):e0212206. https://doi.org/10.1371/journal.pone.0212206

Chuwah C, Noije T, Vuuren D, Hazeleger W, Strunk A, Deetman S et al (2013) Implications of alternative assumptions regarding future air pollution control in scenarios similar to the representative concentration pathways. Atmos Environ 79:787–801. https://doi.org/10.1016/j.atmosenv.2013.07.008

Cisneros R, Brown P, Cameron L, Gaab E, Gonzalez M, Ramondt S et al (2017) Understanding public views about air quality and air pollution sources in the San Joaquin valley, California. J Environ Public Health 2017:1–7. https://doi.org/10.1155/2017/4535142

Croitoru C, Năstase I (2018) A state of the art regarding urban air quality prediction models. E3s Web Conf 32:01010. https://doi.org/10.1051/e3sconf/20183201010

Dehghani S, Vali M, Jafarian A, Oskoei V, Maleki Z, Hoseini M (2022) Ecological study of ambient air pollution exposure and mortality of cardiovascular diseases in elderly. Sci Rep. https://doi.org/10.1038/s41598-022-24653-0

Dube S (2023) Reducing air pollution through smart energy management in industrial clusters of India. Environ Prog Sustain Energy. https://doi.org/10.1002/ep.14169

Fann N, Roman H, Fulcher C, Gentile M, Hubbell B, Wesson K et al (2011) Maximizing health benefits and minimizing inequality: incorporating local-scale data in the design and evaluation of air quality policies. Risk Anal 31(6):908–922. https://doi.org/10.1111/j.1539-6924.2011.01629.x

Fattahi M (2015) The role of urbanization rate in the relationship between air pollution and health expenditures: a dynamic panel data approach. Int Lett Social Human Sci 53:68–72. https://doi.org/10.18052/www.scipress.com/ilshs.53.68

Garg C, Sharma A, Goyal G (2017) A hybrid decision model to evaluate critical factors for successful adoption of GSCM practices under fuzzy environment. Uncertain Supply Chain Manag. https://doi.org/10.5267/j.uscm.2016.7.002

Gómez Y, Estrada-Esquivel H, Rebollar A, Vásquez D (2019) A novel air quality monitoring unit using cloudino and fiware technologies. Math Comput Appl 24(1):15. https://doi.org/10.3390/mca24010015

Guo C, Liu G, Lyu L, Chen C (2020) An unsupervised $PM_{2.5}$ estimation method with different spatio-temporal resolutions based on KIDW-TCGRU. IEEE Access 8:190263–190276. https://doi.org/10.1109/access.2020.3032420

Hao Z, Wu J (2022) Self-assembled zeolitic imidazolate framework/polyimide nanofibers for efficient air pollution control. ACS Appl Nano Mater 5(2):2343–2349. https://doi.org/10.1021/acsanm.1c04021

Hashem I, Usmani R, Almutairi M, Ibrahim A, Zakari A, Alotaibi F et al (2023) Urban computing for sustainable smart cities: recent advances, taxonomy, and open research challenges. Sustainability 15(5):3916. https://doi.org/10.3390/su15053916

He Y, Liu H, Ying W (2024) Constructing stable polyvinyl alcohol/gelatin/cellulose nanocrystals composite electrospun membrane with excellent filtration efficiency for $PM_{2.5}$. Polymers 16(12):1656. https://doi.org/10.3390/polym16121656

Huang X, Jiao T, Liu Q, Zhang L, Zhou J, Li B, Peng Q (2018) Hierarchical electrospun nanofibers treated by solvent vapor annealing as air filtration mat for high-efficiency $PM_{2.5}$ capture. Sci China Mater 62(3):423–436. https://doi.org/10.1007/s40843-018-9320-4

Ikhajiagbe B, Ogwu MC (2020) Hazard quotient, microbial diversity and plant composition of spent crude oil polluted-soil. Beni-Suef Univ J Basic Appl Sci. https://doi.org/10.1186/s43088-020-00052-0

Ikhajiagbe B, Ogwu MC, Ekhator PO, Omoayena II (2022) Implications of soil nitrogen enhancement on the yield performance of soybean (*Glycine max*) in cadmium-polluted soil. J Pollut Effects Control 10(2):1–12. https://doi.org/10.35248/23754-4397.22.10.334

Imansyah M, Putranti T, Mangkurat L (2017) The identification of key sector in CO_2 emissions in production perspective of Indonesia: an input–output analysis. Int J Sustain Fut Human Secur 5(2):21–29. https://doi.org/10.24910/jsustain/5.2/2129

Istiqomah Q (2023) Analysis of distribution of NO_2 pollutants in Pt. X Sanggau Regency, West Kalimantan. J Teknik Sipil 23(3):359. https://doi.org/10.26418/jts.v23i3.64855

Izah SC, Abdelkhalek ST, Ogwu MC, Hamidifar H (2024) Overview and perspectives of biomonitoring in the Global South. In: Izah SC, Ogwu MC, Hamidifar H (eds) Biomonitoring of pollutants in the Global South. Springer, Singapore. https://doi.org/10.1007/978-981-97-1658-6_1

Izah SC, Ogwu MC, Hamidifar H (2024) Biomonitoring of pollutants in the Global South. Springer Nature, Cham. https://doi.org/10.1007/978-981-97-1658-6

Izah SC, Ogwu MC, Akram M (2023) Herbal medicine phytochemistry: applications and trends. Springer, Cham. https://doi.org/10.1007/978-3-031-21973-3

Jacob DE, Nelson IU, Izah SC, Ukpong E, Akpan UU, Ogwu MC (2024) Bioindicators in recreational planning and development: balancing nature and human activities. In: Izah SC, Ogwu MC, Hamidifar H (eds) Biomonitoring of pollutants in the Global South. Springer, Singapore. https://doi.org/10.1007/978-981-97-1658-6_24

Jakfar, Husin H, Zaki M, Mairiza L, Zulrika M, Nasution F, Ahmadi (2023) Optimization study of CO_2 gas absorption with NaOH absorbent continuous system in Raschig Ring packing column using Box–Behnken design. Inventions 8(3):70. https://doi.org/10.3390/inventions8030070

Jiang R, Wu P, Wu C (2022) Driving factors behind energy-related carbon emissions in the U.S. road transport sector: a decomposition analysis. Int J Environ Res Public Health 19(4):2321. https://doi.org/10.3390/ijerph19042321

Johansson D, Rootzén J, Berntsson T, Johnsson F (2012) Assessment of strategies for CO_2 abatement in the European petroleum refining industry. Energy 42(1):375–386. https://doi.org/10.1016/j.energy.2012.03.039

Johnson B (2011) Experience with urban air pollution in Paterson, New Jersey and implications for air pollution communication. Risk Anal 32(1):39–53. https://doi.org/10.1111/j.1539-6924.2011.01669.x

Jordan L, Chui C, Forth M (2018) Child welfare NGOS in Hong Kong: does advocacy work? Int Soc Work 63(2):177–192. https://doi.org/10.1177/0020872818774109

Kencono D (2023) The use of CCTV model analysis for a smart environment in Semarang city. E3s Web Conf 440:07006. https://doi.org/10.1051/e3sconf/202344007006

Kiesewetter G, Borken-Kleefeld J, Schöpp W, Heyes C, Thunis P, Bessagnet B et al (2014) Modelling NO_2 concentrations at the street level in the gains integrated assessment model: projections under current legislation. Atmos Chem Phys 14(2):813–829. https://doi.org/10.5194/acp-14-813-2014

Kosoe EA, Achana GTW, Ogwu MC (2023) Regulations and policies for herbal medicine and practitioners. In: Izah SC, Ogwu MC, Akram M (eds) Herbal medicine phytochemistry. Reference series in phytochemistry. Springer, Cham. https://doi.org/10.1007/978-3-031-21973-3_33-1

Krishna B, Devi R, Priya A (2021) Knowledge, attitude and awareness on effect of air pollution on children among parents. J Pharmaceut Res Int. https://doi.org/10.9734/jpri/2021/v33i47b33134

Kumar A, Gupta I, Brandt J, Kumar R, Dikshit A, Patil R (2016a) Air quality mapping using gis and economic evaluation of health impact for Mumbai city, India. J Air Waste Manag Assoc 66(5):470–481. https://doi.org/10.1080/10962247.2016.1143887

Kumar A, Patil R, Dikshit A, Kumar R (2016b) Comparison of predicted vehicular pollution concentration with air quality standards for different time periods. Clean Technol Environ Policy 18(7):2293–2303. https://doi.org/10.1007/s10098-016-1147-6

Kuo C, Pan R, Wu C, Phan D, Chan C (2018) Application of a time-stratified case-crossover design to explore the effects of air pollution and season on childhood asthma hospitalization in cities of differing urban patterns: big data analytics of government open data. Int J Environ Res Public Health 15(4):647. https://doi.org/10.3390/ijerph15040647

Lefèvre J, Briand Y, Pye S, Tovilla J, Li F, Oshiro K et al (2020) A pathway design framework for sectoral deep decarbonization: the case of passenger transportation. Clim Policy 21(1):93–106. https://doi.org/10.1080/14693062.2020.1804817

Leung A, Kwok E (2012) Evaluation of the nano-confined catalytic oxidation technology for air purification and odor reduction. Adv Mater Res 550–553:607–615. https://doi.org/10.4028/www.scientific.net/amr.550-553.607

Li J, Yang H, Zha S, Yu N, Liu X, Sun R (2021) Effects of covid-19 emergency response levels on air quality in the Guangdong-Hong Kong-Macao greater bay area, China. Aerosol Air Qual Res 21(2):200416. https://doi.org/10.4209/aaqr.2020.07.0416

Li L, Jin X, Li Y, Chen H, Wang Y (2022) The impact of carbon trading policy on regional ecological risk: synergy between market-based environmental policy and government intervention. Front Environ Sci. https://doi.org/10.3389/fenvs.2022.1010522

Liang P, Zhu T, Fang Y, Li Y, Han Y, Wu Y et al (2017a) The role of meteorological conditions and pollution control strategies in reducing air pollution in Beijing during APEC 2014 and victory parade 2015. Atmos Chem Phys 17(22):13921–13940. https://doi.org/10.5194/acp-17-13921-2017

Liang P, Zhu T, Fang Y, Li Y, Wu Y, Hu M, et al (2017) The role of meteorological conditions and pollution control strategies in reducing air pollution in Beijing during APEC 2014 and parade 2015. https://doi.org/10.5194/acp-2017-456

Liao K, Hou X, Strickland M (2016) Resource allocation for mitigating regional air pollution–related mortality: a summertime case study for five cities in the United States. J Air Waste Manag Assoc 66(8):748–757. https://doi.org/10.1080/10962247.2016.1176085

Liu C, Hsu P, Lee H, Ye M, Zheng G, Liu N et al (2015) Transparent air filter for high-efficiency $PM_{2.5}$ capture. Nat Commun. https://doi.org/10.1038/ncomms7205

Liu G, Xiao M, Zhang X, Gál C, Chen X, Li L et al (2017a) A review of air filtration technologies for sustainable and healthy building ventilation. Sustain Cities Soc 32:375–396. https://doi.org/10.1016/j.scs.2017.04.011

Liu M, Huang Y, Ma Z, Zhou J, Liu X, Wang H et al (2017b) Spatial and temporal trends in the mortality burden of air pollution in China: 2004–2012. Environ Int 98:75–81. https://doi.org/10.1016/j.envint.2016.10.003

Liu Y, Yang S, Liu X, Guo P, Zhang K (2021) Driving forces of temporal-spatial differences in CO_2 emissions at the city level for China's transport sector. Environ Sci Pollut Res 28(20):25993–26006. https://doi.org/10.1007/s11356-020-12235-4

Liu Y, Zhou Y, Lu J (2020) Exploring the relationship between air pollution and meteorological conditions in China under environmental governance. Sci Rep. https://doi.org/10.1038/s41598-020-71338-7

Lu T, Liu Y, García A, Wang M, Li Y, Bravo-villasenor G et al (2022) Leveraging citizen science and low-cost sensors to characterize air pollution exposure of disadvantaged communities in Southern California. Int J Environ Res Public Health 19(14):8777. https://doi.org/10.3390/ijerph19148777

Luttrell C, Sills E, Aryani R, Ekaputri A, Evinke M (2017) Beyond opportunity costs: who bears the implementation costs of reducing emissions from deforestation and degradation? Mitig Adapt Strat Glob Change 23(2):291–310. https://doi.org/10.1007/s11027-016-9736-6

Maestas M, Brook R, Ziemba R, Li FRC, Klaver Z, et al (2018) Reduction of personal $PM_{2.5}$ exposure via indoor air filtration systems in Detroit: an intervention study. J Expos Sci Environ Epidemiol 29(4):484–490. https://doi.org/10.1038/s41370-018-0085-2

Masri S, Cox K, Flores L, Rea J, Wu J (2022) Community-engaged use of low-cost sensors to assess the spatial distribution of PM2.5 concentrations across disadvantaged communities: results from a pilot study in Santa Ana, CA. Atmosphere 13(2):304. https://doi.org/10.3390/atmos13020304

Mata TM, Martins AA, Calheiros CSC, Villanueva F, Alonso-Cuevilla NP, Gabriel MF, Silva GV (2022) Indoor air quality: a review of cleaning technologies. Environments 9(9):118. https://doi.org/10.3390/environments9090118

Matte T, Ross Z, Kheirbek I, Eisl H, Johnson S, Gorczynski J et al (2013) Monitoring intraurban spatial patterns of multiple combustion air pollutants in New York city: design and implementation. J Expos Sci Environ Epidemiol 23(3):223–231. https://doi.org/10.1038/jes.2012.126

Meier R, Schindler C, Eeftens M, Aguilera I, Ducret-Stich R, Ineichen A et al (2015) Modeling indoor air pollution of outdoor origin in homes of Sapaldia subjects in Switzerland. Environ Int 82:85–91. https://doi.org/10.1016/j.envint.2015.05.013

Memić B, Džubur A, Avdagić-Golub E (2022) Green IoT: sustainability environment and technologies. Sci Eng Technol 2(1):24–29. https://doi.org/10.54327/set2022/v2.i1.25

Mohraz M, Golbabaei F, Yu I, Mansournia M, Zadeh A, Dehghan S (2018) Preparation and optimization of multifunctional electrospun polyurethane/chitosan nanofibers for air pollution control applications. Int J Environ Sci Technol 16(2):681–694. https://doi.org/10.1007/s13762-018-1649-3

Mora S (2023) Air quality monitoring in coal-centric cities: a hybrid approach. Sustainability 15(16):12624. https://doi.org/10.3390/su151612624

Moreoane L, Mukwevho P, Burger R (2021) The quality of the first and second vaal triangle airshed priority area (vtapa) air quality management plans. Clean Air J. https://doi.org/10.17159/caj/2020/31/2.12178

Mueller S, Tarnay L, O'Neill S, Raffuse S (2020) Apportioning smoke impacts of 2018 wildfires on Eastern Sierra Nevada sites. Atmosphere 11(9):970. https://doi.org/10.3390/atmos11090970

Munsif R, Zubair M, Aziz A, Zafar M (2021) Industrial air emission pollution: potential sources and sustainable mitigation. https://doi.org/10.5772/intechopen.93104

Ngobeni F (2023) Community health risk awareness and knowledge of air pollution in Annadale, Polokwane Local Municipality, Lim-Popo province, South Africa. https://doi.org/10.20944/preprints202305.2149.v1

Niveditha ASM, Arivarasu L (2021) Covid-19 lockdown and its impact on air pollution. J Contemp Issues Bus Govern. https://doi.org/10.47750/cibg.2020.26.02.169

Noël C, Vanroelen C, Gadeyne S (2021) Qualitative research about public health risk perceptions on ambient air pollution. A review study. SSM Popul Health 15:100879. https://doi.org/10.1016/j.ssmph.2021.100879

Ogwu MC, Izah SC (2023) One health implications of agrochemicals and their sustainable alternatives (First Edition). In: Sustainable Development and Biodiversity Series volume 34 (Ramawat K.G.). Springer Nature, Singapore Springer Verlag, Singapore. 826p. Hardcover. https://doi.org/10.1007/978-981-99-3439-3

Ogwu MC, Izah SC, Dessureault-Rompré J, Gasparatos D (2024) Sustainable soil systems in Global South. Springer Nature, Cham. https://doi.org/10.1007/978-981-97-5276-8

Ogwu MC, Ogwu HI, Kosoe EA (2023) Plants used in the management and treatment of cardiovascular diseases: case study of the Benin People of Southern Nigeria. In: Izah SC, Ogwu MC, Akram M (eds) Herbal medicine phytochemistry. Reference series in phytochemistry. Springer, Cham. https://doi.org/10.1007/978-3-031-21973-3_4-1

Pignocchino G, Baldassarre G, Mondino E, Raffetti E (2022) Public risk perception of air pollution in the general population of Italy and Sweden: environmental and socio-demographic drivers. https://doi.org/10.21203/rs.3.rs-1783841/v1

Pope F (2024) Light painting photography makes particulate matter air pollution visible. Commun Earth Environ. https://doi.org/10.1038/s43247-024-01409-4

Quarmby S, Santos G, Mathias M (2019) Air quality strategies and technologies: a rapid review of the international evidence. Sustainability 11(10):2757. https://doi.org/10.3390/su11102757

Raaschou-Nielsen O, Andersen Z, Beelen R, Samoli E, Stafoggia M, Weinmayr G et al (2013) Air pollution and lung cancer incidence in 17 European cohorts: prospective analyses from the European study of cohorts for air pollution effects (escape). Lancet Oncol 14(9):813–822. https://doi.org/10.1016/s1470-2045(13)70279-1

Rahman M, Wang S, Zhao W, Xu X, Zhang W, Arshad A (2022) Investigating the relationship between air pollutants and meteorological parameters using satellite data over Bangladesh. Remote Sens 14(12):2757. https://doi.org/10.3390/rs14122757

Reddy V (2024) Improving the quality of the air and implementing into effective pollution mitigation methods. E3s Web Conf 529:03014. https://doi.org/10.1051/e3sconf/202452903014

Robert J, Wei F, Samson J (2021) Airxorcist: raising awareness about air pollution through narrative storytelling in computer game. Trends Undergrad Res 4(1):c20–27. https://doi.org/10.33736/tur.2808.2021

Roelfsema M, Soest H, Harmsen M, Vuuren D, Bertram C, Elzen M et al (2020) Taking stock of national climate policies to evaluate implementation of the Paris Agreement. Nat Commun. https://doi.org/10.1038/s41467-020-15414-6

Rybarczyk Y, Žalakevičiūtė R (2018) Machine learning approaches for outdoor air quality modelling: a systematic review. Appl Sci 8(12):2570. https://doi.org/10.3390/app8122570

Saadeh R, Khader Y, Malkawi M, Allouh M (2022) Communicating the risks of air pollution to the public: a perspective from Jordan and Lebanon. Environ Health Insights. https://doi.org/10.1177/11786302221127851

Schumacher K, Shandas V (2019) Rescaling air quality management: an assessment of local air quality authorities in the United States. Air Soil Water Res. https://doi.org/10.1177/1178622119842125

Sebos I (2023) Greenhouse gas emissions in Greek agriculture: trends and projections. E3s Web Conf 436:02007. https://doi.org/10.1051/e3sconf/202343602007

Shafique M, Kim R, Kyung-Ho K (2018) Green roof for stormwater management in a highly urbanized area: the case of Seoul, Korea. Sustainability 10(3):584. https://doi.org/10.3390/su10030584

Shao W (2023) Electrospun-SiO_2-nanofiber-reinforced cellulose aerogel loaded with ZIF-67 for air filtration and formaldehyde adsorption. Adv Mater Technol. https://doi.org/10.1002/admt.202301395

Sîrbu A, Becker M, Caminiti S, Baets B, Elen B, Francis L et al (2015) Participatory patterns in an international air quality monitoring initiative. Plos One 10(8):e0136763. https://doi.org/10.1371/journal.pone.0136763

Song J, Jiang L, He Z, Liu Z, Leng X (2022) Characteristics analysis and identification of key sectors of air pollutant emissions in China from the perspective of complex metabolic network. Int J Environ Res Public Health 19(15):9396. https://doi.org/10.3390/ijerph19159396

Souzandeh H, Johnson K, Wang Y, Bhamidipaty K, Zhong W (2016) Soy-protein-based nanofabrics for highly efficient and multifunctional air filtration. ACS Appl Mater Interf 8(31):20023–20031. https://doi.org/10.1021/acsami.6b05339

Srinivasan S (2024) A novel approach integrating IOT and WSN with predictive modeling and optimization for enhancing efficiency and sustainability in smart cities. JES 20(4s):2228–2237. https://doi.org/10.52783/jes.2393

Steffen W, Richardson K, Rockström J, Cornell SE, Fetzer I, Bennett EM et al (2015) Planetary boundaries: guiding human development on a changing planet. Science 347(6223):1259855. https://doi.org/10.1126/science.1259855

Tang J, McNabola A, Misstear B, Caulfield B (2017) An evaluation of the impact of the Dublin port tunnel and HGV management strategy on air pollution emissions. Transp Res Part D Transp Environ 52:1–14. https://doi.org/10.1016/j.trd.2017.02.009

Tsao C, Campbell J, Mena-Carrasco M, Carmichael G, Chen Y (2012) Biofuels that cause land-use change may have much larger non-GHG air quality emissions than fossil fuels. Environ Sci Technol 46(19):10835–10841. https://doi.org/10.1021/es301851x

Tutmez B (2021) Interpreting relationships between pollutants and carbon dioxide emitted into air from industries in Serbia. J Eng Manag Compet 11(2):115–123. https://doi.org/10.5937/jemc2102115t

Ucar Z, Akay A, Bilici E (2020) Towards green smart cities: importance of urban forestry and urban vegetation. Int Archiv Photogram Remote Sens Spatial Inf Sci XLIV-4/W3–2020:399–403. https://doi.org/10.5194/isprs-archives-xliv-4-w3-2020-399-2020

Wai K, Yu K (2023) Application of a machine learning method for prediction of urban neighborhood-scale air pollution. Int J Environ Res Public Health 20(3):2412. https://doi.org/10.3390/ijerph20032412

Waisman H, Bataille C, Winkler H, Jotzo F, Shukla P, Colombier M et al (2019) A pathway design framework for national low greenhouse gas emission development strategies. Nat Clim Chang 9(4):261–268. https://doi.org/10.1038/s41558-019-0442-8

Wang B, Wang Q, Wang Y, Di J, Miao S, Yu J (2019) Flexible multifunctional porous nanofibrous membranes for high-efficiency air filtration. ACS Appl Mater Interfaces 11(46):43409–43415. https://doi.org/10.1021/acsami.9b17205

Wang L, Zhong B, Vardoulakis S, Zhang F, Pilot E, Li Y et al (2016) Air quality strategies on public health and health equity in Europe—a systematic review. Int J Environ Res Public Health 13(12):1196. https://doi.org/10.3390/ijerph13121196

Wang X, Zhu Y, Zeng H, Cheng Q, Zhao X, Xu H et al (2022) Spatialized analysis of air pollution complaints in Beijing using the BERT+CRF model. Atmosphere 13(7):1023. https://doi.org/10.3390/atmos13071023

Wang Z, Zhao C, Pan Z (2015) Porous bead-on-string poly(lactic acid) fibrous membranes for air filtration. J Colloid Interface Sci 441:121–129. https://doi.org/10.1016/j.jcis.2014.11.041

Ward F, Lowther-Payne H, Halliday E, Dooley K, Joseph N, Livesey R et al (2022) Engaging communities in addressing air quality: a scoping review. Environ Health. https://doi.org/10.1186/s12940-022-00896-2

Ward T, Semmens E, Weiler E, Harrar S, Noonan C (2015) Efficacy of interventions targeting household air pollution from residential wood stoves. J Eposure Sci Environ Epidemiol 27(1):64–71. https://doi.org/10.1038/jes.2015.73

Watne Å, Lindén J, Willhelmsson J, Fridén H, Gustafsson M, Castell N (2021) Tackling data quality when using low-cost air quality sensors in citizen science projects. Front Environ Sci. https://doi.org/10.3389/fenvs.2021.733634

Wesseling J, Ruiter H, Blokhuis C, Drukker D, Weijers E, Volten H et al (2019) Development and implementation of a platform for public information on air quality, sensor measurements, and citizen science. Atmosphere 10(8):445. https://doi.org/10.3390/atmos10080445

Wong M, Wilkie A, Garzón-Galvis C, King G, Olmedo L, Bejarano E et al (2020) Community-engaged air monitoring to build resilience near the US-Mexico border. Int J Environ Res Public Health 17(3):1092. https://doi.org/10.3390/ijerph17031092

Xia X, Zhang A, Shi L, Qi Q, Jiang L, Ye Y (2017) The association between air pollution and population health risk for respiratory infection: a case study of Shenzhen, China. Int J Environ Res Public Health 14(9):950. https://doi.org/10.3390/ijerph14090950

Xiao K, Wang Y, Wu G, Fu B, Zhu Y (2018) Spatiotemporal characteristics of air pollutants (PM10, PM2.5, SO_2, NO_2, O_3, and CO) in the inland basin city of Chengdu, Southwest China. Atmosphere 9(2):74. https://doi.org/10.3390/atmos9020074

Yu Y, Jin Z, Li J, Wu Y, Li J (2020) An air pollutant emission reduction path of China's power industry. Atmosphere 11(8):852. https://doi.org/10.3390/atmos11080852

Yuan Y, Wang Y, Chi Y, Jin F (2020) Identification of key carbon emission sectors and analysis of emission effects in China. Sustainability 12(20):8673. https://doi.org/10.3390/su12208673

Zhang F, Luo L, Wang Z, Zhang W, Li C, Qiu Z et al (2019) Estimation of the effects of air pollution on hospitalization expenditures for asthma. Int J Health Serv 50(1):100–109. https://doi.org/10.1177/0020731419874996

Zhao Y, Xu A, Sun Z, Li Y, Hou Q (2022) Identification of health effects of complex air pollution in China. Int J Environ Res Public Health 19(19):12652. https://doi.org/10.3390/ijerph191912652

Zheng X, Wang F, Zheng X, Wu Z, Qiao X, Meng Q et al (2021) Construction and spatio-temporal derivation of hazardous chemical leakage disaster chain. Int J Image Data Fusion 12(4):335–348. https://doi.org/10.1080/19479832.2021.1958929

Zhou A, Zhang J, Wang G (2023) Study on embodied CO_2 emissions and transfer pathways of Chinese industries. Sustainability 15(3):2215. https://doi.org/10.3390/su15032215

Zhu Y, Song X, Wu R, Fang J, Liu L, Wang T et al (2021) A review on reducing indoor particulate matter concentrations from personal-level air filtration intervention under real-world exposure situations. Indoor Air 31(6):1707–1721. https://doi.org/10.1111/ina.12922

Chapter 9
Innovations in Soil Remediation and Land Restoration

Daniel Etim Jacob, Imaobong Daniel Jacob, Koko Sunday Daniel, Idorenyin Kenneth Gideon, and Unyime Ukpong Akpan

Abstract This chapter examines innovative techniques in soil remediation, emphasizing the need for sustainable practices to restore degraded land. It discusses various bioremediation strategies, including phytoremediation, mycoremediation, and microbial bioremediation, which utilize biological organisms to detoxify contaminated soils. Additionally, the chapter highlights advancements in electrokinetic remediation, green chemistry approaches, and biochar applications for carbon sequestration. The importance of policy frameworks and economic incentives in promoting these technologies is also analyzed, emphasizing the role of community engagement in successful remediation projects. Additionally, a promising avenue for further research is the optimisation of soil restoration procedures through the application of neural networks and machine learning. The challenges of scaling up innovative remediation techniques, regulatory barriers, and the need for interdisciplinary collaboration are addressed to pave the way for effective land restoration. By embracing these innovative methods, stakeholders can work towards enhancing soil health, promoting sustainability, and contributing to global environmental goals.

Keywords Soil remediation · Bioremediation · Phytoremediation · Electrokinetic remediation · Green chemistry · Biochar · Sustainable practices

9.1 Introduction

As the demand for sustainable development has grown in importance over the past few decades, soil remediation has gained increased attention worldwide. In addition to protecting ecosystems, effective remediation of soil is essential for reducing the harmful health effects of contamination. The European Commission's Joint Research Centre (JRC) claims that contamination and land degradation pose serious risks to Europe's soil quality. There have been reports of more than 3 million potentially

D. E. Jacob (✉) · I. D. Jacob · K. S. Daniel · I. K. Gideon · U. U. Akpan
Department of Forestry and Wildlife, University of Uyo, Uyo, Nigeria
e-mail: danieljacob@uniuyo.edu.ng

M. C. Ogwu and S. Chibueze Izah (eds.), *Evaluating Environmental Processes and Technologies*, Environmental Science and Engineering,
https://doi.org/10.1007/978-3-031-85327-2_9

polluted locations throughout the continent, many of which need ongoing remediation (Panagos et al. 2020). since of its long-term persistence and ability to penetrate the food chain, soil pollution is especially dangerous since it can endanger human health and food security (Li et al. 2015). Thus, soil contamination remains a pervasive issue globally, posing severe environmental, health, and economic challenges. Soil contamination typically occurs due to industrial activity, improper disposal of hazardous waste, agricultural runoffs, and accidental spills of chemicals and pollutants. Heavy metals, pesticides, organic pollutants, and industrial by-products are common contaminants found in soils. Remediation of contaminated soil is crucial to restore land functionality, protect ecosystems, and prevent adverse health impacts.

Numerous techniques, including physical, chemical, and biological ones, are used in the remediation process. To restore the functionality and health of the soil, these methods seek to eliminate or neutralise pollutants (Ogwu et al. 2024; Ogwu and Izah 2023). Depending on the kind and degree of contamination, common techniques include soil washing, in situ and ex situ treatments, and phytoremediation. Every technique has benefits, drawbacks, and applicability for pollutants and environmental situations (Hou et al. 2019). The cost and sustainability of the methods employed in soil restoration are two of the main issues. Conventional techniques, like the extraction and disposal of soil, are frequently expensive and harmful to the environment, especially when it comes to energy use and greenhouse gas emissions. As a result, sustainable soil remediation techniques like phytoremediation and nanoremediation are becoming more and more popular. Because of their reduced environmental effect and capacity to utilise natural processes for soil recovery, these techniques are becoming more and more popular (Favara and Gamlin 2017; Izah et al. 2023).

For example, phytoremediation uses plants to draw heavy metals and other contaminants from the soil and store them. This approach has become popular since it is economical and environmentally friendly. Plant growth rates, the degree of contamination, and site-specific circumstances, among other things, restrict its use. Phytoremediation is a promising study field because of recent advances in genetic engineering that have made it possible to create plants that are more successful at cleaning up highly polluted soils (Mench et al. 2010). Conversely, nanoremediation employs nanoparticles to degrade or immobilise pollutants at the molecular level. This strategy has demonstrated promise in reducing contamination in heavily contaminated areas and is especially effective against persistent organic pollutants (POPs). However, research is still being done to address worries about the possible toxicity of nanoparticles to human health and soil ecosystems (Fei et al. 2022; Ikhajiagbe et al. 2021, 2022). Furthermore, contemporary soil restoration techniques now include sustainable remediation techniques. These procedures seek to maximise long-term advantages while reducing the environmental impact of cleanup operations. The environmental impact of various remediation strategies has been assessed using life cycle assessment (LCA), which takes into consideration variables like resource consumption, pollution, and overall environmental sustainability (Visentin et al. 2019). Research comparing the environmental effects of thermal desorption and stabilization/solidification methods for mercury-contaminated soils indicated that stabilisation exhibited a reduced environmental footprint, rendering it a more sustainable choice for long-term restoration (Wang et al. 2020; Ikhajiagbe and Ogwu 2020).

In Europe, brownfield redevelopment has become a major component of soil remediation efforts. Brownfields, which are lands previously used for industrial or commercial purposes but are now abandoned due to contamination, represent a significant challenge. Redeveloping these sites requires effective soil remediation to eliminate or mitigate contamination before safe reuse. In situ stabilization/solidification, where contaminants are immobilized within the soil matrix, has emerged as a cost-effective and technically feasible approach for brownfield remediation. This method reduces the mobility of pollutants, thereby preventing their leaching into groundwater (Harbottle et al. 2007). As the global demand for land increases, pressure is mounting to remediate contaminated soils for urban expansion, agriculture, and industrial use. With this growing demand, there has been increased innovation in remediation technologies to improve effectiveness, reduce costs, and enhance environmental sustainability. Research has highlighted the potential of emerging technologies, such as biochar amendments, which can immobilize heavy metals and improve soil fertility simultaneously. Similarly, hybrid remediation methods combining chemical and biological treatments have shown promising results in reducing contamination levels while restoring soil health (Aparicio et al. 2022).

9.2 Emerging Bioremediation Techniques

Bioremediation harnesses the natural abilities of organisms to degrade or detoxify hazardous substances in contaminated environments, including soil and water. Among the various bioremediation strategies, three notable techniques have emerged: phytoremediation, mycoremediation, and microbial bioremediation. Each method utilizes different biological agents to target specific contaminants, providing sustainable and eco-friendly solutions to environmental pollution challenges.

9.2.1 Phytoremediation

Using plants, especially hyperaccumulators, to absorb, accumulate, and detoxify heavy metals and other contaminants from the soil is known as phytoremediation. Because of its affordability and environmental advantages, this method has grown in favour (Shah and Daverey 2020). *Thlaspi caerulescens* and *Euphorbia lagascae* are examples of hyperaccumulator plants that are useful for soil remediation because they have the capacity to absorb large concentrations of heavy metals including cadmium, lead, and zinc from disturbed soils (Dhaliwal et al. 2020). Phytostabilization, phytoextraction, and phytodegradation are some of the ways that phytoremediation operates. By stabilising pollutants in the soil, phytostabilization stops them from leaking into groundwater (Shackira and Puthur 2019). During phytoextraction, metals are absorbed by roots and transported to above-ground plant components, where they can be safely collected and disposed of (Shah and Daverey 2020; Suman et al. 2018). According to Tripathi et al. (2020), phytodegradation is the process by which organic pollutants are broken down by plants and related microbes into less dangerous forms. Transgenic plants with improved phytoremediation capacities have been made possible by recent developments in genetic engineering and molecular biology. To enhance plant genomes' capacity for metal uptake and accumulation, for example, scientists have successfully inserted metal-binding proteins (Shah and Daverey 2020; Suman et al. 2018; Wróbel et al. 2023). By increasing the effectiveness of phytoremediation, this novel technique makes it possible to treat contaminated soils more successfully. A list of prominent hyperaccumulator plant species utilised in phytoremediation, along with information on their absorption capacity and targeted contaminants, is given in Table 9.1. Environmental variables like soil type, pollutant content, and climate can affect how effective phytoremediation is. The potential of phytoremediation as a lasting remediation technique has been demonstrated by its successful application in numerous contaminated sites across the globe, despite these obstacles (Diarra et al. 2022).

9.2.2 Mycoremediation

Mycoremediation employs fungi to degrade or sequester environmental contaminants, leveraging their unique enzymatic capabilities. A wide range of enzymes found in fungi are capable of decomposing complex organic substances, such as heavy metals, herbicides, and hydrocarbons (Chandra and Enespa 2019). For instance as indicated in Table 9.2, white-rot fungi, such as *Phanerochaete chrysosporium*, are particularly effective in degrading lignin and other recalcitrant compounds, making them valuable for remediating contaminated soils (Ijoma and Tekere 2017). The mycoremediation process typically involves the introduction of fungal strains to contaminated sites, where they can colonize and break down pollutants through various metabolic pathways. Fungi can either degrade contaminants directly by secreting extracellular enzymes or sequester toxic metals by forming stable complexes (Kumar et al. 2021). This dual capability makes mycoremediation an effective approach for addressing diverse types of contamination.

Recent research has focused on optimizing mycoremediation techniques by utilizing fungal consortia, which can enhance degradation efficiency compared to single fungal species. Fungal consortia enhance the total breakdown of contaminants by utilising the complementary metabolic pathways of various fungi (Thirumalaivasan et al. 2024). Furthermore, one inventive method for immobilising pollutants and promoting plant growth in remediated areas is the application of fungal

Table 9.1 Phytoremediation plant species for heavy metal absorption

Plant species	Contaminant targeted	Accumulation mechanism	Efficiency of mechanism	Source
Brassica juncea	Lead (Pb), cadmium (Cd)	Hyperaccumulation in roots and shoots	High efficiency for Pb and Cd, with accumulation rates up to 1000 mg/kg in shoots in highly contaminated soils	Bassegio et al. (2020)
Pteris vittata	Arsenic (As)	Phytoextraction	Highly efficient for As, with removal rates exceeding 80% of available arsenic in contaminated soils	Vandana et al. (2020)
Helianthus annuus	Lead (Pb), zinc (Zn)	Rhizofiltration, phytoaccumulation	Moderate efficiency, effective for Pb and Zn up to 200 mg/kg in shoot tissue in moderate contamination scenarios	Shah and Daverey (2020) and Suman et al. (2018)
Populus deltoides	Mercury (Hg), selenium (Se)	Phytostabilization	Moderate efficiency, stabilizes Hg and Se in roots, reducing mobility by approximately 70% in contaminated soils	Akhtar et al. (2023)
Zea mays	Zinc (Zn), nickel (Ni)	Phytostabilization, bioaccumulation	High efficiency, with Zn accumulation of up to 1500 mg/kg in roots and soil stabilization for Zn and Ni	Ruiz-Huerta et al. (2022)
Thlaspi caerulescens	Cadmium (Cd), zinc (Zn)	Hyperaccumulation	High efficiency, hyperaccumulation reaching up to 3000 mg/kg for Zn and Cd, especially effective in areas with high metal concentration	Leitenmaier and Küpper (2011)

(continued)

Table 9.1 (continued)

Plant species	Contaminant targeted	Accumulation mechanism	Efficiency of mechanism	Source
Sorghum bicolor	Uranium (U)	Phytoremediation	Moderate efficiency, with U uptake of approximately 50% from contaminated water sources through root filtration	Phieler et al. (2015)
Typha latifolia	Chromium (Cr), copper (Cu)	Rhizofiltration	High efficiency, with Cr and Cu uptake up to 80% from wastewater in wetlands, effective in low to moderate contamination	Malaviya et al. (2020)
Eichhornia crassipes	Lead (Pb), mercury (Hg)	Phytodegradation	High efficiency, with Pb and Hg removal rates up to 90% in water remediation settings, though less effective in soil	De Souza and Silva (2019)
Salix viminalis	Cadmium (Cd), lead (Pb)	Phytoextraction	High efficiency for Cd and Pb, removing up to 2500 mg/kg in biomass, suitable for long-term remediation in contaminated soils	Mleczek et al. (2018)

biomaterials, such as mycelium-based composites (Geris et al. 2023). The ability of mycoremediation to efficiently treat contaminated soils and streams has been established through its implementation in numerous field investigations. For example, a study by Hamad et al. (2021) shown how a group of fungus can effectively remediate oil-contaminated soils, leading to notable decreases in hydrocarbon levels. This demonstrates that mycoremediation can be a practical and effective solution for addressing environmental contamination.

Table 9.2 Fungal species used in mycoremediation

Fungal species	Contaminant targeted	Degradation mechanism	Efficiency (%)	References
Pleurotus ostreatus	Hydrocarbons, polycyclic aromatic hydrocarbons (PAHs)	Enzymatic degradation via laccase and peroxidase	75–85% for PAHs	Kumar et al. (2021)
Trametes versicolor	Chlorinated compounds, dioxins	Lignin degradation enzymes	70–90% for chlorinated compounds	Ijoma and Tekere (2017)
Phanerochaete chrysosporium	Heavy metals, hydrocarbons	Enzymatic breakdown, biosorption	60–80% for hydrocarbons	Chandra and Enespa (2019)
Aspergillus niger	Heavy metals (Cr, Pb)	Biosorption and bioaccumulation	50–70% for Cr and Pb	Hamad et al. (2021)
Penicillium spp.	Petroleum hydrocarbons	Enzymatic biodegradation	65–78% for petroleum hydrocarbons	Geris et al. (2023)
Lentinula edodes	Polycyclic aromatic hydrocarbons (PAHs)	Enzymatic breakdown via laccase	72–88% for PAHs	Thirumalaivasan et al. (2024)
Cladosporium spp.	Heavy metals (Cu, Zn)	Metal uptake and transformation	55–70% for Cu and Zn	Chandra and Enespa (2019)
Mucor spp.	Heavy metals (Cd, Zn)	Biosorption and enzyme production	60–75% for Cd and Zn	Wróbel et al. (2023)
Ganoderma lucidum	Organic solvents, heavy metals	Laccase-mediated degradation	68–80% for organic solvents	Kumar et al. (2021)
Rhizopus arrhizus	Lead (Pb), chromium (Cr)	Biosorption	58–72% for Pb and Cr	Chandra and Enespa (2019)

9.2.3 Microbial Bioremediation

Using naturally occurring or genetically modified microbes to break down or change environmental pollutants is known as microbial bioremediation. This method breaks down a variety of contaminants, such as hydrocarbons, heavy metals, and pesticides, by taking advantage of the metabolic capacities of bacteria, archaea, and fungus (Maqsood et al. 2022). In situ and ex situ remediation are the two primary categories into which microbial bioremediation falls. While ex situ bioremediation is the extraction of polluted soil or water and subsequent controlled treatment, in situ bioremediation refers to the immediate incorporation of microorganisms in contaminated areas. Each approach has benefits and may be customised for particular contamination situations (Izah et al. 2024a, b; Jacob et al. 2024a, b).

Recent advances in biotechnology have facilitated the development of engineered microorganisms specifically designed to target contaminants. For instance, researchers have successfully introduced genes responsible for pollutant degradation into microbial strains, enhancing their efficiency in bioremediation processes (Maqsood et al. 2022). Microbial consortia, comprising multiple microbial species, have also gained attention for their synergistic effects in degrading complex pollutants more effectively than single-species applications (Islam et al. 2023). The availability of nutrients, oxygen concentrations, and the existence of co-contaminants are some of the variables that affect the effectiveness of microbial bioremediation (Kumar et al. 2021). The results of remediation can be greatly enhanced by augmenting or biostimulating these elements. For instance, bioaugmentation introduces carefully chosen microbial strains to improve degrading capacities, whereas biostimulation uses the addition of nutrients or oxygen to promote the growth of local microbial populations (Islam et al. 2023). As a result, new bioremediation methods such as mycoremediation, microbial bioremediation, and phytoremediation provide creative and practical ways to deal with soil contamination. These techniques will become more and more crucial to environmental restoration initiatives as science and technology develop. By incorporating these bioremediation techniques, land management and pollution remediation become more sustainable, promoting the general well-being of ecosystems and human populations (Jacob et al. 2024c, d; Hait et al. 2024a, b; Abajue and Izah 2024) (Table 9.3)

9.3 Nanotechnology in Soil Remediation

By providing a very efficient and focused method of resolving soil contamination problems, nanotechnology is transforming the field of soil remediation. Nanotechnology offers creative ways to break down pollutants, increase the effectiveness of remediation procedures, and lessen the environmental impact of conventional remediation approaches since it can alter materials at the atomic and molecular levels. Materials with one or more dimensions on the nanometre scale (1–100 nm) are called nanoparticles. They are very useful in environmental applications because of their small size and special chemical and physical characteristics, which include strong reactivity, a large surface area-to-volume ratio, and improved permeability (Table 9.4). Nanoparticles are mostly employed in soil remediation to break down or immobilise pollutants at the molecular level, providing a more accurate and non-invasive substitute for conventional techniques. Zero-valent iron (nZVI) and titanium dioxide (TiO_2) are two of the most widely employed nanoparticles in soil remediation. The capacity of these materials to degrade pollutants such organic molecules, heavy metals, and persistent organic pollutants (POPs) has been well investigated. Utilising these nanoparticles for both in situ and ex situ cleanup procedures has produced encouraging outcomes in recent investigations.

Table 9.3 Microbial species and engineered microbes for bioremediation

Microbial species	Contaminant targeted	Mechanism	Efficiency	Source
Pseudomonas putida	Petroleum hydrocarbons, toluene	Enzymatic breakdown, metabolic pathways	High efficiency in degrading aromatic hydrocarbons (up to 90% reduction in lab settings)	Islam et al. (2023)
Bacillus subtilis	Chromium (Cr), cadmium (Cd)	Metal biosorption and bioaccumulation	Moderate, effective for low-to-moderate metal concentrations, but efficiency decreases at higher levels	Wróbel et al. (2023)
Escherichia coli	Mercury (Hg)	Biotransformation via engineered plasmids	High, with engineered strains showing near 100% mercury removal in controlled environments	Bassegio et al. (2020)
Rhodococcus erythropolis	Hydrocarbons, polycyclic aromatic hydrocarbons (PAHs)	Enzymatic biodegradation	High, especially for PAHs, with up to 85% degradation efficiency	Geris et al. (2023)
Mycobacterium spp.	Hydrocarbons	Enzymatic breakdown of alkanes	Moderate, effective for specific alkanes but efficiency varies by species and alkane structure	Chandra and Enespa (2019)
Geobacter sulfurreducens	Uranium (U), arsenic (As)	Metal reduction, electron transfer	High, particularly effective for uranium and arsenic, with significant reduction rates in anaerobic environments	Vandana et al. (2020)

(continued)

Table 9.3 (continued)

Microbial species	Contaminant targeted	Mechanism	Efficiency	Source
Azotobacter vinelandii	Heavy metals	Bioaccumulation and metal reduction	Moderate to high, effective for a range of metals, with efficiency influenced by environmental conditions	Wróbel et al. (2023)
Deinococcus radiodurans	Radioactive waste	Genetic engineering for radiation resistance	High, shows extreme resistance to radiation and can degrade certain radioactive compounds effectively	Maqsood et al. (2022)
Sphingomonas spp.	Chlorinated solvents	Biodegradation through metabolic enzymes	Moderate to high, with efficiencies ranging from 50 to 80% depending on solvent concentration and exposure time	Ghosh and Singh (2005)
Alcaligenes eutrophus	Heavy metals (Ni, Cd)	Metal resistance and accumulation	High, particularly effective for nickel and cadmium, showing efficient metal uptake and reduction in contaminated soils	Islam et al. (2023)

9.3.1 Zero-Valent Iron (nZVI) in Soil Remediation

One of the most researched and used nanomaterials in soil remediation is zero-valent iron (nZVI). Its main mechanism of action is reductive dechlorination, in which the contaminant receives electrons from the nanoparticles, converting dangerous substances like chlorinated solvents into less hazardous or innocuous molecules (Table 9.5). According to studies, nZVI is an efficient way to remediate heavy metals like lead and arsenic as well as pollutants like trichloroethylene (TCE) and

Table 9.4 Common nanoparticles used in soil remediation

Nanoparticle	Contaminant	Mechanism of action	Efficiency	Challenges	References
Zero-valent iron (nZVI)	Chlorinated solvents (e.g., TCE, PCE), heavy metals (e.g., As, Pb)	Reductive dechlorination, reduction, and immobilization	High efficiency in dechlorination (up to 90%) for chlorinated solvents; moderate for heavy metals	Agglomeration, potential for secondary pollution	Galdames et al. (2020)
Titanium dioxide (TiO_2)	Organic pollutants, hydrocarbons, pesticides	Photocatalytic degradation under UV light	High degradation rates under UV light (up to 85%) for organic pollutants	UV dependence, possible nanoparticle release into environment	Chen et al. (2020)
Silver nanoparticles (AgNPs)	Pathogens, heavy metals	Antibacterial properties, ion exchange	Highly effective in pathogen removal (up to 99%), moderate for heavy metals	Toxicity to non-target organisms	Côa et al. (2020)
Carbon nanotubes (CNTs)	Organic pollutants, heavy metals	Adsorption and sequestration	High adsorption capacity (up to 98%) for various pollutants	Potential ecological risks, high production costs	Khan et al. (2021)
Iron oxide nanoparticles (Fe_3O_4)	Heavy metals, arsenic	Adsorption, reduction, and magnetic separation	Moderate efficiency (~ 70%) in heavy metal removal	Agglomeration, difficulty in recovery	Kumar et al. (2024)
Cerium oxide nanoparticles (CeO_2)	Organic contaminants, heavy metals	Oxidative degradation and sorption	High efficiency for organic pollutants (up to 85%), moderate for heavy metals	Potential release into the environment	Zhang et al. (2024)
Zinc oxide nanoparticles (ZnO)	Organic pollutants, heavy metals	Photocatalytic degradation, ion exchange	High photocatalytic degradation (up to 90%) for organic pollutants	Possible toxicity to aquatic life	Bhattacharjee et al. (2024)

(continued)

Table 9.4 (continued)

Nanoparticle	Contaminant	Mechanism of action	Efficiency	Challenges	References
Manganese oxide nanoparticles (MnO_X)	Chlorinated organics, heavy metals	Oxidation of organic compounds and reduction of metals	Moderate efficiency (~ 75%) for organic compounds and heavy metals	Instability in certain environments	Zhu et al. (2020)
Copper nanoparticles (CuNPs)	Organic pollutants, pathogens	Antimicrobial and catalytic activity	High efficiency for pathogen removal (up to 99%) and moderate for organic pollutants	Toxicity to microorganisms and plants	Manjula et al. (2022)
Palladium nanoparticles (PdNPs)	Persistent organic pollutants (POPs), heavy metals	Catalytic reduction and dechlorination	High efficiency in dechlorination (up to 90%) and heavy metal reduction	High cost, potential toxicity	Wang et al. (2022)

perchloroethylene (PCE). When nZVI is used in situ, the nanoparticles are injected straight into the polluted soil, where they react with the existing contaminants. This method is less disruptive and more economical since it eliminates the requirement for contaminated material to be excavated and transported. One of nZVI's main benefits is its highly reactive nature, which enables pollutants to degrade quickly—often in a matter of weeks or months. To maximise the utilisation of nZVI, however, issues such nanoparticle aggregation, restricted mobility in soil matrices, and possible long-term environmental effects must be resolved. The effectiveness of nZVI for the treatment of soils polluted with chlorinated organic chemicals was shown in a study by Li et al. (2021). According to the study, ZVI greatly outperformed conventional iron-based remediation techniques, achieving 90% dechlorination of TCE in under 24 h. Another work by Ganie et al. (2021) demonstrated nZVI's adaptability in treating both organic and inorganic toxins by demonstrating its capacity to lower the content of arsenic in groundwater.

9.3.2 *Titanium Dioxide (TiO_2) for Photocatalytic Degradation*

Another nanoparticle that has drawn a lot of interest in soil remediation is titanium dioxide (TiO_2), mainly because of its photocatalytic qualities. Reactive oxygen

Table 9.5 Key applications of zero-valent iron (nZVI) in soil remediation

Contaminant	Mechanism of action	Effectiveness	Challenges	References
Trichloroethylene (TCE)	Reductive dechlorination	90% dechlorination within 24 h	Agglomeration, limited mobility in soil	Kumar et al. (2017)
Perchloroethylene (PCE)	Reductive dechlorination	Rapid breakdown to less toxic compounds	Potential groundwater contamination	Dutta et al. (2022)
Arsenic	Reductive precipitation	97% reduction in groundwater concentrations	Long-term stability of iron oxides formed	Sun et al. (2016)
Lead	Reduction and immobilization	High immobilization rates	Limited mobility of nZVI in some soils	Liu et al. (2023)
Chromium (VI)	Reductive transformation	Effective reduction within hours	Requires precise control of environmental conditions	Fang et al. (2021)
Nitrate	Reductive denitrification	Significant reduction under anaerobic conditions	May lead to ammonia formation	Tiso and Schechter (2015)
Mercury	Reduction and immobilization	Effective binding to mercury ions	Environmental persistence of byproducts	Duan et al. (2020)
Zinc	Reduction and sorption	Moderate sorption rates in contaminated soils	Mobility of reduced particles	El-Naggar et al. (2018)
Copper	Adsorption and reduction	High removal rates in mixed contaminants	Adsorptive capacity decreases over time	Zhou et al. (2019)

species (ROS) like hydroxyl radicals are produced when TiO_2 is exposed to ultraviolet (UV) radiation. These ROS can oxidise organic contaminants and convert them into innocuous byproducts like carbon dioxide and water (Table 9.6). Because of this, TiO_2 is very effective at breaking down hydrocarbons, dyes, pesticides, and other organic pollutants in soil. TiO_2 can be administered ex situ, where contaminated soil is treated under controlled conditions, or in situ, when the nanoparticles are dispersed into the soil. The fact that TiO_2 can function in ambient circumstances without requiring high temperatures or pressures is one of its key benefits. TiO_2 is a desirable alternative for extensive soil remediation projects since it is environmentally friendly, non-toxic, and reasonably priced. The efficiency of TiO_2 in breaking down polycyclic aromatic hydrocarbons (PAHs) in polluted soils was shown by research by Liu (2018). According to the study, TiO_2 could completely degrade PAHs under

UV irradiation in 72 h, leaving no traceable remains behind. Over 95% of the target chemicals were broken down within 48 h of treatment, according to a different study by El-Saeid et al. (2022) that investigated the use of TiO_2 in the photocatalytic degradation of pesticides.

Table 9.6 Key applications of titanium dioxide (TiO_2) in soil remediation

Contaminant	Mechanism of action	Effectiveness	Challenges	References
Polycyclic aromatic hydrocarbons (PAHs)	Photocatalytic oxidation	Complete degradation within 72 h	Requires UV light for activation	Liu (2018)
Pesticides	Photocatalytic oxidation	95% degradation within 48 h	Potential release of nanoparticles into the environment	Ahmari et al. (2018)
Dyes	Photocatalytic oxidation	Rapid breakdown to CO_2 and H_2O	Long-term environmental stability unclear	Al Miad et al. (2024)
Hydrocarbons	Photocatalytic oxidation	Effective under ambient conditions	High energy input for UV activation	Dursun and Ayturan (2022)
Volatile organic compounds (VOCs)	Photocatalytic degradation	Significant reduction in air and soil levels	Requires controlled exposure to UV light	Rouhani and Taghipour (2023)
Pharmaceuticals	Photocatalytic degradation	85–90% degradation under UV light	Limited effectiveness without UV enhancement	Ahmadi et al. (2017)
Benzene	Photocatalytic oxidation	Rapid degradation under UV exposure	Limited effectiveness under natural sunlight	Huang et al. (2017)
Toluene	Photocatalytic oxidation	Effective degradation with minimal byproducts	High operational costs for large-scale applications	Shayegan et al. (2019)
Herbicides	Photocatalytic degradation	Complete breakdown of toxic compounds	UV-dependence limits natural application	Mikkelsen et al. (2011)

9.3.3 Risks and Challenges of Nanotechnology in Soil Remediation

Although there are numerous benefits to using nanotechnology for soil remediation, there are also risks and difficulties that need to be properly considered. The possible harm that nanoparticles could do to the environment and human health is one of the main worries. According to studies, some nanoparticles, like carbon nanotubes and nano-silver, can be harmful to aquatic life and could endanger soil biodiversity. Furthermore, because of their small size, nanoparticles can move via groundwater and soil systems, which could unintentionally contaminate nearby ecosystems. Another matter that needs more research is how persistent nanoparticles are in the environment. The long-term effects of the extensive use of compounds like nZVI and TiO_2 in soil remediation are still unknown, despite the fact that they are typically regarded as harmless. The fate and transport of nanoparticles in soil settings, as well as their potential for bioaccumulation in plants and animals, are now being investigated.

Concerns regarding the health and environmental hazards of engineered nanoparticles were brought up in a research by Hristozov et al. (2012), which emphasised the necessity of thorough risk assessments prior to large-scale usage. The possibility of unexpected interactions between nanoparticles and biological systems was also noted by Mu et al. (2014), who suggested that further study is necessary to fully comprehend the scope of these interactions. Notwithstanding these difficulties, there is no denying nanotechnology's potential advantages in soil remediation. Researchers hope to find safer, more effective, and more sustainable solutions for soil health issues by further developing and improving nanomaterials for environmental applications.

9.4 Electrokinetic Remediation Advances

A state-of-the-art method for eliminating a variety of pollutants from soil, particularly heavy metals and organic pollutants, is electrokinetic remediation (EKR). This method overcomes some of the drawbacks of conventional remediation techniques by applying electrical currents to mobilise and extract toxins from the soil matrix. Researchers and practitioners have been paying more attention to EKR's potential as a sustainable and affordable remedy for soil contamination as environmental concerns continue to rise on a worldwide scale. The creation of an electric field across the polluted soil is the basic process of electrokinetic remediation. Electroosmosis, electromigration, and electrophoresis are among the transport phenomena made possible by the electrochemical reactions produced by this electric field (Table 9.7). The process by which pore water in the soil moves in the direction of the electrodes to help carry dissolved pollutants is known as electroosmosis. However, electrophoresis depicts the movement of charged particles under the influence of an electric field, whereas electromigration encompasses the migration of ions that are charged towards oppositely charged electrodes. Electrodes are inserted into the soil in the basic EKR arrangement, with one electrode acting as the anode and the other as the cathode. Contaminants are drawn to the electrodes by the application of a direct current, where they can be removed by collecting techniques or precipitation. According to research, EKR is capable of efficiently treating a range of contaminants, including organic pollutants like pesticides and petroleum hydrocarbons as well as heavy metals like lead, cadmium, and arsenic (Reddy and Cameselle 2009).

Understanding the many mechanisms that control the mobilisation of pollutants is essential, in addition to the fundamentals of electrokinetic remediation. EKR's capacity to address intricate contaminant-soil interactions is one of its main benefits. Depending on the kinds of pollutants and the physicochemical characteristics of the soil, the three mechanisms by which they are mobilized—desorption, ion migration, and electroosmosis—vary greatly. For example, because sand particles have a lower charge density and a higher permeability, electroosmosis usually dominates in sandy soils. On the other hand, because clay particles have a high surface charge and tend to bind pollutants securely, electromigration frequently takes over in clayey soils (Acar and Alshawabkeh 1993). The pH gradient created between the electrodes has

Table 9.7 Mechanisms of electrokinetic remediation

Mechanism	Description	Role in contaminant mobilization	References
Electroosmosis	Movement of pore water towards the electrodes due to the applied electric field	Facilitates the transport of dissolved contaminants	Han et al. (2021)
Electromigration	Movement of charged ions towards the oppositely charged electrode	Essential for mobilizing ionic contaminants like metals	Wang et al. (2021)
Electrophoresis	Movement of charged particles in the electric field	Aids in the transport of colloidal particles and ions	Khair (2022)
pH gradient	Electrolysis generates pH changes, affecting solubility and mobility of contaminants	Alters solubility of heavy metals, enhancing their mobilization	Wang et al. (2021)
Electrode reactions	Oxidation and reduction reactions at the electrodes leading to contaminant degradation	Directly contributes to the breakdown of organic pollutants	Qiao and Xiong (2021)
Anionic and cationic movement	Specific transport of anions and cations based on electrode charge	Helps in the separation and removal of various charged pollutants	Pulkka et al. (2014)
Mass transfer enhancement	Enhanced movement of contaminants through improved fluid dynamics in the pore spaces	Increases the efficiency of contaminant extraction	Essaid et al. (2015)

a major impact on the movement of pollutants during electrokinetic cleanup. Protons are produced at the anode by the electrolysis of water at the electrodes, which lowers pH, while hydroxide ions build up at the cathode, which raises pH. The mobility and solubility of pollutants, particularly heavy metals, can be significantly impacted by this pH change. Changes in pH lead to changed solubility, which enhances the electrokinetic mobilisation of these metals. For example, a drop in pH can help metal ions become soluble and migrate in the direction of the electrodes (Kang et al. 2016).

9.4.1 Applications of Electrokinetic Remediation

Numerous field locations have seen the successful application of electrokinetic remediation, proving its efficacy in a range of contaminated circumstances. Because it can efficiently mobilise elements like lead, cadmium, and arsenic—which are frequently present in industrial locations, mining sites, and areas affected by inappropriate waste disposal—the method has proven very helpful for sites contaminated with heavy metals (Lageman et al. 2005). Because heavy metals are ionic, they can be mobilised by electromigration, which efficiently concentrates them at the extraction electrodes (Table 9.8). Electrokinetic remediation has demonstrated potential for treating organic contaminants in addition to heavy metals. However, because many organic compounds are nonpolar and have low mobility in aqueous solutions, the behaviour of organic pollutants in soil is frequently more complex than that of metals. Techniques like the use of surfactants or co-solvents can be used to improve the efficacy of EKR in treating organic pollutants by increasing contaminant desorption and facilitating electroosmotic transport (Saichek and Reddy 2005). Successful field applications targeting a variety of organic contaminants, such as pesticides and hydrocarbons, have been made possible by these adjustments to the conventional electrokinetic technique. Additionally, because of its adaptability, electrokinetic remediation can be used on multicomponent contamination systems, which is especially useful in many actual contaminated locations. In the vicinity of industrial sites, mixed contamination—which includes both metals and organic compounds—occurs often. Optimising the cleanup of mixed contaminated systems has showed promise thanks to recent developments in electrode design, including the use of improved materials and combinations. These innovations facilitate improved removal efficiencies by enhancing current distribution and contaminant mobilization (Giannis et al. 2010).

9.4.2 Advantages of Electrokinetic Remediation

In the realm of environmental remediation, electrokinetic remediation is becoming more and more popular due to its many benefits. Its capacity to be applied in situ—that is, without the need for excavation or transportation to a treatment facility—is one

Table 9.8 Applications of electrokinetic remediation

Contaminant	Typical sources	Effective techniques	References
Heavy metals	Industrial sites, mining areas, landfills	Electromigration, electroosmosis, electrode modification	Reddy and Cameselle (2009)
Organic pollutants	Agricultural runoff, petroleum spills, waste disposal sites	Surfactant addition, co-solvent use, electrokinetic-enhanced biodegradation	Ngieng et al. (2024)
Mixed contaminants	Landfills, industrial complexes, hazardous waste sites	Integrated EKR with other remediation methods, sequential extraction	Giannis et al. (2010)
Nitrate and phosphate	Agricultural fields, wastewater treatment plants	Electrokinetic stripping and adsorption	Lagum (2021)
Polycyclic aromatic hydrocarbons (PAHs)	Urban runoff, industrial discharges	Enhanced solubilization and degradation	Liu et al. (2022)
Emerging contaminants	Pharmaceuticals, personal care products	Electrokinetic remediation combined with nanotechnology	Sarma (2022)

of its main advantages. This feature minimises environmental disruption and drastically lowers expenses (Table 9.9). Additionally, fine-grained soils like clay, which are notoriously difficult to repair using traditional approaches due to their limited permeability and significant pollutant binding capacity, respond particularly well to electrokinetic remediation (Pamukcu and Kenneth Wittle 1992). The scalability of EKR is another noteworthy benefit. Pilot studies and full-scale implementations have validated the technique's usefulness in real-world circumstances, while laboratory experiments have shown its potential to treat a wide range of pollutants. Both larger, more complicated projects and smaller remediation initiatives can use the technology. EKR can also be used in conjunction with other remediation strategies, including chemical oxidation or phytoremediation, to improve the overall efficacy and efficiency of treatment.

9.4.3 Challenges and Limitations of Electrokinetic Remediation

Electrokinetic remediation has several benefits, but its successful application requires addressing a number of issues and constraints. Energy usage is among the most important problems (Table 9.10). Despite being more energy-efficient than certain traditional techniques, EKR can still have significant energy requirements, especially for large-scale applications. Soil conductivity, electrode spacing, and the type of

Table 9.9 Advantages of electrokinetic remediation

Advantage	Description	References
In situ treatment	Remediation can be conducted without soil excavation, preserving the site integrity	Pamukcu and Kenneth Wittle (1992)
Effective for fine soils	Particularly beneficial for treating clay and silt, which are difficult for conventional methods	Ismaiel (2006)
Scalability	Applicable to both small-scale and large-scale projects, offering flexibility in implementation	Santos and de Carvalho (2022)
Versatility	Can be combined with other remediation technologies (bioremediation, phytoremediation) for enhanced efficacy	Alshawabkeh et al. (1999)
Environmental friendliness	Often utilizes non-toxic reagents, reducing secondary pollution risks	Bhavya et al. (2021)
Cost-effectiveness	Long-term cost savings compared to traditional excavation and landfill methods	Chen and Li (2018)
Speed of remediation	Can accelerate the remediation process, reducing overall project timelines	Suthersan et al. (2016)

contaminants being treated are all intimately related to the process's energy requirements (Lageman and Godschalk 2007). Furthermore, electrokinetic remediation is hampered by the byproducts that are produced during electrochemical processes. Hydrogen and oxygen gases, for instance, are produced when water electrolyses at the electrodes. This might alter the structure of the soil and make it more difficult to remove pollutants.

Additionally, pollutants may precipitate close to the electrodes due to the acid-alkaline front produced during the electrokinetic process, which could lower the removal efficiency overall (Alshawabkeh et al. 1999).

Another major obstacle to electrokinetic remediation is soil heterogeneity. Soil characteristics, such as porosity, permeability, and chemical composition, can vary greatly in natural subsurface settings. Because of this unpredictability, it may be challenging to guarantee full mobilisation and removal of pollutants due to uneven current distribution. Optimising the efficacy of EKR at polluted locations requires addressing soil heterogeneity. The cost of electrokinetic cleanup is still a major deterrent to its widespread use. The initial expenditure of money for equipment and the potential requirement for extended operating durations can make the process

Table 9.10 Challenges and limitations of electrokinetic remediation

Challenge	Description	Potential solutions	References
Energy consumption	High energy demands, particularly for large-scale applications	Optimize current density and electrode design; use renewable energy sources	Kebede et al. (2022)
Formation of byproducts	Electrolysis can produce gases and other byproducts that disrupt soil structure and remediation efforts	Develop gas venting strategies and monitor byproduct formation closely	Aparicio et al. (2022)
Soil heterogeneity	Variability in soil properties can lead to uneven current distribution and inconsistent remediation efficiency	Pre-characterization of soil properties to inform design; adaptive remediation strategies	Huang et al. (2024)
Cost	Initial investment and prolonged operational times may be higher than conventional methods	Conduct pilot studies to demonstrate cost-effectiveness and attract funding	Wuana and Okieimen (2011)
Groundwater contamination	Risk of mobilizing contaminants into groundwater if not properly managed	Implement monitoring and control measures to prevent leaching	Padhye et al. (2023)
Limited depth penetration	Effectiveness may decrease with soil depth or in complex geological formations	Use advanced electrode configurations and materials to enhance penetration	Kiflu et al. (2016)
Regulatory challenges	Compliance with environmental regulations may complicate project approval and execution	Engage with regulators early in the planning process and develop transparent protocols	Fiorino (1996)

more costly than traditional remediation approaches, even though in situ techniques like EKR can lower expenses related to soil extraction and shipping.

9.4.4 Recent Advances in Electrokinetic Remediation

The capabilities and efficacy of electrokinetic remediation are being continuously expanded by recent research and technological advancements. One important area of innovation is the incorporation of nanotechnology into electrokinetic systems; nanoparticles, like carbon nanotubes or zero-valent iron, have demonstrated the ability to improve the degradation of organic contaminants and increase the mobility of metals within the soil matrix (Table 9.11); the introduction of these nanoparticles can be facilitated through electrokinetic transport, leading to improved contaminant removal (Behrouzinia et al. 2022). Another recent development in EKR is the investigation of novel electrode materials that can improve the efficiency of contaminant extraction. Researchers are looking into the use of conductive polymers, modified electrodes, and nanocomposites to improve current distribution and contaminant interaction within the soil. These cutting-edge electrode materials should improve EKR's overall efficacy in removing organic and heavy metal contaminants (Sun et al. 2023). Additionally, current research is concentrating on improving the electrokinetic systems' operating parameters, such as current density, electrode spacing, and treatment duration. Researchers seek to minimise energy usage and any environmental effects while optimising the removal efficiency of different contaminants by methodically modifying these parameters (Bazrafshan et al. 2015).

9.5 Green Chemistry in Soil Treatment

The concepts of green chemistry have been increasingly popular in recent years as essential elements of remediation and soil treatment projects. The design of chemicals and chemical processes that reduce the usage and production of hazardous compounds is the focus of green chemistry. This creative method not only lessens the need for hazardous chemicals to treat soil, but it also encourages environmentally friendly and health-promoting sustainable habits.

9.5.1 Bio-Based Solvents and Eco-Friendly Chemical Agents

A viable substitute for conventional organic solvents, which are frequently made from petrochemicals and can present serious health and environmental hazards, is the use of bio-based solvents. Bio-based solvents come from renewable resources like biomass, plants, and agricultural waste. Their reduced environmental effect, increased biodegradability, and decreased toxicity are their defining characteristics (Table 9.12). The potential of bio-based solvents in soil remediation applications has been emphasised in a number of research. For example, ethyl lactate, glycerol, and terpenes are examples of bio-based solvents that have shown promise in the removal of organic contaminants from contaminated soils. The fermentation of carbohydrates produces ethyl lactate, a solvent that is biodegradable and well-known for its capacity to dissolve a wide range of organic substances. According to research, ethyl lactate is a good green substitute for traditional solvents like toluene or benzene since it can efficiently mobilise and extract hydrocarbons and pesticides from contaminated soils (Foster et al. 2017; Sarkar and Sadhukhan 2022).

Furthermore, glycerol a byproduct of the manufacturing of biodiesel has drawn interest as a possible environmentally benign solvent for soil remediation. Glycerol

Table 9.11 Recent advances in electrokinetic remediation

Innovation	Description	Impact on EKR	References
Integration with renewable energy	Utilizing solar or wind energy to power electrokinetic remediation systems, reducing carbon footprint and operational costs	Increases sustainability and lowers the environmental impact of remediation processes	Vocciante et al. (2021)
Real-time monitoring technologies	Implementation of sensors and IoT devices to monitor contaminant levels and remediation progress in real time	Allows for adaptive management and optimization of remediation strategies	Horst et al. (2022) and Izah et al. (2024a, b)
Bioelectroremediation	Combining electrokinetic remediation with microbial electrochemical systems to enhance biodegradation of organic pollutants	Synergistic effects improve contaminant removal rates and efficiency	Camacho (2021)
Electrode design innovations	Development of advanced electrode geometries (e.g., 3D printed electrodes) that enhance current distribution and reduce energy consumption	Improves effectiveness in challenging subsurface environments	Tan (2023)
Hybrid remediation approaches	Integrating electrokinetic remediation with techniques such as phytoremediation and chemical treatment for enhanced contaminant removal	Provides a comprehensive solution to complex contamination scenarios	Aparicio et al. (2022)
Application of ionic liquids	Utilizing ionic liquids to enhance the solubility of contaminants in electrokinetic systems	Increases the effectiveness of contaminant extraction, particularly for hydrophobic compounds	Boulakradeche et al. (2015)
Modeling and simulation advances	Use of computational models to predict the behavior of contaminants during electrokinetic remediation processes	Enhances planning and design of remediation systems, leading to improved outcomes	Sprocati and Rolle (2021)

Table 9.12 Bio-based solvents for soil treatment

Bio-based solvent	Source/origin	Properties	Advantages	References
Ethyl lactate	Derived from renewable biomass (fermentation)	Low toxicity, biodegradable, effective solvent	Good solubilization of organic pollutants	Pereira and Rodrigues (2014)
Glycerol	Byproduct of biodiesel production	Non-toxic, enhances solubility	Effective in extracting PAHs and pesticides	Cameotra and Makkar (2010)
Terpenes	Plant-derived (essential oils)	Biodegradable, low environmental impact	Effective against a variety of organic compounds	Sil et al. (2020)
Dimethyl sulfoxide (DMSO)	Derived from lignin and wood products	Polar aprotic solvent, high solubility	Non-toxic, enhances the mobilization of contaminants	Aziz et al. (2022)
Ethanol	Fermented plant material	Biodegradable, polar solvent	Reduces toxicity, effective for many contaminants	Gudiña et al. (2015)

is a desirable option for soil treatment applications due to its non-toxic properties and capacity to increase the solubility of organic contaminants. Glycerol has been successfully used in studies to remove persistent organic pollutants, including polycyclic aromatic hydrocarbons (PAHs), from polluted soils (Cameotra and Makkar 2010). Furthermore, a major achievement in green chemistry for soil treatment has been the creation of solvent-based extraction methods that use bio-based solvents. Utilising bio-based solvents, these methods such as solvent extraction and supercritical fluid extraction improve the mobilisation and removal of pollutants while reducing their negative effects on the environment. Advanced extraction techniques combined with environmentally acceptable solvents have the potential to greatly increase the effectiveness of soil remediation procedures.

9.5.2 *Plant-Based Surfactants for Enhanced Biodegradation*

Because they lower water's surface tension and encourage the solubilisation of hydrophobic substances, surfactants are essential for improving the biological breakdown of organic pollutants in soil. However, traditional surfactants might lead to ecological damage and are frequently generated from petrochemical sources. However, a more sustainable option is provided by plant-based surfactants, commonly referred to as biosurfactants, which are made from renewable plant components (Table 9.13). Compared to traditional surfactants, biosurfactants have a number of benefits, such as increased biodegradability, decreased toxicity, and the capacity to efficiently solubilise a variety of organic pollutants. Saponins and other plant-derived surfactants can greatly increase the bioavailability of hydrophobic contaminants, promoting microbial communities' breakdown of them in polluted soils (Pérez et al. 2015). Natural glycosides called saponins are present in a variety of plants and have been demonstrated to have superior surfactant qualities, which makes them useful agents for soil remediation.

The use of plant-based surfactants has been shown to accelerate the breakdown of persistent organic pollutants, such as PAHs and chlorinated chemicals. The application of saponin-based surfactants, for example, greatly accelerated the biodegradation of phenanthrene in contaminated soils, improving the efficacy of pollutant removal, according to a study by Adenutsi et al. (2023). This improvement can be ascribed to biosurfactants' capacity to make hydrophobic pollutants more accessible to microbes, which in turn promotes microbial breakdown processes. Additionally, it has been demonstrated that using plant-based surfactants into bioremediation techniques might enhance the development and activity of advantageous soil microbes. More efficient soil treatment can result from the use of biosurfactants, which can increase microbial diversity and improve microbial metabolic pathways involved in the breakdown of organic contaminants (Gudiña et al. 2015). By encouraging the sorption of contaminants onto soil particles, plant-based surfactants might lessen the

Table 9.13 Plant-based surfactants for soil treatment

Plant-based surfactant	Source	Properties	Advantages	References
Saponins	Derived from various plants	Amphiphilic properties	Enhanced biodegradation of organic pollutants	Pérez et al. (2015)
Lecithin	Phospholipid from soybeans	Biodegradable, non-toxic	Promotes microbial activity in contaminated soils	Sangwan et al. (2022)
Sophorolipids	Produced by yeast (*Candida* spp.)	Low toxicity, high surface activity	Effective in mobilizing hydrophobic compounds	Cameotra and Makkar (2010)
Rhamnolipids	Produced by *Pseudomonas aeruginosa*	Biodegradable, effective surfactant	Enhances degradation of PAHs	Nie et al. (2010)
Plant oil-derived surfactants	Derived from vegetable oils	Biodegradable, amphiphilic	Lower environmental impact compared to synthetic surfactants	Nagtode et al. (2023)

toxicity of polluted soils in addition to improving biodegradation processes. Biosurfactants can reduce hydrophobic pollutants' bioavailability and lessen their harmful effects on soil organisms by binding to them. The promise of plant-based surfactants as efficient agents for environmentally friendly soil treatment is highlighted by their dual action of improving biodegradation and lowering toxicity.

9.5.3 Green Chemistry Application in Large-Scale Remediation Projects

An important development in sustainable environmental management is the incorporation of green chemistry concepts into extensive remediation programs. The demand for creative strategies that minimise secondary pollution and lower energy use has grown as remediation activities become more intricate. A framework for creating and executing remediation plans that put sustainability and environmental safety first is offered by green chemistry (Table 9.14). Using life cycle assessment (LCA) techniques to examine the environmental effects of different remediation technologies is a crucial part of incorporating green chemistry into remediation projects. Stakeholders can choose the most environmentally friendly alternatives by using life cycle assessment (LCA), which enables a thorough evaluation of the economic, social, and environmental effects of remediation techniques (Visentin et al. 2019).

LCA assists in identifying options for minimising environmental harm and lowering resource consumption by taking into account the full life cycle of remediation operations, from resource extraction to disposal. Additionally, using eco-friendly materials and methods through the remediation process is encouraged by the use of green chemistry concepts. This covers the use of energy-efficient technology as well as the use of chemical agents, surfactants, and biodegradable solvents. For

Table 9.14 Integration of green chemistry principles in remediation

Principle	Application	Benefits	References
Life cycle assessment (LCA)	Evaluating environmental impacts of remediation	Informed decision-making for sustainable practices	Visentin et al. (2019)
Use of renewable energy	Integration in remediation processes	Reduces carbon footprint and energy consumption	Klemeš et al. (2019)
Community engagement	Stakeholder involvement in decision-making	Enhances transparency and public trust	Gudiña et al. (2015)
Bioremediation	Utilizing microorganisms for pollutant degradation	Low energy requirements and sustainable outcomes	Bala et al. (2022)

example, bioremediation techniques that take advantage of microorganisms' innate capacity for degradation can drastically lower the energy usage related to conventional remediation techniques. Bioremediation can efficiently treat soil contamination while requiring less energy and chemicals by encouraging the development of native microbial communities (Saeed et al. 2021).

Furthermore, another way to adhere to the principles of green chemistry is by integrating clean sources of energy into remediation initiatives. Utilising energy sources like solar, wind, or biomass can help remediation projects adopt more sustainable techniques and lessen their carbon footprint. Remedial projects can become more energy-efficient, lowering their dependency on fossil fuels and greenhouse gas emissions, by utilising renewable energy. Incorporating green chemistry concepts into remediation initiatives also involves involving the community and stakeholders. Remedial actions can be customised to satisfy local requirements while advancing environmental justice by including local people' knowledge and concerns into the decision-making process (Harclerode et al. 2015). Involving stakeholders promotes openness and trust, which improves community relations and cleanup results.

9.6 Innovative Soil Stabilization and Erosion Control Techniques

A basic technique, soil stabilisation aims to improve the physical and chemical characteristics of soil in order to increase its strength, lessen erosion, and enable sustainable land use, especially in areas that have been damaged or mined. New soil stabilisation and erosion control methods have been developed as a result of growing environmental sustainability consciousness and the need for resilient ecosystems. Some examples of these innovations are the use of natural fibres and recyclable materials in soil reinforcement technologies, the development of bioengineering solutions and vegetative barriers, and the use of geopolymers and bio-based binders.

9.6.1 Geopolymers and Bio-Based Binders

When aluminosilicate minerals react with alkaline activators, inorganic polymers known as geopolymers are created. Because they are strong, long-lasting, and environmentally friendly, they provide a novel approach to soil stabilisation. In contrast to traditional cement, geopolymers are made from industrial waste materials like fly ash and slag, which lowers the carbon footprint of cement manufacturing (Table 9.15). They are appropriate for stabilising contaminated soils due to their capacity to encapsulate heavy metals and other contaminants (ur Rehman et al. 2023). The efficiency of geopolymers in enhancing the mechanical characteristics of soil has been shown in recent research. For example, Hamid and Alnuaim (2024) found that adding geopolymer to sandy soils greatly boosted their compressive strength, which in turn improved their ability to support loads. In a similar vein, Koohestani et al. (2021) discovered that geopolymers may efficiently tie soil particles together, decreasing the likelihood of erosion and enhancing soil stability in general.

Bio-based binders, alongside geopolymers, have become popular as environmentally friendly substitutes for conventional stabilisers. Sustainability, low environmental impact, and biological degradation are some benefits of bio-based binders, which are made from natural materials like cellulose, lignin, and starch (Bumanis et al. 2020). They can decrease fluidity and increase soil cohesiveness, both of which are essential for halting erosion. For example, Yao et al. (2022) showed that lignin-based binders could greatly increase clay soils' shear strength, increasing their resistance to erosion. Additionally, using bio-based binders in soil stabilisation methods helps sequester carbon while also encouraging sustainable land use. Given that soil organic matter is essential for sequestering carbon, this is especially significant in light of climate change.

Table 9.15 Overview of soil stabilization technologies

Technology	Description	Advantages	References
Geopolymers	Inorganic polymer binders formed by the reaction of aluminosilicate materials with alkaline solutions	High compressive strength, low environmental impact, and durability	Ismail et al. (2024)
Bio-based binders	Natural polymers or biodegradable materials used to bind soil particles	Sustainable, renewable, and reduced carbon footprint	Bumanis et al. (2020)
Natural fiber reinforcement	Incorporating fibers from plants like jute, sisal, and coir into soil mixtures	Improves tensile strength and ductility of soil	Gowthaman et al. (2018)
Recycled materials	Use of recycled aggregates or industrial by-products for soil stabilization	Cost-effective and reduces landfill waste	Onyelowe et al. (2019)
Chemical stabilizers	Synthetic chemicals added to soil to improve strength and durability	Quick and effective in enhancing soil properties	Firoozi et al. (2017)
Enzyme-based stabilizers	Use of enzymes to enhance soil structure and stability	Eco-friendly and can improve soil fertility	Shukla et al. (2023)

9.6.2 Natural Fibers and Recycled Materials

An eco-friendly and novel method of improving soil stability is the use of natural fibres for soil reinforcement. Because of their excellent tear strength and biodegradability, natural fibres including hemp, sisal, jute, and coir are appropriate for applications involving soil reinforcement (Sen and Reddy 2011). By adding these fibres to soil matrices, the soil's structure and erosion resistance are improved, which improves performance under a range of loading scenarios. Sharma et al. (2017) discovered, for instance, that incorporating jute fibres into clay soil enhanced its resistance to shear and decreased its plasticity index, suggesting enhanced stability and decreased erosion potential. In a similar vein, Askarinejad et al. (2018) found that adding coir fibres to sandy soils increased their compressive strength and increased their resistance to erosion during periods of high rainfall.

Glass fibres and leftover plastics are two examples of recycled materials that have drawn interest as environmentally friendly alternatives for soil reinforcement. In addition to enhancing mechanical qualities, adding these elements to soil solves environmental issues related to plastic waste. For example, a study by Salimi and Ghazavi (2021) showed that adding shredded plastic trash to soil mixtures improved soil stability by increasing shear strength and decreasing compressibility. Fibres are not the only recycled materials used in soil stabilisation. It has been demonstrated that adding recycled materials, such as asphalt and crushed concrete, can enhance the mechanical qualities of soil (Liu et al. 2020). This strategy helps with resource conservation and waste reduction in addition to encouraging sustainable land use.

9.6.3 Advances in Vegetative Barriers and Bioengineering Solutions

Utilising plants and vegetation to stabilise soil and stop erosion, vegetative barriers are a crucial part of erosion control strategies. By improving soil moisture retention and efficiently lowering surface drainage, these barriers can increase soil stability (Wolka et al. 2018). The goal of recent developments in vegetative barriers has been to maximise their effectiveness by carefully choosing and arranging plants (Table 9.16). Using native vegetation that are well-suited to the local environment is one promising strategy. Native plants are better at halting erosion because of their deep root systems, which improve soil stability and structure (Stokes et al. 2014). Incorporating cover crops into farming methods has also become more popular as a way to enhance soil health and stop erosion.

Cover crops boost nitrogen cycling, increase soil organic matter, and lessen soil compaction. Additionally, bioengineering solutions—which blend ecological principles with conventional engineering techniques—have become more and more popular in erosion control initiatives (Table 9.17). Soil erosion in steep terrain has been successfully prevented by methods like slope bioengineering, which uses living

Table 9.16 Advances in vegetative barriers and bioengineering solutions

Method	Description	Benefits	References
Vegetative buffer strips	Strips of vegetation planted along waterways to reduce runoff	Reduces soil erosion, improves water quality, and enhances biodiversity	Dunn et al. (2022)
Living shorelines	Incorporating native plants and natural materials to stabilize shorelines	Provides habitat, improves water quality, and reduces erosion	Currin (2019)
Bioengineering techniques	Combination of vegetation with engineering techniques to stabilize slopes	Sustainable, cost-effective, and enhances natural beauty	Dorairaj and Osman (2021)
Root reinforcement	Utilization of plant roots to increase soil stability	Enhances soil cohesion and reduces erosion risk	Masi et al. (2021)
Planting strategies	Strategic planting of native vegetation to enhance soil stability	Increases biodiversity and improves soil structure	Lal (2015)
Geosynthetic reinforcements	Use of synthetic materials to provide additional soil stability	Enhances load distribution and reduces erosion	Markiewicz et al. (2024)

plant materials to stabilise slopes (Dorairaj and Osman 2021). For instance, it has been shown that live staking, which is putting native species cuttings onto unstable slopes, effectively increases vegetation development and lowers erosion rates.

Table 9.17 Comparison of traditional versus innovative soil stabilization techniques

Criteria	Traditional techniques	Innovative techniques
Materials used	Cement, lime, and chemical additives	Geopolymers, bio-based binders, natural fibers, recycled materials
Environmental impact	High due to CO_2 emissions and non-renewable resources	Lower, with potential for carbon sequestration and sustainability
Cost	Generally higher due to material and transportation costs	Potentially lower with local materials and reduced energy consumption
Adaptability	Limited to specific soil types and conditions	More versatile, applicable to a broader range of soil conditions
Erosion control	Primarily focuses on short-term stabilization	Long-term stabilization and ecosystem enhancement
Maintenance requirements	Often requires ongoing maintenance and monitoring	More sustainable solutions with lower maintenance needs

9.7 Biochar and Carbon Sequestration in Land Restoration

In the context of land restoration, biochar—a carbon-rich byproduct made from the pyrolysis of organic materials—has shown promise as a means of improving soil health and storing carbon. The following lists its many uses in agricultural and forest restoration projects as well as its several advantages in soil remediation and carbon sequestration, including bettering soil qualities, boosting nutrient retention, and expanding water-holding capacity.

9.7.1 Biochar Production and Characteristics

By pyrolysing biomass in an oxygen-limited atmosphere, organic material is thermally decomposed to produce biochar. Through this process, feedstock such as organic municipal solid waste, forestry waste, and agricultural residues becomes a stable source of carbon (Sarker et al. 2024). Biochar's advantageous qualities in soil applications are influenced by its physical and chemical properties, such as its surface area, permeability, and functional groups (Table 9.18). Biochar's surface area is increased by its porous nature, which creates areas for microbial colonisation and nutrient adsorption (Wong and Ogbonnaya 2021). Its capacity to hold onto moisture and nutrients is further enhanced by the presence of functional groups like carboxyl and hydroxyl, which makes it a useful soil amendment.

9.7.2 Properties of Biochar in Soil Remediation

Biochar is good for soil health because of a number of important characteristics. Its porous nature helps soils retain moisture during dry spells by increasing water retention. This is especially helpful in dry or desertification-prone locations. Because of its high cation exchange capacity (CEC), biochar can retain vital nutrients including phosphorus, potassium, and nitrogen, lowering the requirement for fertiliser and accelerating plant growth (Dey et al. 2023). Additionally, the alkaline properties of biochar can help balance acidic soils, improving their suitability for agricultural use. By enhancing soil aeration and decreasing bulk density, biochar functions as a soil conditioner in terms of physical structure. This characteristic improves root penetration and plant resistance and is especially helpful in compacted or damaged soils. Applications of biochar significantly improve soil texture and productivity, which is important for both food availability and rehabilitation of ecosystems, according to research by Khan et al. (2024).

Table 9.18 Biochar and carbon sequestration in land restoration

Study/project	Location	Feedstock for biochar	Application	Soil/environmental effects	Carbon sequestration impact	References
Brazilian agroforestry systems	Brazil	Agricultural residues	Applied to agroforestry soils to enhance fertility	Increased maize yields by 30%, improved microbial activity, better nutrient availability	Significant soil organic carbon retention	Celentano et al. (2020)
Corn production in North America	Iowa, USA	Corn stalks and forestry residues	Used in crop systems to enhance nutrient retention and water availability	15% increase in corn yield, improved soil moisture retention	Long-term carbon sequestration due to biochar stability	Spokas et al. (2012)
Reforestation in the Amazon Basin	Amazon Basin	Forestry residues	Integrated into planting holes for tree saplings	Improved tree survival and growth, increased soil fertility	Sequestration of carbon in soils, contribution to climate mitigation	Caron (2017)
Post-mining rehabilitation in Australia	Australia	Municipal green waste	Applied to degraded soils from mining activities to restore quality	Improved soil structure, enhanced organic carbon levels, promoted plant growth	Long-term storage of organic carbon in biochar-treated soils	Mercuri et al. (2005)
Agricultural land restoration	China	Rice husks, wheat straw	Applied to intensively farmed agricultural land	Enhanced crop productivity, improved water retention and nutrient availability	Increased carbon sequestration in agricultural soils	Zheng et al. (2013)

(continued)

Table 9.18 (continued)

Study/project	Location	Feedstock for biochar	Application	Soil/environmental effects	Carbon sequestration impact	References
Soil remediation in temperate regions	Germany	Wood chips	Used to remediate contaminated soils, improve structure	Reduced heavy metal bioavailability, enhanced soil fertility	Stable carbon sequestration in remediated soils	Atkinson et al. (2010)
Degraded lands restoration project	Sub-Saharan Africa	Agroforestry residues	Applied to degraded and desertified lands	Enhanced vegetation growth, improved nutrient cycling and soil health	Significant carbon sequestration and improved soil carbon stocks	Lehmann et al. (2015)

9.7.3 Biochar's Role in Carbon Sequestration

The capacity of biochar to store carbon for decades, if not millennia, is one of its most promising features. Because of its strong resistance to decomposition, the carbon in biochar can stay in soil for extended periods of time without releasing CO_2 again. Large-scale biochar use might offset up to 12% of yearly worldwide human greenhouse gas emissions, according to study by Woolf et al. (2018), making it an effective strategy for climate change mitigation. The pyrolysis of organic substances is the first step in the carbon sequestration process in biochar. About half of the carbon in the biomass is trapped in a stable, solid state during this process instead of being released as CO_2 into the atmosphere. Biochar promotes beneficial bacteria and cycle of nutrients by acting as a slow-release carbon source once it is incorporated into the soil. By encouraging higher carbon capture through improved plant development, this establishes a favourable feedback process in which soil health gradually improves (Sarfraz et al. 2019). According to studies, biochar may offer a sustainable way to store carbon. For example, according to a study by Bai (2021), using biochar might raise the amount of soil organic carbon by 1.1–6.8 Mg C ha^{-1} annually on average. Biochar's capacity to sequester carbon makes it a promising tool for both improving soil fertility and reducing the effects of climate change.

9.7.4 Biochar in Agricultural Applications

One of the main industries gaining from biochar additions is agriculture, especially in tropical areas where soils are frequently extremely acidic and lacking in nutrients. Biochar has been demonstrated to boost crop yields in these settings while lowering the demand for chemical fertilisers. For example, research conducted in sub-Saharan Africa has shown that applying biochar significantly boosts the yields of maize, beans, and cassava (Cornelissen et al. 2013). Additionally, using biochar lessens nutrient leaching, which is important in places with a lot of rainfall. There are further advantages of adding biochar to agricultural soils in terms of microbial activity and soil fertility. According to Glaser et al. (2015), biochar offers a home for mycorrhizal fungi and other beneficial soil microbes that develop symbiotic associations with plant roots. These microbes contribute to the overall sustainability of agricultural activities by improving plant health and enhancing nutrient uptake. By lowering soil emissions of the powerful greenhouse gas nitrous oxide (N_2O), biochar can also serve as a climate change mitigation tactic in agriculture. Depending on the kind of soil and the rate of biochar application, Clough et al. (2013) discovered that biochar amendments could lower N_2O emissions by as much as 50%. Because of this, biochar is not only a vital technique for improving soil fertility but also for lessening the environmental impact of agriculture.

In Australia, researchers applied biochar on a significant scale with the goal of improving agricultural productivity in degraded farmlands.

Several wheat and barley plots were treated with biochar at a rate of 10 tonnes per hectare as part of the project. According to the results, crop yields increased by 15–25% when compared to control plots, and retention of soil moisture and intake of nutrients significantly improved as well (Sohi et al. 2010). The project's participating farmers also reported using less fertiliser, which reduced expenses and had a less negative impact on the environment. In a different instance, biochar was used to test its impact on methane emissions in rice paddies in Japan. Flooded rice fields often produce methane, a powerful greenhouse gas. However, studies showed that adding biochar to rice increased yields by 10% and decreased methane emissions by up to 60% (Yamato et al. 2006). This demonstrates how biochar can increase crop yields while lowering agriculture's carbon footprint.

9.7.5 Biochar in Forest Restoration

Biochar has considerable promise for improving tree growth and repairing degraded areas in forest ecosystems. Because of its capacity to enhance soil structure, promote water retention, and trap carbon, biochar is very useful in recovery efforts following wildfires or logging. Biochar additives can speed up forest regrowth by improving soil nitrogen cycling and lowering the demand for external fertilisers, according to findings by Simiele et al. (2022). One of the best examples of biochar's long-term advantages in forest restoration is the massive Terra Preta (Black Earth) biochar project in the Amazon rainforest. Ancient Amazonian civilizations used biochar to create highly fertile soils that remain productive to this day. Modern efforts to replicate this approach have yielded promising results, with biochar-enhanced soils supporting greater biodiversity and forest regeneration.

9.8 Integrated Approaches to Land Restoration

In recent years, integrated approaches to land restoration have gained increasing recognition as effective solutions for reversing land degradation, improving soil health, and fostering sustainable development. These methods acknowledge that restoring degraded land requires more than individual techniques; it involves harmonizing ecological, physical, chemical, and biological processes at multiple scales. This section discusses the main elements and methods that comprise integrated land restoration frameworks, with an emphasis on ecological engineering, landscape-level restoration initiatives, and cooperative strategies between local communities, NGOs, and governments.

9.8.1 Ecological Engineering for Ecosystem and Soil Restoration

By combining engineering techniques with ecological principles, the multidisciplinary field of ecological engineering seeks to restore ecosystems. The major goal is to restore ecosystem processes including water management, carbon sequestration, and nutrient cycling while simultaneously enhancing soil health and reversing degradation. An important component of ecological engineering is the design and implementation of green infrastructure, such as constructed wetlands, riparian buffer zones, and reforested areas (Table 9.19). These solutions have a wide range of applications, including erosion control, biodiversity habitat, and water filtering. Using bio-based products for soil stabilisation is one way ecological engineering can enhance soil health and maintain surrounding ecosystems. According to Lehmann et al. (2015), biochar, a carbon-rich substance made by pyrolysing organic waste, has been shown to be a very successful soil amendment that increases nutrient availability, decreases soil compaction, and improves water retention. Additionally, by encouraging carbon sequestration, its use can help mitigate climate change. Restoring native vegetation to disturbed environments is a crucial component of ecological engineering. Native species can aid in restoring ecological processes that have been disturbed by human activity and are frequently better suited to the local environmental circumstances. For instance, drought-resistant plants that stabilise soils and enhance moisture retention have been the focus of initiatives in arid and semi-arid environments (Muller et al. 2017). In these regions, the utilisation of deep-rooted plants and perennial grasses improves the amount of organic matter in the soil while reducing erosion from wind and water.

9.8.2 Landscape-Level Restoration Projects

At the landscape level, integrated approaches involve restoring large tracts of land with a view toward balancing land use for agriculture, biodiversity conservation, and human settlement. Landscape ecology emphasizes the importance of spatial heterogeneity, meaning that different parts of a landscape may require tailored interventions depending on their specific ecological conditions and land use patterns. Restoration at this scale is crucial for achieving connectivity between habitats, allowing wildlife to move freely across ecosystems, which is essential for maintaining biodiversity. Agroforestry systems, which incorporate trees and shrubs into agricultural landscapes, are a popular approach in landscape-level restoration (Table 9.20). By increasing organic matter content, decreasing erosion, and boosting biodiversity, this technique has been demonstrated to promote soil health (García-Tejero et al. 2020). Because different plant species offer a variety of ecosystem services, such as carbon storage and nitrogen cycling, agroforestry also improves ecosystem resilience. Additionally, landscape-level restoration initiatives heavily rely on watershed management.

Table 9.19 Ecological engineering techniques for soil restoration

Technique	Description	Benefits	Example projects	References
Green infrastructure	Use of natural and semi-natural systems such as wetlands	Water filtration, habitat creation, erosion control	Constructed wetlands in the Netherlands	Štrbac et al. (2023)
Biochar application	Bio-based carbon material used as a soil amendment	Increases water retention, nutrient availability, carbon sequestration	Amazon Basin reforestation project	Mou (2024)
Native vegetation reintroduction	Planting of locally adapted species to restore ecosystem functions	Enhances biodiversity, improves soil health, reduces erosion	Arid land restoration in sub-Saharan Africa	Muller et al. (2017)
Perennial grasses	Use of grasses with deep root systems in degraded soils	Soil stabilization, moisture retention, organic matter improvement	Grassland rehabilitation in Australia	Reed (2014)

Reduced water quality, sedimentation of water bodies, and soil erosion are frequently the results of poor land management in watersheds. Integrated watershed management approaches seek to restore degraded watersheds by combining afforestation, terracing, and riverbank stabilization with community-based agricultural practices (Lal 2019). For example, restoration projects in the Loess Plateau of China have successfully restored degraded landscapes by integrating erosion control measures with sustainable agriculture.

9.8.3 Collaborative Approaches: Government, NGOs, and Local Communities

Effective cooperation between different stakeholders, including as governments, non-governmental organisations, and local populations, is frequently essential to ensuring the effectiveness of comprehensive land restoration programs. In order to ensure that indigenous understanding and needs are taken into consideration while planning and carrying out restoration activities, these projects usually employ a participatory approach. The Great Green Wall Initiative in Africa is among the most prominent examples; it attempts to improve water management, plant trees, and support sustainable agriculture in order to repair degraded land throughout the Sahel region (Goffner

Table 9.20 Landscape-level restoration techniques

Technique	Description	Benefits	Example projects	References
Agroforestry systems	Integration of trees into agricultural systems	Improved soil fertility, reduced erosion, enhanced biodiversity	Reforestation in Brazil's Atlantic forest	Oliveira et al. (2024)
Watershed management	Restoring watersheds through afforestation, terracing, and riverbank stabilization	Improved water quality, reduced erosion, enhanced agricultural productivity	Loess Plateau restoration project, China	Lal (2019)
Habitat connectivity	Restoring ecological corridors between fragmented habitats	Maintains biodiversity, enhances ecosystem resilience	European Green Belt Initiative	Ortega et al. (2023)
Terracing	Creating stepped landforms to reduce water runoff and soil erosion	Reduces erosion, improves water retention, supports sustainable agriculture	Ethiopian highlands rehabilitation	Tamene et al. (2017)

et al. 2019). In many cases, local communities are essential to the upkeep and administration of regenerated landscapes. For rehabilitation efforts to be successful in the long run, it has been crucial to empower these communities through financial assistance and capacity building. For instance, community-based forest management is emphasised by the UN's REDD + Program (Reducing Emissions from Deforestation and Forest Degradation) as a means of restoring damaged forests and providing financial advantages to the local populace (Muttaqin et al. 2019). These programs frequently incorporate systems for ecosystem service payment, which incentivise communities and landowners to maintain or restore ecological functioning. Additionally, NGOs frequently act as intermediaries, offering financial backing, lobbying assistance, and technical know-how for land restoration initiatives. Governments and non-governmental organisations can work together to fill policy gaps and make sure that restoration initiatives align with both national and global objectives, such the UN Decade on Ecosystem Restoration (2021–2030). By fostering partnerships between multiple stakeholders, integrated approaches can scale up restoration activities and promote long-term sustainability.

9.8.4 Challenges and Opportunities

Despite the many advantages of integrated approaches, a number of issues need to be resolved to guarantee their long-term viability. Lack of funds for extensive restoration initiatives, particularly in developing nations, is a significant obstacle. The extent of degradation frequently surpasses the financial resources made available by international programs such as the Green Climate Fund and the Global Environment Facility (GEF). In order to guarantee that restoration initiatives are in line with sustainable development objectives and climate mitigation plans, national policy consistency is also required. Technical difficulties with monitoring and evaluation also exist. It can be challenging to evaluate the overall efficacy of integrated approaches since they frequently entail intricate relationships between different ecological processes. The use of sophisticated monitoring tools, such remote sensing and geospatial analysis, to monitor restoration efforts is growing, but in situations with limited resources, these tools are not always available (Jacob et al. 2012a, b, 2013; Jacob and Olajide 2011; Kosoe et al. 2023).

There are plenty of chances to expand integrated land restoration initiatives despite these obstacles. More funding for these strategies is strongly encouraged by the growing understanding of the significance of repairing damaged ecosystems in relation to biodiversity preservation, climate change mitigation, and sustainable development. Furthermore, the Paris Agreement and other international policy frameworks' increasing emphasis on nature-based solutions emphasises how important land restoration may be to reaching environmental goals worldwide (IPBES 2019).

A growing knowledge of how nature-based solutions (NbS) like forest restoration, agroecology, and wetland restoration not only restore ecosystems but also aid in addressing global issues like food security, water scarcity, and climate change is reflected in the emphasis on NbS. In this regard, land restoration becomes a crucial component of comprehensive plans for accomplishing the Sustainable Development Goals (SDGs) of the UN, especially those pertaining to life on land (SDG 15), climate action (SDG 13), and ecologically sound communities and cities (SDG 11).

As part of a more concerted effort to reduce degradation of the environment and desertification while advancing sustainable development, several national and international projects are increasing the use of integrated restoration approaches. For example, nations in Asia, Africa, and Latin America have committed to the Bonn Challenge, which seeks to restore 350 million hectares of degraded landscapes worldwide by 2030. Restoring productivity to degraded lands is the challenge's main focus, as this would improve ecosystem services and livelihoods. Another illustration is the United Nations Convention to Combat Desertification's (UNCCD) Land Degradation Neutrality (LDN) project, which encourages integrated restoration techniques that strike a balance between ecosystem health and land use in order to eliminate net land degradation by 2030 (Chasek et al. 2019).

9.9 Policy and Economic Considerations in Soil Remediation

A major global issue, soil deterioration has an impact on the economy, society, and environment. Innovative solutions are needed to address this problem, backed by strong legislative frameworks and financial incentives that motivate investors, particularly in the private sector, to spend money on soil remediation. The economics of soil restoration methods, current international soil health legislation, and possible incentives to promote participation from the public and commercial sectors are all covered in this section. It explores the functions of global frameworks such as the Sustainable Development Goals (SDGs) of the UN, the European Union's soil regulations, and new tools like carbon credits in funding soil restoration initiatives.

9.9.1 *International Policies on Soil Health and Remediation*

International policy frameworks, especially in areas like the European Union (EU), are entwined with efforts to preserve and restore soil health. In advancing policies for soil conservation, the EU has taken the lead. The *EU Soil Thematic Strategy* was introduced as an overarching framework aimed at preventing soil degradation, preserving soil functions, and restoring soils affected by various forms of pollution (Loj 2009). Although the strategy was not legally binding, it significantly influenced member states to develop national soil protection strategies. Additionally, the *European Green Deal*, launched in 2020, emphasizes soil health as part of a larger agenda toward sustainability. The *Farm to Fork Strategy* and *Biodiversity Strategy* under the Green Deal focus on sustainable agriculture, land use practices, and biodiversity protection, which collectively contribute to maintaining healthy soils. This reflects the understanding that soil health is crucial to food security and climate action.

Globally, the Sustainable Development Goals (SDGs) of the UN offer a thorough framework for addressing social and environmental issues, such as soil degradation. Protecting, restoring, and using terrestrial ecosystems sustainably are all part of SDG 15, "Life on Land," which also calls for stopping land degradation and fighting desertification Owuor (2021). The SDGs encourage global initiatives and support for soil remediation projects by establishing goals for environmentally friendly land use and soil restoration. Furthermore, the preservation of healthy soils is emphasised as a crucial element of sustainable land management in frameworks such as the United Nations Convention to Combat Desertification (UNCCD). The UNCCD promotes land restoration and encourages countries to implement Land Degradation Neutrality (LDN) targets as part of their national action plans. Countries participating in the convention have committed to avoiding net land degradation by balancing land degradation and restoration efforts. However, despite these initiatives, soil protection legislation remains fragmented, and its effectiveness depends largely on national-level implementation and enforcement. For example, some countries have robust soil protection laws, while others lack dedicated legislation addressing soil health. The lack of global, legally binding frameworks on soil protection continues to be a critical barrier to widespread remediation efforts.

9.9.2 Economic Aspects of Soil Remediation Techniques

One important consideration in the choice and application of soil remediation methods is cost-effectiveness. The kind of contamination, the project's size, the remediation technique employed, and the land's post-restoration worth all affect how economically feasible soil restoration is. Different methods have varying costs and effects on the environment, such as bioremediation, electrokinetic remediation, and phytoremediation. To find out if the financial benefits of cleanup outweigh the expenses, a cost–benefit analysis is frequently performed. This assessment typically considers several factors, including:

i. **Direct Costs**: These include the costs associated with the remediation technology, labor, materials, equipment, and monitoring. Traditional methods like excavation and landfill disposal tend to have high direct costs due to the need for transportation and landfilling fees. Innovative techniques, such as bioremediation and phytoremediation, often have lower upfront costs but may require longer timeframes for results.
ii. **Indirect Costs**: These involve potential losses from reduced land value or ongoing health risks if contaminated land remains untreated. Indirect costs may also include ecosystem service losses, such as reduced water filtration and biodiversity.
iii. **Long-Term Benefits**: The long-term benefits of soil remediation can be substantial, including restored ecosystem services, enhanced agricultural productivity, and improved public health outcomes. Additionally, rehabilitated land may increase in value, providing economic returns through development or agriculture. Furthermore, effective soil remediation can prevent further environmental degradation, which would otherwise incur additional costs.

9.9.3 Carbon Credits and Financing Soil Restoration

As global efforts to combat climate change intensify, carbon credits are becoming a key financial mechanism to support soil remediation and land restoration projects. One metric tonne of carbon dioxide (CO_2) or its equivalent in other greenhouse gases can be reduced by purchasing carbon credits, which are marketable certifications. Agroforestry, regenerative agriculture, and the application of biochar are examples of soil restoration initiatives that can produce carbon credits, giving investors and landowners access to an extra source of income. Soil carbon sequestration has been identified as one of the most scalable and economical ways to combat climate change. A substantial contribution to global carbon reduction efforts might be made by restoring land to increase soil organic carbon reserves, which could offset up to 5.3 gigatons of CO_2 yearly (Lal 2019). Carbon markets, both voluntary and regulated, are increasingly incorporating soil carbon sequestration projects, allowing landowners and companies engaged in soil restoration to sell carbon credits on the market.

Incentives for soil carbon capture and storage through carbon credits were first introduced by nations like Australia through the Carbon Farming Initiative (CFI). Through the CFI, landowners and farmers can obtain carbon credits by using land management techniques like reforestation and no-till farming that increase soil carbon. Businesses wishing to offset their carbon footprint can then purchase these credits. However, especially in developing nations, the prospect of carbon credits to fund soil remediation is still not fully utilised. The difficulty of precisely quantifying and confirming soil carbon sequestration, which varies depending on soil type, climate, and land use patterns, is one of the difficulties. New methods, such soil carbon modelling and remote sensing, are being explored to solve this problem by increasing the precision and openness of soil carbon assessments.

9.9.4 Incentivizing Private Sector Involvement

Private sector participation is crucial to scaling up soil remediation and land restoration projects. Governments can play a key role by offering financial incentives, such as tax breaks, subsidies, and grants, to encourage companies to invest in soil health. Public–private partnerships (PPPs) can also make it easier for private businesses and government agencies to share resources, expertise, and risks. The Dow Chemical Company's partnership with the Nature Conservancy to carry out extensive soil remediation operations in damaged industrial areas is a good illustration of private sector involvement. Along with improving environmental results, this collaboration raised the company's corporate social responsibility (CSR) profile and demonstrated the need of soil remediation as a component of a larger sustainability plan (Hawkins and Prickett 2016). Private landowners may also be persuaded to participate in soil restoration by financial incentives like payment for ecosystem services (PES). PES programs compensate landowners that implement soil-healthy practices like agroforestry, cover crops, and decreased tillage. These programs help to internalize the environmental benefits of soil conservation, which are often undervalued in traditional markets.

9.10 Future Directions and Challenges

To improve the scalability and efficacy of novel soil remediation approaches, several future research areas and problems must be addressed as the field of soil remediation continues to develop. The following discusses possible future paths for soil remediation, technological developments, social acceptance, and legal considerations while considering how artificial intelligence (AI) and neural networks might improve these procedures (Amabie et al. 2024).

9.10.1 Research Areas for Innovation

The complexities associated with contaminated soil necessitate ongoing research into innovative remediation techniques. One promising area is the integration of nano-remediation technologies, which utilize nanoparticles to target specific contaminants at a molecular level. According to research, nanoparticles can improve the effectiveness of remediation campaigns. For example, using nano-iron for in situ dechlorination of chlorinated solvents has shown notable rates of contaminant removal (Hannum and Pamukcu 2019). Additionally, there is potential for studying biosurfactants, which are biologically generated surfactants that increase the bioavailability of hydrophobic contaminants. These substances, which come from microorganisms, can enhance the breakdown of organic contaminants in soil (Makkar et al. 2011). Climate resilience in remediation solutions is another important area of research. Knowing how climate change affects soil condition and pollutant movement is crucial given the rise in extreme weather events. It will be more crucial than ever to incorporate climate-resilient techniques, such as choosing plant species for phytoremediation that can tolerate drought or flooding (Majhi et al. 2022). This area of study has the potential to produce more environmentally friendly remediation techniques that improve ecosystem resilience while simultaneously addressing contaminants.

9.10.2 Technological Barriers

Despite the advancements in soil remediation technologies, several technological barriers hinder their widespread application. To determine their efficacy and safety, for example, a lot of novel approaches need to undergo thorough field testing and validation, which can be an expensive and time-consuming procedure. Furthermore, current soil characteristics like low pH or excessive salinity may restrict the use of some technologies, requiring more investigation into how well these techniques work in various soil types and contamination situations (Hopmans et al. 2021). Another difficulty with novel approaches is their scalability. Numerous effective remediation initiatives have been carried out on a local scale; nevertheless, it can be challenging to translate these outcomes into large-scale applications. Creating pilot projects that show how beneficial these methods are on a bigger scale is essential to winning over regulators and the general public.

9.10.3 Regulatory Considerations

The viability of introducing novel cleanup technology is largely determined by regulatory frameworks. Current regulations may not be well-suited to accommodate emerging technologies, leading to lengthy approval processes that can delay the deployment of innovative solutions. To overcome this challenge, regulatory agencies need to collaborate with researchers and practitioners to develop adaptive frameworks that encourage the integration of innovative technologies while ensuring environmental safety (Watson et al. 2018). Furthermore, harmonization of regulations across different jurisdictions can facilitate the broader adoption of effective remediation techniques. Variability in regulations can create barriers for practitioners operating in multiple regions, emphasizing the need for international collaboration to establish best practices and standards.

9.10.4 Social Acceptance and Stakeholder Engagement

The success of soil remediation methods depends on how the public views them and how socially acceptable they are. Engaging stakeholders—such as local communities, environmental organizations, and industry representatives—early in the remediation process can foster trust and cooperation. Effective communication about the benefits and safety of innovative techniques is essential to alleviate concerns and resistance (Jacob and Owolabi 2011; Jacob and Ogogo 2011). Additionally, community involvement in monitoring and decision-making can enhance the legitimacy of remediation projects and empower local populations. Incorporating local knowledge and priorities into remediation planning can lead to more effective and socially accepted outcomes.

9.10.5 Machine Learning and Artificial Intelligence

The application of machine learning (ML) and artificial intelligence (AI) to soil remediation offers a revolutionary chance to maximise restoration tactics. In order to forecast the results of different procedures and improve cleanup plans, these systems can evaluate enormous volumes of data from soil samples, weather trends, and prior remediation attempts. For instance, machine learning algorithms can identify patterns in contamination data, leading to improved targeting of specific contaminants and more effective remediation strategies (Janga et al. 2023). Moreover, AI can assist in monitoring soil health post-remediation. By employing sensor technologies and satellite imagery, AI systems can continuously assess soil conditions and the effectiveness of remediation efforts, allowing for adaptive management strategies that respond to changing environmental conditions (Himeur et al. 2022). The real-time data collected through these technologies can facilitate prompt interventions, maximizing the success of remediation initiatives.

9.11 Conclusion

Investigating cutting-edge soil rehabilitation methods highlights the urgent need for long-term solutions to the widespread problem of soil degradation. Modern remediation techniques are complex, as evidenced by the incorporation of cutting-edge technologies like biochar applications, geoengineering techniques, and artificial intelligence possibilities. Each technique discussed highlights not only the advancements in scientific understanding but also the imperative to restore soil health as a foundation for resilient ecosystems and communities.

Moreover, the role of strong policy frameworks and economic incentives cannot be overstated. Effective regulations and incentives are vital for fostering an environment conducive to the adoption of innovative practices. International policies, such as the European Union's directives and the Sustainable Development Goals, serve as essential guides that encourage countries to prioritize soil health and restoration efforts. By aligning remediation strategies with these frameworks, nations can ensure that efforts are cohesive and beneficial on a global scale.

Another essential component of successful remediation projects is community engagement. Stakeholders can guarantee that the techniques used are contextually appropriate and meet the particular requirements of the communities affected by soil degradation by incorporating indigenous communities in the development and implementation of soil restoration initiatives. In the end, collaborative techniques produce more sustainable results by fostering a sense of shared responsibility among community members and enhancing the legitimacy of cleanup efforts.

Looking ahead, the potential for incorporating machine learning and artificial intelligence into soil remediation techniques provides a window into a more effective future. A more proactive approach to issues with soil health is made possible by these technologies' ability to support continuous surveillance, forecasting, and adaptive management. The capacity to evaluate enormous volumes of data will allow professionals to maximise the use of resources and customise remedial activities to particular environmental circumstances.

References

Abajue MC, Izah SC (2024) Arthropods as a sustainable biomonitoring tool for environmental pollution. In: Izah SC, Ogwu MC, Hamidifar H (eds) Biomonitoring of pollutants in the global south. Springer, Singapore, pp 535–567. https://doi.org/10.1007/978-981-97-1658-6_15

Acar YB, Alshawabkeh AN (1993) Principles of electrokinetic remediation. Environ Sci Technol 27(13):2638–2647

Adenutsi CD, Turkson JN, Wang L, Zhao G, Zhang T, Quaye JA, Erzuah S, Sokama-Neuyam YA (2023) Review on potential application of saponin-based natural surfactants for green chemical enhanced oil recovery: perspectives and progresses. Energy Fuels 37(13):8781–8823

Ahmadi M, Motlagh HR, Jaafarzadeh N, Mostoufi A, Saeedi R, Barzegar G, Jorfi S (2017) Enhanced photocatalytic degradation of tetracycline and real pharmaceutical wastewater using MWCNT/TiO_2 nano-composite. J Environ Manage 186:55–63

Ahmari H, Heris SZ, Khayyat MH (2018) The effect of titanium dioxide nanoparticles and UV irradiation on photocatalytic degradation of Imidaclopride. Environ Technol 39(4):536–547

Akhtar MS, Hameed A, Aslam S, Ullah R, Kashif A (2023) Phytoremediation of metal-contaminated soils and water in Pakistan: a review. Water Air Soil Pollut 234(1):11

Al Miad A, Saikatb SP, Alam MK, Hossain MS, Bahadur NM, Ahmed S (2024) Metal oxides-based photocatalyst for the efficient degradation of organic pollutants for a sustainable environment: a review. Nanoscale Adv

Alshawabkeh AN, Gale RJ, Ozsu-Acar E, Bricka RM (1999) Optimization of 2-D electrode configuration for electrokinetic remediation. J Soil Contam 8(6):617–635

Amabie T, Izah SC, Ogwu MC, Hait M (2024) Harmonizing tradition and technology: the synergy of artificial intelligence in traditional medicine. In: Izah SC, Ogwu MC, Akram M (eds) Herbal medicine phytochemistry. Reference series in phytochemistry. Springer, Cham. https://doi.org/10.1007/978-3-031-21973-3_70-1

Aparicio JD, Raimondo EE, Saez JM, Costa-Gutierrez SB, Alvarez A, Benimeli CS, Polti MA (2022) The current approach to soil remediation: a review of physicochemical and biological technologies, and the potential of their strategic combination. J Environ Chem Eng 10(2):107141

Askarinejad A, Akca D, Springman SM (2018) Precursors of instability in a natural slope due to rainfall: a full-scale experiment. Landslides 15:1745–1759

Atkinson CJ, Fitzgerald JD, Hipps NA (2010) Potential mechanisms for achieving agricultural benefits from biochar application to temperate soils: a review. Plant Soil 337:1–18

Aziz T, Farid A, Haq F, Kiran M, Ullah A, Zhang K, Li C, Ghazanfar S, Sun H, Ullah R, Ali A (2022) A review on the modification of cellulose and Its applications. Polymers 14:3206

Bai X (2021) Soil organic carbon in Maine, its responses to biochar addition, and climate change. The University of Maine

Bala S, Garg D, Thirumalesh BV, Sharma M, Sridhar K, Inbaraj BS, Tripathi M (2022) Recent strategies for bioremediation of emerging pollutants: a review for a green and sustainable environment. Toxics 10(8):484

Bassegio C, Campagnolo MA, Schwantes D, Gonçalves Junior AC, Manfrin J, Schiller ADP, Bassegio D (2020) Growth and accumulation of Pb by roots and shoots of *Brassica juncea* L. Int J Phytorem 22(2):134–139

Bazrafshan E, Mohammadi L, Ansari-Moghaddam A, Mahvi AH (2015) Heavy metals removal from aqueous environments by electrocoagulation process—a systematic review. J Environ Health Sci Eng 13:1–16

Behrouzinia S, Ahmadi H, Abbasi N, Javadi AA (2022) Insights into enhanced electrokinetic remediation of copper-contaminated soil using a novel conductive membrane based on nanoparticles. Environ Geochem Health 44(3):1015–1032

Bhattacharjee N, Som I, Saha R, Mondal S (2024) A critical review on novel eco-friendly green approach to synthesize zinc oxide nanoparticles for photocatalytic degradation of water pollutants. Int J Environ Anal Chem 104(3):489–516

Bhavya G, Belorkar SA, Mythili R, Geetha N, Shetty HS, Udikeri SS, Jogaiah S (2021) Remediation of emerging environmental pollutants: a review based on advances in the uses of eco-friendly biofabricated nanomaterials. Chemosphere 275:129975

Boulakradeche MO, Akretche DE, Cameselle C, Hamidi N (2015) Enhanced electrokinetic remediation of hydrophobic organics contaminated soils by the combination of non-ionic and ionic surfactants. Electrochim Acta 174:1057–1066

Bumanis G, Vitola L, Pundiene I, Sinka M, Bajare D (2020) Gypsum, geopolymers, and starch—alternative binders for bio-based building materials: a review and life-cycle assessment. Sustainability 12(14):5666

Camacho JV (2021) Electrobioremediation of polluted soils. In: Electrochemically assisted remediation of contaminated soils. Springer, Cham, pp 297–313

Cameotra SS, Makkar RS (2010) Biosurfactant-enhanced bioremediation of hydrophobic pollutants. Pure Appl Chem 82(1):97–116

Caron AS (2017) Stand diversification and tropical ecosystems: community ecology of beetles and ecosystem functioning in Sardinilla. McGill University (Canada), Panama

Celentano D, Rousseau GX, Paixão LS, Lourenço F, Cardozo EG, Rodrigues TO, e Silva HR, Medina J, de Sousa TMC, Rocha AE, de Oliveira Reis F (2020) Carbon sequestration and nutrient cycling in agroforestry systems on degraded soils of Eastern Amazon, Brazil. Agrofor Syst 94:1781–1792

Chandra P, Enespa (2019) Fungal enzymes for bioremediation of contaminated soil. In: Recent advancement in white biotechnology through fungi: volume 3: perspective for sustainable environments, pp 189–215

Chasek P, Akhtar-Schuster M, Orr BJ, Luise A, Ratsimba HR, Safriel U (2019) Land degradation neutrality: the science-policy interface from the UNCCD to national implementation. Environ Sci Policy 92:182–190

Chen W, Li H (2018) Cost-effectiveness analysis for soil heavy metal contamination treatments. Water Air Soil Pollut 229:1–13

Chen D, Cheng Y, Zhou N, Chen P, Wang Y, Li K, Huo S, Cheng P, Peng P, Zhang R, Wang L (2020) Photocatalytic degradation of organic pollutants using TiO_2-based photocatalysts: a review. J Clean Prod 268:121725

Clough TJ, Condron LM, Kammann C, Müller C (2013) A review of biochar and soil nitrogen dynamics. Agronomy 3(2):275–293

Côa F, Bortolozzo LS, Petry R, Da Silva GH, Martins CH, de Medeiros AM, Sabino CM, Costa RS, Khan LU, Delite FS, Martinez DST (2020) Environmental toxicity of nanopesticides against non-target organisms: the state of the art. In: Nanopesticides: from research and development to mechanisms of action and sustainable use in agriculture, pp 227–279

Cornelissen G, Martinsen V, Shitumbanuma V, Alling V, Breedveld GD, Rutherford DW, Sparrevik M, Hale SE, Obia A, Mulder J (2013) Biochar effect on maize yield and soil characteristics in five conservation farming sites in Zambia. Agronomy 3(2):256–274

Currin CA (2019) Living shorelines for coastal resilience. In: Coastal wetlands. Elsevier, pp 1023–1053

De Souza CB, Silva GR (2019) Phytoremediation of effluents contaminated with heavy metals by floating aquatic macrophytes species. In: Biotechnology and bioengineering. IntechOpen, pp 137–153

Dey S, Purakayastha TJ, Sarkar B, Rinklebe J, Kumar S, Chakraborty R, Datta A, Lal K, Shivay YS (2023) Enhancing cation and anion exchange capacity of rice straw biochar by chemical modification for increased plant nutrient retention. Sci Total Environ 886:163681

Dhaliwal SS, Singh J, Taneja PK, Mandal A (2020) Remediation techniques for removal of heavy metals from the soil contaminated through different sources: a review. Environ Sci Pollut Res 27(2):1319–1333

Diarra I, Kotra KK, Prasad S (2022) Application of phytoremediation for heavy metal contaminated sites in the South Pacific: strategies, current challenges and future prospects. Appl Spectrosc Rev 57(6):490–512

Dorairaj D, Osman N (2021) Present practices and emerging opportunities in bioengineering for slope stabilization in Malaysia: an overview. PeerJ 9:e10477

Duan P, Khan S, Ali N, Shereen MA, Siddique R, Ali B, Iqbal HM, Nabi G, Sajjad W, Bilal M (2020) Biotransformation fate and sustainable mitigation of a potentially toxic element of mercury from environmental matrices. Arab J Chem 13(9):6949–6965

Dunn RM, Hawkins JM, Blackwell MS, Zhang Y, Collins AL (2022) Impacts of different vegetation in riparian buffer strips on runoff and sediment loss. Hydrol Process 36(11):e14733

Dursun S, Ayturan ZC (2022) Simultaneous removal of gaseous benzene and toluene with photocatalytic oxidation process at high temperatures under UVC irradiation. Environ Sci Pollut Res 29(25):38232–38247

Dutta N, Usman M, Ashraf MA, Luo G, Zhang S (2022) Efficacy of emerging technologies in addressing reductive dechlorination for environmental bioremediation: a review. J Hazard Mater Lett 3:100065

El-Naggar A, Shaheen SM, Ok YS, Rinklebe J (2018) Biochar affects the dissolved and colloidal concentrations of Cd, Cu, Ni, and Zn and their phytoavailability and potential mobility in a mining soil under dynamic redox-conditions. Sci Total Environ 624:1059–1071

El-Saeid MH, BaQais A, Alshabanat M (2022) Study of the photocatalytic degradation of highly abundant pesticides in agricultural soils. Molecules 27(3):634

Essaid HI, Bekins BA, Cozzarelli IM (2015) Organic contaminant transport and fate in the subsurface: evolution of knowledge and understanding. Water Resour Res 51(7):4861–4902

Fang Y, Wu X, Dai M, Lopez-Valdivieso A, Raza S, Ali I, Peng C, Li J, Naz I (2021) The sequestration of aqueous Cr(VI) by zero valent iron-based materials: from synthesis to practical application. J Clean Prod 312:127678

Favara P, Gamlin J (2017) Utilization of waste materials, non-refined materials, and renewable energy in in situ remediation and their sustainability benefits. J Environ Manage 204:730–737

Fei L, Bilal M, Qamar SA, Imran HM, Riasat A, Jahangeer M, Ghafoor M, Ali N, Iqbal HM (2022) Nano-remediation technologies for the sustainable mitigation of persistent organic pollutants. Environ Res 211:113060

Fiorino DJ (1996) Toward a new system of environmental regulation: the case for an industry sector approach. Envtl L 26:457

Firoozi AA, Guney Olgun C, Firoozi AA, Baghini MS (2017) Fundamentals of soil stabilization. Int J Geo-Eng. 8:1–16

Foster WG, Evans JA, Little J, Arbour L, Moore A, Sauve R, Andrés León J, Luo W (2017) Human exposure to environmental contaminants and congenital anomalies: a critical review. Crit Rev Toxicol 47(1):59–84

Galdames A, Ruiz-Rubio L, Orueta M, Sánchez-Arzalluz M, Vilas-Vilela JL (2020) Zero-valent iron nanoparticles for soil and groundwater remediation. Int J Environ Res Public Health 17(16):5817

Ganie AS, Bano S, Khan N, Sultana S, Rehman Z, Rahman MM, Sabir S, Coulon F, Khan MZ (2021) Nanoremediation technologies for sustainable remediation of contaminated environments: recent advances and challenges. Chemosphere 275:130065

García-Tejero IF, Carbonell R, Ordoñez R, Torres FP, Durán Zuazo VH (2020) Conservation agriculture practices to improve the soil water management and soil carbon storage in Mediterranean rainfed agro-ecosystems. In: Soil health restoration and management, pp 203–230

Geris R, Calil S, Rigoli IC, Vidal RRL, da Silva AF, Malta M (2023). Advanced applications of lignocellulosic fibers and mycelium-based composites for a sustainable world. In: Trends and innovations in energetic sources, functional compounds and biotechnology: science, simulation, experiments. Springer Nature Switzerland, Cham, pp 367–400

Ghosh M, Singh SP (2005) A review on phytoremediation of heavy metals and utilization of it's by products. Asian J Energy Environ 6(4):18

Giannis A, Pentari D, Wang JY, Gidarakos E (2010) Application of sequential extraction analysis to electrokinetic remediation of cadmium, nickel and zinc from contaminated soils. J Hazard Mater 184(1–3):547–554

Glaser B, Wiedner K, Seelig S, Schmidt HP, Gerber H (2015) Biochar organic fertilizers from natural resources as substitute for mineral fertilizers. Agron Sustain Dev 35:667–678
Goffner D, Sinare H, Gordon LJ (2019) The Great Green Wall for the Sahara and the Sahel Initiative as an opportunity to enhance resilience in Sahelian landscapes and livelihoods. Reg Environ Change 19:1417–1428
Gowthaman S, Nakashima K, Kawasaki S (2018) A state-of-the-art review on soil reinforcement technology using natural plant fiber materials: past findings, present trends and future directions. Materials 11(4):553
Gudiña EJ, Rodrigues AI, Alves E, Domingues MR, Teixeira JA, Rodrigues LR (2015) Bioconversion of agro-industrial by-products in rhamnolipids toward applications in enhanced oil recovery and bioremediation. Bioresour Technol 177:87–93
Hait M, Patel D, Izah SC (2024a) Molecular techniques and technologies in biomonitoring for environmental sustainability. In: Izah SC, Ogwu MC, Hamidifar H (eds) Biomonitoring of pollutants in the global south. Springer, Singapore, pp 605–637. https://doi.org/10.1007/978-981-97-1658-6_17
Hait M, Sahu P, Biswas S, Izah SC (2024b) Sustainable application of artificial intelligence in biomonitoring for environmental sustainability: challenges and prospects. In: Izah SC, Ogwu MC, Hamidifar H (eds) Biomonitoring of pollutants in the global south. Springer, Singapore, pp 747–778. https://doi.org/10.1007/978-981-97-1658-6_22
Hamad AA, Moubasher HA, Moustafa YM, Mohamed NH, Abd-El Rhim EH (2021) Petroleum hydrocarbon bioremediation using native fungal isolates and consortia. Sci World J 2021(1):6641533
Hamid W, Alnuaim A (2024) Experimental study on improving the bearing capacity of Sabkha soil using geopolymer. Int J Geosynth Ground Eng 10(5):84
Han D, Wu X, Li R, Tang X, Xiao S, Scholz M (2021) Critical review of electro-kinetic remediation of contaminated soils and sediments: mechanisms, performances and technologies. Water Air Soil Pollut 232(8):335
Hannum L, Pamukcu S (2019) Aided transport of nano-iron in clay soils using direct electric field. In: Technical proceedings of the 2007 cleantech conference and trade show, Aug 2019. CRC Press, pp 144–147
Harbottle MJ, Al-Tabbaa A, Evans CW (2007) A comparison of the technical sustainability of in situ stabilisation/solidification with disposal to landfill. J Hazard Mater 141(2):430–440
Harclerode M, Ridsdale DR, Darmendrail D, Bardos P, Alexandrescu F, Nathanail P, Pizzol L, Rizzo E (2015) Integrating the social dimension in remediation decision-making: state of the practice and way forward. Remediat J 26(1):11–42
Hawkins N, Prickett G (2016) The case for green infrastructure. In: Turbulence: a corporate perspective on collaborating for resilience, pp 87–99
Himeur Y, Rimal B, Tiwary A, Amira A (2022) Using artificial intelligence and data fusion for environmental monitoring: a review and future perspectives. Inf Fusion 86:44–75
Hopmans JW, Qureshi AS, Kisekka I, Munns R, Grattan SR, Rengasamy P, Ben-Gal A, Assouline S, Javaux M, Minhas PS, Raats PAC (2021) Critical knowledge gaps and research priorities in global soil salinity. Adv Agron 169:1–191
Horst J, Welty N, Gallegos J, Dunham D, Adams A (2022) The rapid advancement of environmental sensors in remediation. Ground Water Monit Remediat 42(2)
Hou S, Wu B, Peng D, Wang Z, Wang Y, Xu H (2019) Remediation performance and mechanism of hexavalent chromium in alkaline soil using multi-layer loaded nano-zero-valent iron. Environ Pollut 252:553–561
Hristozov DR, Gottardo S, Critto A, Marcomini A (2012) Risk assessment of engineered nanomaterials: a review of available data and approaches from a regulatory perspective. Nanotoxicology 6(8):880–898
Huang G, Wu Y, Cheng L, Zhou D, Wang X, Ding M et al (2024) Spatial heterogeneity of soil moisture caused by drainage and its effects on cadmium variation in rice grain within individual fields. Sci Total Environ 946:174500

Huang H, Huang H, Feng Q, Liu G, Zhan Y, Wu M, Lu H, Shu Y, Leung DY (2017) Catalytic oxidation of benzene over Mn modified TiO_2/ZSM-5 under vacuum UV irradiation. Appl Catal B 203:870–878

Ijoma GN, Tekere M (2017) Potential microbial applications of co-cultures involving ligninolytic fungi in the bioremediation of recalcitrant xenobiotic compounds. Int J Environ Sci Technol 14:1787–1806

Ikhajiagbe B, Ogwu MC (2020) Hazard quotient, microbial diversity and plant composition of spent crude oil polluted-soil. In: Beni-Suef Univ J Basic Appl Sci 9(26). https://doi.org/10.1186/s43088-020-00052-0

Ikhajiagbe B, Igiebor FA, Ogwu MC (2021) Growth and yield performances of rice (*Oryza sativa* var. nerica) after exposure to biosynthesized nanoparticles. Bull Natl Res Cent 45. Article number 62. https://doi.org/10.1186/s42269-021-00508-y

Ikhajiagbe B, Ogwu MC, Ekhator PO, Omoayena II (2022) Implications of soil nitrogen enhancement on the yield performance of soybean (*Glycine max*) in cadmium-polluted soil. J Pollut Eff Control 10(2):1–12. https://doi.org/10.35248/23754-4397.22.10.334

IPBES W (2019) Intergovernmental science-policy platform on biodiversity and ecosystem services. In: Summary for policy makers of the global assessment report on biodiversity and ecosystem services of the intergovernmental science-policy platform on biodiversity and ecosystem services. IPBES Secretariat, Bonn, Germany

Islam MA, Karim A, Ethiraj B, Raihan T, Khan MMR, Kadier A, Al Nadhari S, Al-Masri AA, Ameen F (2023) Interspecies microbial interactions in bioelectrochemical system and biodegradation: a state of the art review. Sci Total Environ 891:164623

Ismaiel HAH (2006) Treatment and improvement of the geotechnical properties of different soft fine-grained soils using chemical stabilization. Shaker, Aachen, Germany

Ismail AH, Kusbiantoro A, Tajunnisa Y, Ekaputrc JJ, Laory I (2024) A review of aluminosilicate sources from inorganic waste for geopolymer production: sustainable approach for hydrocarbon waste disposal. Clean Mater 100259

Izah SC, Ogwu MC, Akram M (2023) Herbal medicine phytochemistry: applications and trends. Springer, Cham, Switzerland. https://doi.org/10.1007/978-3-031-21973-3

Izah SC, Abdelkhalek ST, Ogwu MC, Hamidifar H (2024a) Overview and perspectives of biomonitoring in the global south. In: Izah SC, Ogwu MC, Hamidifar H (eds) Biomonitoring of pollutants in the global south. Springer, Singapore, pp 3–36. https://doi.org/10.1007/978-981-97-1658-6_1

Izah SC, Ogwu MC, Hamidifar H (2024b) Biomonitoring of pollutants in the global south. Springer Nature, Switzerland. https://doi.org/10.1007/978-981-97-1658-6

Jacob DE, Ogogo AU (2011) Community participation in protected area management: a case study of Cross River National Park. In: Popoola L, Ogunsanwo K, Idumah F (eds) Forestry in the context of the millennium development goals. Proceedings of the 34th annual conference of the forestry association of Nigeria, Osogbo, Osun State, Nigeria, vol 1, pp 412–415

Jacob DE, Olajide O (2011) Relevance and challenges of geographic information system (GIS) in the management of protected forest in Nigeria. Niger J Agric Food Environ 7(2):63–66

Jacob DE, Owolabi JT (2011) The need for community participation in protected area management: a case study of Cross River National Park. J Geogr Environ Plan (JOGEP) 7(2):61–69

Jacob DE, Ufot IN, Udeagha AU (2012a) The significance of satellite remote sensing and GIS applications in environmental meteorology. J Environ Manage Saf 3(4):1–10

Jacob DE, Udeagha AU, Ufot IN (2012b) The potentials of participatory geographic information system (PGIS) in natural resources management in Nigeria. In: Onyekwelu JC, Agbeja BO, Adekunle VAJ, Lameed GA, Adesoye PO, Omole AO (eds) De-reservation, encroachment and deforestation: implications for the future of Nigerian forest estate and carbon emission reduction. Proceedings of the 3rd biennial national conference of the forest and forest products society, Ibadan, pp 51–55

Jacob DE, Udoakpan UI, Daniel KS, Nelson IU, Okon KE (2013) Strategies for ICT application in sustainable forest management in Nigeria. Niger J Agric Food Environ 9(1):56–62

Jacob DE, Nelson IU, Okweche SI, Oko PA, Izah SC (2024a) Suitability of mammals indigenous to the global south as bioindicator species for assessing environmental health. In: Izah SC, Ogwu MC, Hamidifar H (eds) Biomonitoring of pollutants in the global south. Springer, Singapore, pp 451–484. https://doi.org/10.1007/978-981-97-1658-6_13

Jacob DE, Nelson IU, Efenakpo OD, Izah SC (2024b) Reptiles as environmental sentinels: exploring their significance. In: Izah SC, Ogwu MC, Hamidifar H (eds) Biomonitoring of pollutants in the global south. Springer, Singapore, pp 485–533. https://doi.org/10.1007/978-981-97-1658-6_14

Jacob DE, Nelson IU, Izah SC, Ukpong E, Akpan UU, Ogwu MC (2024c) Bioindicators in recreational planning and development: balancing nature and human activities. In: Izah SC, Ogwu MC, Hamidifar H (eds) Biomonitoring of pollutants in the global south. Springer, Singapore, pp 835–878. https://doi.org/10.1007/978-981-97-1658-6_24

Jacob DE, Nelson IU, Oluwafemi OJ, Izah SC, Ogwu MC (2024d) Environmental stewardship: safeguarding biodiversity in protected landscapes and recreational parks using biosecurity. In: Izah SC, Ogwu MC, Hamidifar H (eds) Biomonitoring of pollutants in the global south. Springer, Singapore, pp 779–834. https://doi.org/10.1007/978-981-97-1658-6_23

Janga JK, Reddy KR, Kvns R (2023) Integrating artificial intelligence, machine learning, and deep learning approaches into remediation of contaminated sites: a review. Chemosphere 140476

Kang CH, Kwon YJ, So JS (2016) Bioremediation of heavy metals by using bacterial mixtures. Ecol Eng 89:64–69

Kebede AA, Kalogiannis T, Van Mierlo J, Berecibar M (2022) A comprehensive review of stationary energy storage devices for large scale renewable energy sources grid integration. Renew Sustain Energy Rev 159:112213

Khair AS (2022) Nonlinear electrophoresis of colloidal particles. Curr Opin Colloid Interface Sci 59:101587

Khan FSA, Mubarak NM, Tan YH, Khalid M, Karri RR, Walvekar R, Abdullah EC, Nizamuddin S, Mazari SA (2021) A comprehensive review on magnetic carbon nanotubes and carbon nanotube-based buckypaper for removal of heavy metals and dyes. J Hazard Mater 413:125375

Khan S, Irshad S, Mehmood K, Hasnain Z, Nawaz M, Rais A, Gul S, Wahid MA, Hashem A, Abd_Allah EF, Ibrar D (2024) Biochar production and characteristics, its impacts on soil health, crop production, and yield enhancement: a review. Plants 13(2):166

Kiflu H, Kruse S, Loke MH, Wilkinson PB, Harro D (2016) Improving resistivity survey resolution at sites with limited spatial extent using buried electrode arrays. J Appl Geophys 135:338–355

Klemeš JJ, Varbanov PS, Ocłoń P, Chin HH (2019) Towards efficient and clean process integration: utilisation of renewable resources and energy-saving technologies. Energies 12(21):4092

Koohestani B, Darban AK, Mokhtari P, Darezereshki E, Yilmaz EROL (2021) Geopolymerization of soil by sodium silicate as an approach to control wind erosion. Int J Environ Sci Technol 18:1837–1848

Kosoe EA, Achana GTW, Ogwu MC (2023) Regulations and policies for herbal medicine and practitioners. In: Izah SC, Ogwu MC, Akram M (eds) Herbal medicine phytochemistry. Reference series in phytochemistry. Springer, Cham. https://doi.org/10.1007/978-3-031-21973-3_33-1

Kumar MA, Bae S, Han S, Chang Y, Lee W (2017) Reductive dechlorination of trichloroethylene by polyvinylpyrrolidone stabilized nanoscale zerovalent iron particles with Ni. J Hazard Mater 340:399–406

Kumar A, Yadav AN, Mondal R, Kour D, Subrahmanyam G, Shabnam AA, Khan SA, Yadav KK, Sharma GK, Cabral-Pinto M, Fagodiya RK (2021) Myco-remediation: a mechanistic understanding of contaminants alleviation from natural environment and future prospect. Chemosphere 284:131325

Kumar M, Rany M, Ansari JR (2024) Chemical methods for producing iron oxide magnetic nanoparticles: a Review. Iran J Chem Chem Eng (IJCCE) 43(5)

Lageman R, Godschalk MS (2007) Electro-bioreclamation: a combination of in situ remediation techniques proves successful at a site in Zeist, the Netherlands. Electrochim Acta 52(10):3449–3453

Lageman R, Clarke RL, Pool W (2005) Electro-reclamation, a versatile soil remediation solution. Eng Geol 77(3–4):191–201
Lagum AA (2021) Integrating electrochemical and biological phosphorus removal processes via electrokinetic-based technology. J Environ Chem Eng 9(6):106609
Lal R (2015) Restoring soil quality to mitigate soil degradation. Sustainability 7(5):5875–5895
Lal R (2019) Conceptual basis of managing soil carbon: inspired by nature and driven by science. J Soil Water Conserv 74(2):29A-34A
Lehmann J, Kuzyakov Y, Pan G, Ok YS (2015) Biochars and the plant-soil interface. Plant Soil 395:1–5
Leitenmaier B, Küpper H (2011) Cadmium uptake and sequestration kinetics in individual leaf cell protoplasts of the Cd/Zn hyperaccumulator *Thlaspi caerulescens*. Plant Cell Environ 34(2):208–219
Li P, Lin C, Cheng H, Duan X, Lei K (2015) Contamination and health risks of soil heavy metals around a lead/zinc smelter in southwestern China. Ecotoxicol Environ Saf 113:391–399
Li Y, Zhao HP, Zhu L (2021) Remediation of soil contaminated with organic compounds by nanoscale zero-valent iron: a review. Sci Total Environ 760:143413
Liu B (2018) Enhanced photocatalytic oxidation of polycyclic aromatic hydrocarbons in offshore produced water. Doctoral dissertation, Memorial University of Newfoundland
Liu L, Cai G, Zhang J, Liu X, Liu K (2020) Evaluation of engineering properties and environmental effect of recycled waste tire-sand/soil in geotechnical engineering: a compressive review. Renew Sustain Energy Rev 126:109831
Liu FS, Lockett BR, Sorichetti RJ, Watmough SA, Eimers MC (2022) Agricultural intensification leads to higher nitrate levels in Lake Ontario tributaries. Sci Total Environ. 830:154534
Liu Y, Qiao J, Sun Y (2023) Enhanced immobilization of lead, cadmium, and arsenic in smelter-contaminated soil by sulfidated zero-valent iron. J Hazard Mater 447:130783
Loj G (2009) The new thematic strategy for protecting soils in Europe. In: Desertification and risk analysis using high and medium resolution satellite data: training workshop on mapping desertification. Springer Netherlands, pp 83–88
Majhi PK, Raza B, Behera PP, Singh SK, Shiv A, Mogali SC, Bhoi TK, Patra B, Behera B (2022) Future-proofing plants against climate change: a path to ensure sustainable food systems. Biodiversity, functional ecosystems and sustainable food production. Springer International Publishing, Cham, pp 73–116
Makkar RS, Cameotra SS, Banat IM (2011) Advances in utilization of renewable substrates for biosurfactant production. AMB Express 1:1–19
Malaviya P, Singh A, Anderson TA (2020) Aquatic phytoremediation strategies for chromium removal. Rev Environ Sci Biotechnol 19:897–944
Manjula NG, Sarma G, Shilpa BM, Suresh Kumar K (2022) Environmental applications of green engineered copper nanoparticles. In: Phytonanotechnology. Springer Nature Singapore, Singapore, pp 255–276
Maqsood Q, Hussain N, Mumtaz M, Bilal M, Iqbal HM (2022) Novel strategies and advancement in reducing heavy metals from the contaminated environment. Arch Microbiol 204(8):478
Markiewicz A, Koda E, Kiraga M, Wrzesiński G, Kozanka K, Naliwajko M, Vaverková MD (2024) Polymeric products in erosion control applications: a review. Polymers 16(17):2490
Masi EB, Segoni S, Tofani V (2021) Root reinforcement in slope stability models: a review. Geosciences 11(5):212
Mench M, Lepp N, Bert V, Schwitzguébel JP, Gawronski SW, Schröder P, Vangronsveld J (2010) Successes and limitations of phytotechnologies at field scale: outcomes, assessment and outlook from COST Action 859. J Soils Sediments 10:1039–1070
Mercuri AM, Duggin JA, Grant CD (2005) The use of saline mine water and municipal wastes to establish plantations on rehabilitated open-cut coal mines, Upper Hunter Valley NSW, Australia. For Ecol Manage 204(2–3):195–207

Mikkelsen SH, Hansen E, Christensen TB, Baun A, Hansen SF, Binderup ML (2011) Survey on basic knowledge about exposure and potential environmental and health risks for selected nanomaterials. Danish Ministry of the Environment

Mleczek M, Gąsecka M, Waliszewska B, Magdziak Z, Szostek M, Rutkowski P, Kaniuczak J, Zborowska M, Budzyńska S, Mleczek P, Niedzielski P (2018) *Salix viminalis* L.—a highly effective plant in phytoextraction of elements. Chemosphere 212:67–78

Mou M (2024) Innovation potential of wood-based biochar for soil improvement in forestry. Master's thesis, Itä-Suomen yliopisto

Mu Q, Jiang G, Chen L, Zhou H, Fourches D, Tropsha A, Yan B (2014) Chemical basis of interactions between engineered nanoparticles and biological systems. Chem Rev 114(15):7740–7781

Muller A, Schader C, El-Hage Scialabba N, Brüggemann J, Isensee A, Erb KH, Smith P, Klocke P, Leiber F, Stolze M, Niggli U (2017) Strategies for feeding the world more sustainably with organic agriculture. Nat Commun 8(1):1–13

Muttaqin MZ, Alviya I, Lugina M, Hamdani FAU (2019) Developing community-based forest ecosystem service management to reduce emissions from deforestation and forest degradation. Forest Policy Econ 108:101938

Nagtode VS, Cardoza C, Yasin HKA, Mali SN, Tambe SM, Roy P, Singh K, Goel A, Amin PD, Thorat BR, Cruz JN (2023) Green surfactants (biosurfactants): a petroleum-free substitute for sustainability—comparison, applications, market, and future prospects. ACS Omega 8(13):11674–11699

Ngieng HY, Jusoh MNH, Ahmad N, Al Masud MA, Samaraweera H, Mohamed MM (2024) Enhanced soil decontamination via electrokinetic removal of organic pollutants. Ind Domestic Waste Manage 4(1):42–55

Nie M, Yin X, Ren C, Wang Y, Xu F, Shen Q (2010) Novel rhamnolipid biosurfactants produced by a polycyclic aromatic hydrocarbon-degrading bacterium *Pseudomonas aeruginosa* strain NY3. Biotechnol Adv 28(5):635–643

Ogwu MC, Izah SC (2023) One health implications of agrochemicals and their sustainable alternatives. In: Ramawat KG (ed) Sustainable development and biodiversity series, 1st edn, vol 34. Springer Nature, Singapore, Springer Verlag, Singapore, 826 pp. Hardcover ISBN 978-981-99-3438-6. https://doi.org/10.1007/978-981-99-3439-3

Ogwu MC, Izah SC, Dessureault-Rompré J, Gasparatos D (2024) Sustainable soil systems in global south. Springer Nature, Switzerland. https://doi.org/10.1007/978-981-97-5276-8

Oliveira PPA, Bernardi ACC, Pezzopane JRM, Bosi C, Júnior FP, Tadini AM, Martin-Neto L, Rodrigues PHM (2024) Potential of integrated trees-pasture-based systems for GHG emission mitigation and improving soil carbon dynamics in the Atlantic forest biome, Southeastern of Brazil. Eur J Agron 158:127219

Onyelowe KC, Van Bui D, Ubachukwu O, Ezugwu C, Salahudeen B, Van Nguyen M, Ikeagwuani C, Amhadi T, Sosa F, Wu W, Ta Duc T (2019) Recycling and reuse of solid wastes; a hub for ecofriendly, ecoefficient and sustainable soil, concrete, wastewater and pavement reengineering. Int J Low-Carbon Technol 14(3):440–451

Ortega U, Ametzaga-Arregi I, Sertutxa U, Peña L (2023) Identifying a green infrastructure to prioritise areas for restoration to enhance the landscape connectivity and the provision of ecosystem services. Landsc Ecol 38(12):3751–3765

Owuor GO (2021) Monitoring land degradation neutrality using geospatial techniques in support of sustainable land management: a case study of narok county (Doctoral dissertation, University of Nairobi)

Padhye LP, Srivastava P, Jasemizad T, Bolan S, Hou D, Shaheen SM, Rinklebe J, O'Connor D, Lamb D, Wang H, Siddique KH (2023) Contaminant containment for sustainable remediation of persistent contaminants in soil and groundwater. J Hazard Mater 455:131575

Pamukcu S, Kenneth Wittle J (1992) Electrokinetic removal of selected heavy metals from soil. Environ Prog 11(3):241–250

Panagos P, Jones A, Van Liedekerke M, Orgiazzi A, Lugato E, Montanarella L (2020) JRC support to the European Joint Programme for soil (EJP SOIL). Technical report by the Joint Research Centre (JRC), EUR 30450 EN, JRC122248

Pereira CS, Rodrigues AE (2014) Ethyl lactate main properties, production processes, and applications. In: Alternative solvents for natural products extraction. Springer Berlin Heidelberg, Berlin, Heidelberg, pp 107–125

Pérez AJ, Hassan EM, Pecio Ł, Omer EA, Kucinska M, Murias M, Stochmal A (2015) Triterpenoid saponins and C-glycosyl flavones from stem bark of *Erythrina abyssinica* Lam and their cytotoxic effects. Phytochem Lett 13:59–67

Phieler R, Merten D, Roth M, Büchel G, Kothe E (2015) Phytoremediation using microbially mediated metal accumulation in *Sorghum bicolor*. Environ Sci Pollut Res 22:19408–19416

Pulkka S, Martikainen M, Bhatnagar A, Sillanpää M (2014) Electrochemical methods for the removal of anionic contaminants from water—a review. Sep Purif Technol 132:252–271

Qiao J, Xiong Y (2021) Electrochemical oxidation technology: a review of its application in high-efficiency treatment of wastewater containing persistent organic pollutants. J Water Process Eng 44:102308

Reddy KR, Cameselle C (2009) Electrochemical remediation technologies for polluted soils, sediments and groundwater. Wiley

Reed KFM (2014) Perennial pasture grasses—an historical review of their introduction, use and development for southern Australia. Crop Pasture Sci 65(8):691–712

Rouhani S, Taghipour F (2023) Photocatalytic oxidation of volatile organic compounds (VOCs) in air using ultraviolet light-emitting diodes (UV-LEDs). Chem Eng Sci 272:118617

Ruiz-Huerta EA, Armienta-Hernández MA, Dubrovsky JG, Gómez-Bernal JM (2022) Bioaccumulation of heavy metals and As in maize (*Zea mays* L) grown close to mine tailings strongly impacts plant development. Ecotoxicology 31(3):447–467

Saeed Q, Xiukang W, Haider FU, Kučerik J, Mumtaz MZ, Holatko J, Naseem M, Kintl A, Ejaz M, Naveed M, Brtnicky M (2021) Rhizosphere bacteria in plant growth promotion, biocontrol, and bioremediation of contaminated sites: a comprehensive review of effects and mechanisms. Int J Mol Sci 22(19):10529

Saichek RE, Reddy KR (2005) Electrokinetically enhanced remediation of hydrophobic organic compounds in soils: a review. Crit Rev Environ Sci Technol 35(2):115–192

Salimi K, Ghazavi M (2021) Soil reinforcement and slope stabilisation using recycled waste plastic sheets. Geomech Geoeng 16(6):497–508

Sangwan S, Kaur H, Sharma P, Sindhu M, Wati L (2022) Routing microbial biosurfactants to agriculture for revitalization of soil and plant growth. In: New and future developments in microbial biotechnology and bioengineering. Elsevier, pp 313–338

Santos PDO, de Carvalho MM (2022) Exploring the challenges and benefits for scaling agile project management to large projects: a review. Requirements Eng 27(1):117–134

Sarfraz R, Hussain A, Sabir A, Ben Fekih I, Ditta A, Xing S (2019) Role of biochar and plant growth promoting rhizobacteria to enhance soil carbon sequestration—a review. Environ Monit Assess 191:1–13

Sarkar AK, Sadhukhan S (2022) Bioremediation of salt-affected soil through plant-based strategies. In: Advances in bioremediation and phytoremediation for sustainable soil management: principles, monitoring and remediation, pp 81–100

Sarker TR, Ethen DZ, Asha HH, Islam S, Ali MR (2024) Transformation of municipal solid waste to biofuel and bio-chemicals—a review. Int J Environ Sci Technol 1–22

Sarma H (2022) Understanding emerging contaminants in soil and water: current perspectives on integrated remediation approaches. In: Emerging contaminants in the environment, pp 1–38

Sen T, Reddy HJ (2011) Application of sisal, bamboo, coir and jute natural composites in structural upgradation. Int J Innov Manage Technol 2(3):186

Shackira AM, Puthur JT (2019) Phytostabilization of heavy metals: understanding of principles and practices. In: Plant-metal interactions, pp 263–282

Shah V, Daverey A (2020) Phytoremediation: a multidisciplinary approach to clean up heavy metal contaminated soil. Environ Technol Innov 18:100774

Sharma Y, Purohit DGM, Sharma S (2017) Improvement of soil properties by using jute fibre as soil stabilizer. Am J Eng Res 6(10):123–129

Shayegan Z, Haghighat F, Lee CS (2019) Photocatalytic oxidation of volatile organic compounds for indoor environment applications: three different scaled setups. Chem Eng J 357:533–546

Shukla BK, Saraswat R, Bharti G, Sharma PK, Singh D, Srivastava PK, Yadav P, Mishra S (2023) Soil stabilization through microbial bioenzymes: a comprehensive review of recent developments in eco-friendly approaches. In: International conference on environmental geotechnology, recycled waste materials and sustainable engineering. Springer Nature Singapore, Singapore, pp 75–95

Sil A, Pramanik K, Samantaray P, Firoz M, Yadav V (2020) Essential oils: a boon towards eco-friendly management of phytopathogenic fungi. J Entomol Zool Stud 8(4):1884–1891

Simiele M, De Zio E, Montagnoli A, Terzaghi M, Chiatante D, Scippa GS, Trupiano D (2022) Biochar and/or compost to enhance nursery-produced seedling performance: a potential tool for forest restoration programs. Forests 13(4):550

Sohi SP, Krull E, Lopez-Capel E, Bol R (2010) A review of biochar and its use and function in soil. Adv Agron 105:47–82

Spokas KA, Cantrell KB, Novak JM, Archer DW, Ippolito JA, Collins HP, Boateng AA, Lima IM, Lamb MC, McAloon AJ, Lentz RD (2012) Biochar: a synthesis of its agronomic impact beyond carbon sequestration. J Environ Qual 41(4):973–989

Sprocati R, Rolle M (2021) Electrokinetic remediation of contaminated sites: advances from process-based simulations and machine learning. In: 15th annual water research conference: Danish water forum. Danish Water Forum, pp 56–56

Stokes A, Douglas GB, Fourcaud T, Giadrossich F, Gillies C, Hubble T, Kim JH, Loades KW, Mao Z, McIvor IR, Mickovski SB (2014) Ecological mitigation of hillslope instability: ten key issues facing researchers and practitioners. Plant Soil 377:1–23

Štrbac S, Kašanin-Grubin M, Pezo L, Stojić N, Lončar B, Ćurčić L, Pucarević M (2023) Green infrastructure designed through nature-based solutions for sustainable urban development. Int J Environ Res Public Health 20(2):1102

Suman J, Uhlik O, Viktorova J, Macek T (2018) Phytoextraction of heavy metals: a promising tool for clean-up of polluted environment? Front Plant Sci 9:1476

Sun J, Chillrud SN, Mailloux BJ, Stute M, Singh R, Dong H, Lepre CJ, Bostick BC (2016) Enhanced and stabilized arsenic retention in microcosms through the microbial oxidation of ferrous iron by nitrate. Chemosphere 144:1106–1115

Sun Z, Zhao M, Chen L, Gong Z, Hu J, Ma D (2023) Electrokinetic remediation for the removal of heavy metals in soil: limitations, solutions and prospection. Sci Total Environ 165970

Suthersan SS, Horst J, Schnobrich M, Welty N, McDonough J (2016) Remediation engineering: design concepts. CRC Press

Tamene L, Adimassu Z, Ellison J, Yaekob T, Woldearegay K, Mekonnen K, Thorne P, Le QB (2017) Mapping soil erosion hotspots and assessing the potential impacts of land management practices in the highlands of Ethiopia. Geomorphology 292:153–163

Tan MY (2023) An overview of progresses and challenges of electrochemically integrated multi-electrode arrays for probing localised corrosion in complex environmental conditions. Corros Eng Sci Technol 58(8):819–837

Thirumalaivasan N, Gnanasekaran L, Kumar S, Durvasulu R, Sundaram T, Rajendran S, Nangan S, Kanagaraj K (2024) Utilization of fungal and bacterial bioremediation techniques for the treatment of toxic waste and biowaste. Front Mater 11:1416445

Tiso M, Schechter AN (2015) Nitrate reduction to nitrite, nitric oxide and ammonia by gut bacteria under physiological conditions. PLoS ONE 10(3):e0119712

Tripathi S, Singh VK, Srivastava P, Singh R, Devi RS, Kumar A, Bhadouria R (2020) Phytoremediation of organic pollutants: current status and future directions. In: Abatement of environmental pollutants, pp 81–105 ur Rehman Z, Junaid MF, Ijaz N, Khalid U, Ijaz Z (2023) Remediation methods of heavy metal contaminated soils from environmental and geotechnical standpoints. Sci Total Environ 867:161468

Vandana UK, Gulzar ABM, Singha LP, Bhattacharjee A, Mazumder PB, Pandey P (2020) Hyperaccumulation of arsenic by *Pteris vittata*, a potential strategy for phytoremediation of arsenic-contaminated soil. Environ Sustain 3:169–178

Visentin C, da Silva Trentin AW, Braun AB, Thomé A (2019) Application of life cycle assessment as a tool for evaluating the sustainability of contaminated sites remediation: a systematic and bibliographic analysis. Sci Total Environ 672:893–905

Vocciante M, Dovì VG, Ferro S (2021) Sustainability in electrokinetic remediation processes: a critical analysis. Sustainability 13(2):770

Wang L, Chen L, Tsang DC (2020) Green and sustainable stabilization/solidification. In: Soil and groundwater remediation technologies. CRC Press, pp 213–222

Wang Y, Li A, Cui C (2021) Remediation of heavy metal-contaminated soils by electrokinetic technology: mechanisms and applicability. Chemosphere 265:129071

Wang W, Nadagouda MN, Mukhopadhyay SM (2022) Advances in matrix-supported palladium nanocatalysts for water treatment. Nanomaterials 12(20):3593

Watson R, Wilson HN, Smart P, Macdonald EK (2018) Harnessing difference: a capability-based framework for stakeholder engagement in environmental innovation. J Prod Innov Manag 35(2):254–279

Wolka K, Mulder J, Biazin B (2018) Effects of soil and water conservation techniques on crop yield, runoff and soil loss in sub-Saharan Africa: a review. Agric Water Manag 207:67–79

Woolf D, Lehmann J, Cowie A, Cayuela ML, Whitman T, Sohi S (2018) Biochar for climate change mitigation. In: Soil and climate, pp 219–248

Wong JW, Ogbonnaya UO (2021) Biochar porosity: a nature-based dependent parameter to deliver microorganisms to soils for land restoration. Environ. Sci. Pollut. Res. 28:46894–46909

Wróbel M, Śliwakowski W, Kowalczyk P, Kramkowski K, Dobrzyński J (2023) Bioremediation of heavy metals by the genus *Bacillus*. Int J Environ Res Public Health 20(6):4964

Wuana RA, Okieimen FE (2011) Heavy metals in contaminated soils: a review of sources, chemistry, risks and best available strategies for remediation. Int Sch Res Notices 2011(1):402647

Yamato M, Okimori Y, Wibowo IF, Anshori S, Ogawa M (2006) Effects of the application of charred bark of *Acacia mangium* on the yield of maize, cowpea and peanut, and soil chemical properties in South Sumatra, Indonesia. Soil Sci Plant Nutr 52(4):489–495

Yao H, Wang Y, Liu J, Xu M, Ma P, Ji J, You Z (2022) Review on applications of lignin in pavement engineering: a recent survey. Front Mater 8:803524

Zhang X, Sathiyaseelan A, Zhang L, Lu Y, Jin T, Wang MH (2024) Zirconium and cerium dioxide fabricated activated carbon-based nanocomposites for enhanced adsorption and photocatalytic removal of methylene blue and tetracycline hydrochloride. Environ Res 261:119720

Zheng R, Chen Z, Cai C, Wang X, Huang Y, Xiao B, Sun G (2013) Effect of biochars from rice husk, bran, and straw on heavy metal uptake by pot-grown wheat seedling in a historically contaminated soil. BioResources 8(4):5965–5982

Zhou L, Li N, Owens G, Chen Z (2019) Simultaneous removal of mixed contaminants, copper and norfloxacin, from aqueous solution by ZIF-8. Chem Eng J 362:628–637

Zhu S, Ho SH, Jin C, Duan X, Wang S (2020) Nanostructured manganese oxides: natural/artificial formation and their induced catalysis for wastewater remediation. Environ Sci Nano 7(2):368–396

Chapter 10
Advancements in Waste-to-Energy Conversion and Recycling Technologies

Odangowei Inetiminebi Ogidi and Sylvester Chibueze Izah

Abstract The rapid growth of the global population and corporate interests has exacerbated issues associated with increasing energy consumption, insufficient waste management, and environmental degradation. The 5Rs (Reduce, Reuse, Repair, Recover, and Recycle) form the basis of contemporary methods that address these challenges effectively. There is a practical solution for materials that cannot be recycled: transformation of waste into energy, or WtE. In WtE, the system processes include biological conversion (anaerobic digestion, fermentation) and thermal conversion (incineration, gasification, and pyrolysis) and physicochemical conversion (transesterification). While the ecological processes involve in WtE includes microbial decomposition (breakdown of organic waste by microorganisms), thermophilic digestion (high-temperature decomposition of organic waste) and photodegradation (breakdown of organic waste by light). Therefore, this chapter systematically evaluates various waste-to-energy processes, emphasizing their capacity to reduce waste. The processes encompass pyrolysis, petrol generation, anaerobic digestion, and combustion. Sustainable materials management, improved landfill mining, and integrated waste management represent effective complementary strategies for environmentally responsible waste disposal. A thorough review of current research and technology developments has highlighted the following environmental impacts of waste-to-energy transformations: a shift towards more sustainable energy sources, less waste in landfills, and lower emissions of greenhouse gases. The results show how important it is to generate Waste-to-Energy (WtE) to enhance energy efficiency, reduce the environmental impact of manufacturing, and advance the circular economy. This chapter concludes by proposing directions for additional

O. I. Ogidi (✉)
Department of Biochemistry, Faculty of Basic Medical Sciences, Bayelsa Medical University, Yenagoa, Bayelsa State, Nigeria
e-mail: ogidiodangowei@gmail.com

S. C. Izah
Department of Community Medicine, Faculty of Clinical Sciences, Bayelsa Medical University, Yenagoa, Bayelsa State, Nigeria

Department of Microbiology, Faculty of Science, Bayelsa Medical University, Yenagoa, Bayelsa State, Nigeria

M. C. Ogwu and S. Chibueze Izah (eds.), *Evaluating Environmental Processes and Technologies*, Environmental Science and Engineering,
https://doi.org/10.1007/978-3-031-85327-2_10

research and media focus that may enhance the ecological benefits of waste-to-energy (WtE) projects.

Keywords Waste-to-energy · Technology · Recycling · Waste · Incineration · Microbial fuel cell · Pyrolysis

10.1 Introduction

The rapid industrialisation and advancement that have increased the global population have given rise to many issues, such as the increasing need for energy, massive amounts of garbage, and environmental damage. The 5R concept, as shown in the waste-to-energy connection, is crucial for addressing these multifaceted problems. A country's energy supply is vital to its economy. It is a fundamental component of every country's living standard. Fossil fuels have been a world priority for energy generation since the beginning of the industrial revolution (Sharma et al. 2020). Their restocking will take millions of years and is bad for the environment. These resources cannot be replenished. Emissions of greenhouse gases, primarily carbon dioxide (CO_2), linked to the combustion of fossil fuels have increased by more than 40% since the beginning of the industrial period. Changing trends in precipitation and global temperatures are two examples of how rising CO_2 emissions negatively impact human life. A person's ability to eat is fundamental to their well-being. There is a lot of food that goes to waste or becomes useless somewhere between collecting it and eating it. The bulk of food waste is organic material from various places, such as farms, restaurants, processing plants, and homes (Alao et al. 2022).

Conventional disposal options for food scraps include landfills or incinerators used with other types of municipal waste. In contrast, dioxins and other harmful gases are slowly released into the air due to this process. According to reports, the food business is responsible for 26% of the greenhouse gas emissions causing climate change (Girish et al. 2015). The processes that convert garbage into usable energy. Waste-to-energy (WtE) technologies are currently demonstrating significant potential, particularly in underdeveloped countries, where they have the potential to convert municipal solid waste (MSW) into viable electricity. They are significant because the current state of sustainable development necessitates the availability of modern energy services that are ecologically friendly, cheap, dependable, and efficient (Ogidi and Izah 2024; Ogidi 2024). The extensive solid waste management systems of industrialised nations include WtETs. As per Damodharan et al. (2019), the top nations in the world for investments in WtE plants are the US, China, Germany, France, Italy, Japan, the Netherlands, Singapore, Sweden, Turkey, and the UK (Damodharan et al. 2019). More than ever before, WtE technologies are helping with sustainable solid waste management and reducing environmental problems. Turning solid waste into usable energy and/or heat is the job of a waste-to-energy (WtE) facility. Aside from being an efficient and economical way to recover energy, it is also kind to the environment. On top of that, it's a clean, efficient energy alternative that might replace fossil fuels and cut down on carbon emissions. Dessie et al. (2020) have estimated that 4 and 2 metric tonnes of solid waste, respectively, could generate the same amount of energy as 1 and 1½ metric tonnes of oil and coal.

The objective of this chapter was to discuss the thermal treatment methods of gasification, pyrolysis, and incineration, as well as anaerobic digestion and other biological (non-thermal) treatment approaches. Also covered are technologies for using landfill gas, ways to produce hydrogen biologically, and microbial fuel cells. In addition, the present risks and difficulties linked to WtE technologies are also investigated.

10.2 The Five R Principles in Waste Management

The Five R principles in waste management (Fig. 10.1)—Refuse, Reduce, Reuse, Repurpose, and Recycle—represent a hierarchical approach aimed at minimizing waste and promoting sustainable waste management practices. These principles serve as the foundation for waste management systems globally, guiding efforts to reduce the environmental impact of waste. In the context of WTE conversion, the Five R principles provide a framework that emphasizes the importance of minimizing waste generation while harnessing residual waste as a resource. Studies show that refusing items like single-use plastics, non-recyclable materials, and hazardous products significantly reduces the waste stream, thus decreasing the demand for waste processing and energy conversion facilities (Papargyropoulou et al. 2014). This principle is particularly relevant in WTE contexts as it lessens the amount of waste requiring disposal or conversion, optimizing energy output from unavoidable waste (Arena 2012). In the waste-to-energy sphere, refuse efforts can reduce operational costs and emissions. For instance, in jurisdictions where refuse policies are enforced—such as bans on non-essential plastic products—the volume of waste suitable for energy recovery is often lower but of higher caloric value, enhancing WTE conversion efficiency (Kumar and Samadder 2017).

However, the success of refuse efforts relies heavily on regulatory frameworks, consumer awareness, and active participation in reducing consumable waste. "Reduce" emphasizes minimizing resource use and waste generation through efficiency and sustainable consumption practices (Okosodo and Ogidi 2023). Research highlights that reducing material use across industries and households directly impacts the volume of waste generated and the need for waste-to-energy solutions (Singh and Ordoñez 2016). By reducing waste volume at the source, the remaining waste can be allocated for recycling or energy recovery, enabling a more streamlined and effective WTE process. Reducing waste is critical in increasing WTE efficiency. When communities and industries adopt waste reduction practices, WTE facilities operate on a smaller, more manageable scale, which can reduce emissions, energy input, and financial costs (Tan et al. 2015). For example, studies on urban waste show that reduction in packaging waste has allowed WTE plants to focus on

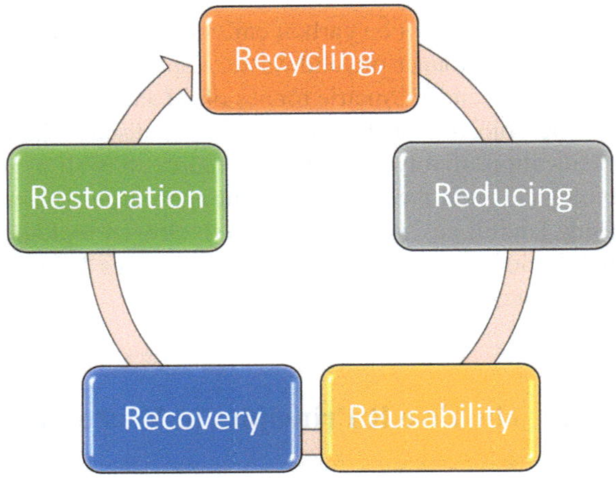

Fig. 10.1 The 5R principle of waste management

higher-energy-yield materials, thus maximizing energy conversion rates (Tsai and Kuo 2010).

"Reuse" involves repurposing products multiple times before they are disposed of. This principle is foundational in minimizing waste that ultimately enters landfills or incineration plants, thereby reducing the load on waste-to-energy systems (Velenturf and Purnell 2017). Reusing products lessens the immediate demand for waste disposal and energy recovery by delaying the disposal timeline. Research has demonstrated that reusing materials like glass, metal, and certain plastics decreases the volume of waste destined for energy conversion facilities, allowing these facilities to focus on non-reusable waste with a high caloric value. Studies indicate that reuse initiatives, particularly in construction and manufacturing, significantly reduce the burden on WTE systems. Consequently, WTE conversion can focus on truly non-reusable waste, achieving better energy recovery efficiency. Repurposing, or finding alternative uses for waste products, transforms discarded materials into valuable resources, circumventing the need for immediate disposal. This principle aligns with the concept of WTE by providing an alternative route to utilizing waste. For example, organic waste can be repurposed into biofertilizers and fuels, thereby reducing waste input into WTE systems (Iqbal et al. 2020; Nizzy and Ogwu 2024). Repurposing can improve waste-to-energy processes by reducing the volume and enhancing the quality of waste feedstock. Studies suggest that materials repurposed for construction, agricultural, and artisanal applications lower the volume of waste requiring incineration, which in turn improves the operational efficiency of WTE facilities (Di Maria and Micale 2015). By prioritizing repurposing, WTE plants can focus on waste streams that yield higher energy returns, such as biomass or synthetic organic compounds, improving the calorific efficiency of energy recovery. Recycling stands as a key process in waste management, converting waste materials into new products to reduce the consumption of fresh resources. In the context of waste-to-energy, recycling mitigates the volume of waste directed to energy recovery, ensuring that only non-recyclable waste enters WTE facilities. Recycling reduces emissions and enhances the efficiency of WTE systems by sorting out materials that do not contribute positively to energy conversion, such as metals and non-combustible materials (Pires et al. 2011).

Integrating recycling with WTE processes can lead to a more balanced and sustainable waste management system. In Sweden and Japan, for instance, stringent recycling programs have reduced waste volumes, allowing WTE plants to focus on truly non-recyclable waste, optimizing energy outputs and lowering emissions (Johansson 2013). Recycling also lowers the environmental footprint of WTE facilities by reducing the amount of non-combustible waste incinerated, which could otherwise contribute to higher emissions. The Five R principles collectively influence the effectiveness and efficiency of waste-to-energy conversion. WTE facilities that operate within a framework that emphasizes the Five R hierarchy typically experience optimized energy outputs, as they process higher-quality waste streams with enhanced caloric values (Parkes et al. 2015). Countries that implement comprehensive Five R strategies have achieved higher WTE efficiencies, as observed in Germany and the Netherlands, where extensive recycling and reduction initiatives complement WTE facilities. Incorporating the Five R principles into WTE systems also promotes a circular economy by ensuring that waste is minimized, reused, repurposed, and recycled to the greatest extent possible before entering WTE facilities. This approach not only preserves resources but also mitigates the environmental impact of waste incineration, which has traditionally been criticized for its emissions and land use (Tan et al. 2015).

10.3 Energy Supply and Waste Management

The management of waste and the provision of a consistent energy supply have been longstanding challenges for humanity. Despite significant advancements, these issues remain pertinent today. Transitioning to an atom economy, wherein each atom is utilised to its maximum potential, is essential for addressing these issues. Achieving this goal necessitates mastering fundamental processes and mechanisms involved in energy and waste production. There has been a significant rise in energy demand in the last century. This will persist as long as individuals strive to improve their living standards, resulting in increased consumption rates. Energy consumption has evolved through stages, beginning with wood (biomass), then coal oil, and culminating in natural gas. Wood requires cultivation, harvesting, drying, and processing before utilisation. Coal requires extraction, transportation, and further processing before utilisation. In contrast, oil requires refining and transportation before its utilization (Alwahaibi and Al Muslahi 2024).

Conversely, gaseous fuels are readily available. With the growth of human populations, it is anticipated that energy consumption patterns will persist in this manner, prioritising distributed energy sources that necessitate minimal processing time before utilisation. Efficient waste management has been essential for humanity throughout history and will remain necessary for the foreseeable future. Inadequate treatment or management of waste is the primary factor contributing to most social health issues. Initially, waste management's primary focus was sewage and human waste. It has evolved into municipal solid waste (MSW), which encompasses a significant proportion of non-hazardous solid waste and emphasizes gaseous emissions, specifically CO_2, CH_4, and other compounds such as NO_x and SO_x. CO_2 emissions pose a significant threat to global warming, leading to increasing concerns regarding the social implications of waste. Energy and waste exhibit a significant interdependence. Energy consumption produces gaseous and thermal byproducts, which represent a considerable issue. The Carnot cycle establishes the maximum efficiency of converting thermal energy into work, resulting in inevitable thermal waste (Muhammad et al. 2023).

Gaseous waste emissions, including particulate matter, CO_2, NO_x, and SO_x, are generated through combustion or gasification, constituting 80% of global energy generation. More and more landfill gas is being converted into fuels and energy (LFGTE), alongside the expansion of waste-to-energy (WTE) facilities, which have seen the addition of approximately 100 new plants in recent years. This trend indicates a convergence of energy production and waste management. Technological advancement aims to discover the most efficient methods for energy extraction from various sources while minimising waste production. Countries lacking long-term electricity generation plans urgently seek to utilise available local resources, especially biomass and waste materials. Comprehending the fundamental reaction sequences (mechanisms) involved is essential for optimising energy production from these 'alternative' sources under ideal conditions. Devices that maximise energy extraction and minimize waste can be developed more effectively. Effective management of waste, in relation to its generation rate, is primarily achieved through two methods: recycling and reuse. Municipal solid waste may be disposed of in landfills or subjected to thermal conversion in specialised chambers at elevated temperatures, significantly decreasing the volume of waste to a fraction of its original size while facilitating the recovery of energy and resources. Turbines transform heat from combustion or gasification into steam, subsequently producing power. A waste-to-energy (WTE) process is what this method is known as. Energy and metals can be recovered through the recycling of MSW, with most of the by-product ash serving beneficial roles in landfill maintenance in the United States, construction, and other infrastructure initiatives in the European Union and Japan. Repurposing residual ash presents a near-zero waste solution, as trash-to-energy processes decrease the volume of municipal solid waste by 90% (Naresh and Munaswamy 2019). The interconnection between energy supply and waste management is shown in Fig. 10.2.

10.4 Waste-to-Energy Technologies

The three different types of waste-to-energy technologies are depicted in Fig. 10.3. These technologies include physico-chemical, thermo-chemical, and biological treatment methods (World Energy Council 2016; Amer et al. 2014). Table 10.1 provides

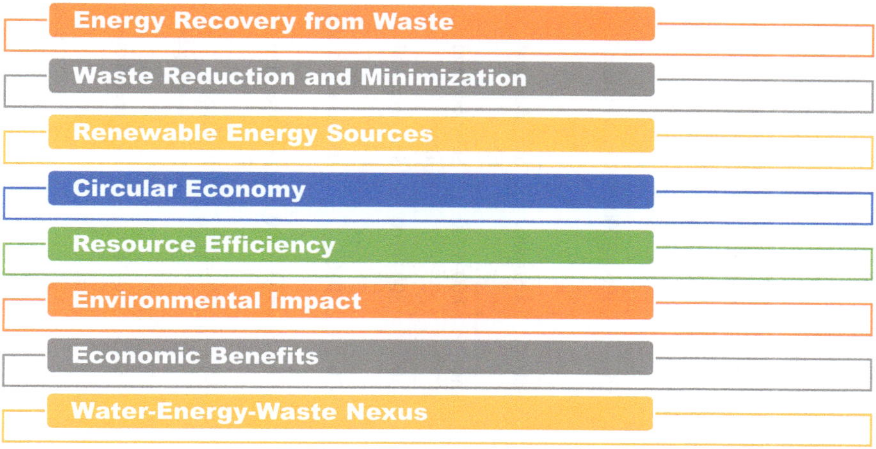

Fig. 10.2 Interconnections between energy supply and waste management

an overview of the many types of waste-to-energy technologies, as well as the operational parameters associated with each of these technologies. While Table 10.2 shows the advantages and disadvantages of these technologies.

10.4.1 Biological Treatment Processes

Food scraps, paper, green debris, and kitchen refuse comprise over 60% of the preponderance of biodegradable waste (Angelonidi and Smith 2015). This is why waste is a highly appealing biomass material for power facilities (Hamad et al. 2014). The quantity of biodegradable organic components in municipal solid waste (MSW) makes energy recovery from MSW an ecologically appropriate process, as noted by Kumar and Samadder (2017). Microbial decomposition of MSW organic components is necessary for biomass thermal power plants (BTPs). Biochemical treatment plants (BTPs) can be either aerobic or anaerobic. While aerobic BTPs require oxygen for proper operation, anaerobic BTPs can handle MSW without it. Anaerobic treatment, which uses less energy than aerobic treatment but still produces a flammable gas with carbon dioxide and methane components, is this study's focus. A valuable byproduct of this process is biogas, which may be burned to generate energy.

10.4.1.1 The Anaerobic Digestion Process

Researchers have shown that bacteria undergo anaerobic degradation of organic matter in a specific sequence known as the ADP (Veluchamy and Kalamdhad 2017; Braguglia et al. 2015; Zhang et al. 2016; Fedailaine et al. 2015). Biogas has multiple

Table 10.1 Different types of waste to energy technologies and their operational parameters

S. No.	Technology	Description	Operational parameters				
			Feedstock	Capacity range (tph)	Efficiency (%)	Temperature range (°C)	Emission control
1	Incineration	Combustion of waste	Municipal solid waste	100–1000	20–30	800–1300	Scrubber, and filters
2	Gasification	Partial oxidation of waste	MSW, biomass	10–500	30–40	800–1400	Scrubber, and filters
3	Plasma gasification	High temperature gasification	MSW, hazardous waste	10–100	50–70	1000–5000	None required
4	Pyrolysis	Thermal decomposition of waste	MSW, biomass	2–200	40–60	400–800	Scrubber, and filters
5	Fluidized bed combustion	Combustion in fluidized bed	MSW, biomass	10–500	30–40	800–1200	Scrubber, and filters
6	Circulating fluidized bed	Combustion in circulating bed	MSW, biomass	10–500	30–40	800–1200	Scrubber, and filters
7	Anaerobic digestion	Microbial decomposition of organic waste	Organic waste	1–100	50–70	30–60	Biogas scrubbing
8	Advanced thermal conversion	Various thermal processes	MSW, biomass	5–200	40–70	400–1200	Scrubber, and filters
9	Microbial fuel cells	Bio-electrochemical systems that harness energy from microorganisms breaking down organic waste	MSW, biomass, hydrocarbons	1–100	50–90	10–60	Reduced greenhouse gas
10	Biological hydrogen production	Microorganisms are used in breaking down organic waste	MSW, organic waste, biomass	100–1000	50–90	10–60	Zero greenhouse

(continued)

Table 10.1 (continued)

S. No.	Technology	Description	Operational parameters				
			Feedstock	Capacity range (tph)	Efficiency (%)	Temperature range (°C)	Emission control
11	Landfill gas utilization	Capturing and processing methane-rich gas emitted from landfills	MSW, biomass	1000–10,000	60–80	38–93	Greenhouse gas reduction
12	Trans-esterification	Physicochemical process converting triglycerides in biomass into biodiesel and glycerol	Biomass	10,000–100,000	70–98	150–200	Reduced greenhouse gas

Table 10.2 Advantages and disadvantages of waste-to-energy (WtE) technologies

Technology	Description	Advantages	Disadvantages
Incineration	Burning waste at high temperatures to produce steam/electricity	High energy conversion efficiency, volume reduction, and hazardous waste management	Air pollution, toxic ash, high capital and operating costs
Gasification	Heating waste in low-oxygen conditions to produce synthetic gas (syngas)	Higher energy efficiency than incineration, fewer emissions, and flexible fuel output	Complex process, high capital costs, and tar formation
Pyrolysis	Thermal decomposition of waste in absence of oxygen to produce oil, gas, and char	Low emissions, flexible fuel output, and moderate capital costs	Lower energy efficiency, char disposal issues
Anaerobic digestion	Microbial breakdown of organic waste to produce biogas (CH_4, CO_2)	Low-temperature process, high organic waste conversion, and nutrient-rich digestate	Slow process, sensitive to contaminants, and biogas cleaning required
Plasma arc gasification	High-temperature plasma arc breaks down waste into syngas	High energy efficiency, minimal emissions, and vitrified ash	Extremely high capital costs, complex operation
Circulating fluidized bed (CFB)	Combustion of waste in a fluidized bed of sand or ash	High combustion efficiency, flexible fuel input, and lower emissions	High capital costs, erosion issues
Refuse-derived fuel (RDF)	Processing waste into fuel pellets/briquettes for combustion	Low capital costs, flexible fuel output, and waste diversion	Quality variability, combustion emissions
Biological hydrogen production	Microbial conversion of organic waste into hydrogen gas	Renewable energy source, low emissions, and potential for carbon sequestration	Low efficiency, complex process control, and high capital costs
Microbial fuel cells (MFCs)	Microorganisms break down waste, generating electricity	Low-temperature process, potential for wastewater treatment, and direct electricity generation	Low power density, high capital costs, and membrane fouling

(continued)

Table 10.2 (continued)

Technology	Description	Advantages	Disadvantages
Landfill gas utilization	Capture and utilization of methane emissions from landfills	Reduces greenhouse gas emissions, energy generation, and waste management	Dependent on landfill conditions, gas cleaning required, and potential for explosions
Trans-esterification	Chemical conversion of waste fats/oils into biodiesel	Renewable energy source, low emissions, and compatible with existing infrastructure	High feedstock costs, catalyst requirements, and glycerol byproduct management

potential uses; it can be converted into sustainable natural gas and fuels for vehicles or burned to provide energy (Arneil et al. 2013). The ADP undergoes four distinct phases: hydrolysis, acidogenesis, acetogenesis, and methanogenesis, as indicated by numerous studies (Divya et al. 2015; Manjusha and Sajeena Beevi 2016; Rafieenia et al. 2017; Ren et al. 2018; Wang et al. 2018). Nielfa and Cano (2015) and Kamali et al. (2016) have proposed that the ADP process can be comprehended by examining the chemical reaction that takes place when anaerobic microorganisms, including *Listeria* and *Escherichia coli*, facilitate the biochemical conversion of organic compounds, such as glucose, into carbon dioxide and methane. Biogas generation systems are classified as "wet" or "dry" by the ADP according to the concentration of total solids (TS) employed (Angelonidi and Smith 2015). The two types of anaerobic digestion systems are different. Wet systems work at low solids levels, whereas dry systems work at high solids levels (Cesaro et al. 2010). Reducing landfill waste is one potential advantage of biogas generation. Biogas can produce heat and electricity, used in various business and residential areas and for transportation purposes (Moya et al. 2017).

There are a variety of temperature ranges in which anaerobic digestion processes (ADP) can take place; however, the optimal conditions for producing biogas and methane are found between 35 and 60 °C (Friedmann 2015; Arikan and Mulbry 2015). According to Vindis et al. (2009), thermophilic and mesophilic anaerobic digestions are two ways biogas can be produced. Different temperatures can be used to conduct anaerobic digestion processes, as Moya et al. (2017) observed. One illustration is that the most prevalent type of anaerobic digesters in use today, LS systems, are capable of processing solids only 15% of the time. These can be used to treat a variety of feedstocks, including industrial sludge, culinary waste slurries, FOG, and manure (Zhang et al. 2014). Yeqing et al. (2017) discovered that HS anaerobic digestion systems can process solid content inputs exceeding 15%. Their assistance is required to effectively manage waste materials with high solids concentrations, including lignocellulose, farm waste, organic domestic waste, green waste, energy crops, and agricultural byproducts (Iyiola et al. 2023; Paul and Dutta 2018; Chen et al. 2017). Compared to LS anaerobic digestion systems, these digesters frequently generate a more solid final product while utilising less heating energy. Nevertheless, their output necessitates a prolonged retention time and additional mixing and material-moving equipment. Anaerobic digestion systems employ numerous operational parameters. These include a carbon ratio of 20–30, a retention time of 10–40 days for thermophilic digesters and 14 days for mesophilic digesters.

10.4.1.2 Types of Anaerobic Digestion Systems

Batch, multi-stage, and single-stage procedures are all viable options for managing anaerobic municipal solid waste (Fig. 10.4). Total solids (TS) from the digester reactor slurries can be used for further classification. MS displays TS of about 15–20%, LS a little more than 10%, and HS 22–40%, as indicated earlier. In addition to single-stage and multi-stage systems, there are three subtypes: multi-stage low solids (MSLS), multi-stage high solids (MSHS), and single-stage high solids (SSLS). The process in a batch reactor begins with the addition of feedstock, continues with the reaction, and ends with the disposal of the used feedstock. Due to two critical distinctions, batch systems can produce biogas at rates that are 50–100 times higher than those of in-vessel landfills and can accomplish significantly higher reaction rates (Jouhara et al. 2017).

The leachate is initially reused on an ongoing basis. Subsequently, they operate at elevated temperatures in comparison to landfills. The three primary types of batch systems are sequential batch systems, up-flow anaerobic sludge blanket reactors, and single-stage batch systems. The high-water consumption of LS necessitates implementing more costly post-treatment techniques and a larger reactor, which are also its

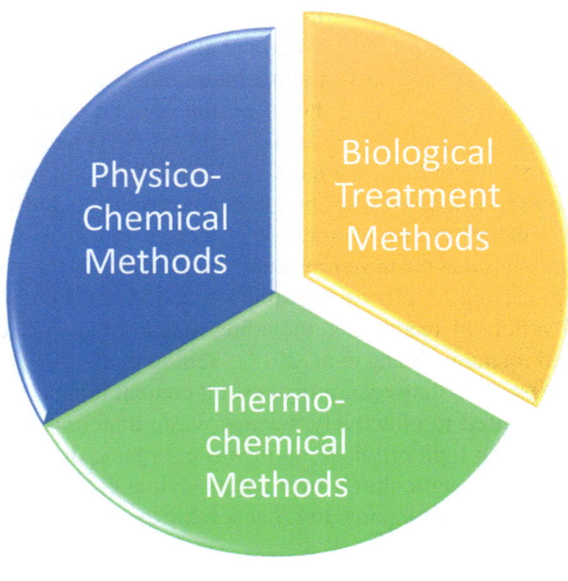

Fig. 10.3 Waste-to-energy technologies

primary drawbacks. Dewatering is a critical phase after digestion, which is one of the reasons post-treatment is expensive (Feodorov 2016). Although LS systems provide a greater reactor capacity per unit of production, the higher cost of HS systems compensates for this disadvantage. Regarding health, HS reactors are technically superior and have increased organic loading rates. In contrast to the preponderance of low-solids anaerobic digestion facilities that were constructed in the 1980s, a growing number of high-solids methods have been implemented in recent years (Cesaro et al. 2010). The results of this study suggest that high-solids plants are rapidly superseding other options as the gold standard in anaerobic digestion.

The ADP has multiple challenges. Digesters designed for this purpose take days to weeks to break down organic matter; nevertheless, substrate characteristics impact digestion efficiency (Xu et al. 2016). Anaerobic digestion is highly variable due to the difficulty of balancing acidogenic and methanogenic bacteria, which are different in physiology, dietary needs, growth rates, and sensitivity to environmental factors (Chen et al. 2008). System instability, harmful intermediate chemical synthesis, and reduced methane production could occur without transparent processes and optimisation controls (Xua et al. 2017). Perrot and Subiantoro (2018) state anaerobic systems are not very good at treating waste and do not produce much energy.

10.4.1.3 Biological Hydrogen Production Methodologies

A subtype of waste-to-energy technology called biological hydrogen production processes (BHPPs) converts renewable energy sources like biomass into hydrogen (Das et al. 2008). According to Beyene et al. (2018), hydrogen's abundance makes it the most ecologically friendly energy source. According to Beyene et al. (2018) and Das et al. (2008), it can be readily transformed into electricity and has a high energy density, but it does not exist in its basic form and does not become a greenhouse gas. Energy-intensive physical–chemical methods of waste treatment add to greenhouse gas emissions; nonetheless, these qualities outperform them (Beyene et al. 2018). With their ability to function at typical air pressure and temperature, BHPPs have earned a good reputation as a method for producing renewable energy.

The presence or absence of light requirements allows for the categorisation of BHPPs. Photosynthesis, MECs, and fermentation are the three processes that can be used to create BHPPs (Lee et al. 2010). However, some obstacles come with putting the BHPP into action. Even though water and sunlight can be used directly to produce hydrogen gas through photosynthesis, this process is vulnerable to oxygen (Foyer 2018; Izah et al. 2024a, b). Despite MECs' youth, they have great promise for significantly improving conversion efficiency. The output rate is rapid, but the conversion efficiency is subpar during fermentation (Łukajtis et al. 2018). Furthermore, numerous impediments could render this process unsuitable for industrial applications. These include the fact that photobioreactors and MECs are susceptible to fluctuations in natural light levels during photofermentation (Vinod et al. 2018), that they consume a significant amount of energy (Xia et al. 2016), and that they occupy a substantial amount of space.

10.4.1.4 Microbial Fuel cells (MFCs)

According to Kook et al. (2016), MFCs can produce bioelectricity from various waste materials. This technique uses electrochemically active microorganisms to generate energy from substrates (Ogidi et al. 2024d), regardless of whether oxygen is present (Fischer 2018). The chemical energy that is wasted can be converted into power through anodic oxidation pathways in MFCs (Hu et al. 2018). Modern fuel cells (MFCs) are an improvement over older technology since they can function at lower temperatures and without the use of chemical catalysts (Nastro 2014). Microbes act as electron acceptors in microbial fuel cells, which break down organic and inorganic compounds to generate electricity without burning (Jannelli et al. 2016). Kook et al. (2016) demonstrated the efficacy of the microbial fuel cell approach by generating energy and removing COD from MSW using dual-chamber microbial cells. To reduce expenses, MFCs can be used to produce energy from MSW (Hu et al. 2018). Multiple chemical substances are present in high concentrations in the anodic chamber of MSW, making it an ideal fuel for microbial fuel cells (MFCs). The process of organic compound breakdown involves the transfer of electrons from the cathode to the anode by bacteria. The cathode then gets electrons (Ogidi 2022).

Meanwhile, oxygen helps move protons from the anode to the cathode by acting as an electron acceptor. In an electric circuit, the movement of electrons produces electric energy. Felt carbon and distribution According to Chiu et al. (2016), microbial fuel cells with two chambers can make more power than those with one. Many people are looking forward to MFC technology since it is an affordable and eco-friendly low-energy treatment solution. Extensive use of this technology raises concerns due to its immaturity and the limited power output of MFCs (Kook et al. 2016). Energy recovery and the preservation of MFC stability are greatly affected by the substrate concentration in the anode chamber, which is challenging because of the different physicochemical characteristics of MSW. Researchers and scientists also face obstacles because of the wide variety of fuel types and operating circumstances that impact MFC efficiency (Nastro 2014).

10.4.1.5 Landfill Gas Utilization

Remediation is necessary for the safe disposal of municipal solid waste, which comprises garbage from both homes and businesses. Liquid waste management facilities are one option for dealing with MSW. In most cases, gas emissions during landfill restoration are caused by the breakdown of organic materials (Scarlat et al. 2015). Particulate matter, carbon dioxide, and natural gas's principal component, methane, along with trace amounts of non-methane organic chemicals, make up generated gas (De Gioannis et al. 2017; US EPA 2017). But, if not handled correctly, the greenhouse gas composition of these waste products could have worldwide environmental effects (Singh et al. 2016; Moya et al. 2017). Landfill temperature, humidity, oxygen levels, waste type, and age are just a few of the environmental elements that affect how quickly and how much gas is produced by landfills; other site-specific characteristics include these.

In their analysis, Njoku et al. (2018) found that many elements affect how well landfill gas (LFG) utilisation technology works. First, the availability of qualified operators; second, the sum of all investment costs (including capital, expenditure, and operational expenses); third, the suggested ideal system for LFG utilisation; fourth, the consistency of the MSW supply; and fifth, the existence of LFG end users. Governments, landfill owners, and enterprises should collect, repurpose, and put landfill gas (LFG) to use for energy instead of letting it escape into the environment. Industries like transportation, manufacturing, and real estate would reap these benefits. Cogeneration facilities can produce heat and electricity from the abundant methane in LFG (Pavlas et al. 2010; Yi et al. 2018). The Global Methane Initiative reports that LFG helps reduce emissions of methane (2011). According to Achinas et al. (2016), there are several ways in which biogas production could improve agriculture. These include lowering energy costs, increasing efficiency and productivity, and offering services that prioritise environmental compliance. In addition, local communities benefit from LFG efforts through greater job possibilities and higher incomes.

According to Xiaoli et al. (2016), China and the US, the two biggest economies in the world, produce a disproportionate amount of municipal solid waste (28% of total emissions) and greenhouse gas emissions (14% of total emissions), respectively. There were 2343 landfills in the US, but they only managed to handle 63% of the 250 million metric tonnes of MSW that was generated that year. Landfills only hold about 45% of China's municipal solid waste (MSW), even though the country produces and maintains more than 500 MSW landfills (Aniekan and Ikechukwu 2016; Brink et al. 2013). Approximately 2.6 million metric tonnes of methane may be captured each year by modern controlled landfills in the US, which can generate more than 50 MW of thermal and electrical energy (Moya et al. 2017). China has the potential to capture 15 million metric tonnes of landfill methane, according to Brink et al. (2013). With a daily processing capacity of 18,800–20,000 metric tonnes of municipal solid garbage and an electrical generation capacity of up to 50 MW, the Sudokwon landfill in South Korea has been officially certified as the giant landfill in the world.

While there may be some financial benefits to using landfill gas for energy, there are also significant costs to consider when managing leachate and disposing of condensate. Landfill gas escapes from most sites because there aren't enough segregation devices, and the MSF isn't covered quickly enough. Some nations struggle to upgrade their landfill gas because of harmful substances such as furan derivatives, dioxins, benzene, cadmium, and mercury (Kashyap et al. 2016). Landfill gas has a lower combustion efficiency than natural gas because of its higher CO_2 concentration (40–50% vol.), its effects of CO_2 on flame temperatures and combustion rates, and its narrowing of the flammable range (Anggono et al. 2016). The planet's warming is attributed to a significant volume of CO_2 by Qin et al. (2001). An additional drawback is that landfill decomposition processes are lengthy, ranging from years to decades.

Consequently, it is challenging to construct landfill projects in regions with restricted land availability (Xu et al. 2016).

10.4.2 Thermal Treatment Processes

According to Aleluia and Ferrão (2016), thermal treatment is a term that describes any waste management approach that utilises increased temperatures to control garbage. To maximise the energy that may be generated, municipal solid waste management uses cutting-edge thermal treatment technologies. Several different techniques have been utilised, including gasification, incineration, and pyrolysis (Astrup et al. 2015; Gurgul et al. 2017; Pape Thomsen et al. 2017; Samolada and Zabaniotou 2014; Dong et al. 2018). These techniques have been utilised in the past. Gasification and pyrolysis have been utilised for a considerable amount of time to produce various fuels, including charcoal, ash, and petrol (Bartocci et al. 2018). According to Heidenreich and Foscolo (2015), the production of producer gas, which functions as a combustible gas, is accomplished by the gasification of coal byproducts using steam and oxygen.

10.4.2.1 Incineration

According to Wang et al. (2017a, b), incineration is a method for treating garbage that uses burning to produce energy from organic waste components. Desiccating and incinerating the processed waste at temperatures approaching 1000 °C can be utilised to produce power. Bottom ash, vapour, particles, and heat are the byproducts of this process. Based on what Vaida and Lelea (2017) have found, this is the gold standard for MSW therapy. According to Wang et al. (2018), it is the most effective way to treat MSW since it lessens the negative impact on the environment and decreases the need for landfill space. The technique can significantly reduce the quantity of waste and weight by 90% and 70%, respectively, according to Ma et al. (2016). This results in less handling time and fewer toxic consequences.

10.4.2.2 Pyrolysis

In an oxygen-free or inert gas-surrounded environment, organic molecules undergo thermochemical decomposition at high temperatures (300–800 °C), a process known as pyrolysis (Moya et al. 2017; Czajczyńska et al. 2017). As a result of its controllable solid-to-liquid-to-gas byproduct ratios as a function of operating parameters like temperature and heating rate, pyrolysis is now being considered as an MSW treatment option (Czajczyńska et al. 2017). According to Grycová et al. (2016), it can recover more value molecules and transform low-density materials into high-density biofuels. According to Yang et al. (2018), charcoal has been made for centuries by pyrolysis. The main reactions dictate the pyrolysis process's product yields—the decomposition of solids—and the secondary reactions—the reactivity of primary volatiles (Zhan et al. 2018). The experimental apparatus and operating parameters determine the efficiency of secondary reactions. Heat processes rely on pyrolysis, which converts raw materials into reactive intermediates (Kwon et al. 2010). Wu et al. (2017) found that pyrolysis might avoid dioxin contamination, in contrast to the incineration method. The excellent energy conversion efficiency, optimal operating costs, and large processing capacity of pyrolysis make it an attractive and viable option for managing low-resource materials.

The pre-treatment of trash is essential for the efficient operation of pyrolysis technology. Heating organic materials to 300 °C in oxygen-poor chambers initiates their thermal degradation. Then, in a controlled setting, the temperature is raised to 800 °C to produce solid, liquid, and gaseous debris. The waste product of this process is called syngas. Hydrogen, methane, and carbon dioxide make up this substance, which could be utilised as a biofuel (Moya et al. 2017; Ogidi and Otele 2022). According to Vijay and Hawkes (2017), syngas can be converted into energy via boilers, engines, fuel cells, heat pumps, and turbines.

The oily liquids generated by municipal solid waste pyrolysis are not suitable for energy recovery due to their high oxygen content, corrosive nature, and instability when heated or subjected to time (Zhe et al. 2018). There is a direct correlation between the efficacy of the process and the quality of the pyrolysis outputs. Consequently, it is imperative to enhance the volatile products. Pyrolysis has certain advantages; however, it also generates carbon monoxide, which is highly thermally reactive. Consequently, this waste treatment method has no successful industrial-scale applications. Furthermore, the energy generated and maintained through pyrolysis, which is employed to decompose internal waste in MSW, incurs substantial expenses (Rafati et al. 2016).

10.4.2.3 Gasification

The process of gasification entails the partial oxidation of carbonaceous materials, whether in a solid or liquid state, at high temperatures (500–1400 °C) and variable atmospheric pressures (approximately 1–33 bars). Syngas is the outcome. This process leads to the thermal decomposition of waste materials, producing gas and various by-products such as ash, char, and creosote (Thakare and Nandi 2016).

One relatively new approach to recycling waste into usable energy is gasification. According to Moya et al. (2017) and Thakare and Nandi (2016), this approach dramatically decreases landfilling and emissions of greenhouse gases. Gasification is necessary to transform organic materials into gases that are rich in hydrogen and can be used for renewable energy, as opposed to pyrolysis, which relies only on temperature to produce gaseous species (Gurgul et al. 2018; Moya et al. 2017). It is also possible to use steam, oxygen, or carbon dioxide. Experts have begun using new reactor technologies with various plant designs to gasify municipal solid waste. Gasifiers can be grouped into different types, as shown in Fig. 10.5.

Fig. 10.4 Types of anaerobic digestion systems

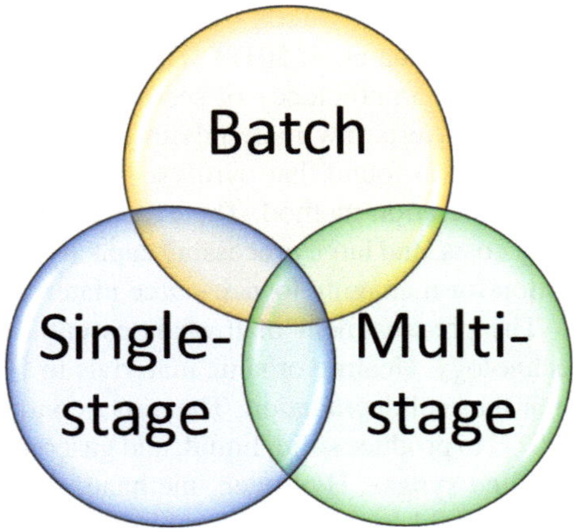

Gasification requires pre-processing of MSW to create fuel from garbage, which is the main drawback of the process. According to Panepinto et al. (2015), tar can be formed during this process. It can clog systems that convert gas into electricity, hamper systems that remove compressors, sulphur, ceramic filters, and erode heat exchangers, and harm engines and gas turbines. On top of that, gasification is an intricate and sensitive process (Soni and Naik 2016). Consequently, there is a significant amount of uncertainty regarding its economic feasibility, reliability, and performance due to the lack of sufficient reference data for full-scale application. In contrast to the utilisation of syngas recovery in industrial boilers or petrol engines, the direct combustion of municipal solid refuse appears to be a more efficient method of power generation.

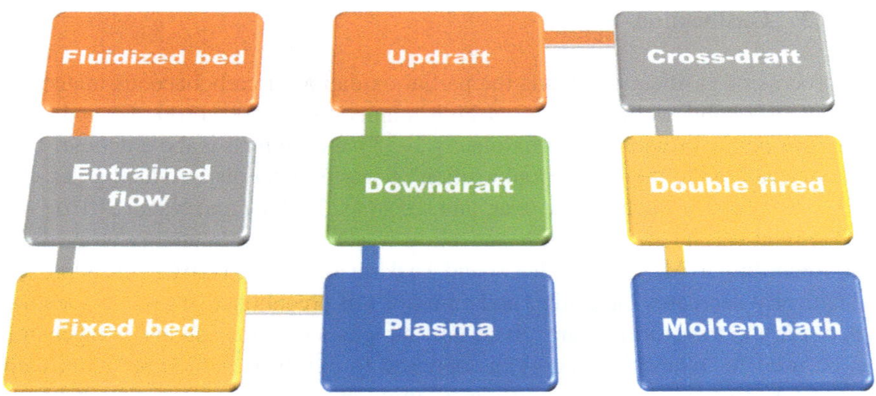

Fig. 10.5 Types of gasifiers

10.4.3 Physicochemical Technologies

10.4.3.1 Trans-esterification

Trans-esterification is one of the most widely recognised methods for converting refuse into energy. This process converts used cooking oils and animal lipids into liquid fuels through a physicochemical reaction (Shahnazari et al. 2020). The transesterification process, which is a component of the MSW stream, is used to produce biodiesel from the fatty components of culinary waste. The fat fractions are composed of inedible animal components, such as blood, fat, and internal organs, that are obtained from slaughterhouses, as well as residual cooking oil from families and restaurants, as per Shahzad et al. (2017) and Li et al. (2012). Before the trans-esterification process commences, the gathered used cooking oil (UCO) is initially screened and subsequently dehydrated to eliminate moisture and impurities. Conversely, rendering facilities utilise inedible animal components to generate lard and another valuable byproduct, meat and bone meal, from abattoir waste. The trans-esterification reaction between the processed fat and monoalkyl alcohols, such as ethyl or methyl alcohol, is facilitated by an acid or base catalyst, leading to biodiesel production. Potassium hydroxide (KOH) or carbonic acid (H_2CO_3) are frequently employed as catalysts in the trans-esterification reaction (Gardy et al. 2014; Nizami et al. 2016).

The catalyst selection is contingent upon the concentration of free fatty acids (FFAs) in the feedstock. In contrast to used culinary oil, which typically contains only 15% free fatty acids, rendered animal fat may contain up to 20%. Acid-catalyzed reactions require 1–8 h to complete, as per Sharma et al. (2008) and Thamsiriroj and Murphy (2010), whereas alkane-catalyzed reactions require only 30 min. Nevertheless, the alkane-catalyzed approach is outperformed by the acid-catalyzed trans-esterification process in terms of the ultimate product yield. However, the reactor is subjected to a corrosive environment for an extended period, which may result in increased maintenance costs. A two-stage transesterification process is the solution because it can process feedstocks with a high concentration of free fatty acids. The utilisation of an acid catalyst in an esterification reaction is one potential initial step in the conversion of free fatty acids into biodiesel. The alkane catalyst may convert the remaining triglycerides into biodiesel in a transesterification process (Gerpen 2005). Biodiesel and glycerol are the primary byproducts of the transesterification process. The quality and quantity of biodiesel produced are contingent upon the amount of unsaturated fatty acids in the feedstock. The purity and yield of biodiesel are both improved when the feedstock contains a higher concentration of unsaturated fatty acids. The quality and production of biodiesel are adversely affected when the feedstock contains significant levels of saturated fatty acids, as per Shahzad et al. (2017).

10.5 Recycling and Waste-to-Energy

Systematic separation of recyclable and compostable materials from municipal solid waste (MSW) is crucial for their effective management. To equitably allocate the time and resources necessary for distinguishing recyclables from non-recyclables, as well as the costs borne by municipalities for separate collection and processing systems, it is essential to segregate these materials at the source, specifically within homes, businesses, and institutions. However, if the source-separated materials are not viable, they will ultimately be disposed of in a landfill. Despite significant initiatives by the petrochemical sector and various governments to enhance plastic recycling, more than 80% of plastic waste generated in the US is disposed of in landfills, highlighting the absence of suitable markets for particular products (Themelis 2007). Six individuals believe that waste-to-energy (WTE) negatively impacts recycling behaviour. Towns that prioritise recycling identify the materials and financial constraints that render specific types of municipal solid waste less recyclable. Subsequently, they examine energy recovery via waste-to-energy (WTE) technologies as a viable method to minimise landfill usage. An excellent example of this is the Earth Engineering Centre's Sustainable Waste Management Ladder (Yue et al. 2020).

Many countries have successfully reduced or eliminated landfilling through the implementation of waste-to-energy and recycling programs (Table 10.3). Unchecked combustion was historically a significant contributor to harmful air pollution. Recent advancements in technology have enhanced waste-to-energy operations, enabling them to exceed the Maximum Achievable Control Technology (MACT) standard. Following one year of operation, a waste-to-energy (WTE) facility emits an equivalent quantity of dioxin as that produced during a 15-min fireworks display. There is a significant reduction in pollutants and fuel consumption linked to transporting municipal solid waste (MSW) to further away landfills when using waste-to-energy (WTE) processing instead of the more traditional ways (Ogidi et al. 2024a, b, c). Fewer trucks use the nation's highways (Zhang et al. 2024; Kosoe and Ogwu 2024). According to US Department of Transportation traffic statistics, seven people die and forty-plus sustain serious injuries annually because of the large number of trucks required to transport 2 million metric tonnes of excess municipal solid waste from New Jersey to landfills in Pennsylvania, Virginia, and Ohio. An article by Elizabeth Rosenthal in the New York Times indicates that waste-to-energy (WTE) method is extensively utilised in Europe, particularly in Denmark, while the United States lags significantly in its implementation. Analysis of the 'Room for Debate' discourse revealed that 57% of participants expressed support, while only 24% opposed. WTE may be gaining recognition as a significant solution to global waste management and energy supply challenges (The Public Acceptability of Incineration 2010).

Table 10.3 Case studies of some countries with effective waste-to-energy and recycling integration

Country	WtE technology	Recycling rate (%)	Energy generation (TWh/year)	Greenhouse gas reductionmillion tons CO_2-eq/year	Notable initiatives
Sweden	Gasification, grate incineration	52	17	2.5	District heating, carbon neutral by 2045
Denmark	Gasification, incineration	45	10	1.5	Biogas production, waste reduction target 20% by 2025
Japan	Advanced gasification, high-temperature incineration	60	12	1.8	Recycling-oriented society by 2030, waste-to-product technologies
Germany	Gasification, incineration	65	15	2.2	Closed-loop recycling, biogas production, green energy promotion
Singapore	Advanced incineration	60	2	0.5	Zero waste nation by 2035, waste-to-resource technologies
Austria	Gasification, incineration	55	8	1.2	High-quality recycling, biogas production, waste reduction target 50% by 2025
Netherlands	Gasification, incineration	50	6	0.9	Circular economy framework, biogas production, waste reduction target 65% by 2025
Switzerland	Gasification, incineration	55	4	0.6	High-quality recycling, biogas production, waste reduction target 50% by 2025
Norway	Gasification, incineration	45	3	0.5	Carbon capture and storage, waste-to-energy production
Finland	Gasification, incineration	40	2	0.4	Advanced biofuel production, waste reduction target 30% by 2025
Canada	Gasification, incineration	30	1	0.2	Waste-to-energy production, circular economy framework

(continued)

Table 10.3 (continued)

Country	WtE technology	Recycling rate (%)	Energy generation (TWh/year)	Greenhouse gas reductionmillion tons CO_2-eq/year	Notable initiatives
Australia	Advanced incineration	35	1.5	0.3	Waste-to-energy production, recycling target 80% by 2030
New Zealand	Gasification, incineration	25	0.5	0.1	Waste reduction target 50% by 2025, biogas production
United States	Gasification, incineration	35	20	4	Waste-to-energy production, recycling target 50% by 2030
South Korea	Advanced gasification, incineration	50	5	1	Waste-to-energy production, recycling target 70% by 2030
China	Gasification, incineration	20	10	2	Waste-to-energy production, recycling target 50% by 2025
India	Gasification, incineration	15	2	0.5	Waste-to-energy production, recycling target 50% by 2030
Brazil	Gasification, incineration	20	1	0.3	Waste-to-energy production, recycling target 50% by 2030
Russia	Gasification, incineration	10	1	0.2	Waste-to-energy production, recycling target 30% by 2025
South Africa	Gasification, incineration	15	0.5	0.8	Integrated waste management plans, waste-to-energy tariffs, recycling targets

10.6 Environmental Impacts of Waste-to-Energy Conversion

Presently, combating climate change and finding replacements for non-renewable energy sources are the top two global concerns. Energy production from non-renewable sources has damaged the environment in several ways. Furthermore, it harms public health due to the sulphur, nitrogen compounds, and heavy metals released into the air as toxic fumes. Moreover, fossil fuels (coal, oil, gas) and other non-renewable resources will run out eventually (Raj et al. 2020; Ahmed et al. 2024).

Waste-to-energy (WtE) technologies significantly reduce landfill waste, which helps preserve land and reduces landfill-related environmental problems, including methane emissions and leachate. Power generation from renewable sources is made possible by waste-to-energy (WtE) technology, which converts garbage into fuels, heat, and electricity, among other types of energy (Ramprasad et al. 2022). This lessens the need for fossil fuels and the emissions of greenhouse gases by increasing the generation of renewable energy (Hussain et al. 2020). By making it easier to recover metals and other materials for recycling, certain waste-to-energy (WtE) technologies enable resource restoration and recycling, which in turn supports the circular economy and conservation of resources (Goud et al. 2022; Hou et al. 2024; Yue et al. 2020; Zhang et al. 2024). By removing organic waste from landfills, Waste-to-Energy (WtE) technology can reduce emissions of methane, a powerful greenhouse gas. There is a further reduction in CO_2 emissions because the generated power is used instead of fuels that come from fossil fuels (Indira et al. 2022). To achieve sustainable growth, we must meet the energy demands of our growing population. This may include making better use of existing resources as well as developing and refining novel storage materials (Jaidass et al. 2018; Shukla et al. 2024; Lakshmi et al. 2021). Some of the many tests available for evaluating material properties are X-ray diffraction and X-ray spectroscopy (EDS) (Sambiani et al. 2023). Combinations and energy storage device development and evaluation often benefit from mathematical modelling techniques in artificial intelligence and soft computing (Spandana and Rao 2018). To evaluate the efficiency and performance of storage devices, it is necessary to know their electrical properties, and predictive approaches can help with this (Kumar et al. 2022). The environmental impacts of waste to energy are illustrated in Fig. 10.6.

Point-by-point environmental impacts of Waste-to-Energy (WtE) conversion are illustrated below:

Positive Environmental Impacts

1. **Greenhouse Gas (GHG) Reduction**: WtE conversion reduces GHG emissions by offsetting fossil fuels and utilizing methane from landfills.
2. **Waste Volume Reduction**: WtE technologies decrease waste volume, minimizing landfilling and associated environmental issues.
3. **Renewable Energy Generation**: WtE produces renewable energy, contributing to a cleaner energy mix.

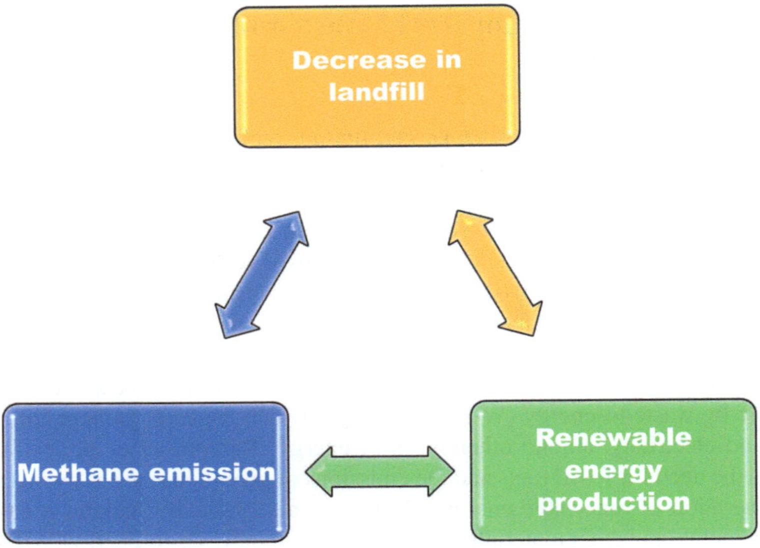

Fig. 10.6 Environmental impacts of waste-to-energy

4. **Conservation of Natural Resources**: WtE helps conserve natural resources by reducing the need for fossil fuels and landfills.
5. **Pollution Control**: Modern WtE facilities employ advanced pollution control systems, minimizing air and water pollution.

Negative Environmental Impacts

1. **Air Pollution**: WtE facilities emit:pollutants like particulate matter (PM), nitrogen oxides (NO_x), sulfur dioxide (SO_2), and volatile organic compounds (VOCs).
2. **Ash and Residue Generation**: WtE processes produce ash and residues, requiring proper disposal and management.
3. **Water Pollution**: WtE facilities may generate wastewater, potentially contaminating surface and groundwater.
4. **Soil Contamination**: Improper ash and residue disposal can lead to soil contamination.
5. **Noise Pollution**: WtE facilities can generate noise, affecting nearby communities.

10.7 Challenges and Future Research Direction

Both Mohan et al. (2020) and Dessie et al. (2020) concluded that insufficient garbage recovery and recycling methods are a barrier to the generation of electricity from waste. The waste-to-energy business has been confronted with many technological, political, and financial challenges, such as insufficient funding, improper data collecting and analysis, and legislation conflicting with one another. Khosravani and Reinicke (2020), Pandey et al. (2021) and Rajmohan et al. (2020) agree that more excellent research into these constraints would benefit various organisations. It is of the utmost need to place effective waste management, public education, and socio-demographic variables at the forefront of priorities. The findings of studies carried out by Ren et al. (2021), Varjani et al. (2020) and Hannan et al. (2020), pointed out that the level of public awareness has a considerable impact on the efficiency of a waste management system. Improved awareness is a crucial part of the solid waste management cycle (Kumar and Samadder 2017; Agoh et al. 2024). Through improved awareness, achieving the best possible results in municipal solid waste management (MSWM) is possible. According to Ravi et al. (2020) and Munir et al. (2018), the current global scenario is causing various environmental difficulties, including failing to separate garbage in the proper regions worldwide.

According to Ye et al. (2021) and Rodrigo-Comino et al. (2020), this state is the outcome of inadequate social coordination and planning. This is the conclusion reached by the researchers. According to Varjani et al. (2017) and Sulaiman et al. (2015), the most successful methods of information distribution were public knowledge surveys, seminars on solid waste management, and advertising the benefits of efficient solid waste disposal. These methods were shown to be the most effective. According to Agabo-García et al. (2019), the critical focus of future studies should be on the constraints and challenges associated with waste-to-energy technology implementation. Recent developments in technology have led to the development of microbial fuel cells, which can potentially reduce the adverse effects of human activities on the environment. Yin et al. (2020) researched this technique and used microorganisms to produce hydrogen gas and other high-energy leftovers from municipal solid trash. According to Kainthola et al. (2019), the generation of biogas is tricky because of the complicated composition of numerous feedstocks and the delayed biodegradability of these feedstocks. Many different pre-treatment procedures can be utilised to fix this issue successfully. In their conclusion to the pre-treatment study, Yaashikaa et al. (2019) emphasise the significance of pre-treatment strategies and the potential for future developments.

10.8 Conclusion

The waste-to-energy (WtE) conversion process is a critical strategy for addressing the significant issues associated with renewable energy generation and refuse disposal. Worldwide initiatives have demonstrated that waste-to-energy (WtE) is essential to improved energy efficiency and cleaner manufacturing. Waste-to-energy (WtE) generation should be the primary focus of future research initiatives to increase efficiency, decrease emissions, and broaden its applicability across various waste streams. Lawmakers, business leaders, and academics must collaborate continuously to guarantee that WtE conversion is a seamless component of comprehensive sustainable development strategies. The Waste-to-Energy (WtE) conversion strategy suggests that sustainable energy can be generated using thermodynamic combustion, anaerobic digestion, gasification, and pyrolysis, among other eco-friendly technologies. This method also endeavours to resolve the issue of waste disposal. Greenhouse gas emissions can be reduced, and renewable electricity can be generated through waste-to-energy by implementing sustainable materials management methods, such as refuse control and beneficial landfill mining. This approach reduces the amount of refuse that is disposed of in landfills. Waste-to-energy (WtE) technology remains indispensable for the advancement of energy performance and manufacturing efficiency. Improving waste-to-energy (WtE) systems ought to be the focus of future studies which is to enhance their efficiency, reduce emissions, and more effectively manage various categories of waste. Legislators, stakeholders, and academic researchers must work together to incorporate waste-to-energy into sustainable development programs. To realise sustainable development, the future of WtETs is contingent upon the political will and commitment of governments and stakeholders in the energy and environmental sectors, including commercial corporations, in a future where worries about pollution have less of an effect. The utilisation of environmental economic instruments for the purpose of garbage creation, collection, transportation, treatment, and disposal could be advantageous to the contemporary waste management system.

References

Achinas S, Jan G, Euverink W (2016) Theoretical analysis of biogas potential prediction from agricultural waste. Resour-Effic Technol 2:143–147

Agabo-García C, Pérez M, Rodríguez-Morgado B, Parrado J, Solera R (2019) Biomethane production improvement by enzymatic pre-treatments and enhancers of sewage sludge anaerobic digestion. Fuel 255:115713. https://doi.org/10.1016/j.fuel.2019.115713

Agoh EC, Ogwu MC, Chukwuemeka OS, Ekeledo PI (2024) Environmental and human health effects of cassava processing and processing waste. In: Ogwu MC, Izah SC, Cunha Alves AA, Babu SC (eds) Plant biology, sustainability and climate change, sustainable cassava. Academic Press, pp 203—219. https://doi.org/10.1016/B978-0-443-21747-0.00001-1

Ahmed AA, Qahtan TF, Owolabi TO, Agunloye AO, Rashid M, Ali MSM (2024) Waste to sustainable energy based on TENG technology: a comprehensive review. J Clean Prod 4:141354

Alao MA, Popoola OM, Ayodele TR (2022) Waste-to-energy nexus: an overview of technologies and implementation for sustainable development. Clean Energy Syst 3:100034

Aleluia J, Ferrão P (2016) Characterization of urban waste management practices in developing Asian countries: a new analytical framework based on waste characteristics and urban dimension. Waste Manage 58:415–429

Alwahaibi AM, Al Muslahi N (2024) Integrated waste management system for Oman: a sustainable approach leveraging IoT and web application simulation. Tuijin Jishu/J Propul Technol 45(2):20–38

Amer A, Agll A, Hamad YM, Hamad TA, Sheffield JW (2014) Case studies in thermal engineering study of energy recovery and power generation from alternative energy source. Case Stud Therm Eng 4:92–98

Angelonidi E, Smith SR (2015) A comparison of wet and dry anaerobic digestion processes for the treatment of municipal solid waste and food waste. Water Environ J 29(4):549–557

Anggono W, Wardana ING, Lawes M, Hughes KJ, Wahyudi S, Hamidi N, Hayakawa A (2016). The influence of CO_2 in biogas flammability limit and laminar burning velocity in spark ignited premix combustion at various pressures. In: Sustainable energy & advanced materials: proceeding of the international conference & exhibition on sustainable energy & advanced materials, vol 1717, pp 1–7. https://doi.org/10.1063/1.4943425

Aniekan I, Ikechukwu O (2016) Review Of Municipal solid waste management technologies and its practices in China and Germany. Int J Technol Enhanc Emerg Eng Res 4(5):1–7

Arena U (2012) Process and technological aspects of municipal solid waste gasification. A review. Waste Manage 32(4):625–639

Arikan O, Mulbry W (2015) Effect of temperature on methane production from field-scale anaerobic digesters treating dairy manure. Available from http://articles.extension.org/pages/72729/effect-of-temperature-on-methane-productionfrom-field-scale-anaerobic-digesters-treating-dairy-manu. Accessed 16 Mar 2042

Arneil R, Arancon D, Sze C, Lin K, Chan KM, Kwan TH, Luque R (2013) Advances on waste valorization: new horizons for a more sustainable society. Energy Sci Eng 1(2):53–71

Astrup TF, Tonini D, Turconi R, Boldrin A (2015) Life cycle assessment of thermal waste-to-energy technologies: review and recommendations. Waste Manage 37:104–115

Bartocci P, Zampilli M, Bidini G, Fantozzi F (2018) Hydrogen-rich gas production through steam gasification of charcoal pellet. Appl Therm Eng 132:817–823

Beyene HD, Werkneh AA, Ambaye TG (2018) Current updates on waste to energy (WtE) technologies: a review. Renew Energy Focus 24:1–11. https://doi.org/10.1016/j.ref.2017.11.001

Braguglia CM, Gianico A, Gallipoli A, Mininni G (2015) The impact of sludge pretreatments on mesophilic and thermophilic anaerobic digestion efficiency: role of the organic load. Chem Eng J 270:362–371

Brink S, Godfrey H, Kang M, Lyser S, Majkut J, Mignotte S, Singer L (2013) The Woodrow Wilson School's graduate policy workshop: methane mitigation opportunities in China. Available from https://www.princeton.edu/~mauzeral/teaching/WWS591e_Methane_Workshop_FinalReport2013.pdf. Accessed 19 Mar 2024

Cesaro A, Belgiorno V, Naddeo V (2010) Comparative technology assessment of anaerobic digestion of organic fraction of MSW. WIT Trans Ecol Environ 142:355–366

Chen Y, Cheng JJ, Creamer KS (2008) Inhibition of anaerobic digestion process: a review. Bioresour Technol 99(10):4044–4064. https://doi.org/10.1016/j.biortech.2007.01.057

Chen H, Liu J, Chang X, Chen D, Xue Y, Liu P, Han S (2017) A review on the pretreatment of lignocellulose for high-value chemicals. Fuel Process Technol 160:196–206

Chiu H, Pai T, Liu M, Chang C, Lo F, Chang T, Chu Y (2016) Electricity production from municipal solid waste using microbial fuel cells. Waste Manage Res 34(7):619–629

Czajczyńska D, Anguilano L, Ghazal H, Krzy R, Reynolds AJ, Jouhara H (2017) Potential of pyrolysis processes in the waste management sector. Therm Sci Eng Progr 3:171–197. https://doi.org/10.1016/j.tsep.2017.06.003

Damodharan D, Rajesh Kumar B, Gopal K, De Poures MV, Sethuramasamyraja B (2019) Utilization of waste plastic oil in diesel engines: a review. Rev Environ Sci Bio/technol 18(4):681–697
Das D, Khanna N, Veziroğlu TN (2008) Recent developments in biological sequence databases. Chem Ind Chem Eng Q 14(2):57–64
De Gioannis G, Muntoni A, Polettini A, Pomi R, Spiga D (2017) Energy recovery from one- and two-stage anaerobic digestion of food waste. Waste Manage 68:595–602
Dessie W, Luo X, Wang M, Feng L, Liao Y, Wang Z, Yong Z, Qin Z (2020) Current advances on waste biomass transformation into value-added products. Appl Microbiol Biotechnol 104(11):4757–4770. https://doi.org/10.1007/s00253-020-10567-2
Di Maria F, Micale C (2015) Recycling and recovery routes of plastic solid waste (PSW): a review. Waste Manage 30(8):1395–1402
Divya D, Gopinath LR, Christy PM (2015) A review on current aspects and diverse prospects for enhancing biogas production in sustainable means. Renew Sustain Energy Rev 42:690–699
Dong J, Tang Y, Nzihou A, Chi Y, Weiss-Hortala E, Ni M (2018) Life cycle assessment of pyrolysis, gasification and incineration waste-to-energy technologies: theoretical analysis and case study of commercial plants. Sci Total Environ 626:744–753
Fedailaine M, Moussi K, Khitous M, Abada S, Saber M, Tirichine N (2015) Modeling of the anaerobic digestion of organic waste for biogas production. Procedia Comput Sci 52:730–737
Feodorov V (2016) Modern technologies of treatment and stabilization for sewage sludge from water treatment plant. Agric Agric Sci Procedia 10:417–430
Fischer F (2018) Photoelectrode, photovoltaic and photosynthetic microbial fuel cells. Renew Sustain Energy Rev 90:16–27
Foyer CH (2018) Reactive oxygen species, oxidative signaling and the regulation of photosynthessis. Environ Exp Bot 154:134–142
Friedmann H (2015) The influence of temperature on the biogas production. Available from http://friedmann-biogaspraxis.com/en/the-influence-of-temperature-on-the-biogasproduction/. Accessed 16 Mar 2024
Gardy J, Hassanpour A, Lai X, Cunliffe A, Rehan M (2014) The influence of blending process on the quality of rapeseed oil-used cooking oil biodiesels. Int Sci J Environ Sci 3:1–8
Gerpen JV (2005) Biodiesel processing and production. Fuel Process Technol 86:1097–1107. https://doi.org/10.1016/j.fuproc.2004.11.005
Girish KM, Naik R, Prashantha SC, Nagabhushana H, Nagaswarupa HP, Raju KA et al (2015) Zn_2TiO_4: Eu^{3+} nanophosphor: self explosive route and its near UV excited photoluminescence properties for WLEDs. Spectrochim Acta Part A Mol Biomol Spectrosc 138:857–865
Global Methane Initiative (2011) Landfill methane: reducing emissions, advancing recovery and use opportunities. Available from https://www.globalmethane.org/documents/landfill_fs_eng.pdf. Accessed 17 Mar 2024
Goud JS, Srilatha P, Kumar RV, Kumar KT, Khan U, Raizah Z et al (2022) Role of ternary hybrid nanofluid in the thermal distribution of a dovetail fin with the internal generation of heat. Case Stud Therm Eng 35:102113
Grycová B, Koutník I, Pryszcz A (2016) Pyrolysis process for the treatment of food waste. Bioresour Technol 218:1203–1207
Gurgul A, Szczepaniak W, Zabłocka-Malicka M (2017) Incineration and pyrolysis vs. steam gasification of electronic waste. Sci Total Environ 624:1119–1124
Gurgul A, Szczepaniak W, Zabłocka-Malicka M (2018) Incineration and pyrolysis vs. steam gasification of electronic waste. Sci Total Environ 624:1119–1124
Hamad TA, Agll AA, Hamad YM, Sheffield JW (2014) Solid waste as renewable source of energy: current and future possibility in Libya. Case Stud Therm Eng 4:144–152
Hannan MA, Hossain Lipu MS, Akhtar M, Begum RA, Al Mamun MA, Hussain A, Mia MS, Basri H (2020) Solid waste collection optimization objectives, constraints, modeling approaches, and their challenges toward achieving sustainable development goals. J Clean Prod 277:123557. https://doi.org/10.1016/j.jclepro.2020.123557

Heidenreich S, Foscolo PU (2015) New concepts in biomass gasification. Prog Energy Combust Sci 46:72–95

Hou Y, Wang Q, Zhou K, Zhang L, Tan T (2024) Integrated machine learning methods with oversampling technique for regional suitability prediction of waste-to-energy incineration projects. Waste Manage 174:251–262

Hu J, Zhang Q, Lee DJ, Ngo HH (2018) Feasible use of microbial fuel cells for pollution treatment. Renew Energy 129:824–829

Hussain Z, Mishra J, Vanacore E (2020) Waste to energy and circular economy: the case of anaerobic digestion. J Enterp Inf Manag 33(4):817–838

Indira DNVSLS, Ganiya RK, Babu PA, Xavier AJ, Kavisankar L, Hemalatha S et al (2022) Improved artificial neural network with state order dataset estimation for brain cancer cell diagnosis. BioMed Res Int 2022

Iqbal M, Basit A, Shahid M, Niazi NK, Majeed A, Ullah S (2020) Sustainable bioeconomy: development and future of biomass conversion strategies. Renew Sustain Energy Rev 117:109474. https://doi.org/10.1016/j.rser.2019.109474

Iyiola AO, Kolawole AS, Ajayi FO, Ogidi OI, Ogwu MC (2023) Sustainable water use and management for agricultural transformation in Africa, chap 22. In: Water resources management for rural development: challenges and mitigation, vol 1. Elsevier, pp 287–300. https://doi.org/10.1016/B978-0-443-18778-0.00001-5

Izah SC, Ogwu MC, Hamidifar H (2024a) Biomonitoring of pollutants in the global south. Springer Nature, Switzerland. https://doi.org/10.1007/978-981-97-1658-6

Izah SC, Ogwu MC, Loukas A, Hamidifar H (2024b) Water related challenges and management strategies. Springer Nature, Singapore. https://doi.org/10.1007/978-981-97-4966-9

Jaidass N, Moorthi CK, Babu AM, Babu MR (2018) Luminescence properties of Dy^{3+} doped lithium zinc borosilicate glasses for photonic applications. Heliyon 4(3):1–19

Jannelli N, Nastro RA, Cigolotti V, Minutillo M, Falcucci G, Scarl AT (2016) Low pH, high salinity: too much for microbial fuel cells? Appl Energy. https://doi.org/10.1016/j.apenergy.2016.07.079

Johansson N (2013) Waste-to-energy in Sweden: adaptation to the circular economy. Environ Innov Soc Trans 9:1–16

Jouhara H, Czajczynska D, Ghazal H, Krzyzynska R, Anguilano L, Reynolds AJ, Spencer N (2017) Municipal waste management systems for domestic use. Energy 139:485–506

Kainthola J, Kalamdhad AS, Goud VV (2019) A review on enhanced biogas production from anaerobic digestion of lignocellulosic biomass by different enhancement techniques. Process Biochem 84:81–90. https://doi.org/10.1016/j.procbio.2019.05.023

Kamali M, Gameiro T, Elisabete M, Costa V, Capela I (2016) Anaerobic digestion of pulp and paper mill wastes. An overview of the developments and improvement opportunities. Chem Eng J 298:162–182

Kashyap RK, Chugh P, Nandakumar T (2016) Opportunities and challenges in capturing landfill gas from an active and un-scientifically managed land fill site—a case study. Procedia Environ Sci 35:348–367

Khosravani MR, Reinicke T (2020) On the environmental impacts of 3D printing technology. Appl Mater Today 20. https://doi.org/10.1016/j.apmt.2020.100689

Kook L, Rozsenberski T, Nemestothy N, Belafi-Bako K, Bakonyi P (2016) Bioelectrochemical treatment of municipal waste liquor in microbial fuel cells for energy valorization. J Clean Prod 112:4406–4412

Kosoe EA, Ogwu MC (2024) Sustainable cassava processing: processes, techniques, wastes, and waste streams. In: Ogwu MC, Izah SC, Cunha Alves AA, Babu SC (eds) Plant biology, sustainability and climate change, sustainable cassava. Academic Press, pp 257–272. https://doi.org/10.1016/B978-0-443-21747-0.00011-4

Kumar A, Samadder SR (2017) A review on technological options of waste to energy for effective management of municipal solid waste. Waste Manage 69:407–422. https://doi.org/10.1016/j.wasman.2017.08.046

Kumar KU, Babu P, Basavapoornima C, Praveena R, Rani DS, Jayasankar CK (2022) Spectroscopic properties of Nd^{3+}-doped boro-bismuth glasses for laser applications. Physica B 646:414327
Kwon E, Westby KJ, Castaldi MJ (2010) Transforming municipal solid waste (MSW) into fuel via the gasification/pyrolysis process. In: Proceedings of the 18th annual North American waste to energy conference, Orlando, Florida, USA, 11–13 May 2010, pp 1–7
Lakshmi L, Reddy MP, Santhaiah C, Reddy UJ (2021) Smart phishing detection in web pages using supervised deep learning classification and optimization technique ADAM. Wireless Pers Commun 118(4):3549–3564
Lee H, Vermaas WFJ, Rittmann BE (2010) Biological hydrogen production: prospects and challenges. Trends Biotechnol 28(5):262–271
Li J, Zhou H, Cao Y (2012) Transesterification of waste cooking oil to produce biodiesel using acid and alkaline catalyst. Adv Mater Res 518–523:3566–3572. https://doi.org/10.4028/www.scientific.net/AMR.518-523.3566
Łukajtis R, Hołowacz I, Kucharska K, Glinka M, Rybarczyk P, Przyjazny A, Kamiński M (2018) Hydrogen production from biomass using dark fermentation. Renew Sustain Energy Rev 91:665–694
Ma H, Cao Y, Lu X, Ding Z, Zhou W (2016) Review of typical municipal solid waste disposal status and energy technology. Energy Procedia 88:589–594
Manjusha C, Sajeena Beevi B (2016) Mathematical modeling and simulation of anaerobic digestion of solid waste. Procedia Technol 24:654–660
Mohan SV, Varjani S, Pant P, Sauer M, Chang JS (2020) Circular bioeconomy approaches for sustainability. Bioresour Technol 318:124084. https://doi.org/10.1016/j.biortech.2020.124084
Moya D, Aldás C, López G, Kaparaju P (2017) Municipal solid waste as a valuable renewable energy resource: a worldwide opportunity of energy recovery by using waste-to-energy technologies. Energy Procedia 134:286–295
Muhammad A, Naidu BM, Sundram VPK, Hussain MZSM, Chew LL, Amirrudin FA (2023) Sustainable waste management in Malaysia: leveraging supply chain solutions for a greener future. Inf Manage Bus Rev 15(3 (SI)):147–154
Munir MT, Mansouri SS, Udugama IA, Baroutian S, Gernaey KV, Young BR (2018) Resource recovery from organic solid waste using hydrothermal processing: opportunities and challenges. Renew Sustain. Energy Rev 96(2017):64–75. https://doi.org/10.1016/j.rser.2018.07.039
Naresh M, Munaswamy P (2019) Smart agriculture system using IoT technology. Int J Recent Technol Eng 7(5):98–102
Nastro RA (2014) Microbial fuel cells in waste treatment: recent advances. Int J Perform Eng 10(4):367–376
Nielfa A, Cano R (2015) Theoretical methane production generated by the co-digestion of organic fraction municipal solid waste and biological sludge. Biotechnol Rep 5:14–21
Nizami AS, Mohanakrishna G, Mishra U, Pant D (2016) Trends and sustainability criteria for liquid biofuels. In: Singh RS, Pandey A, Gnansounou E (eds) Biofuels production and future perspective. Taylor and Francis Group, London, pp 59–88. https://doi.org/10.1201/9781315370743-5
Nizzy AM, Ogwu MC (2024) Valorization of cassava processing by-products into biofuel for a sustainable environment. In: Ogwu MC, Izah SC, Cunha Alves AA, Babu SC (eds) Plant biology, sustainability and climate change, sustainable cassava. Academic Press, pp 291–309. https://doi.org/10.1016/B978-0-443-21747-0.00010-2
Njoku PO, Odiyo JO, Durowoju OS, Edokpayi JN (2018) A review of landfill gas generation and utilisation in Africa. Open Environ Sci 10:1–15
Ogidi OI (2022) Impact of pharmaceutical compounds on the microbial ecology of surface water resources, chap 12. In: Biodiversity in Africa: potentials, threats and conservation. Springer Nature Singapore Pte Ltd. https://doi.org/10.1007/978-981-19-3326-4_12
Ogidi OI (2024) Strategies of sustainable management of water resources in the global south. In: Water crisis and sustainable management in the global south. Springer Nature Singapore Pte Ltd.

Ogidi OI, Izah SC (2024) Water contamination by municipal solid wastes and sustainable management strategies. In: Water crisis and sustainable management in the global south. Springer Nature Singapore Pte Ltd.

Ogidi OI, Otele A (2022) An overview on the production of biofuel from water hyacinth (*Eichhornia crassipes*). Ekowe J Sci Technol (EJOST) 2:1–15

Ogidi OI, Izah SC, Tobia PS (2024a) Biosurfactant production from cassava processing wastes. In: Sustainable cassava: strategies from production through waste management. Elsevier

Ogidi OI, Ajoko IT, Tawariowei AM (2024b) Sustainable application of genetic ecotoxicological techniques in biomonitoring for environmental sustainability. In: Biomonitoring of pollutants in the global south. Springer Nature Singapore Pte Ltd.

Ogidi OI, Onwuagba CG, Richard-Nwachukwu N (2024c) Biomonitoring tools, techniques and approaches for environmental assessments. In: Biomonitoring of pollutants in the global south. Springer Nature Singapore Pte Ltd.

Ogidi OI, Angaye TCN, Carbom HE (2024d) Microorganisms in the monitoring of terrestrial and aquatic ecosystems: applications, challenges and future prospects. In: Biomonitoring of pollutants in the global south. Springer Nature Singapore Pte Ltd.

Okosodo EF, Ogidi OI (2023) Biodiversity conservation strategies and sustainability, chap 3. In: Sustainable utilization and conservation of Africa's biological resources and environment, sustainable development and biodiversity. Springer Nature Singapore Pte Ltd.

Pandey AK, Gaur VK, Udayan A, Varjani S, Kim SH, Wong JWC (2021) Biocatalytic remediation of industrial pollutants for environmental sustainability: research needs and opportunities. Chemoshpere 272:129936. https://doi.org/10.1016/j.chemosphere.2021.129936

Panepinto D, Tedesco V, Brizio E, Genon G (2015) Environmental performances and energy efficiency for MSW gasification treatment. Waste Biomass Valorization 6(1):123–135

Papargyropoulou E, Lozano R, Steinberger JK, Wright N, Ujang ZB (2014) The food waste hierarchy as a framework for the management of food surplus and food waste. J Clean Prod 76:106–115. https://doi.org/10.1016/j.jclepro.2014.04.020

Pape Thomsen T, Arossy ZS, Ahrenfeldt J, Henriksen UB, Frandsen FJ, Müller-St DS, Over E (2017) Changes imposed by pyrolysis, thermal gasification and incineration on composition and phosphorus fertilizer quality of municipal sewage sludge. J Environ Manage 198:308–318

Parkes O, Lettieri P, Bogle IDL (2015) Life cycle assessment of integrated waste management systems for alternative waste processing facilities. Waste Manage 36:350–368

Paul S, Dutta A (2018) Challenges and opportunities of lignocellulosic biomass for anaerobic digestion. Resour Conserv Recycl 130:164–174

Pavlas M, Tou M, Bébar L, Stehlík P (2010) Waste to energy—an evaluation of the environmental impact. Appl Therm Eng 30:2326–2332

Perrot JF, Subiantoro A (2018) Municipal waste management strategy review and waste-to-energy potentials in New Zealand. Sustainability 10(3114):1–12

Pires A, Martinho G, Chang NB (2011) Solid waste management in European countries: a review of systems analysis techniques. J Environ Manage 92(4):1033–1050

Qin W, Egolfopoulos FN, Tsotsis TT (2001) Fundamental and environmental aspects of landfill gas utilization for power generation. Chem Eng J 82(1):157–172

Rafati L, Rahmani Boldaji M, Khodadadi M, Atafar Z, Ehrampoush MH, Mojtaba Momtaz S, Mokhtari M (2016) Waste to energy: challenges and opportunities in Iran. J Environ Health Sustain Dev 1(3):175–184

Rafieenia R, Girotto F, Peng W, Cossu R, Pivato A, Raga R, Lavagnolo MC (2017) Effect of aerobic pre-treatment on hydrogen and methane production in a two-stage anaerobic digestion process using food waste with different compositions. Waste Manage 59:194–199

Raj TV, Hoskeri PA, Muralidhara HB, Manjunatha CR, Kumar KY, Raghu MS (2020) Facile synthesis of perovskite lanthanum aluminate and its green reduced graphene oxide composite for high performance supercapacitors. J Electroanal Chem 858:113830

Rajmohan KS, Chandrasekaran R, Varjani S (2020) A review on occurrence of pesticides in environment and current technologies for their remediation and management. Indian J Appl Microbiol 60(2):125–138. https://doi.org/10.1007/s12088-019-00841-x

Ramprasad P, Basavapoornima C, Depuru SR, Jayasankar CK (2022) Spectral investigations of Nd^{3+}: $Ba(PO_3)_2 + La_2O_3$ glasses for infrared laser gain media applications. Opt Mater 129:112482

Ravi A, Gurunathan B, Rajendiran N, Varjani S, Gnansounou E, Pandey A, You S, Raman JK, Ramanujam P (2020) Contemporary approaches towards augmentation of distinctive heterogeneous catalyst for sustainable biodiesel production. Environ Technol Innov 19:100906. https://doi.org/10.1016/j.eti.2020.100906

Ren Y, Yu M, Wu C, Wang Q, Gao M, Huang Q (2018) A comprehensive review on food waste anaerobic digestion: research updates and tendencies. Bioresour Technol 247:1069–1076

Ren X, Liu T, Awasthi MK, Varjani S, Pandey A, Zhang Z (2021) Municipal solid waste biorefineries: a case study in China. In: Bhaskar T, Varjani S, Pandey A, Rene ER (eds) Waste biorefinery: value addition through resources utilization. Elsevier, Amsterdam, Netherlands

Rodrigo-Comino J, López-Vicente M, Kumar V, Rodríguez-Seijo A, Valkó O, Rojas C, Pourghasemi HR, Salvati L, Bakr N, Vaudour E, Brevik EC, Radziemska M, Pulido M, Di Prima S, Dondini M, de Vries W, Santos ES, de Mendonça-Santos ML, Yu Y, Panagos P (2020) Soil science challenges in a new era: a transdisciplinary overview of relevant topics. Air Soil Water Res 13. https://doi.org/10.1177/1178622120977491

Sambiani K, Lare Y, Zanguina A, Narra S (2023) Location-allocation combining fuzzy analytical hierarchy process for waste to energy facilities siting in developing urban areas: the case study of Lomé, Togo. Heliyon 9(9):1–19

Samolada MC, Zabaniotou AA (2014) Comparative assessment of municipal sewage sludge incineration, gasification and pyrolysis for a sustainable sludge-to-energy management in Greece. Waste Manage 34:411–420

Scarlat N, Motola V, Dallemand JF, Monforti-ferrario F, Mofor L (2015) Evaluation of energy potential of municipal solid waste from African urban areas. Renew Sustain Energy Rev 50:1269–1286

Shahnazari A, Rafiee M, Rohani A, Nagar BB, Ebrahiminik MA, Aghkhani MH (2020) Identification of effective factors to select energy recovery technologies from municipal solid waste using multi-criteria decision making (MCDM): a review of thermochemical technologies. Sustain Energy Technol Assess 40. https://doi.org/10.1016/j.seta.2020.100737

Shahzad K, Nizami AS, Sagir M, Rehan M, Maier S, Khan MZ et al (2017) Biodiesel production potential from fat fraction of municipal waste in Makkah. PLoS ONE 12:e0171297. https://doi.org/10.1371/journal.pone.0171297

Sharma YC, Singh B, Upadhyay SN (2008) Advancements in development and characterization of biodiesel: a review. Fuel 87:2355–2373. https://doi.org/10.1016/j.fuel.2008.01.014

Sharma S, Soumen B, Nagaraj P, Shetti MK, Pavan W, Tejraj MA (2020) Waste-to-energy nexus: a sustainable development. Environ Pollut 267:115501

Shukla KA, Sofian ADABA, Singh A, Chen WH, Show PL, Chan YJ (2024) Food waste management and sustainable waste to energy: current efforts, anaerobic digestion, incinerator and hydrothermal carbonization with a focus in Malaysia. J Clean Prod 448:141457

Singh J, Ordoñez I (2016) Resource recovery from post-consumer waste: important lessons for the upcoming circular economy. J Clean Prod 134:342–353

Singh S, Anunay G, Rohit G, Shivangi G, Vipul V (2016) Greenhouse gas emissions from landfills: a case of NCT of Delhi, India. J Climatol Weather Forecast 4(1):1–6

Soni V, Naik V (2016) Gasification—a process for energy recovery and disposal of municipal solid waste. Am J Mod Energy 2(6):38–42

Spandana K, Rao VS (2018) Internet of Things (IoT) based smart water quality monitoring system. Int J Eng Technol (UAE) 7(3):259–262

Sulaiman N, Hannan MA, Mohamed A, Majlan EH, Wan Daud WR (2015) A review on energy management system for fuel cell hybrid electric vehicle: issues and challenges. Renew Sustain Energy Rev 52:802–814. https://doi.org/10.1016/j.rser.2015.07.132

Tan Y, Wang L, Xu X (2015) Urban waste management: waste-to-energy technologies as a sustainable alternative. Renew Sustain Energy Rev 50:1480–1493. https://doi.org/10.1016/j.rser.2015.05.054

Thakare S, Nandi S (2016) Study on potential of gasification technology for municipal solid waste (MSW) in Pune city. Energy Procedia 90:509–517

Thamsiriroj T, Murphy JD (2010) How much of the target for biofuels can be met by biodiesel generated from residues in Ireland? Fuel 89:3579–3589. https://doi.org/10.1016/j.fuel.2010.06.009

The Public Acceptability of Incineration (2010) Research undertaken by National Society for Clean Air and Environmental Protection. ISBN 0903 474 51 4. www.nsca.org.uk. Rosenthal E. Europe finds clean energy in trash, but US lags. New York Times, environment section, 12 Apr 2010

Themelis NJ (2007) Thermal treatment review: global growth of traditional and novel thermal treatment technologies. Waste Manage World 2007:37–45

Tsai WT, Kuo KK (2010) Waste-to-energy in Taiwan: a review. Renew Sustain Energy Rev 14(5):1417–1421

US EPA (2017) Basic information about landfill gas. Available from https://www.epa.gov/lmop/basic-information-about-landfill-gas. Accessed 17 Mar 2018

Vaida D, Lelea D (2017) Municipal solid waste incineration: recovery or disposal. case study of city Timisoara, Romania. Procedia Eng 181:378–384

Varjani S, Rakholiya P, Ng HY, Taherzadeh MJ, Ngo HH, Chang JS, Wong JWC, You S, Teixeira JA, Bui XT (2020) Bio-based rhamnolipids production and recovery from waste streams: status and perspectives. Bioresour Technol 319:124213. https://doi.org/10.1016/j.biortech.2020.124213

Velenturf AP, Purnell P (2017) Resource recovery from waste: restoring the balance between resource scarcity and waste overload. Sustainability 9(9):1603

Veluchamy C, Kalamdhad AS (2017) Influence of pretreatment techniques on anaerobic digestion of pulp and paper mill sludge: a review. Bioresour Technol 245:1206–1219

Vijay A, Hawkes A (2017) The techno-economics of small-scale residential heating in low carbon futures. Energies 10(1915):1–23

Vindis P, Mursec B, Janzekovic M, Cus F (2009) The impact of mesophilic and thermophilic anaerobic digestion on biogas production. J Achiev Mater Manuf Eng 36(2):192–198

Vinod SY, Vinoth R, Dharmesh Y (2018) Bio-hydrogen production from waste materials: a review. MATEC Web Conf 192:1–5. https://doi.org/10.1051/matecconf/201819202020

Wang X, Zhao J, Yang Q, Sun J, Peng C, Chen F, Zeng G (2017a) Evaluating the potential impact of hydrochar on the production of short-chain fatty acid from sludge anaerobic digestion. Bioresour Technol 246:234–241

Wang Y, Yan Y, Chen G, Zuo J, Yan B, Yin P (2017b) Effectiveness of waste-to-energy approaches in China: from the perspective of greenhouse gas emission reduction. J Clean Prod 163:99–105

Wang L, Øye B, Skreiberg Ø, Becidan M, Vatland PS, Fossum M, Stuen J (2018) Characterization of ash deposits from municipal solid waste (MSW) incineration plants. Energy Procedia 142:630–635

World Energy Council (2016) World energy resources: waste to energy. Available from https://www.worldenergy.org/wpcontent/uploads/2017/03/WEResources_Waste_to_Energy_2016.pdf. Accessed 17 Feb 2024

Wu D, Zhang A, Xiao L, Ba Y, Ren H, Liu L (2017) Pyrolysis characteristics of municipal solid waste in oxygen-free circumstance. Energy Procedia 105:1255–1262

Xia A, Jacob A, Herrmann C, Murphy JD (2016) Fermentative bio-hydrogen production from galactose. Energy 96:346–354. https://doi.org/10.1016/j.energy.2015.12.087

Xiaoli C, Tonjes DJ, Mahajan D (2016) Methane emissions as energy reservoir: context, scope, causes and mitigation strategies. Prog Energy Combust Sci 56:33–70. https://doi.org/10.1016/j.pecs.2016.05.001

Xu S, He H, Luo L (2016) Recycling of solid waste for biofuels and bio-chemicals. https://doi.org/10.1007/978-981-10-0150-5. Accessed 13 Jan 2024

Xua F, Lia Y, Gea X, Yang L, Lia Y (2017) Anaerobic digestion of food waste—challenges and opportunities. Bioresour Technol 247:1047–1058

Yaashikaa PR, Senthil Kumar P, Varjani SJ, Saravanan A (2019) Advances in production and application of biochar from lignocellulosic feedstocks for remediation of environmental pollutants. Bioresour Technol 292:122030. https://doi.org/10.1016/j.biortech.2019.122030

Yang Y, Heaven S, Venetsaneas N, Banks CJ, Bridgwater AV (2018) Slow pyrolysis of organic fraction of municipal solid waste (OFMSW): characterisation of products and screening of the aqueous liquid product for anaerobic digestion. Appl Energy 213:158–168

Ye Y, Ngo HH, Guo W, Chang SW, Nguyen DD, Varjani A, Ding A, Bui XT, Nguyen DP (2021) Bio-membrane based integrated systems for nitrogen recovery in wastewater treatment: current applications and future perspectives. Chemosphere 265:129076

Yi S, Jang YC, An AK (2018) Potential for energy recovery and greenhouse gas reduction through waste-to-energy technologies. J Clean Prod 176:503–511

Yin Z, Zhu L, Li S, Hu T, Chu R, Mo F, Hu D, Liu C, Li B (2020) A comprehensive review on cultivation and harvesting of microalgae for biodiesel production: environmental pollution control and future directions. Bioresour Technol 301(2019):122804. https://doi.org/10.1016/j.biortech.2020.122804

Yue L, Jayapal M, Cheng X, Zhang T, Chen J, Ma X, Zhang W (2020) Highly dispersed ultra-small nano Sn-SnSb nanoparticles anchored on N-doped graphene sheets as high performance anode for sodium ion batteries. Appl Surf Sci 512:145686

Zhan H, Zhuang X, Song Y, Yin X, Cao J, Shen Z, Wu C (2018) Step pyrolysis of N-rich industrial biowastes: regulatory mechanism of NO_x precursor formation via exploring decisive reaction pathways. Chem Eng J 344:320–331

Zhang C, Su H, Baeyens J, Tan T (2014) Reviewing the anaerobic digestion of food waste for biogas production. Renew Sustain Energy Rev 38:383–392

Zhang Q, Hu J, Lee D (2016) Biogas from anaerobic digestion processes: research updates. Renew Energy 98:108–119

Zhang Z, Chen Z, Zhang J, Liu Y, Chen L, Yang M, Osman AI, Farghali M, Liu E, Hassan D, Ihara I, Lu K, Rooney DW, Yap PS (2024) Municipal solid waste management challenges in developing regions: a comprehensive review and future perspectives for Asia and Africa. Sci Total Environ 930:172794. https://doi.org/10.1016/j.scitotenv.2024.172794

Zhe X, Syed-Hassan SSA, Xu J, Song H, Su S, Zhang S, Xiang J (2018) Evolution of coke structures during the pyrolysis of bio-oil at various temperatures and heating rates. J Anal Appl Pyrol 134:336–342

Chapter 11
Carbon Capture, Storage, and Utilization Technologies

Amit Kumar Chaturwedi, Nand Kumar Kashyap, Archana Shukla, Shailendra Yadav, Gourisankar Roymahapatra, and Milan Hait

Abstract This chapter explores the transformative potential of Carbon Capture, Storage, and Utilization (CCSU) technologies, a pioneering field that addresses climate change by turning carbon waste into valuable resources. As industrial and energy sectors remain significant carbon dioxide (CO_2) emitters, CCSU offers a strategic pathway to capture these emissions directly at the source, minimizing their environmental footprint. The chapter delves into advanced capture methods such as chemical absorption, membrane separation, and cryogenic distillation, which isolate CO_2 with high efficiency. Once captured, the CO_2 undergoes secure geological storage, utilizing natural rock formations to trap emissions underground and prevent atmospheric release. Beyond storage, the chapter highlights utilization technologies that transform CO_2 into products like alternative fuels, building materials, and chemicals, fostering a circular economy where waste becomes an asset. The economic and environmental impacts of CCSU, coupled with its potential to reshape energy and industrial systems, underscore its role in sustainable development and decarbonization efforts. Through detailed insights, this chapter provides a comprehensive view of CCSU's current advancements, challenges, and outlook. It is an essential technology in the global transition towards a low-carbon economy.

The original version of the chapter has been revised. A correction to this chapter can be found at https://doi.org/10.1007/978-3-031-85327-2_17

A. K. Chaturwedi · N. K. Kashyap · M. Hait (✉)
Department of Chemistry, Dr. C. V. Raman University, Kota, Bilaspur, C.G., India
e-mail: haitmilan@gmail.com

A. Shukla
Department of Chemistry, LCIT College of Commerce and Science, Bodri, Bilaspur, C.G., India

S. Yadav
Department of Chemistry, AKS University, Satna, M.P., India

G. Roymahapatra
Department of Applied Science, Haldia Institute of Technology, Haldia, W.B., India

M. C. Ogwu and S. Chibueze Izah (eds.), *Evaluating Environmental Processes and Technologies*, Environmental Science and Engineering,
https://doi.org/10.1007/978-3-031-85327-2_11

Keywords Carbon circularity · Emission transformation · CO_2 valorization · Sustainable sequestration · Decarbonization pathways · Eco-innovative utilization

11.1 Introduction

The urgent need to reduce carbon emissions has become evident as the world grapples with climate change (IPCC 2022). Carbon dioxide (CO_2) levels in the atmosphere have increased dramatically throughout the last hundred years, primarily because of transportation, energy generation, and industrial operations. Since CO_2 is a significant greenhouse gas, researchers in the scientific community, environmental movement, and engineering community have been looking for new ways to cut down on, recycle, and reuse carbon emissions. Carbon Capture, Storage, and Utilization (CCSU) technologies are one example of a solution that has recently shown enormous promise. These innovations have the potential to tackle emissions directly, turning waste (Gupta et al. 2023) into resources and opening new avenues for a more sustainable future (Tan et al. 2016). This section dives into these complex technologies, examining how they change global environmental policies, economies, and industries. The core concept of CCSU is a bold but basic one. These technologies can capture, store, and even convert CO_2 emissions into valuable goods rather than allowing them to enter our environment. For big industrial emitters, carbon capture is like a filter that pulls CO_2 out of the air before it can do any damage. Industries like cement mills and power plants may use CCSU to lessen their environmental impact (Hait et al. 2019a) without having to make significant changes to how they make things. With emissions laws tightening worldwide, this skill is vital for maintaining the economic sustainability of carbon-intensive industries and combating climate change. Technologies that store and use CO_2 provide long-term environmental advantages beyond just capturing the gas (Hait et al. 2018a). Carbon emissions may be saved by converting it into alternative fuels or construction supplies or safely storing it in natural formations. What is the rationale for concentrating on carbon capture? While important, traditional approaches to lowering emissions—like increasing energy efficiency or moving to renewable energy sources—are inadequate when used alone. Green energy is making significant progress, but it's challenging to decarbonize businesses like steel, cement, and chemical production because they use high-temperature processes and complicated chemical reactions that make CO_2 (Belhaj and Bera 2017). This is the so-called "hard-to-abate" area where CCSU can potentially have a huge impact. CCSU bridges climate objectives with industrial productivity by collecting emissions from these businesses and redirecting them onto safer or more profitable paths. CCSU may also reduce emissions when paired with biofuels in a BECCS system. CCSU may actively decrease the amount of carbon dioxide in the air by removing CO_2 from bioenergy sources. Meeting the ambitious objectives set out in global climate accords, such as the Paris Agreement, which aims to limit global temperature increases to well below 2 °C, is increasingly dependent on these techniques, which are feasible and considered crucial (Bickle 2009).

Capturing CO_2, the first stage of the CCSU process, is an area of outstanding innovation in and of itself. Chemical absorption, in which solvents bond preferentially to CO_2, was once the method of choice for carbon capture. This approach is still in use but is now more efficient and cheaper because of developments in capture materials, including solid sorbents, membrane technology, and cryogenic procedures. Every approach has its own set of benefits. Chemical absorption, for example, works very well for big factories. On the other hand, membrane separation and cold methods could be helpful in locations that require incredibly pure CO_2 (Roh et al. 2018). Thanks to the variety of collection techniques, a wide range of sectors, regions, and operating sizes may benefit from CCSU technology. Carbon capture is a dynamic field that is always changing because of new materials being investigated for potential improvements in capture efficiency and reductions in operating costs. For instance, scientists have a great deal of interest in porous materials such as metal–organic frameworks (MOFs) because of their exceptional CO_2 collection capability. In addition, technologies that remove (Hait et al. 2019b) CO_2 straight into the air, known as direct air capture (DAC) (Kramer and Haigh 2009), are a huge step forward in addressing scattered emissions, which is especially important when dealing with more minor, more dispersed sources of pollution (Chaturwedi and Hait 2023).

There are two main options for what happens to collected CO_2 after transportation: storage or use. One of the most talked-about options is geological storage (Bickle 2009), which involves inserting CO_2 hooked on deep constructions similar salt aquifers or abandoned oil and gas sites. This method, like turning back the clock on natural carbon sequestration, offers a stable, long-term answer to emissions by keeping CO_2 isolated from the air for lengthy periods (Solomon et al. 2009). The rocks' geological stability has been thoroughly investigated, guaranteeing that CO_2 would stay contained with little chance of leaking once captured. In terms of use, CO_2 is finding more and more new homes in a wide variety of products, including building supplies, alternative fuels, and chemicals. Industries have the potential to foster a circular economy by utilizing trash fumes to produce products such as fizzy concrete, synthetic fuels (Thakur et al. 2018), and even plastics. These CO_2-based goods provide a long-term answer to garbage disposal problems while simultaneously decreasing use of fossil fuels. Sustainability and economically feasible emissions control are key to CCSU's purpose; hence, the circular economy transition is crucial. The environmental benefits of CCSU are apparent, but the economic ramifications are just as substantial. CCSU technologies may stimulate investment in environmentally friendly infrastructure and technology, creating new CO-based goods and carbon credits markets. In the face of mounting decarbonization pressures (Sanna et al. 2014), CCSU offers a solution that satisfies regulations without sacrificing productivity for affected sectors. Furthermore, CCSU provides financial incentives, which make emissions reductions economically appealing when governments implement carbon pricing schemes. One of the most critical problems we face today is climate change, and technology that can absorb, store, and utilize carbon provides a complete solution. These technologies, through the process of CO capture, storage, and reutilization, meet not only future environmental (Hait et al. 2018b) objectives but also foster economic development and sustainability. A low-carbon economy is within reach, thanks to CCSU's ever-expanding influence, which will have far-reaching consequences for people, places, and things. This chapter delves into the complexities of CCSU, allowing readers to uncover its possibilities, obstacles, and groundbreaking developments that contribute to its role as a foundation for resilience to climate change and sustainable development (Aminu et al. 2017).

This chapter describes the challenges and opportunities of CO_2 recycling as they relate to CCSU research. It will discuss carbon dioxide storage units CCSUs and provide the techniques for CO_2 mineralization. The latter part of this chapter introduces and explains the process of simecting CO_2 and industrial garbage into the deep goaf for oxidation.

11.2 Current CO_2 Scenario Overview

Current environmental and climatic issues include rising CO_2 emissions from industrial operations, energy generation, and transportation requirements globally (Shivanna 2022). CO_2 is a major conservatory fume that is causing Ground's atmosphere to warm, which in turn is making climate change worse, the world's temperature rise, and ecosystems to shift. Concentrations of atmospheric CO_2 have risen unprecedentedly in recent decades, exceeding pre-industrial levels by more than 40%. Human-caused emissions are causing polar caps to melt, rising sea levels, extreme weather events to become more frequent and destructive, and declining biodiversity (Fawzy et al. 2020; Nunes 2023). A significant source of carbon dioxide emissions comes from the combustion of fossil fuels in power stations and other heavy sectors. This includes companies that make cement, steel, and chemicals. Deforestation, agricultural practices, and waste management all contribute to the complicated profile of CO_2, making it even more difficult for the world to reduce emissions. As soon as possible, governments, researchers, and businesses are looking into and implementing plans like CCSU to lower and reuse CO_2 pollution and boost the routine of green energy. This highlights the need for improved global collaboration and policy frameworks since meeting significant reduction objectives is still a worldwide problem. The shift toward low-carbon economies, which aim to preserve natural habitats and guarantee a stable future climate, has made CO_2 reduction an integral part of sustainable development objectives (Peter 2018).

11.3 Carbon Capture and Storage

Carbon Capture and Storage, often known as CCS, is an important technology that assists in reducing greenhouse gas emissions. This method involves the capture of CO_2 from industrial operations or power production and its subsequent subterranean storage. The two parts of this procedure are the carbon capture technique and the scope and goal (Singh et al. 2023).

11.3.1 Method of Carbon Capture

There are three main techniques employed in CCSU: pre-combustion, oxyfuel combustion, and post-combustion. The following are included:

Pre-combustion

Synthesis gas, a byproduct of burning fossil fuels, consists of hydrogen (H_2) and carbon monoxide (CO). This decarbonization process involves the interaction of fossil fuels, including coal or natural gas, with air, oxygen (O_2), and steam (or lack thereof), before combustion. These Eqs. (11.1) and (11.2) show the main reactions that happen during partial oxide formation and steam reforming, which create syngas (Jansen et al. 2015).

$$C_nH_m + \frac{n}{2}O_2 \rightarrow nCO + \left(\frac{m}{2}\right)H_2 \quad \Delta H_{CH_4} = -36\,kJ\,mol^{-1} \tag{11.1}$$

$$C_nH_m + nH_2O \rightarrow nCO + \left(\frac{n+m}{2}\right)H_2 \quad \Delta H_{CH_4} = 206\,kJ\,mol^{-1} \tag{11.2}$$

The usual reformer products from steam reform are 43% hydrogen, 11% carbon monoxide, 21% water, and 6% carbon dioxide (Osman et al. 2021). "Auto-thermal reforming" describes a pre-combustion process combining steam reforming with partial oxidation. In this setup, the exothermic heat created during partial oxidation powers any endothermic steam reform reaction. HCl, Hg, carbonyl sulphide, and H_2S are among the contaminants detached subsequently the syngas mix after it has cooled (Cormos et al. 2017). After purifying the syngas, we react CO with steam H_2O to heat it to a higher temperature, as shown in Eq. (11.3). This increases % of CO_2 and makes subsequent CO_2 separation easier. Additionally, it produces hydrogen fuel, which is decarbonized and only releases hydrogen gas when burned.

$$CO + H_2O \rightarrow CO_2 + H_2 \quad \Delta H = -41\,kJ\,mol^{-1} \tag{11.3}$$

The chemical industry stores or uses CO_2 segregated by physical and chemical absorption. The pre-combustion method has captured CO_2 for more than ninety-five years. A Rankine + Brayton combination cycle plant may generate electricity from hydrogen-rich fuels. This approach simplifies energy recovery, air separation, and reforming. When using oil or coal, further purification steps are necessary to remove contaminants, ash, and compounds covering sulfur (Kumar et al. 2018; Li et al. 2019; Osman et al. 2021). In the first version of the integrated gasification combined cycle (IGCC), the WGSR stage was responsible for 44% of the overall efficiency loss.

This was because it needed a lot of energy to make steam and heat. High-pressure pre-combustion CO_2 enables its compression and liquefaction for storage or transit, requiring minimal energy. Fuel cells can utilize hydrogen as a fuel for transportation or as a component in creating compounds with added value, thereby encouraging its production. The flexible outputs of the pre-combustion route allow you to quickly choose between producing hydrogen gas or electricity, depending on your needs. You can separate H_2 and CO_2 mixtures using chemical and physical absorption methods. One example is syngas scrubbing, which uses an acidic liquid solvent to both carbon dioxide and hydrogen sulfide (Osman et al. 2021).

Physical solvents pose several challenges, including their inability to absorb CO_2 or H_2, high thickness, poor heat reactivity, poisonous nature, and potential for fire. Dilute CO_2 streams are unsuitable for low-temperature CO_2 separation methods used in the food sector, such as chilling, compression, condensation, fishing, and cryogenic range distillation (Nizzy and Ogwu 2024; Kosoe and Ogwu 2024). It is critical to enhance separation technology for very pure H_2 and CO_2. Reactors that combine reaction and separation, such as adsorptive and membrane reactors, show enormous promise for reducing energy needs and boosting efficiency (Portillo et al. 2023). Adsorptive reactor technology employs a selective solid CO adsorbent to remove CO from the stream, shifting the equilibrium reaction toward H_2 generation. Activated carbons, zeolites, and metal–organic frameworks are inappropriate because their CO_2 adsorption ability decreases at high temperatures. This method makes use of a wide variety of engineered adsorbent systems. Membrane reactors use palladium or an alloy of it; nonetheless, even at reduced reaction temperatures, they are susceptible to deactivation and sulfur poisoning. We use dense polymeric membranes, which are inexpensive, thermally unstable, and insensitive to hydrogen. CMSMs operate well because of their sulfur resistance and durability (Li et al. 2019; Krishnan et al. 2023). Although there are obstacles to enhancing the overall efficiency of pre-combustion technology, it shows promise in carbon collection, storage, and usage. Avoid solvent reduction by lowering the solvent regeneration temperature. We are addressing this problem by using ionic liquids known for their very low volatility. However, the multitude of potential configurations resulting from various combinations of anions and cations presents a challenge in selecting the appropriate ionic liquid Zhou et al. 2021b). We now use computer-aided molecular design (CAMD) to design ionic liquids architecturally. When put through industrial pre-combustion conditions of 40 °C and 30 bars, the hydroxyl (OH)-ammonium (NH_3) and hydroxyl-imidazolium bis (trifluoromethyl sulphonyl) amide from the ionic liquid solvent work best for this technology. There is some theoretical evidence that the pre-combustion method might be 38–45% less expensive than the post-combustion route and 21–24% cheaper than the oxyfuel combustion approach. The difficulty of the setup procedure, additional costs, and the need to convert existing facilities have hindered its commercialization (McDonald et al. 2015).

Oxyfuel Combustion

Power systems may increase boiler efficiency and decrease emissions from carbon-based fuels by using the oxyfuel combustion approach (Chen 2018). Despite having a modest efficiency penalty, the high cost of producing and separating O_2 makes it challenging to market. Researchers are exploring novel avenues for air separation, such as chemical looping techniques, ion-transport membranes, and oxygen-transport membranes, to address this problem. The energy requirement for cryogenic air separation has prompted the development of oxygen-transport membranes, also known as mixed ionic-electronic conducting membranes (MIEC) (Chen et al. 2024). Compared to technologies implemented after combustion, these membranes may reduce energy expenditure by 19–50%. There has been a meteoric rise in the number of papers focusing on oxygen-transport membranes in recent years, going from an average of 30 in 1985 to 200 in 2012 (Portillo et al. 2022; Kotowicz and Balicki 2014: Shin and Park 2023; Aguilar-Martinez et al. 2023).

Utilizing the chemical looping approach in oxyfuel in conjunction with IGCC instead of the air-separating unit has the potential to increase the net efficiency of the power plant by three percent. It is also possible to reduce power expenses by 7–12% and capital costs by 10–18%. Current or future power plants may use oxyfuel combustion routes with other fuels, such as lignocellulosic biomass or municipal solid waste, which is beneficial. The BioCCS or BECCS method for reducing greenhouse gas emissions combines bioenergy with carbon capture and storage. Total emissions from the net generation of power from oxyfuel burning of lignocellulosic biomass were $-$ 0.27 kg CO_2 $MJel^{-1}$. Carbon capture and burning of city solid waste together have reduced CO_2 emissions by 0.70 kg CO_2 equivalent per kg of wet trash (Carboa et al. 2009; Wienchol et al. 2020; Cormos et al. 2017). Ultimately, the oxyfuel combustion method presents an opportunity to reduce carbon emissions and the impact of climate change via the use of negative carbon technology.

Twenty BECCS initiatives throughout the globe center on efforts to convert waste into energy, produce ethanol, gasify biomass, and build biogas facilities. However, problems persist, such as inferior efficiency, higher power prices, and higher biomass costs compared to fossil fuels (Shi et al. 2020; Gladysz and Ziebik 2015; Yang et al. 2019). Including air separation, CO_2 purification, and compression units reduces the efficiency of the oxyfuel combustion system to 9–13%. Incorporating heat into the process is essential for increasing its economic viability. Here, heat from steam cycles and compressor cooling systems has shown to be useful (Stanger et al. 2015; Hong et al. 20102010). Pressurized oxyfuel combustion cycles outperform air cycles by 3%. Although there is a rising interest in BECCS on a worldwide scale, its current capabilities are typically below 100 MW. Supercritical CO_2-cycle biomass oxyfuel combustion reduces CO_2 by 3.7 megatons per year. Under some conditions, such as a carbon tax exceeding $28.3 per tonne of CO_2, BECCS technology may prove to be more cost-effective than fossil fuels (Wei et al. 2020; Cook 2009).

Post-combustion

The process of capturing and separating fuel gas from a combustible system from diluted CO_2 in an oxidizing environment is known as the post-combustion pathway (Zhang et al. 2014). Fossil fuel engines undergo denitrification, desulfurization, dust removal, and chilling to stop solvent degradation and capture carbon dioxide in their emissions (Wu et al. 2020). Figure 11.1 shows the subsequent steps, which include feeding the fuel gas, which mostly consists of CO_2, H_2O, and N_2, to the absorber, which houses the solvent. Regeneration of CO_2 follows a water wash of the scrubbed gas. We store the compressed CO_2 gas in geological reservoirs or saltwater aquifers. The primary technique of solvent regeneration in this process is monoethanolamine absorption, which is energy-demanding. Additionally, N-methyldiethanolamine and 2-amino-2-methyl-1-propanol are absorbents. We use adsorption routes such as calcium looping, temperature swing adsorption, or pressure swing adsorption after combustion. Since amine solutions are very selective for acidic gases and have a large ability to absorb carbon dioxide, they are the solvents of choice. The downsides include evaporation, solvent loss, corrosion, deterioration, and a large energy footprint during regeneration. One possible solution to the high initial investment required by post-combustion technologies is membrane separation (Karaszova et al. 2020). It is simple to adapt and scale up with existing power plants, has a low operating cost, and consumes little energy. It also leaves a little carbon imprint. Some problems that can arise include water condensation, the quick loss of selectivity and permeability after operation, and the passage of emissions through the membrane. Metal–organic frameworks (MOFs) have interesting properties, like functionalized hole shapes and custom structures that could work well for capturing CO_2. Despite the existence of over 75,000 potential structures for MOF materials, their industrial application remains unattainable due to their high energy requirements for regeneration and rapid instability. The addition of polar groups or the loading of exposed metal sites inside the structure of MOFs are two possible modifications to these materials (Gkotsis et al. 2023). Among the hundreds of adsorbents, including MOFs, computational screening modeling approaches may help identify the best-performing materials. MOF materials, such as diarylethene and azobenzene, need more energy to undergo desorption than sunlight, which acts as an external stimulus. Park et al. achieved the maximum CO adsorption capacity among photoresponsive MOFs reported in literature, 89.6 cm^3 g^{-1}. Though more costly, pre-combustion technology outperforms post-combustion technology in terms of efficiency (Pfister et al. 2017). Identifying a more effective absorption solution is critical for lowering precombustion expenses. The post-combustion pathway has been around the longest and is the most utilized of the three primary methods for capturing and storing carbon (Zhou et al. 2021; Jiang et al. 2022).

Chemisorption, a ready-made method that uses amine-based solvents to retrofit existing power plants, is available. Pilot-scale power plants have shown a CO_2 absorption capability of 80 metric tons per day. The first coal-fired power plants that collect and store carbon will likely be open for business by 2020 or 2025. This will enable its use in other large-scale economic processes that consume significant amounts of carbon (Farmahini et al. 2020; Park et al. 2018). In the field of carbon capture and storage, researchers have compared both conventional and innovative technologies, such as post-combustion and partial oxy-combustion, in the pilot-scale absorption pathway. Novel solvents provide a number of advantageous characteristics, including reduced environmental impact, cheap manufacturing cost, minimal corrosiveness, degradation, and by-products, and high cycle capabilities (Stephenne 2014). The Shell business has developed a novel CO capture technique, Shell Cansolv™. It

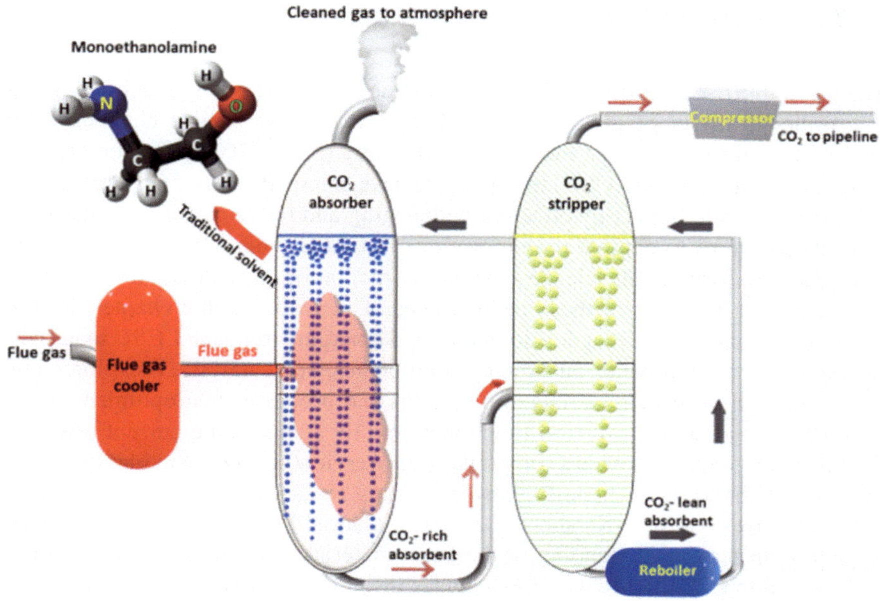

Fig. 11.1 Post-combustion technology involves cooling hot flue gas, sending it to a CO_2-absorber unit with monoethanolamine solvent

depends on amine solvent technologies and is applicable to a wide range of sectors, including chemical, mining, energy production, and refineries. Compared to the most popular solvent, monoethanolamine, the regeneration energy needed to capture one tonne of CO_2 is lower with the Cansolv™ solvent, ranging from 2.5 to 2.9 GJ per tonne of CO_2. This is one benefit of the solvent (Qazvini and Telfer 2020).

An evaluation of the technical readiness level (TRL) of various carbon capture, storage, and use methods is necessary for the decarbonization of the industrial sector. Only pre-combustion technologies, which treat natural gas, have achieved commercial scale (TRL9). The demonstration scale (TRL7) includes adsorption post-combustion, oxyfuel combustion, IGCC, BECCS technology, membrane polymers, direct air capture, and other capture methods (Krishnamurthy et al. 2021). Only three carbon storage solutions have achieved commercial scale: amine post-combustion in power plants, salinity deposits, and CO_2-enhanced recovery of oils (TRL7). There are a number of storage technologies that are still in the early phases of development, including TRL2 for ocean storage and TRL3 for mineral storage, as well as TRL7 for CO_2-enhanced gas recovery in depleting oil and gas fields (Gulzar et al. 2020; Singh and Stéphenne 2014; Yun et al. 2020). In order to consolidate various CO_2 sources and take advantage of carbon capture's economies of scale, the building of suitable infrastructure is essential during the early phases of CCSU implementation.

11.3.2 Scope and Goal

The inevitable rise in atmospheric carbon dioxide (CO_2) and other greenhouse gases is a principal driver of global climate change, presenting critical challenges for environmental sustainability and human well-being (Nunes 2023). Carbon capture, storage, and utilization (CCSU) technologies have emerged as essential strategies in mitigating climate impact by effectively managing and reducing carbon emissions. This chapter aims to explore the innovations in CCSU, with a particular focus on advancements in materials science and process integration that enhance the efficiency and economic viability of these technologies. The scope of this chapter encompasses examination of recent developments and emerging trends in CCSU materials, including novel absorbents, adsorbents, membranes, and catalysts. It explores into the integration of these materials into systems and processes that optimize carbon capture, storage, and conversion functionalities. By synthesizing current research, this chapter intends to provide a comprehensive overview of state-of-the-art technologies and methodologies, identify gaps in existing research, and suggest directions for future investigations (Ketabchi et al. 2023; Gabrielli et al. 2020). It is expected to gain a nuanced understanding of the latest material innovations and their applications within CCSU systems. This includes insights into how these advancements can be scaled from laboratory settings to industrial applications, the economic implications of these technologies, and their potential impact on policy and regulatory frameworks. Furthermore, this chapter seeks to inspire interdisciplinary collaboration among scientists, engineers, and policymakers to foster the development and deployment of more effective carbon management strategies, ultimately contributing to global climate mitigation efforts (Bhavsar et al. 2023; Kosoe et al. 2023).

11.4 CO_2 Storage Technology

This discussion extensively examines the literature, developments, and arguments surrounding various CCSU techniques. Figure 11.2 shows the most prevalent CCS and use CCSU systems for CO_2 storage. They can reduce emissions of carbon dioxide. Unfortunately, their current global deployment is not sufficient to achieve the goal of zero net CO_2 emissions by 2050 (Wan 2022). Scientists are looking into different ways to store CO_2, such as physically, biologically, and chemically (Chen et al. 2016; Liu et al. 2021). This is due to the need to store the captured CO_2 to prevent its effects. Biological storage is the transformation of atmospheric CO_2 into organic matter by means of photosynthesis, in which living creatures take in and store carbon. Carbon cycle regulation and climatic stability depend on this mechanism. Plants and soil both store carbon (Masson-Delmotte et al. 2018; Lal 2008; Ogwu 2019, 2023), respectively; this is known as biological storage. By transforming CO_2 into organic molecules like carbohydrates and proteins, photosynthetic organisms like plants and algae are vital to biological storage (Zhu et al. 2010). The organism's

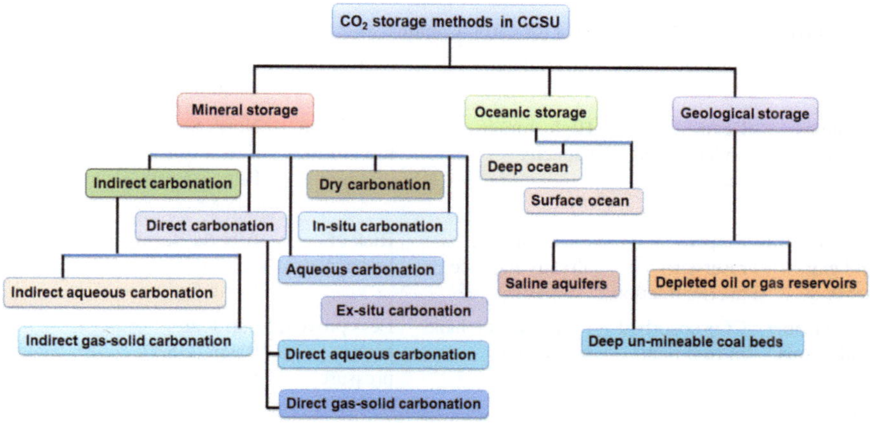

Fig. 11.2 CO_2 storage methods in CCSU

tissues can either retain these chemicals or discharge them into the soil, where they undergo further breakdown to become organic matter (Paustian et al. 2016). The earth's crust (Bachu 2000; Lehmann 2007) and water (Metz et al. 2005) contain various forms of physical storage. The chemical storage process known as mineral carbonation (Bachu 2000) involves the interaction of carbon dioxide with minerals. The next chapters will go into detail on chemical and physical storage. There are three primary methods to store carbon dioxide: There are three primary methods for storing carbon dioxide: (i) geological storage in deep geological formations, (ii) in the thick water of the ocean, and (iii) in mineral carbonates (Bachu 2000).

11.4.1 Geological Storage Technology

Similar to the natural storage of fossil fuels in the ground, we can store CO_2 by injecting it into the appropriate deep rock formation or layer at the appropriate depth. Research on geological CO_2 storage has seen a dramatic increase in published accounts during the last decade (Gale et al. 2015). A total of fourteen locations throughout the globe are now storing more than one million metric tons of carbon dioxide (Bui et al. 2008). The estimated global CO_2 storage capacity varies from 100 to 20,000 gigatonnes CO_2, depending on where the study is conducted (Chunbao Charles and Cang 2010). Most scientists concur that salty groundwater, unmineable deep coal beds, and depleted oil and gas reserves are the optimal locations for underground CO_2 storage (Bachu 2000). Geological storage sites for carbon capture entail securing CO_2 underground and mitigating emissions by drawing on deep salt aquifers, depleted gas and oil reserves, and unmineable coal seams (Table 11.1).

Table 11.1 Geological storage sites for carbon capture

Storage site type	Characteristics	Advantages	Challenges	Example locations	References
Saline aquifers	Deep underground saline water	High storage capacity	Long-term monitoring required	North Sea (Norway), Illinois Basin (USA)	Celia et al. (2015)
Depleted oil and gas fields	Empty reservoirs post-extraction	Infrastructure already in place	Risk of leakage	Alberta Basin (Canada), North Sea	
Unmineable coal seams	CO_2 absorbed in coal matrix	Enhances methane production	Less proven, high pressure needed	San Juan Basin (USA), Surat Basin (Australia)	
Basalt formations	Reacts chemically with CO_2	Permanent mineralization	Limited accessibility	Iceland, Pacific Northwest (USA)	

Depleted Oil or Gas Reservoirs

The feasibility of geological storage as a large-scale storage option makes it a popular EOR/EGR approach (Kumar et al. 2017; Bai et al. 2016). Indeed, experts view CO_2 storage in depleted oil and gas reservoirs as a highly efficient option for energy recovery. This is due to several factors, including: (i) the ability to determine storage capacity thanks to thorough study and exploration throughout the hydrocarbon exploration stages; (ii) the availability of both underground and surface infrastructure, including pipelines and injection wells, which can be used for storage processes with little to no modification required (Wildgust et al. 2013; Ryerson et al. 2013; Hawkes and Gardner 2013); and (iii) the widespread use of CO_2 injection as an EOR technology by the oil and gas industry, which can be utilized for storage purposes (Li et al. 2006).

In order to get the last bit of oil out of operating wells, enhanced oil recovery (EOR) uses CO_2 to raise the reservoir pressure. Additionally, by injecting CO_2, we can recover natural gas (CH_4) from coal seams. The main idea of this approach (White et al. 2003) is that adding CO_2 to the coal bed has the dual benefit of dispersing CH_4 and storing CO_2 in the coal's porous structure. There are established technologies that back the use of CO_2 injection for EOR. Research has focused on a few process areas, including migration simulation (Chiaramonte et al. 2011), geochemical modeling (Cantucci et al. 2009), and leakage/risk assessment (Klusman 2003). One environmental concern associated with EOR is the production of large amounts of water, potentially containing radioactive elements and dangerous heavy metals (Igunnu and Chen 2014).

Deep Unmineable Coal Beds

Coal bed methane (CBM) reservoirs are natural coal deposits with substantial methane gas pockets. This method releases large quantities of CO_2 into the air when it decreases pressure inside the coal seam during extraction. Nevertheless, CO_2 sequestration is an alternative to methane gas that allows coal beds to retain huge quantities of CO_2. This process involves injecting CO_2 into unmineable coal seams and allowing it to adsorb onto the coal surface. Following this process, a coal matrix binds the carbon dioxide, allowing it to remain underground for an extended period, potentially lowering its emissions (Jia et al. 2018). Globally, deep, unmineable coal beds could potentially store an estimated 500 gigatonnes of CO_2. In deep, unmineable coal beds, the stability of CO_2 storage over the long term is contingent on a number of parameters, including the kind of coal, the rank of the coal, the depth, and the pressure. Because of their high sorption capacity and limited permeability, high-rank coal seams are more stable and can store more CO_2. Deep, unmineable coal seams can effectively sequester CO_2 geologically over an extended period (Hu and Cheng 2017; Liu et al. 2017; Bao et al. 2019; Baran et al. 2014). Coal bed methane has already established infrastructure and accumulated a wealth of expertise, which is a major technological advantage when it comes to CO_2 sequestration in deep, underground coal beds. CO_2 injection can improve methane production, partially offset CO_2 sequestration expenses, and potentially avoid coal mining-related environmental problems. However, there are numerous challenges associated with CO_2 sequestration in deep, unmineable coal beds. One is that there could be CO_2 leakage from nearby rock formations or faults. Another is that CO_2 injection, monitoring, and confirmation are expensive. Lastly, policies and regulatory frameworks should ensure the safety of the process (White et al. 2013).

Saline Aquifers

Deep saline aquifers, located at depths ranging from 700 to 1000 m below ground level, are known to contain high-salinity formation brines (Singh 2008). These saline aquifers may not have any economic value; nonetheless, they may be of use as a storage location for the injected carbon dioxide that is collected during the carbon capture and storage process. Saline aquifers are considered an essential choice for CO_2 storage due to their enormous capacity to store carbon dioxide. They have the potential to absorb 10,000 Gt of carbon dioxide, which is the same as the emissions from major stationary sources for more than a century, according to estimates (De Silva et al. 2015; Celia et al. 2015).

Saltwater aquifers may store ten thousand gigatonnes of carbon dioxide, or twenty to five hundred percent the amount of emissions predicted by the year 2050. Their ability to store CO_2 over the long term is about 30 kg/m^3. Most current CO_2 collecting stations have simple access to them, which makes the procedure more cost-effective (Singh 2008; De Silva et al. 2015; Celia et al. 2015). Their high mineralization makes them unfit for human consumption, but they provide a beneficial alternative for CO_2 storage that won't deplete freshwater supplies. Rocks must possess sufficient porosity to facilitate the displacement of brine or gas from their pore structures, thereby enabling the injection and storage of CO_2. The best places to store carbon dioxide are deep saline aquifers because of their high porosity and permeability. However, compared to other natural storage places such as oil areas and coal seams, there is less knowledge about their storage mechanisms (Davison et al. 2001). Yang et al. (2010) examined the features of CO_2 sequestration in saltwater aquifers, including the gas's behavior in various phases, its relationships with water and rock, and the processes that trap CO_2. Elements impacting CO_2 mineral entrapment during brine sequestration have been the subject of much research (Pruess et al. 2003; Gunter et al. 2004; Rackley 2010). Due to a lack of knowledge about the geochemical behavior of salty aquifers, global estimations of CO_2 storage capacity are still not exact. Many saltwater aquifers are not considered the most attractive storage options from an economic perspective due to their high capital expenditures and lack of infrastructure (Gunter 2000; Reichle et al. 1999). Although there are a number of disadvantages, geological CO_2 storage does have the ability to drastically cut emissions of greenhouse gases. Natural flaws or fissures in rock formations pose a significant risk of CO_2 leakage, which is a major concern. Because of this, trapped carbon dioxide gas might leak to the surface, endangering both people and the planet (Liu et al. 2012; Szulczewski et al. 2014). Increased earthquake frequency, acidity of subsurface water, and displacement of geological structures are further dangers. Concerns with geological storage stem from the fact that it is both costly and energy-intensive, necessitating constant site monitoring and maintenance. Regardless of these obstacles, geological storage may still be a beneficial way to cut down on emissions of greenhouse gases (Herzog 2009; Lee and Choi 2018).

11.4.2 Oceanic Storage Technology

The water, an essential natural carbon sink, holds a significant amount of extra carbon dioxide. The ocean's thermohaline circulation dissolves and re-dissolves carbon as it exchanges CO_2 between the air and sea. Physical factors such as pressure, salinity, and temperature influence ocean storage. These variables affect both the solubility of carbon dioxide in saltwater and the pace of its transfer to the deep ocean. Generally, colder, saltier water dissolves CO_2 more easily than warmer, fresher water. Because their water is colder and saltier than the rest of the ocean, the polar areas are ideal locations for ocean storage (White et al. 2003). Because pressure influences both the solubility and the rate of transit of carbon dioxide to the deep ocean, it is an essential component of ocean storage (McNeil and Matear 2008). Furthermore, organic matter, such as phytoplankton, sinks via the biological pump and carries CO_2 to the deep waters of the marine environment (Liu et al. 2021).

Researchers have studied two oceanic systems to mimic natural carbon sequestration processes. The first one skips the mixed layer altogether and pumps carbon dioxide directly into the ocean floor. There is now little hope for attributing carbon that is trapped in the water, despite discussions between academics and businesses.

Injecting carbon dioxide (CO_2) into the deep ocean creates liquid CO_2 lakes in a supercritical, high-pressure environment, a process akin to geological storage. It is possible to release captured CO_2 into the ocean or seabed by means of a pipeline or a ship. Because of the immense CO_2 storage capability of the world's deepest ocean depths, oceanic storage offers a substantial theoretical capacity for CO_2 storage. For instance, the Sunda trench, located below 6 km, has the capacity to store up to 19,000 Gt of liquid CO_2, surpassing the CO_2 production from all global fossil fuel reserves combined. Similarly, the Puerto Rico trench, situated deeper than 7 km, can store up to 24,000 Gt of liquid CO_2. The possibility of the stored CO_2 escaping back into the atmosphere, however, has caused some to express alarm (Goldthorpe 2017). Therefore, it is crucial to monitor the situation to ensure that the CO_2 does not escape back into the atmosphere (Ha-Duong and Keith 2003). In the second scenario, we activate the biological pump by adding nutrients to the ocean's surface. A key component of ocean fertilization is the addition of nutrients that foster the growth of phytoplankton. Phytoplankton are tiny algae that store carbon dioxide in their cells via photosynthesis and subsequently vent it into the ocean (Khan et al. 2023; Lovelock and Rapley 2007).

11.4.3 Mineral Storage Technology

One potential method for CCS is mineral carbonation (MC), which is based on the same principles as natural weathering. An exothermic process forms stable carbonate minerals by combining carbon dioxide with alkaline earth metal-containing minerals and waste materials (Seifritz 1990; Kandji et al. 2017; Lackner 2002). Thermodynamic stability of carbonates is higher than that of CO_2, making CO_2 storage in minerals a safe and long-term option that does not need constant monitoring. A number of distinct benefits distinguish MC from competing carbon storage technologies. These include the capacity to produce goods with additional value, the possibility of in-situ deployment by a range of businesses, and the very long-term stability of CO_2 (Bobicki et al. 2012; Surampalli et al. 2015). The literature on MC includes numerous investigations and evaluations (Galina et al. 2019). Review articles have covered the early work up to 2006; Torróntegui et al. addressed important research up to 2010, and Bodor et al. provided a synopsis of the expansion of MC into industrial waste (Sanna and Maroto-Valer 2017; Wang et al. 2018). We have extensively reviewed the history and current state of MC, as well as its application to various feedstock materials. Ultramafic mining deposits, fly ash, steel manufacturing waste, and pH swing operations have all been the subject of recent assessments that have focused on MC. The techno-economic components of several prior studies, however, have given rise to questions (Li et al. 2018). Table 11.2 presents the estimated CO_2 storage capacities of the methods explained above.

Table 11.2 Estimated CO_2 storage capacity

Option	Gigatons of carbon (GTC)	References
Depleted oil or gas reservoirs	180–250	Anonymous (2020)
Deep unmineable coal beds	1–55	
Saline aquifers	275–2750	
Ocean storage	> 5000	
Mineral carbonation	Very large	

CO_2 Mineralization Methods

Together in situ and ex situ methods for conducting mineralization reactions are applicable, depending on specifics of the reaction site. An in-situ process injects CO_2 into subsurface reservoirs, causing a reaction between CO_2 and alkaline minerals, instead of mining and transporting feedstock as part of an ex-situ process (Saran et al. 2018; Yadav and Mehra 2021). Because it doesn't need a chemical plant, in situ carbonation is cheaper. The dry method of mineral carbonation has poor conversion efficiency and slower kinetics compared to the aqueous approach (Huijgen 2007; Lackner et al. 1997). However, it can produce solid sorbents that absorb CO_2, like calcined limestone. In mineral carbonation, the aqueous approach converts carbonate most efficiently, and this new MC process is the most commonly used method. Minerals can be carbonated in two main ways: directly carbonation reacts magnesium and calcium-containing minerals directly with carbon dioxide; and indirectly carbonation takes magnesium or calcium-containing minerals out of a solution using CO_2 and then adds them to the carbonation solution (Naraharisetti et al. 2017).

Direct Carbonation

One simple way to carbonate anything is using direct carbonation, which just requires one step (Liu et al. 2018). Direct aqueous carbonation and gas–solid carbonation are the two main forms of this process. Direct carbonation in gas–solid systems occurs when minerals react with CO_2 gas, whereas in direct aqueous systems, slurries undergo the same reaction. In most cases, procedures with a moisture content below 0.2 are considered to be on the dry path (Lim et al. 2010). The method's inefficiency and delayed carbonation kinetics, however, make it unfit for use on a grand scale. As mentioned before, aqueous carbonation is superior for carbonating minerals (Malayan 2021).

Direct Gas–Solid Carbonation

The simple process of direct carbonation, known as gas–solid carbonation, entails reacting gaseous carbon dioxide with solid minerals. Lackner et al. first described this approach (Lackner et al. 1997). Its advantages include straightforward process design and the possibility of heat recovery at elevated temperatures (Zevenhoven and Kavaliauskaite 2004). But without optimization to speed up the carbonation process, this approach is not practicable because of the sluggish response rate. Even when subjected to very high temperatures and pressures, the carbonation rate of dry rock minerals during direct carbonation remains negligible, according to experimental results (Bobicki et al. 2012). Kwon et al. (2011) indicated that adding moisture to the flue gas can significantly accelerate dry carbonation processes. Despite this, the technique isn't very effective at capturing CO_2—it requires eight tons of olivine to capture one ton of CO_2, making it unsuitable for large-scale use.

Equation (11.4) describes the process of directly carbonating silicates using the dry method. When the temperature and pressure are the same, the Gibbs free energy value for wollastonite ($CaSiO_3$) is − 44.6 kJ/mol, for forsterite (Mg_2SiO_4) is − 43.1 kJ/mol, and for serpentine ($Mg_3Si_2O_5(OH)_4$) is − 16.9 kJ/mol. This means that the reactions are likely to happen on they're. This procedure lengthens reaction durations because of the slow reaction kinetics and low conversion rate associated with it (Wang et al. 2019; Myers et al. 2019).

$$\begin{aligned}&(\mathrm{Ca, Mg})\mathrm{xSiyOx} + 2\mathrm{y} + \mathrm{zH}_{2\mathrm{z}}(\mathrm{s}) + \mathrm{xCO}_2(\mathrm{g}) \\ &\quad \rightarrow \mathrm{x}(\mathrm{Ca/Mg})\mathrm{CO}_3(\mathrm{s}) + \mathrm{ySiO}_2(\mathrm{s}) + \mathrm{zH}_2\mathrm{O}(\mathrm{g/l})\end{aligned} \tag{11.4}$$

Direct Aqueous Carbonation

When carbonation is done directly in water, minerals with a high reactivity, such as serpentine, wollastonite, and olivine, are used. Additionally, the shape and size of these particles influence the effectiveness and speed of the reaction. Although they may provide quicker response speeds, finely powdered particles pose challenges with clogging and separation (Onwudiegwu et al. 2024). Particle size optimization achieves the ideal reaction efficiency. Further significance lies in the control of the reaction's byproducts, particularly the solid carbonates and the liquid waste (Li et al. 2019). While pre-disposal treatment is necessary to limit the environmental consequences of liquid effluent, processing solid carbonates may increase storage capacity or make them usable in building materials. In comparison to the dry method, direct aqueous carbonation has many benefits, such as faster reaction rates, greater ability to absorb CO_2, and reduced energy needs. But you have to think about process parameters, mineral choice, effluent, and reaction product control very carefully (Gadikota et al. 2014).

Indirect Carbonation

Multiple steps are involved in the indirect carbonation of minerals. The first step is to remove the magnesium or calcium ions from the mineral ore using acids or other solvents. They subsequently generate their corresponding hydroxides or oxides. In the second step, oxides or hydroxides react with CO_2 to form carbonates (Lim et al. 2010).

Indirect Gas–Solid Carbonation

When magnesium oxides or calcium hydroxides are carbonated, it happens at a much faster rate than when the corresponding silicates are carbonated directly.

Ca/Mg silicates produce Ca/Mg oxides/hydroxides due to their rare occurrence in nature (Teir 2008). Therefore, the indirect gas–solid carbonation process first transforms the extracted Ca or Mg into their respective oxides or hydroxides. Carbonation then occurs in an airless environment with these hydroxides or oxides of minerals. Research has established that under certain circumstances, the transformation of magnesium hydroxide, $Mg(OH)_2$ into magnesium carbonate ($MgCO_3$) occurs quickly. For instance, Lackner et al. (1997) reported a complete transformation of $Mg(OH)_2$ into $MgCO_3$ in just two hours at 500 °C and 340 bar of pressure. Similarly, Butt et al. (1998) reported a 90% conversion rate from magnesium hydroxide, $Mg(OH)_2$ to magnesium carbonate ($MgCO_3$) at a pressure of 53 MPa and temperatures of 565 °C over a thirty-minute period, with particles measuring 0.1 μm.

Indirect Aqueous Carbonation

The indirect aqueous approach first chemically extracts magnesium and calcium from minerals in an acidic environment, followed by carbonation in an alkaline environment. This approach allows for the extraction of nearly pure minerals by eliminating contaminants from their natural core. Leaching using acids, molten salts, caustic soda, or bioleaching are some of the methods used to extract minerals (Eloneva et al. 2008; Sipilä et al. 2008). However, the use of acid leaching makes recovering the alkali earth metals an energy-intensive process. Although they are more corrosive, molten salts regenerate with less energy. Another indirect carbonation technique is the pH swing method, which first extracts the solid matrix from the mineral carbonate under low pH conditions, and then increases the pH to allow carbonate production in the second stage (Park and Fan 2004; Azdarpour et al. 2015). Since indirect carbonation involves an additional manufacturing step and extraction agents, direct carbonation is simpler and cheaper. Carbonation of minerals may be accomplished by a variety of techniques, including direct gas–solid carbonation, indirect gas–solid carbonation, and indirect aqueous carbonation methods.

11.5 Carbon Utilization

The term "carbon utilization" refers to the processes that contribute to sustainability and reduce emissions by converting collected CO_2 into useful goods. The alternative to releasing CO_2 into the environment is to transform it into other forms of energy, such as chemicals, construction materials, and alternative fuels. This method transforms emissions into assets with numerous industrial uses, thereby contributing to the establishment of a cyclical carbon economy (Styring et al. 2023).

11.5.1 Applications of Carbon Capture

There are applications for captured carbon, each harnessing CO_2 in unique ways that drive both sustainability and innovation. There are shown in Table 11.3.

This table highlights how captured CO_2 applications support both industrial innovation and environmental sustainability by transforming emissions into valuable products and resources.

Table 11.3 Diverse applications of carbon capture

Application	Description	Benefits	References
Enhanced oil recovery	CO_2 is inserted into exhausted grease reservoirs to intensification pressure and enhance oil extraction. This process also stores CO_2 underground, preventing it from entering the atmosphere	Boosts grease yield, securely stores CO_2 underground, reduces atmospheric CO_2	Shaw and Mukherjee (2022)
Building materials production	Captured CO_2 is mineralized and incorporated into materials like concrete, aggregates, and bricks, forming stable carbonates	Reduces CO_2 emissions, increases material strength, and improves durability for construction purposes	
Alternative fuels synthesis	CO_2 is converted into synthetic fuels like methanol and ethanol through renewable energy-powered chemical reactions	Provides sustainable, low-carbon energy alternatives for transportation and industry	
Chemical feedstock for industry	CO_2 is used as a feedstock in manufacturing chemicals like urea, methanol, and polymers, replacing fossil-based raw materials	Reduces fossil fuel dependency, enables production of fertilizers, plastics, and synthetic fibers from CO_2	
Algae and biomass production	CO_2 is fed to algae in controlled environments, producing biomass that can be converted into biofuels, animal feed, and food supplements	Supports sustainable bio-based products, transforms emissions into valuable resources for various industries	
Carbonated beverages and dry ice	Captured CO_2 is used to carbonate beverages and produce dry ice for shipping and storage in the food and beverage industry	Provides a practical application of CO_2, turning emissions into consumer and industrial products	

11.5.2 Comparison of Carbon Capture and Storage (CCS) and Carbon Capture, Storage, and Utilization (CCSU)

CCS is a promising way to reduce atmospheric CO_2 (Tan et al. 2016). Storage is a prerequisite for the development of carbon capture and storage technologies. Previous reviews (Riaz and Cinar 2014; Belhaj and Bera 2017; Aminu et al. 2017; Thakur et al. 2018) laid out the several physicochemical approaches to efficient CO_2 storage and highlighted the difficulties presented by these many methods and projects. Mineral carbonation (MC), offshore, and geological CO_2 storage methods have been studied (Sanna et al. 2014; Pham et al. 2016; Bickle 2009). Michael Economides, an expert in energy, argues that there are insurmountable technical and financial obstacles that make carbon capture and storage (CCS) an unrealistic method for limiting CO_2 (Ehlig-Economides and Economides 2010). As a potential method for reducing emissions of greenhouse gases, CCSU uses a similar concept. The CCSU method emphasizes the exploitation of collected carbon dioxide as a vital factor, while all three components capture, storage and utilization are significant. There are financial incentives to use CCSU technology since carbon exploitation not only lowers atmospheric CO_2 concentrations but also produces goods with additional value (Roh et al. 2018). Table 11.4 compares the CO collection capacities of CCS and CCSU, providing a brief overview of their benefits and drawbacks. It is important to note that these estimations are generic and may vary based on the technology and implementation employed.

Table 11.4 Comparing carbon capture and storage and carbon capture, storage, and utilization

Method	Advantage	Disadvantage	CO_2 capture value	References
CCS	It reduces carbon emissions from significant emitters such as power plants and factories	Gas collection, transportation, and storage use a lot of energy and resources; preventing leakage and ensuring the CO_2 remains stable over time are major problems	It can eliminate 90% of CO_2 emissions at the source level	Zhao and Itakura (2023)
CCSU	Utilization may lower the total cost of carbon capture; capturing CO_2 has several benefits, including lowering emissions and making money via its usage in goods like fuels and chemicals	Analyze the cost implications of different utilization methods and their impact on economic viability	It captures nearly all carbon dioxide emissions, reaching up to 99%	

11.6 Economical Beneficial of Carbon Capture Storage and Utilization

To ensure that CCSU prioritizes ecosystem conservation, community benefits, and cost-effectiveness, it is important to include ecological, public, and economic implications in the process (Fig. 11.6). This will provide a maintainable way. This comprehensive strategy promotes sustainable development by bringing CCSU in line with long-term climate objectives, efficiency in resource use, and beneficial social consequences.

When considering CCS as a potential large-scale option to reduce emissions of greenhouse gases, its economic feasibility is of the utmost importance. Many believe that carbon capture and storage (CCS) is an essential tactic for achieving global climate targets, particularly in polluting industries and power plants that rely on fossil fuels (Chen et al. 2015). A major obstacle, meanwhile, is the astronomical price tag of CCS, which includes everything from capture to storage. The three primary components of these costs are carbon collection, transportation, and storage. Since the extraction of CO from industrial emissions requires sophisticated technology, the cost of capture alone accounts for around 70% of the total cost. Although the latter stages of transport and storage are less costly, they nevertheless need substantial expenditures in storage facilities, tanker systems, and pipelines (Guo and Huang 2019). Because CCS projects need a large amount of finance, market mechanisms on their own are not always enough to encourage their adoption. For CCS to be more economically viable, governments and governing bodies have put a price on carbon, set up tax credits, and given direct funding. By attaching a monetary penalty to carbon emissions, carbon pricing mechanisms like cap-and-trade systems or carbon taxes may incentivize businesses to use CCS (Onyeaka et al. 2021). Companies that participate in carbon capture and storage also get direct financial assistance in the form of tax credits, such as the U.S. 45Q incentive. Scaling CCS still prioritizes cost

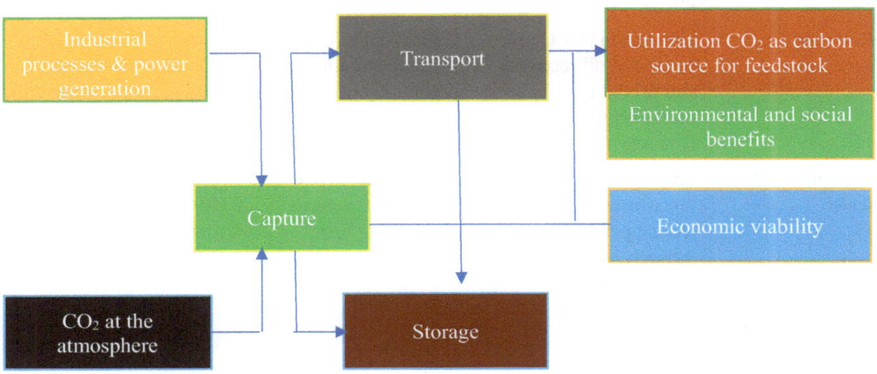

Fig. 11.6 The carbon capture storage and utilization process incorporate ecological, public, and economic implications

reduction, even with these incentives (Nagireddi et al. 2024). Cost savings may be possible because of technological developments that increase capture efficiency and economies of scale (Table 11.5). To further reduce CCS costs, it is possible to discover revenue-generating uses for collected CO_2, such as EOR or product conversion. To make CCS a viable, long-term solution for widespread implementation, political will, technical advancement, and favorable economic circumstances must all come together.

Table 11.5 Economic impact of technological developments and their challenges

Factor	Description	Economic impact	Challenges
Capture costs	Advanced technology is needed to separate CO_2 from emissions, contributing up to 70% of the total cost	Major cost component in CCS projects; impacts feasibility	High initial investment in R&D and technology required
Transport costs	CO_2 must be transported to storage sites, usually via pipelines or tankers	Adds to overall cost; requires infrastructure development	Requires extensive pipeline networks or specialized tankers, especially in remote areas
Storage costs	Long-term storage sites, such as saline aquifers or oil reservoirs, are necessary for sequestration	Important for project sustainability; requires maintenance	Limited availability of suitable storage sites and regulatory approval needed
Carbon pricing	Various mechanisms exist to encourage the adoption of CCS	Helps offset CCS costs, making projects economically feasible	Pricing volatility and policy changes can impact consistency
Tax credits and incentives	Examples like the U.S. 45Q tax credit provide direct financial support for CCS projects	Encourages CCS adoption and attracts private investment	Limited scope in some regions; complex application processes
Revenue generation	Utilization of CO_2 in enhanced oil recovery and product manufacturing (e.g., concrete)	Creates additional revenue streams, potentially offsetting costs	Limited markets for CO_2-based products and regulatory barriers
Economies of scale	Scaling CCS infrastructure can reduce per-unit costs through larger installations and network effects	Reduces average costs, making CCS more viable long-term	High initial costs and long payback periods deter investment

11.7 Potential Application and Opportunities of Carbon Capture Storage and Utilization

Technology for CCSU has enormous promise, with the ability to transform industries as diverse as energy and building. CCSU technologies help cut down on greenhouse gas pollution right away, especially in hard-to-decarbonize industries like steel, cement, and fossil fuel energy (Bachu et al. 2007). The CCSU system prevents dangerous amounts of CO_2 from getting into the air by removing it directly from sources of emission, including power stations or industrial sites. After being collected, CO_2 may be safely stored in geological formations or used creatively to make useful goods (Cormos et al. 2017).

There is a wide range of potential uses for CO_2, such as converting it into artificial fuels, polymers, and construction materials (such as carbon-negative concrete). These applications promote a regenerative economy by repurposing waste materials and cutting down on the need for raw materials. The immediate objective is to decrease dependence on fossil fuels, but CCSU also helps with the recovery of enhanced oil by increasing the efficiency of oil extraction via the use of CO_2 (Yasemi et al. 2023).

Additionally, the CCSU sector paves the way for sustainable economic transition by creating many new jobs in the emerging technology sector. Around the world, the importance of achieving net-zero goals is growing, with CCSU emerging as a practical solution to balance industrial activity with environmental responsibility (Osman et al. 2021). Table 11.6 highlighting the potential applications and opportunities of CCSU across different industries.

11.8 Challenges of Carbon Capture Storage and Utilization Adoption

CCSU technology has the probable to significantly decrease greenhouse gas releases, but it is facing formidable obstacles to widespread adoption. Costly CCSU infrastructure and operating needs, such as securely transporting and storing CO_2, are one of the main barriers. Without considerable subsidies from governments or policy incentives, the high prices of CCSU technologies prevent their widespread adoption, especially in businesses with already slim profit margins (Dalei and Joshi 2022; Obahiagbon and Ogwu 2023, 2024).

Technical challenges also complicate CCSU implementation (Table 11.7). Sophisticated machinery is needed for efficient CO_2 collecting, and evaluating the stability and leaking potential of CO_2 stored in geological formations is a difficult process. Further complicating logistics might be the transportation of CO_2 from collection locations to storage or usage facilities, particularly in areas lacking well-established pipeline networks or secure storage choices. Public opinion and regulatory considerations provide a further obstacle (Park 2018). Many locals are wary of CCSU projects, particularly those involving storage sites, since they are worried about the

Table 11.6 Potential applications and opportunities of carbon capture storage and utilization

Application area	Description	Opportunities
Enhanced oil recovery	Injection of CO_2 into depleted oil fields to enhance oil recovery	Increased oil production and reduced CO_2 emissions; economic incentive to offset capture costs
Cement and concrete production	Utilization of CO_2 in cement curing or as an ingredient in concrete	Reduces overall emissions in cement manufacturing and creates stronger, more durable building materials
Synthetic fuels	Conversion of captured CO_2 into methanol, ethanol, or synthetic hydrocarbons	Provides low-carbon alternatives to fossil fuels; supports energy transition in transportation and power sectors
Chemicals and polymers	Transformation of CO_2 into chemicals like methanol or carbonates used in polymers	Enables production of sustainable chemicals; integrates CO_2 into manufacturing chains, reducing dependency on virgin fossil resources
Agriculture and algae cultivation	CO_2 utilization in greenhouses and algae growth for biofuel and biomass	Enhances agricultural productivity; supports bio-based energy and material production, potentially replacing high-carbon inputs
Energy storage and batteries	CO_2 conversion into electrochemical energy storage materials	Supports the development of sustainable energy storage solutions, crucial for renewable energy integration
Carbon nanomaterials	CO_2 transformation into nanomaterials like graphene and carbon fibers	High-value applications in electronics, composites, and construction materials, promoting a circular economy

potential for CO_2 leaks. On top of that, there is a lack of uniformity in regulatory frameworks, and different foreign standards make it much more difficult for businesses to implement CCSU on a worldwide scale. For CCSU to fulfill its mission of promoting sustainable industrial practices and reaching net-zero objectives, it is crucial to address these social, technological, and economic obstacles (Selma et al. 2014).

Table 11.7 Challenges in carbon capture storage and utilization implementation

Challenge	Description	Impacts	Potential solutions	References
High cost	Expensive technology and operations	Limits adoption	Subsidies, carbon credits	Dziejarski et al. (2023)
Infrastructure needs	Requires specialized pipelines	Storage and transport limitations	Investment in infrastructure	
Monitoring and safety	Ensuring long-term containment	Environmental risks	Advanced monitoring technologies	
Public acceptance	Skepticism about safety	Delays project approvals	Community education	

11.9 Upcoming Scope of Carbon Capture Storage and Utilization

Technological advancements, regulatory backing, and the pressing need to address environmental concerns are fueling worldwide interest in CCSU, which holds enormous promise for its future scope. The capacity of CCSU to convert CO_2 emissions into resources creates enormous prospects in many different industries, leading to a more sustainable and circular economy. With governments and businesses alike setting lofty net-zero emissions goals, CCSU is a vital tool for cutting emissions, particularly in industries where regulations are notoriously difficult to change, such as chemical manufacturing, cement production, and steel production (Bose et al. 2024). CCSU's role in producing useful goods from CO_2, such as synthetic chemicals, alternative fuels, and concrete, while also reducing emissions, aligns well with the worldwide movement towards sustainable production techniques (Richard et al. 2024). New developments in CO_2 conversion have made it possible to reintroduce captured carbon into supply chains, which increases economic value and reduces reliance on virgin resources. CCSU enhances the effectiveness of renewable energy sources by converting CO_2 produced by burning biomass or biogas plants into a climate-neutral product. This will allow these sources to have an even greater influence on achieving climate targets (Nwabueze and Leggett 2024).

There is anticipation that ongoing research and development will lead to even greater CCSU efficiency, cheaper prices, and a wider range of applications, opening it up to a wider range of sectors. Government incentives, carbon pricing systems, and multinational cooperation are all working to make large-scale CCSU deployment more likely. Thus, the future of CCSU has enormous promise to transform sectors, strengthen economies, and have a significant impact on reaching global sustainability targets (Bose et al. 2024).

Figure 11.3 highlights that there could be a significant difference between the levels of transformations required in the energy system with CCS and without CCS

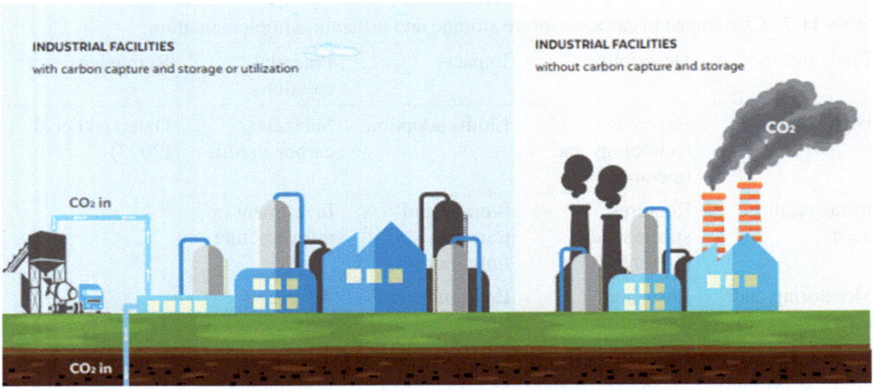

Fig. 11.3 Industrial facilities with carbon capture and storage or utilization and without carbon capture and storage

in the economy. To achieve net-zero carbon by 2050 (IEA 2021), the proportion of renewable energy sources other than hydropower might fall by 13% if CCS becomes practical for power production from fossil fuels. With CCS technology, fossil fuels would contribute 30% of energy, up from 5% without it. Implementing CCS in refineries could boost biofuel sales and slow down the transition to electric freight trucks, leading to a 21% decrease in sales by 2050 compared to 67% without it. CCS has a mitigating potential of about 780 Mt/year below 60 USD/t CO_2 in a 2 °C scenario and 250 Mt/year up to 75 USD/t CO_2 in a far below 2 °C scenario, according to deep decarbonization scenarios in India (Das et al. 2023). Nevertheless, CCS feasibility is an important assumption to consider, and there has been little progress in India over the last decade about its economic and policy viability The CCS government must address issues such as high liability, sanctions, inadequate public knowledge and engagement, poor long-term planning, and stewardship (Clarke et al. 2022). To grasp the evolving story of CCSU in India, it is necessary to review papers produced by the Indian government, literature from across the world, and important announcements made by Indian businesses (Anika et al. 2022). We cover the assessment's important findings, conclusions, and next steps, highlighting areas that require attention to maintain the ongoing CCS/CCSU discussion in India.

11.10 Conclusion

CCSU technologies are an important part of the world's strategy to combat climate change since they reduce emissions of carbon while simultaneously improving resource efficiency and creating new economic opportunities. The energy-intensive businesses and sectors that rely on fossil fuels continue to produce significant amounts of emissions. CCSU offers a complete solution to this problem by capturing and storing carbon, but also by using it as an asset. As CCSU decarbonizes established industries while simultaneously paving the way for future low-carbon ones, it becomes clear how important it is to the shift to a low-carbon economy. The potential of CCSU has been substantially enhanced because of developments in capture technologies such as chemical absorption, membrane technology, and cryogenic separation, which have reduced prices and improved efficiency. To manage atmospheric carbon concentrations and lessen environmental effects in the long run, storage in safe geological formations is an attractive option. The usage component of CCSU changes the way we think about emissions and resource consumption by transforming CO_2 into fuels, building materials, and chemical feedstocks. This establishes a model for a circular economy that turns carbon from waste into a resource. These developments point to a change in perspective, moving away from CO_2 being seen as a byproduct and toward its important role in a sustainable manufacturing cycle. Nevertheless, CCSU still faces substantial obstacles on the road to realizing its maximum potential. Challenges that can prevent broad use include high starting costs, complicated infrastructure needs, and high energy input requirements. For CCSU to be economically viable, scaling technologies are essential for making them inexpensive. This is especially true in areas that restrict decarbonization choices. Furthermore, the widespread use of CCSU requires strict regulatory frameworks, public backing, and incentives to address societal concerns about CO_2 transportation and storage, all while ensuring safety. The environmental and economic advantages of CCSU are substantial, since it can create jobs, propel technical advancement, and encourage the growth of environmentally conscious businesses. To make CCSU financially feasible, government measures are crucial, including carbon pricing, tax incentives, and R&D subsidies. To construct the required infrastructure and to exchange technological knowledge, it is crucial to get the private sector involved and for different sectors to work together.

In the future, CCSU's involvement will likely grow in response to the more ambitious global climate objectives, especially in industries with few viable alternatives that are difficult to change. To defeat economic and technological obstacles and make solutions more accessible and efficient, greater funding for CCSU's research and development is essential. Combining CCSU with renewable energy, bioenergy, and hydrogen technologies could potentially achieve even more effective ways to reduce emissions and manage resources sustainably. In essence, CCSU has transformative potential for reducing emissions and promoting a robust, sustainable economic paradigm. As we strive to reduce our reliance on fossil fuels, CCSU offers a practical solution for controlling and recycling CO_2 emissions. This will allow companies that produce a lot of carbon to switch to more sustainable methods without negatively impacting their bottom lines. Global efforts to address climate change and achieve sustainable development may use CCSU as a core pillar by promoting innovation, encouraging cooperation, and enacting supportive legislation. We must all work together to make CCSU a mainstream option and ensure its success so it can contribute significantly to the low-carbon future.

References

Aguilar-Martínez O, Santolalla-Vargas CE, Tzompantzi F, González F, Romero-Ibarra IC, de los Reyes JA, Díaz-Barriga Arceo LG, Santes V (2023) Improvement of the photocatalytic activity of CeO_2 on the degradation of chlorophenolic compounds obtained by solution combustion synthesis (SCS). Effect of urea as fuel. Solid State Sci 143:107274. https://doi.org/10.1016/j.solidstatesciences.2023.107274

Bermeo M, Vega LF, Abu-Zahra KM (2022) Critical assessment of the performance of next-generation carbon-based adsorbents for CO_2 capture focused on their structural properties. Sci Total Environ 810:151720. https://doi.org/10.1016/j.scitotenv.2021.151720

Carbo MC, Jansen D, Hendriks C, de Visser E, Ruijg GJ, Davison J (2009) Opportunities for CO_2 capture through oxygen conducting membranes at medium-scale oxyfuel coal boilers. Energy Procedia 1:487–494

Chen S (2018) Fundamentals of oxy-fuel combustion. In: Zheng C, Liu Z (eds) Oxy-fuel combustion. Academic Press, pp 13–30. https://doi.org/10.1016/B978-0-12-812145-0.00002-5

Chen Z, Wang L, Huang Z, Zhuang P, Shi Y, Evrendilek F, Huang S, He Y, Liu J (2024) Dynamic and optimal ash-to-gas responses of oxy-fuel and air combustions of soil remediation biomass. Renew Energy 225:120299. https://doi.org/10.1016/j.renene.2024.120299

Cook PJ (2009) Demonstration and deployment of carbon dioxide capture and storage in Australia. Energy Procedia 1:3859–3866. https://doi.org/10.1016/j.egypro.2009.02.188

Farmahini H, Friedrich M, Brandani S, Sarkisov L (2020) Exploring new sources of efficiency in process-driven materials screening for post-combustion carbon capture. Energy Environ Sci 13:2020. https://doi.org/10.1039/C9EE03977E

Gkotsis P, Peleka E, Zouboulis A (2023) Membrane-based technologies for post-combustion CO_2 capture from flue gases: recent progress in commonly employed membrane materials. Membranes 13(12):898. https://doi.org/10.3390/membranes13120898

Gładysz P, Ziębik A (2015) Life cycle assessment of an integrated oxy-fuel combustion power plant with CO_2 capture, transport and storage—Poland case study. Energy 92:328–340. https://doi.org/10.1016/j.energy.2015.07.052

Goldthorpe S (2017) Potential for very deep ocean storage of CO_2 without ocean acidification: a discussion paper. Energy Procedia 114:5417–5429. https://doi.org/10.1016/j.egypro.2017.03.1686

Ha-Duong M, Keith DW (2003) Carbon storage: the economic efficiency of storing CO_2 in leaky reservoirs. Clean Techn Environ Policy 5:181–189

Hong J, Field R, Gazzino M, Ghoniem AM (2010) Operating pressure dependence of the pressurized oxy-fuel combustion power cycle. Energy 35:5391–5399. https://doi.org/10.1016/j.energy.2010.07.016

James AM, Reynolds J, Reed DG, Styring P, Dawson R (2021) A pressure swing approach to selective CO_2 sequestration using functionalized hypercrosslinked polymers. Materials 14(7):1605. https://doi.org/10.3390/ma14071605

Karaszova M, Zach B, Petrusová Z, Červenka V, Bobák M, Šyc M, Izák P (2020) Post-combustion carbon capture by membrane separation, review. Sep Purif Technol 238:116448. https://doi.org/10.1016/j.seppur.2019.116448

Kotowicz J, Balicki A (2014) Enhancing the overall efficiency of a lignite-fired oxyfuel power plant with CFB boiler and membrane-based air separation unit. Energy Convers Manag 80:20–31. https://doi.org/10.1016/j.enconman.2013.12.069

Krishnamurthy S, Lind A, Bouzga A, Pierchala J, Blom R (2021) Post combustion carbon capture with supported amine sorbents: from adsorbent characterization to process simulation and optimization. Chem Eng J 406:127121. https://doi.org/10.1016/j.cej.2020.127121

Krishnan A, Nighojkar A, Kandasubramanian B (2023) Emerging towards zero carbon footprint via carbon dioxide capturing and sequestration. Carbon Capture Sci Technol 9:100137. https://doi.org/10.1016/j.ccst.2023.100137

Kumar M, Sundaram S, Gnansounou E, Larroche C, Thakur IS (2018) Carbon dioxide capture, storage and production of biofuel and biomaterials by bacteria: a review. Bioresour Technol 247:1059–1968. https://doi.org/10.1016/j.biortech.2017.09.050

Lovelock JE, Rapley CG (2007) Ocean pipes could help the earth to cure itself. Nature 449(7161):403. https://doi.org/10.1038/449403a

McDonald TM, Mason JA, Kong X, Bloch ED, Gygi D, Dani A, Crocellà V, Giordanino F, Odoh SO, Drisdell WS, Vlaisavljevich B, Dzubak AL, Poloni R, Schnell SK, Planas N, Lee K, Pascal T, Wan LF, Prendergast D, Neaton JB, Smit B, Kortright JB, Gagliardi L, Bordiga S, Reimer JA, Long JR (2015) Cooperative insertion of CO_2 in diamine-appended metal-organic frameworks. Nature 519(7543):303–308. https://doi.org/10.1038/nature14327

McNeil BI, Matear RJ (2008) Southern ocean acidification: a tipping point at 450-ppm atmospheric CO_2. Biol Sci 105:18860–18864. https://doi.org/10.1073/pnas.0806318105

Pfister M, Belaissaoui B, Favre E (2017) Membrane gas separation processes from wet postcombustion flue gases for carbon capture and use: a critical reassessment. Ind Eng Chem Res 56:591–602. https://doi.org/10.1021/acs.iecr.6b03969

Portillo E, Gallego Fernández LM, Cano M, Alonso-Farinas B, Navarrete B (2022) Techno-economic comparison of integration options for an oxygen transport membrane unit into a coal oxy-fired circulating fluidized bed power plant. Membranes (Basel) 12(12):1224. https://doi.org/10.3390/membranes12121224

Portillo E, Gandara-Loe J, Reina TR, Pastor-Pérez L (2023) Is the RWGS a viable route for CO_2 conversion to added value products? A techno-economic study to understand the optimal RWGS conditions. Sci Total Environ 857:159394. https://doi.org/10.1016/j.scitotenv.2022.159394

Qazvini OT, Telfer SG (2020) A robust metal–organic framework for post-combustion carbon dioxide capture. J Mater Chem A 8:12028–12034. https://doi.org/10.1039/D0TA04121A

Schneider LJ, Bralower TJ, Kump LR (2011) Response of nannoplankton to early Eocene ocean destratification. Palaeogeogr Palaeoclimatol Palaeoecol 310:152–162. https://doi.org/10.1016/j.palaeo.2011.06.018

Shi Y, Liu Q, Shao Y, Zhong W (2020) Energy and exergy analysis of oxy-fuel combustion based on circulating fluidized bed power plant firing coal, lignite and biomass. Fuel 269:117424. https://doi.org/10.1016/j.fuel.2020.117424

Shin J, Park S (2023) Numerical analysis for optimizing combustion strategy in an ammonia-diesel dual-fuel engine. Energy Convers Manag 284:116980. https://doi.org/10.1016/j.enconman.2023.116980

Singh A, Stéphenne K (2014) Shell Cansolv CO_2 capture technology: achievement from first commercial plant. Energy Procedia 63:1678–1685

Stanger R, Wall T, Sporl R, Paneru M, Grathwohl S, Weidmann M, Scheffknecht G, McDonald D, Myohanen K, Ritvanen J, Rahiala S, Hyppanen T, Mletzko J, Kather A, Santos S (2015) Oxyfuel combustion for CO_2 capture in power plants. Int J Greenhouse Gas Control 40:55–125. https://doi.org/10.1016/j.ijggc.2015.06.010

Stephenne K (2014) Start-up of world's first commercial post-combustion coal fired CCS project: contribution of Shell Cansolv to SaskPower Boundary Dam ICCS project. Energy Procedia 63:6106–6110. https://doi.org/10.1016/j.egypro.2014.11.642

Wei X, Manovic V, Hanak DP (2020) Techno-economic assessment of coal- or biomass-fired oxy-combustion power plants with supercritical carbon dioxide cycle. Energy Convers Manag 221:113143. https://doi.org/10.1016/j.enconman.2020.113143

White CM, Strazisar BR, Granite EJ, Hoffman JS, Pennline HW (2003) Separation and capture of CO_2 from large stationary sources and sequestration in geological formations—coalbeds and deep saline aquifers. J Air Waste Manage Assoc 53:645–715

Wienchol P, Szlek A, Ditaranto M (2020) Waste-to-energy technology integrated with carbon capture—challenges and opportunities. Energy 198:117352. https://doi.org/10.1016/j.energy.2020.117352

Wu X, Wang M, Liao P, Shen J, Li Y (2020) Solvent-based post-combustion CO_2 capture for power plants: a critical review and perspective on dynamic modelling, system identification, process control and flexible operation. Appl Energy 257:113941

Yang X, Szuhanszki J, Tian Y, Ingham D, Ma L, Pourkashanian M (2019) Understanding the effects of oxyfuel combustion and furnace scale on biomass ash deposition. Fuel 247:36–46. https://doi.org/10.1016/j.fuel.2019.03.031

Yun S, Oh SY, Kim JK (2020) Techno-economic assessment of absorption-based CO_2 capture process based on novel solvent for coal-fired power plant. Appl Energy 268:114933. https://doi.org/10.1016/j.apenergy.2020.114933

Zhang X, Singh B, He X, Gundersen T, Deng L, Zhang S (2014) Post-combustion carbon capture technologies: energetic analysis and life cycle assessment. Int J Greenhouse Gas Control 27:289–298. https://doi.org/10.1016/j.ijggc.2014.06.016

Zhou T, Shi H, Ding X, Zhou Y (2021a) Thermodynamic modeling and rational design of ionic liquids for pre-combustion carbon capture. Chem Eng Sci 229:116076. https://doi.org/10.1016/j.ces.2020.116076

Zhou X, Zang X, Long Z, Liang D (2021b) Multiscale analysis of the hydrate based carbon capture from gas mixtures containing carbon dioxide. Sci Rep 11:9197. https://doi.org/10.1038/s41598-021-88531-x

Aminu MD, Nabavi SA, Rochelle CA, Manovic V (2017) A review of developments in carbon dioxide storage. Appl Energy 208:1389–1419. https://doi.org/10.1016/j.apenergy.2017.09.015

Anika OC, Nnabuife SG, Bello A, Okoroafor ER, Kuang B, Villa R (2022) Prospects of low and zero-carbon renewable fuels in 1.5-degree net zero emission actualisation by 2050: a critical review. Carbon Capture Sci Technol 5:100072. https://doi.org/10.1016/j.ccst.2022.100072

Anonymous (2020) Global CCS institute, the global status of CCS: 2020. Global CCS Institute, Melbourne, Australia. Available online: https://www.globalccsinstitute.com/wp-content/uploads/2021/03/Global-Status-of-CCS-Report-English.pdf

Azdarpour A, Asadullah M, Mohammadian E, Hamidi H, Junin R, Karaei MA (2015) A review on carbon dioxide mineral carbonation through pH-swing process. Chem Eng J 279:615–630

Bachu S (2000) Overview of CO_2 storage in geological media. Energy Convers Manag 41:633–640

Bachu S, Bonijoly D, Bradshaw J, Burruss R, Holloway S, Christensen NP, Mathiassen OM (2007) CO_2 storage capacity estimation: methodology and gaps. Int J Greenh Gas Control 1:430–443

Bai M, Zhang Z, Fu X (2016) A review on well integrity issues for CO_2 geological storage and enhanced gas recovery. Renew Sustain Energy Rev 59:920–926

Bao L, Liu D, Liu J, Lu Y, Wei Y (2019) CO_2 sequestration potential in deep coal seams: a review. Fuel 235:1037–1050

Baran P, Zarębska K, Krzystolik P, Hadro J, Nunn A (2014) CO_2-ECBM and CO_2 sequestration in Polish coal seam—experimental study. J Sustain Min 13:22–29

Belhaj H, Bera A (2017) A brief review of mechanisms for carbon dioxide sequestration into aquifer reservoirs. Int J Pet Eng 3(2):49–66. https://doi.org/10.1504/IJPE.2017.089315

Bhavsar A, Hingar D, Ostwal S, Thakkar I, Jadeja S, Shah M (2023) The current scope and stand of carbon capture storage and utilization—a comprehensive review. Case Stud Chem Environ Eng 8:100368. https://doi.org/10.1016/j.cscee.2023.100368

Bickle MJ (2009) Geological carbon storage. Nat Geosci 2(12):815–818. https://doi.org/10.1038/ngeo687

Bobicki ER, Liu Q, Xu Z, Zeng H (2012) Carbon capture and storage using alkaline industrial wastes. Prog Energy Combust Sci 38:302–320

Bose D, Bhattacharya R, Kaur T, Pandya R, Sarkar A, Ray A, Mondal S, Mondal A, Ghosh P, Chemudupati R (2024) Innovative approaches for carbon capture and storage as crucial measures for emission reduction within industrial sectors. Carbon Capture Sci Technol 12:100238. https://doi.org/10.1016/j.ccst.2024.100238

Bui M, Adjiman CS, Bardow A, Anthony EJ, Boston A, Brown S, Fennell PS, Fuss S, Galindo A, Hackett LA et al (2008) Carbon capture and storage (CCS): the way forward. Energy Environ Sci 11:1062–1176

Butt DP, Lackner KS, Wendt CH, Vaidya R, Pile DL, Park Y, Holesinger T, Harradine DM, Koji N (1998) The kinetics of binding carbon dioxide in magnesium carbonate. In: Proceedings of the 23th international technical conference on coal utilization and fuel systems, Clearwater, FL, USA, pp 9–13

Cantucci B, Montegrossi G, Vaselli O, Tassi F, Quattrocchi F, Perkins EH (2009) Geochemical modeling of CO_2 storage in deep reservoirs: the Weyburn Project (Canada) case study. Chem Geol 265:181–197

Celia MA, Bachu S, Nordbotten JM, Bandilla KW (2015) Status of CO_2 storage in deep saline aquifers with emphasis on modeling approaches and practical simulations. Water Resour Res 6846–6892. https://doi.org/10.1002/2015WR017609

Chaturwedi AK, Hait M (2023) Hydrochemical investigation of water bodies from the effluents of Sirgitti industrial area of Bilaspur (C.G.). Bull Environ Pharmacol Life Sci 12(10):177–183. http://www.bepls.com

Chen GL, Song XF, Sun SY, Xu YX, Yu JG (2016) Solubility and diffusivity of CO_2 in n-butanol + N235 system and absorption mechanism of CO_2 in a coupled reaction–extraction process. Front Chem Sci Eng 10:480–489

Chen Y-P, Bashir S, Liu J (2015) Carbon capture and storage, chap 7. In: Liu JL, Bashir S (eds) Advanced nanomaterials and their applications in renewable energy. Elsevier, pp 329–366. https://doi.org/10.1016/B978-0-12-801528-5.00007-5

Chiaramonte L, Zoback M, Friedmann J, Stamp V, Zahm C (2011) Fracture characterization and fluid flow simulation with geomechanical constraints for a CO_2–EOR and sequestration project Teapot Dome Oil Field, Wyoming, USA. Energy Procedia 4:3973–3980

Chunbao Charles XU, Cang DQ (2010) A brief overview of low CO_2 emission technologies for iron and steel making. J Iron Steel Res Int 17:1–7

Clarke L, Wei Y-M, De La Vega Navarro A, Garg A, Hahmann AN, Khennas S, Azevedo IML, Löschel A, Singh AK, Steg L, Strbac G, Wada K (2022) Energy systems. In: Shukla PR, Skea J, Slade R, Al Khourdajie A, van Diemen R, McCollum D, Pathak M, Some S, Vyas P, Fradera R, Belkacemi M, Hasija A, Lisboa G, Luz S, Malley J, (eds) IPCC, 2022: climate change 2022: mitigation of climate change. Contribution of working group III to the sixth assessment report of the Intergovernmental Panel on Climate Change. Cambridge University Press, Cambridge, UK, New York, NY, USA. https://doi.org/10.1017/9781009157926.008

Cormos A-M, Dinca C, Petrescu L, Chisăliţă D, Szima S, Cormos C-C (2017) Carbon capture and utilisation technologies applied to energy conversion systems and other energy-intensive industrial applications. Fuel 211. https://doi.org/10.1016/j.fuel.2017.09.104

Dalei NN, Joshi J (2022) Potential matching of carbon capture storage and utilization (CCSU) as enhanced oil recovery in perspective to Indian oil refineries. Clean Technol Environ Policy 24(9):2701–2717. https://doi.org/10.1007/s10098-022-02359-1

Das A, Saini V, Parikh K, Parikh J, Ghosh P, Tot M (2023) Pathways to net zero emissions for the Indian power sector. Energy Strat Rev 45:101042. https://doi.org/10.1016/j.esr.2022.101042

Davison J, Freund P, Smith A (2001) Putting carbon back in the ground; IEA greenhouse gas R&D programme. Cheltenham, UK. ISBN 1898373280

De Silva GPD, Ranjith PG, Perera MSA (2015) Geochemical aspects of CO_2 sequestration in deep saline aquifers: a review. Fuel 155:128–143

Dziejarski B, Krzyżyńska R, Andersson K (2023) Current status of carbon capture, utilization, and storage technologies in the global economy: a survey of technical assessment. Fuel 342:127776. https://doi.org/10.1016/j.fuel.2023.127776

Ehlig-Economides C, Economides MJ (2010) Sequestering carbon dioxide in a closed underground volume. J Pet Sci Eng 70:123–130

Eloneva S, Teir S, Salminen J, Fogelholm CJ, Zevenhoven R (2008) Fixation of CO_2 by carbonating calcium derived from blast furnace slag. Energy 33:1461–1467

Fawzy S, Osman AI, Doran J et al (2020) Strategies for mitigation of climate change: a review. Environ Chem Lett 18:2069–2094. https://doi.org/10.1007/s10311-020-01059-w

Gabrielli P, Gazzani M, Mazzotti M (2020) The role of carbon capture and utilization, carbon capture and storage, and biomass to enable a net-zero-CO_2 emissions chemical industry. Ind Eng Chem Res 59(15):7033–7045. https://doi.org/10.1021/acs.iecr.9b06579
Gadikota G, Matter J, Kelemen P, Park AHA (2014) Chemical and morphological changes during olivine carbonation for CO_2 storage in the presence of NaCl and $NaHCO_3$. Phys Chem Chem Phys 16:4679–4693
Gale J, Abanades JC, Bachu S, Jenkins C (2015) Special issue commemorating the 10th year anniversary of the publication of the Intergovernmental Panel on Climate Change special report on CO_2 capture and storage. Int J Greenh Gas Control 40:1–5
Galina NR, Arce GLAF, Ávila I (2019) Evolution of carbon capture and storage by mineral carbonation: data analysis and relevance of the theme. Miner Eng 142:105879
Gulzar A, Gulzar A, Ansari MB, He F, Gai S, Yang P (2020) Carbon dioxide utilization: a paradigm shifts with CO_2 economy. Chem Eng J Adv 3:100013. https://doi.org/10.1016/j.ceja.2020.100013
Gunter WD (2000) CO_2 sequestration in deep unmineable coal seams. Alberta Research Council, Edmonton, AB, Canada
Gunter WD, Bachu S, Benson S (2004) The role of hydrogeological and geochemical trapping in sedimentary basins for secure geological storage of carbon dioxide. Geol Soc Lond Spec Publ 233:129–145
Guo J, Huang C (2019) Feasible roadmap for CCS retrofit of coal-based power plants to reduce Chinese carbon emissions by 2050. Appl Energy 259:114112. https://doi.org/10.1016/j.apenergy.2019.114112
Gupta P, Biswas S, Chaturwedi AK, Janghel U, Tamrakar G, Verma R, Hait M (2023) Health impact of ground water sample and its effect on flora of Kanker District, Chhattisgarh, India. ES Food Agrof 13:940
Hait M, Chandran J, Chaturwedi AK, Vaishnav MM (2019a) Treatment of water bodies from the effluent of Sirgitti industrial area Bilaspur using bioadsorbents. Pharma Sci Mon 10(2):112–120
Hait M, Gir ND, Dey A, Chaturwedi A, Kashyap NK (2018a) Removal of water pollutants from the effluent of Tifra industrial area in Bilaspur (C.G.) using bio-adsorbents. Pharma Innov J 6(3):15–20
Hait M, Gir ND, Kashyap NK, Dey A, Chaturwedi A (2019b) Bioremoval of water pollutants from the effluents of Tifra industrial area in Bilaspur (C.G.). Int J Curr Eng Sci Res 6(1):399–401
Hait M, Kashyap NK, Chaturwedi AK (2018b) Physicochemical investigation of water bodies of Arpa River at Bilaspur urban areas. Int J Chem Stud 6(1):1773–177. ISSN 2321-4902
Hawkes CD, Gardner C (2013) Pressure transient testing for assessment of wellbore integrity in the IEAGHG Weyburn–Midale CO_2 monitoring and storage project. Int J Greenh Gas Control 16:S50–S61
Herzog HJ (2009) Carbon capture and storage from fossil fuel use. Encycl Energy 1:429–437
Hu Y, Cheng L (2017) Carbon dioxide sequestration in unmineable coal seams: a review. Int J Coal Geol 173:118–134
Huijgen WJJ (2007) Carbon dioxide sequestration by mineral carbonation. Energy Research Centre of the Netherlands, Petten, The Netherlands. ISBN 90-8504-573-8
IEA (2021) Net zero by 2050. IEA, Paris https://www.iea.org/reports/net-zero-by-2050. Licence: CC BY 4.0
Igunnu ET, Chen GZ (2014) Produced water treatment technologies. Int J Low-Carbon Technol 9:157
IPCC global warming of 1.5 °C (2018) In: Masson-Delmotte V, Zhai P, Potner H-O, Roberts D, Skea J, Shukla PR, Pirani A, Moufouma-Okia W, Péan C, Pidcock R et al (eds) An IPCC special report on the impacts of global warming of 1.5 °C above pre–industrial levels and related global greenhouse gas emission pathways, in the context of strengthening the global response to the threat of climate change, sustainable development, and efforts to eradicate poverty. Cambridge University Press, Cambridge, UK, New York, NY, USA

IPCC (2022) Climate change (2022): mitigation of climate change. In: Shukla PR, Skea J, Slade R, Al Khourdajie A, van Diemen R, McCollum D, Pathak M, Some S, Vyas P, Fradera R et al (eds) Contribution of working group III to the sixth assessment report of the Intergovernmental Panel on Climate Change. Cambridge University Press, Cambridge, UK, New York, NY, USA. https://doi.org/10.1017/9781009157926

IPCC (2005) IPCC special report on carbon dioxide capture and storage. In: Metz B, Davidson O, de Coninck HC, Loos M, Meyer LA (eds) Prepared by working group III of the Intergovernmental Panel on Climate Change. Cambridge University Press, Cambridge, UK, New York, NY, USA

Jia Y, Li X, Zhang L, Chen Y, Sun X, Gao F (2018) Enhanced coal bed methane recovery and CO_2 storage potential in coal seams: a review. Fuel 215:251–263

Jiang J-M, Jin L, Huang L, Wang W-T (2022) The future climate under different CO_2 emission scenarios significantly influences the potential distribution of *Achnatherum inebrians* in China. Sustainability 14:4806. https://doi.org/10.3390/su14084806

Kandji EHB, Plante B, Bussière B, Beaudoin G, Dupont P (2017) Geochemical behavior of ultramafic waste rocks with carbon sequestration potential: a case study of the Dumont Nickel Project, Amos, Québec. Environ Sci Pollut Res 24:11734–11751

Ketabchi MR, Babamohammadi S, Davies WG, Gorbounov M, Soltani SM (2023) Latest advances and challenges in carbon capture using bio-based sorbents: a state-of-the-art review. Carbon Capture Sci Technol 6:10008. https://doi.org/10.1016/j.ccst.2022.100087

Khan U, Ogbaga CC, Abiodun O-A, Adeleke O, Ikubanni AA, Okoye PP, Okolie PU, Jude A (2023) Assessing absorption-based CO_2 capture: research progress and techno-economic assessment overview. Carbon Capture Sci Technol 8:100125. https://doi.org/10.1016/j.ccst.2023.100125

Klusman RW (2003) Evaluation of leakage potential from a carbon dioxide EOR/sequestration project. Energy Convers Manag 44:1921–1940

Kosoe EA, Achana GTW, Ogwu MC (2023) Regulations and policies for herbal medicine and practitioners. In: Izah SC, Ogwu MC, Akram M (eds) Herbal medicine phytochemistry. Reference series in phytochemistry. Springer, Cham. https://doi.org/10.1007/978-3-031-21973-3_33-1

Kosoe EA, Ogwu MC (2024) Sustainable cassava processing: processes, techniques, wastes, and waste streams. In: Ogwu MC, Izah SC, Cunha Alves AA, Babu SC (eds) Plant biology, sustainability and climate change, sustainable cassava. Academic Press, pp 257–272. https://doi.org/10.1016/B978-0-443-21747-0.00011-4

Kramer GJ, Haigh M (2009) No quick switch to low-carbon energy. Nature 462(7273):568–569. https://doi.org/10.1038/462568a

Kumar S, Kumar A, Mandal A (2017) Characterizations of surfactant synthesized from Jatropha oil and its application in enhanced oil recovery. AIChE J 63:2731–2741

Kwon S, Fan M, Dacosta HFM, Russell AG (2011) Factors affecting the direct mineralization of CO_2 with olivine. J Environ Sci 23:1233–1239

Lackner KS (2002) Carbonate chemistry for sequestering fossil carbon. Annu Rev Energy Environ 27:193–232

Lackner KS, Butt DP, Wendt CH (1997) Progress on binding CO_2 in mineral substrates. Energy Convers Manag 38:S259–S264

Lal R (2008) Sequestration of atmospheric CO_2 in global carbon pools. Energy Environ Sci 1:86–100

Lee JS, Choi EC (2018) CO_2 leakage environmental damage cost—a CCS project in South Korea. Renew Sustain Energy Rev 93:753–758

Lehmann A (2007) Ocean carbon dioxide storage: geological and chemical aspects. Encycl Ocean Sci 3:1795–1803

Li JJ, Hitch M, Power I, Pan Y (2018) Integrated mineral carbonation of ultramafic mine deposits—a review. Minerals 8:147

Li JJ, Jacobs AD, Hitch M (2019) Direct aqueous carbonation on olivine at a CO_2 partial pressure of 6.5 MPa. Energy 173:902–910

Li ZW, Dong MZ, Li SL, Huang S (2006) CO_2 sequestration in depleted oil and gas reservoirs–caprock characterization and storage capacity. Energy Convers Manag 47:1372–1382

Lim M, Han GC, Ahn JW, You KS (2010) Environmental remediation and conversion of carbon dioxide (CO_2) into useful green products by accelerated carbonation technology. Int J Environ Res Public Health 7:203–228

Liu D, Fernandez Y, Ola O, Mackintosh S, Maroto-Valer MM, Parlett CMA, Lee AF, Wu JCS (2012) On the impact of Cu dispersion on CO_2 photoreduction over Cu/TiO_2. Catal Commun 25:78–82

Liu J, Qin Y, Zhang J, Zhou L (2017) Carbon dioxide sequestration in deep unmineable coal seams: a review. Energy Sci Eng 5:157–172

Liu W, Su S, Xu K, Chen Q, Xu J, Sun ZJ, Wang Y, Hu S, Wang XL, Xue YT et al (2018) CO_2 sequestration by direct gas–solid carbonation of fly ash with steam addition. J Clean Prod 178:98–107

Liu WZ, Teng LM, Rohani S, Qin ZF, Zhao B, Xu CC, Ren S, Liu QC, Liang B (2021) CO_2 mineral carbonation using industrial solid wastes: a review of recent developments. Chem Eng J 416:129093

Malyan C (2021) Carbon capture, utilization, and storage (CCSU) in India. https://www.ceew.in/sites/default/files/ceew-study-on-the-role-of-carbon-capture-utilization-and-storage-in-india.pdf

Myers CA, Nakagaki T, Akutsu K (2019) Quantification of the CO_2 mineralization potential of ironmaking and steelmaking slags under direct gas–solid reactions in flue gas. Int J Greenh Gas Control 87:100–111

Nagireddi S, Agarwal JR, Vedapuri D (2024) Carbon dioxide capture, utilization, and sequestration: current status, challenges, and future prospects for global decarbonization. ACS Eng Au 4(1):22–48. https://doi.org/10.1021/acsengineeringau.3c00049

Naraharisetti PK, Yeo TY, Bu J (2017) Factors influencing CO_2 and energy penalties of CO_2 mineralization processes. Chem Phys Chem 18:3189–3202

Nizzy AM, Ogwu MC (2024) Valorization of cassava processing by-products into biofuel for a sustainable environment. In: Ogwu MC, Izah SC, Cunha Alves AA, Babu SC (eds) Plant biology, sustainability and climate change, sustainable cassava. Academic Press, pp 291–309. https://doi.org/10.1016/B978-0-443-21747-0.00010-2

Nunes LJR (2023) The rising threat of atmospheric CO_2: a review on the causes, impacts, and mitigation strategies. Environments 10:66. https://doi.org/10.3390/environments10040066

Nwabueze QA, Leggett S (2024) Advancements in the application of CO_2 capture and utilization technologies—a comprehensive review. Fuels 5:508–532. https://doi.org/10.3390/fuels5030028

Obahiagbon EG, Ogwu MC (2023) Sustainable supply chain management in the herbal medicine industry. In: Izah SC, Ogwu MC, Akram M (eds) Herbal medicine phytochemistry. Reference series in phytochemistry. Springer, Cham. https://doi.org/10.1007/978-3-031-21973-3_64-1

Obahiagbon EG, Ogwu MC (2024) The nexus of business, sustainability, and herbal medicine. In: Izah SC, Ogwu MC, Akram M (eds) Herbal medicine phytochemistry. Reference series in phytochemistry. Springer, Cham. https://doi.org/10.1007/978-3-031-21973-3_67-1

Ogwu MC (2019) Towards sustainable development in Africa: the challenge of urbanization and climate change adaptation. In: Cobbinah PB, Addaney M (eds) The geography of climate change adaptation in urban Africa. Springer Nature, Switzerland, pp 29–55. https://doi.org/10.1007/978-3-030-04873-0_2

Ogwu MC (2023) Plants as monitors and managers of pollution. In: Srivastav AL, Grewal AS, Tiwari M, Pham TD (eds) Role of green chemistry in ecosystem restoration to achieve environmental sustainability. Elsevier, Amsterdam, Netherlands, pp 51–60. ISBN 978-0-443-15291-7. https://doi.org/10.1016/B978-0-443-15291-7.00022-5

Onwudiegwu C, Sylva L, Aigberua AO, Hait M (2024) Biological monitoring of air pollutants. In: The handbook of environmental chemistry. Springer, Berlin, Heidelberg, pp 1–28. https://doi.org/10.1007/698_2024_1139

Onyeaka H, Miri T, Obileke KC, Hart A, Anumudu C, Al-Sharify ZT (2021) Minimizing carbon footprint via microalgae as a biological capture. Carbon Capture Sci Technol 1:100007. https://doi.org/10.1016/j.ccst.2021.100007

Osman AI, Hefny M, Abdel Maksoud MIA, Elgarahy AM, Rooney DW (2021) Recent advances in carbon capture storage and utilisation technologies: a review. Environ Chem Lett 19(2):797–849. https://doi.org/10.1007/s10311-020-01133-3

Park AHA, Fan LS (2004) CO_2 mineral sequestration: physically activated dissolution of serpentine and pH swing process. Chem Eng Sci 59:5241–5247

Park S (2018) CO_2 reduction-conversion to precipitates and morphological control through the application of the mineral carbonation mechanism. Energy 153:413–421

Paustian K, Lehmann J, Ogle S, Reay D, Robertson GP, Smith P (2016) Climate-smart soils. Nature 532:49–57

Peter SC (2018) Reduction of CO_2 to chemicals and fuels: a solution to global warming and energy crisis. ACS Energy Lett 3(7):1557–1561. https://doi.org/10.1021/acsenergylett.8b00878

Pham LHHP, Rusli R, Keong LK (2016) Consequence study of CO_2 leakage from ocean storage. Procedia Eng 148:1081–1088

Pruess K, Xu T, Apps J, Garcia J (2003) Numerical modeling of aquifer disposal of CO_2. SPE J 8:49–60

Rackley SA (2010) Carbon capture and storage. Butterworth-Heinemann, Oxford, UK; Elsevier, Burlington, VT, USA. ISBN 9781856176361

Reichle D, Houghton J, Kane B, Ekmann J, Benson S, Clarke J, Dahlman R, Hendrey G, Herzog H, Hunter-Cevera J et al (1999) Carbon sequestration research and development. Report. National Technical Information Service, Springfield, VA, USA. Available online: https://digital.library.unt.edu/ark:/67531/metadc739790/. Accessed 1 Feb 2023

Riaz A, Cinar Y (2014) Carbon dioxide sequestration in saline formations: part I—review of the modeling of solubility trapping. J Pet Sci Eng 124:367–380

Richard G, Hait M, Ibrahim M (2024) Emerging chemical air contaminants and their associated human health effects. In: The handbook of environmental chemistry. Springer, Berlin, Heidelberg, pp 1–24. https://doi.org/10.1007/698_2024_1130

Roh K, Lim H, Chung W, Oh J, Yoo H, Al-Hunaidy AS, Imran H, Lee JH (2018) Sustainability analysis of CO_2 capture and utilization processes using a computer-aided tool. J CO_2 Util 26:60–69. https://doi.org/10.1016/j.jcou.2018.05.015

Ryerson FJ, Lake J, Whittaker SD, Johnson JW (2013) Natural CO_2 accumulations in the western Williston Basin: a mineralogical analog for CO_2 injection at the Weyburn site. Int J Greenh Gas Control 16:S25–S34

Sanna A, Maroto-Valer M (2017) CO_2 sequestration by ex-situ mineral carbonation. World Scientific Publishing, Singapore. ISBN 978-1-78634-160-0

Sanna A, Uibu M, Caramanna G, Kuusik R, Maroto-Valer M (2014) A review of mineral carbonation technologies to sequester CO_2. Chem Soc Rev 43(23):8049–8080. https://doi.org/10.1039/C4CS00035H

Saran RK, Arora V, Yadav S (2018) CO_2 sequestration by mineral carbonation: a review. Glob NEST J 20:497–503

Seifritz W (1990) CO_2 disposal by means of silicates. Nature 345:486

Selma LS, Dohle, Michael SS (2014) Public perception of carbon capture and storage (CCS): a review. Renew Sustain Energy Rev 38:848–863. https://doi.org/10.1016/j.rser.2014.07.017

Shaw R, Mukherjee S (2022) The development of carbon capture and storage (CCS) in India: a critical review. Carbon Capture Sci Technol 2:100036. https://doi.org/10.1016/j.ccst.2022.100036

Shivanna KR (2022) Climate change and its impact on biodiversity and human welfare. Proc Indian Natl Sci Acad Part A Phys Sci 88(2):160–171. https://doi.org/10.1007/s43538-022-00073-6

Singh N (2008) Deep saline aquifiers for sequestration of carbon dioxide. In: Proceedings of the international geological congress, Oslo, Norway, pp 6–14

Singh N, Farina I, Petrillo A, Colangelo F, De Felice F (2023) Carbon capture, sequestration, and usage for clean and green environment: challenges and opportunities. Int J Sustain Eng 16(1):248–268. https://doi.org/10.1080/19397038.2023.2256379

Sipilä J, Teir S, Zevenhoven R (2008) Carbon dioxide sequestration by mineral carbonation literature review update 2005–2007. Report Vt1. Åbo Akademi University, Faculty of Technology, Heat Engineering Laboratory, Turku, Finland. ISBN 978-952-12-2036-4

Solomon S, Plattner GK, Knutti R, Friedlingstein P (2009) Irreversible climate change due to carbon dioxide emissions. Proc Natl Acad Sci USA 106(6):1704–1709. https://doi.org/10.1073/pnas.0812721106

Styring P, McCord S, Rackley S (2023) Carbon dioxide utilization, chap 17. In: Rackley S, Andrews G, Clery D, De Richter R, Dowson G, Knops P, Li W, Mccord S, Ming T, Sewel A, Styring P, Tyka M (eds) Negative emissions technologies for climate change mitigation. Elsevier, pp 391–413. https://doi.org/10.1016/B978-0-12-819663-2.00005-8

Surampalli RY, Zhang TC, Tyagi RD, Naidu R, Gurjar BR, Ojha CSP, Yan S, Brar SK, Ramakrishnan A, Kao CM (2015) Carbon capture and storage: physical, chemical, and biological methods. American Society of Civil Engineers, Reston, VA, USA. ISBN 9780784478912

Szulczewski ML, MacMinn CW, Juanes R (2014) Theoretical analysis of how pressure buildup and CO_2 migration can both constrain storage capacity in deep saline aquifers. Int J Greenh Gas Control 23:113–118

Tan Y, Nookuea W, Li H, Thorin E, Yan J (2016) Property impacts on carbon capture and storage (CCS) processes: a review. Energy Convers Manag 118:204–222. https://doi.org/10.1016/j.enconman.2016.04.019

Teir S (2008) Fixation of carbon dioxide by producting carbonates from minerals and steelmaking slags. Ph.D. thesis, Department of Energy Technology, Helsinki University of Technology, Helsinki, Finland

Thakur IS, Kumar M, Varjani SJ, Wu Y, Gnansounou E, Ravindran S (2018) Sequestration and utilization of carbon dioxide by chemical and biological methods for biofuels and biomaterials by chemoautotrophs: opportunities and challenges. Bioresour Technol 256:478–490. https://doi.org/10.1016/j.biortech.2018.01.095

Wan YH (2022) A techno-economic review on carbon capture, utilization and storage systems for achieving a net-zero CO_2 emissions future. Carbon Capture Sci Technol 3:100044

Wang F, Dreisinger D, Jarvis M, Hitchins T (2019) Quantifying kinetics of mineralization of carbon dioxide by olivine under moderate conditions. Chem Eng J 360:452–463

Wang F, Dreisinger DB, Jarvis M, Hitchins T (2018) The technology of CO_2 sequestration by mineral carbonation: current status and future prospects. Can Metall Q 57:46–58

White CM, Strazisar BR, Granite EJ, Hoffman JS, Pennline HW (2003) Separation and capture of CO_2 from large stationary sources and sequestration in geological formations—coalbeds and deep saline aquifers. J Air Waste Manag Assoc 53:645–715

Wildgust N, Gilboy C, Tontiwachwuthikul P (2013) Introduction to a decade of research by the IEAGHG Weyburn–Midale CO_2 monitoring and storage project. Int J Greenh Gas Control 16:S1–S4

Yadav S, Mehra A (2021) A review on ex situ mineral carbonation. Environ Sci Pollut Res 28:12202–12231

Yang F, Bai BJ, Tang DZ, Dunn-Norman S, Wronkiewicz D (2010) Characteristics of CO_2 sequestration in saline aquifers. Pet Sci 7:83–92

Yasemi S, Khalili Y, Sanati A, Bagheri M (2023) Carbon capture and storage: application in the oil and gas industry. Sustainability 15:14486. https://doi.org/10.3390/su151914486

Zevenhoven R, Kavaliauskaite I (2004) Mineral carbonation for long-term CO_2 storage: an exergy analysis. Int J Thermodyn 7:23–31. https://dergipark.org.tr/en/pub/ijot/issue/5751/76664

Zhao Y, Itakura K (2023) A state-of-the-art review on technology for carbon utilization and storage. Energies 16:3992. https://doi.org/10.3390/en16103992

Zhu XG, Long SP, Ort DR (2010) Improving photosynthetic efficiency for greater yield. Annu Rev Plant Biol 61:235–261.

Chapter 12
Emerging Technologies for Climate Change Mitigation and Adaptation

Milan Hait, Amit Kumar Chaturwedi, Jayati Chatterjee Mitra, Rashmi Verma, and Nand Kumar Kashyap

Abstract Climate change reflects an enormous risk to ecological systems, human health, and worldwide economic growth. With the increase in global temperatures and the growing frequency of catastrophic storms, the urgency for mitigation (reducing greenhouse gas emissions) and adaptation (adjusting to the changing climate) has reached a critical level. Emerging technologies are crucial in addressing these challenges, providing innovative solutions for reducing emissions, enhancing resilience, and enabling sustainable development. This chapter explores key emerging technologies in climate change mitigation and adaptation, including renewable energy advancements, carbon capture and storage, artificial intelligence for climate modeling and decision-making, climate-resilient agriculture, and smart infrastructure. These technologies not only offer the potential to reduce the carbon footprint but also provide novel approaches to safeguarding communities and ecosystems against the impacts of climate change. The chapter also discusses the challenges and opportunities these technologies present in terms of scalability, policy frameworks, and global cooperation.

Keywords Climate change · Emerging technology · Mitigation · Adaptation · Carbon capture and storage · Global warming

M. Hait · A. K. Chaturwedi · J. C. Mitra · R. Verma · N. K. Kashyap (✉)
Department of Chemistry, Dr. C. V. Raman University, Bilaspur, Chhattisgarh, India
e-mail: nandk79@gmail.com

M. Hait
e-mail: haitmilan@gmail.com

A. K. Chaturwedi
e-mail: chaturwedia12@gmial.com

J. C. Mitra
e-mail: jayatimitra@cvru.ac.in

R. Verma
e-mail: rashmiverma@cvru.ac.in

M. C. Ogwu and S. Chibueze Izah (eds.), *Evaluating Environmental Processes and Technologies*, Environmental Science and Engineering,
https://doi.org/10.1007/978-3-031-85327-2_12

12.1 Introduction

Climate change represents the major obstacle of the twenty-first century, posing an imminent risk to the numerous growing achievements realized globally, particularly in Asia and the Pacific. Both the severity and frequency of extreme weather conditions are expected to increase significantly. The region encompasses seven of the ten nations, the majority at risk globally, and the potential economic repercussions could be significant (Ebi et al. 2021). According to research studies conducted by the Asian Development Bank (ADB), under a business-as-usual scenario, annual economic losses could reach 5.28% in East Asia and 12.68% in the Pacific by the year 2100. The estimated yearly monetary expenses of catastrophes in the Asia and Pacific region could reach $53.79 billion. In 2013, developing Asia accounted for 35% of energy-related international carbon dioxide emissions; the percentage keeps increasing, further intensifying the region's vulnerability. Given the present scenario, ADB is strengthening its commitment to assist its developing member countries (DMCs) in tackling the challenges caused by climate change. Economics is vital in effectively addressing climate change, and technological advancements will also be essential (Mimura 2013). Climate change presents significant challenges to the global community.

The most recent Intergovernmental Panel on Climate Change (IPCC) report indicates that the world may experience reduced water supply and agricultural output across various regions, heightened risks for coastal areas, and a greater vulnerability to more severe weather. The climate is changing and is highly likely to persist in this trend. Significant attempts to reduce greenhouse gas emissions (GHGE) will only mitigate climate change, not entirely prevent it (Bolan et al. 2024). Therefore, adaptation is essential. Technology is a crucial element of adaptation. Technology has the potential to safeguard society against evolving climate conditions, enhance efficiency, and facilitate more effective utilization of endangered resources like water (Kaur et al. 2023). This report outlines various existing and emerging technologies that can assist the world in adapting to climate change (IPCC 2014). Climate change has affected ecosystems, including global warming, elevated ocean levels, glacier melting, and increased severe weather. These changes have resulted in negative consequences for human wellness, agricultural output, water accessibility, and biodiversity, among other areas (Nguyen et al. 2023; Ogwu et al. 2014; Ogwu 2019). The Asia–Pacific region accounts for 50% of GHGE, which has doubled since 2000. The Economic and Social Commission for Asia and the Pacific (ESCAP) Sustainable Development Goal (SDG) 2021 progress report indicates that it is improbable for the region to meet the targets set under SDG 13 (Climate Action) by the year 2030. Addressing these problems necessitates an increased implementation of technologies for climate change adaptation and mitigation across various sectors (Bolan et al. 2024).

Emerging technologies encompass novel, revolutionary, inventive, and frequently disruptive advancements, such as artificial intelligence (AI), the Internet of Things (IoT), 5G, digital twins, robotics, Space 2.0 technologies, and big data. These technologies provide numerous novel applications to tackle the difficulties climate change poses across different sectors. For instance, advanced 3D printing reduces carbon emissions and minimizes waste in manufacturing processes, while big data identifies adverse emissions throughout the supply chain (Licardo et al. 2024). Artificial intelligence can enhance energy efficiency by utilizing data from smart meters to estimate energy demand and assist utilities in optimizing energy production. 5G-enabled smart grids facilitate the connection of multiple data points over extensive distances, encompassing wind turbines, rooftop solar panels, and electric vehicle batteries. There are sophisticated technological solutions available to decrease dependence on fossil fuel energy, lower GHGE, reduce the carbon footprint of energy sources, adopt renewable energy technologies, and enhance carbon capture and storage. These technologies are crucial enablers of sustainability and environmental durability, offering significant possibilities to assess, tackle, and adapt to climate change (Zafar et al. 2024). Realizing the long-term objectives of the Paris Agreement to address climate change necessitates the integration of both innovative and developing technologies, alongside establishing novel industries and marketplaces to ensure their practical application. The emerging technologies must be affordable, sustainably sourced, ethnically appropriate, and society-suitable while addressing the complexities associated with climate change (Richard et al. 2024).

The effectiveness of technology transfer and dispersion relies on supportive policy and regulatory frameworks, the ability to absorb, implement, and enhance technologies suited to local conditions, financial resources, and investment, as well as collaboration across all levels—international, regional, national, and sub-national (Furszyfer Del Rio et al. 2021). Increase understanding of policy instruments and approaches aimed at facilitating the diffusion and adoption of emerging technologies for climate change adaptation and mitigation, the Yunnan Academy of Scientific and Technical Information (YASTI) in China, in collaboration with the Chinese Academy of Science Innovation Cooperation Center, Bangkok (CASICCB) and the United Nations Economic and Social Commission for Asia and the Pacific (ESCAP), hosted the Regional Workshop on Emerging Technologies to Address Climate Change from September 14–17, 2021 (Yao et al. 2023). The swift advancement of climate change poses a critical challenge of the twenty-first century, endangering ecosystems, human societies, and global economies. In response, emerging technologies have taken a pivotal role in alleviating climate change's effects and adjusting to its unavoidable outcomes.

The novel approaches, which include renewable energy technologies, advanced materials, and climate-smart agriculture, are transforming our approach to the climate crisis and providing new avenues for sustainability (Hait et al. 2024a). As the global community strives to achieve the ambitious objectives outlined in the Paris Agreement and other international frameworks, emerging technologies offer essential tools to decrease greenhouse gas emissions, improve resilience to climate impacts, and promote low-carbon, sustainable development (Hait et al. 2024b). This extensive range of technologies includes established innovations like solar and wind energy alongside advanced solutions such as carbon capture and storage (CCS), geoengineering, artificial intelligence (AI) for climate modeling, and smart infrastructure aimed at enhancing climate resilience (Strengers and Nicholls 2017). The demand for these technologies is increasingly critical, as they provide technical solutions and play a significant role in fostering economic transformation, generating green jobs, and promoting environmental equity, particularly in vulnerable communities (Kashyap et al. 2024). Integrating emerging technologies in climate change mitigation and adaptation is essential for tackling the diverse challenges a warming planet presents, ranging from carbon–neutral cities to improved agricultural practices.

This chapter describes the significant emerging technologies that offer potential solutions for addressing the impacts of climate change and adapting to its unavoidable consequences, emphasizing their capabilities, challenges, and the current initiatives aimed at scaling them effectively. Utilizing these innovations allows for reducing the causes of climate change while enhancing our capacity to adapt to its consequences, fostering a more sustainable and resilient future.

12.2 Climate Change and Its Impact

Climate change involves the long-term alterations in temperature, precipitation patterns, and various atmospheric conditions that transpire over decades or centuries. Anthropogenic actions such as the burning of fossil fuels (coal, oil, and natural gas), the destruction of forests, and various industrial operations that emit greenhouse gases (GHGs) into the surrounding environment are the main contributing factors to this event (Patel et al. 2024). These gases accumulate and retain heat, leading to increased global warming, also called greenhouse impact (Malhi et al. 2021). IPCC (2012) predicts ongoing increases in temperature, precipitation, severe events, and sea levels due to climate change, both worldwide and in Asia. Also, these changes may accelerate; across the twenty-first century, global temperature rise is expected to intensify, with seasonal and regional variability across Asia (Cruz et al. 2007). The warming is expected to be slower in Southeast Asia and faster in Central, West, and North Asia. Annual rainfall may rise in Asia, with North and East Asia seeing the most significant rises. Central Asia may see a drop in mean precipitation. South, East, and Southeast Asia may see severe weather, including heat waves and intense rains (WHO 2021). With a 2–5 °C rise in sea surface temperature, we expect tropical storm strength to increase by 10–20%. The global sea level is projected to increase by 3.8 mm/year by the end of the twenty-first century (Christiansen et al. 2011). The effects of climate change are extensive and complex, influencing ecosystems, economies, and communities. The significant repercussions of climate change involve the following.

12.2.1 *Biodiversity*

Climate change significantly impacts global biodiversity, leading to species disadvantage. Studies have demonstrated a significant link between various climate events and the dynamics of species on a large scale (Hait et al. 2024c). The rate and extent of climate change are modifying the suitable habitat ranges for organisms in aquatic and terrestrial environments. Alterations in overall climate patterns affect the sustainability of ecological systems in various ways, including moves in species diversity, adjustments in range, modifications in the timing of activities, and adjustments in microhabitat utilization (Hait et al. 2024d; Egboduku et al. 2024). A species' capacity to withstand adverse conditions, engage in ecological relationships, and navigate distribution limitations frequently influences its location. Therefore, the local species must accept, adapt, relocate, or confront the possibility of extinction (Ortiz-Bobea et al. 2021). The most successful species demonstrate a superior ability to adapt to novel environments, or exhibit reduced resilience in their current ones. A significant factor to consider is the insufficient habitat connectivity and accessibility to microclimates, vital in increasing vulnerability to global warming and excessive heatwave events. The carbon sequestration rates are experiencing fluctuations due to climate-driven development in the distribution worldwide of mangrove forests (Manes et al. 2021).

12.2.2 Changing Weather Patterns

Climate change increases the frequency and strength of catastrophic storms, including hurricanes, cyclones, and typhoons. Increased ocean temperatures contribute additional energy to severe weather, resulting in greater strength and damaging hurricanes. Certain regions undergo extended periods of drought due to changes in rainfall patterns. In contrast, others may encounter increased frequency and intensity of floods, particularly in areas with low elevations (Sugg et al. 2023). The escalating impacts of climate change are unmistakably reshaping our weather patterns, manifesting in the increasingly frequent and intense extreme weather events that communities around the globe must now confront (Sugg et al. 2023). As ocean waters warm, they catalyze more powerful hurricanes, devastating coastal regions and altering the very fabric of local economies. Concurrently, prolonged droughts are becoming a grim reality for many areas, straining water resources and impacting agriculture. In stark contrast, low-lying regions face the dire consequences of frequent and severe flooding, forcing populations to adapt or relocate in search of safer ground. These interconnected phenomena highlight the urgency of addressing climate change and the necessity for comprehensive strategies to mitigate its far-reaching impacts on our planet's weather systems (Robinson et al. 2021).

12.2.3 Tourism

Tourism is a multifaceted economic sector that serves as an effective tool for employment, earnings development, and the accumulation of significant foreign exchange. It fosters intercultural communication and collaboration, provides employment opportunities for innovators, and ultimately contributes to the overall development of a nation (Arshad et al. 2018). By promoting cultural exchange and collaboration, tourism enriches the experiences of visitors and locals alike and drives innovation in various sectors. This synergy ultimately enhances a nation's economic growth and global standing (Scott 2021; Jacob et al. 2024). Climate change significantly impacts the tourism sector, among other sectors, as the climate plays a crucial role in facilitating tourism in specific geographic areas. Various locations at distinct times of the year draw domestic and international tourists, influenced by the suitability and alignment of particular weather conditions. Consequently, the significant fluctuations in these weather trends caused by climate change will ultimately pose challenges to the regional and national economies. These challenges may include decreased tourist arrivals in affected areas, leading to reduced revenue for local businesses and potential job losses (Obahiagbon and Ogwu 2023, 2024). This could jeopardize the overall economic stability of regions that rely on tourism (Bujosa et al. 2015). According to the Intergovernmental Panel on Climate Change (IPCC) report, there has been a significant reduction in the length of the skiing season worldwide due to the depletion of certain skiing areas and notable changes in the environmental conditions of tourist attractions. The lack of sufficient socioeconomic and sociopolitical capital increasingly puts highly demanding tourist sites at risk of vulnerability (IPCC 2013). Tourism's vulnerability relies on various elements, including the degree of exposure, sensitivity, essential sectors, and capacity evaluation factors. Sectors such as nutrition, healthcare, habitats, amenities, water supply, and the general accessibility of a specific area are susceptible to climate change. This susceptibility highlights the interconnectedness of environmental health and tourism, where adverse climate impacts can directly affect the resources and infrastructure supporting local communities and visitors. As these essential sectors face challenges, tourist attractions' overall appeal and sustainability may diminish, further exacerbating their vulnerability (Balsalobre-Lorente et al. 2022; Abbass et al. 2022).

12.2.4 Ecosystem Disruption

Plant and animal species cannot adapt quickly enough to changing climates, leading to ecosystem disruptions. Coral reefs, for example, are susceptible to temperature changes and ocean acidification and suffer significant bleaching events. The ongoing impacts of climate change represent substantial challenges to the delicate equilibrium of environments, with flora and fauna facing difficulties adapting to new circumstances (FAO 2018). The plight of coral reefs is a poignant example; these vibrant ecosystems, already under siege from rising temperatures and ocean acidification, are enduring devastating bleaching events that compromise their survival. As species struggle to keep pace with the rapidly changing environment, the ripple effects on biodiversity and ecological stability become more pronounced. Without concerted efforts to mitigate climate change and protect vulnerable habitats, the delicate interconnections that sustain life on our planet risk unraveling, highlighting an urgent call to action for conservation and sustainability initiatives (Ferreira et al. 2020). Temperature rise forces species to migrate to cooler areas, disrupting established ecosystems. This can lead to imbalances, such as the decline of species that cannot adapt or migrate quickly enough. The rising temperature phenomenon is a challenge for individual species and a catalyst for broader ecological upheaval. The forced migration of organisms toward cooler habitats throws the intricate balance of ecosystems into disarray (Ishikawa-Ishiwata and Furuya 2022). This forced relocation disrupts established relationships among species, leading to imbalances that can precipitate declines in specific populations. The cascading effects of these changes underscore the urgent need for comprehensive strategies that address climate change and its far-reaching impacts on biodiversity. By recognizing the interconnectedness of these issues, we can better advocate for sustainable practices that protect both our planet and its diverse inhabitants (Rehman et al. 2021).

12.2.5 Ocean Changes

The intricate relationship between ocean changes and climate impact reveals a profound interconnectedness that is impossible to overlook. As rising temperatures contribute to coral bleaching and the disruption of marine ecosystems, the implications extend far beyond the water's surface, affecting global weather patterns and food security. The alarming increase in sea levels poses a significant threat to coastal communities, underscoring the urgency for collective action against climate change. Moreover, the acidification of oceans endangers marine life and jeopardizes the livelihoods of millions who depend on these resources (Hoegh-Guldberg et al. 2017). Addressing these challenges requires a multifaceted approach integrating scientific research, sustainable practices, and policy reforms. By recognizing the vital role oceans play in our climate system, we can foster a more profound commitment to preserving their health for future generations, ensuring ecological balance and human prosperity.

12.2.5.1 Ocean Acidification

Elevated carbon dioxide emissions in the environment result in elevated levels of carbon dioxide in the oceans, contributing to greater acidity in the water. This poses a significant risk to marine ecosystems, especially for organisms dependent on $CaCO_3$ for their shells and skeletons, including corals, mollusks, and certain plankton. The continuous increase of greenhouse gases in the atmosphere is not just a climate issue; it carries significant consequences for aquatic life, especially for species reliant on calcium to form their outer shells and skeletal systems. The absorption of increased carbon dioxide by seawater leads to acidity, which interferes with the complex equilibrium of aquatic environments and poses significant challenges to the survival of these living things. The repercussions of this phenomenon extend beyond individual species, potentially unraveling entire food webs and jeopardizing the livelihoods of communities that rely on healthy oceans. Addressing ocean acidification is imperative for preserving biodiversity and ensuring the resilience of our planet's marine environments for future generations (Yamano et al. 2011).

12.2.5.2 Disruption of Marine Ecosystems

As ocean temperatures rise, the distribution of fish and other marine species changes, which can have severe economic impacts on fisheries and food security. The repercussions of increasing ocean temperatures extend beyond mere environmental shifts; they fundamentally alter the dynamics of marine ecosystems and human livelihoods. As fish distribution patterns adjust in response to warmer waters, many species find themselves in unfamiliar territories, disrupting established ecological balances. This upheaval threatens biodiversity and poses significant economic challenges for fisheries that rely on stable populations and predictable catch locations. Consequently, food security becomes increasingly precarious as communities dependent on these resources face dwindling supplies and rising costs. The intricate connections between marine health and human welfare underscore the urgency of addressing climate change, highlighting the need for sustainable practices that can safeguard both our oceans and the communities that rely upon them (Saunders et al. 2014).

12.2.5.3 Melting Ice Caps and Glaciers

Global warming leads to melting polar ice caps and glaciers, contributing to the increase in ocean levels. Coastal towns and tiny islands face significant risks from floods, severe storms, and the intrusion of saline ocean water into freshwater supplies. This intrusion poses a substantial risk to the availability of safe water for consumption, disrupts ecological systems, and hinders economic growth, complicating these populations' ability to maintain their traditional lifestyles. Consequently, the intrusion may force numerous individuals to relocate, exacerbating sustainability and social issues. This increased vulnerability poses significant risks to infrastructure, ecosystems, and local economies, making it essential for communities to develop adaptive strategies. As a result, addressing climate change is critical for environmental sustainability and the protection of human livelihoods (Ricke et al. 2013).

12.2.5.4 Coastal Erosion

Rising sea levels and stronger storms increase coastal erosion, threatening infrastructure and ecosystems, particularly in densely populated coastal regions. As coastal erosion intensifies due to rising sea levels and increasingly severe storms, the impact on infrastructure and ecosystems becomes alarmingly clear (Qi et al. 2017). Densely populated coastal regions face a dual threat: not only are homes and businesses at risk, but vital ecosystems that provide natural barriers and habitat for diverse species are also under siege. This pressing issue necessitates immediate attention and action to mitigate the consequences of erosion, highlighting the urgent need for comprehensive strategies that balance human development with environmental preservation. The future of our coastlines depends on our ability to address these challenges head-on, ensuring safety for communities while safeguarding the delicate ecosystems (Precht and Aronson 2004).

12.2.6 Agricultural Impact

Climate change impacts agriculture in many ways. Reduced precipitation, snowfall, glacier melting, and saltwater seepage into reservoirs, lakes, and rivers may reduce production by depleting freshwater resources. Increased rainfall, snowstorms, and melting ice may impair agricultural productivity due to floods. The typical temperatures, especially their increases, along with the period and location of rainfall, may significantly affect crop yields. They may also change insect predator dispersion, population density, and disease or fungal infection rates (Betts et al. 2018). Climate change has a direct effect on agricultural productivity. Increased temperatures, changed rainfall patterns, and a rise in severe weather, such as droughts and floods, can lead to a decrease in harvest rates. This can lead to food shortages, higher prices, and increased pressure on food systems. As a result, these challenges can exacerbate food insecurity and disproportionately affect vulnerable populations who rely heavily on stable food supplies (Baul, and McDonald 2015; Ogwu and Izah 2023; Kosoe et al. 2023). Addressing these impacts is crucial for ensuring sustainable agricultural practices and food availability for all. Warmer temperatures may also create favorable conditions for pests and plant diseases, threatening global food supplies. The increasing temperatures due to climate change significantly threaten the global food supply, creating an environment where pests and diseases can thrive. As warmer conditions become more prevalent, crops are increasingly vulnerable to infestations and infections that can devastate yields (Fargnoli and Lombardi 2020; Imarhiagbe et al. 2023). This interconnection between rising temperatures and agricultural stability underscores the urgency of addressing climate change, as the repercussions extend beyond mere food production to encompass economic stability and food security for millions worldwide. By recognizing the delicate balance of our ecosystems and the impact of climate fluctuations on agriculture, it becomes clear that proactive measures are essential to safeguarding our food systems against these escalating challenges. Ensuring a resilient future for global agriculture will require concerted efforts in sustainable practices, innovation in pest management, and a commitment to reducing greenhouse gas emissions (Malhi et al. 2021). A study using CERES (crop estimation through resources and environmental synthesis) showed that in northwest India, wheat and rice yields could rise by 27.9% and 14.8%, respectively, if carbon dioxide levels were doubled.

Nevertheless, the heightened heat stress from elevated GHGE levels significantly undermines the advantage. Adhering to the current irrigation schedule will improve rice and wheat yields by 20% and 5%, respectively, even under higher CO_2 and temperature stress. In extreme water constraints and temperature stress, we expect rice and wheat yields to drop even with increased CO_2 (Deryng et al. 2016). Increased CO_2 concentration may counteract agricultural output losses from rising temperatures and soil moisture (Long et al. 2006a, b). High CO_2 concentrations lower worldwide output losses by reducing agricultural water consumption (4–17%). Regional variability in agrarian yields is attributable mainly to crop growth circumstances (Uddling et al. 2018). High CO_2 levels reduce protein-based nutrients (N, Fe, Zn, and S) in non-leguminous crops. Higher CO_2 levels in rice crops at 29 °C led to increased vegetative and reproductive development and increased seed production despite reducing the seed set (Madan et al. 2012; Table 12.1).

12.2.7 Human Health Risks

Climate change significantly impacts human health, as widely recognized. The World Health Organization (WHO) indicates that climate change could lead to an estimated 250,000 additional deaths annually from 2030 to 2050 (Watts et al. 2015). Severe weather events cause mortality and morbidity, and vector-transmitted illnesses are on the rise globally, contributing to these deaths (Lemery et al. 2021). Global health landscape is continually evolving, presenting novel challenges and pressing concerns that demand a coordinated response. As the world becomes increasingly interconnected, pandemics, chronic disease prevalence, and healthcare inequities transcend national borders, necessitating collaborative strategies for effective management (Meierrieks 2021). Furthermore, emerging health issues influenced by environmental shifts, technological advancements, and changing population dynamics require innovative solutions to safeguard public health (Izah et al. 2023). Nations can better navigate these complex challenges by fostering international partnerships and embracing a holistic approach to health policy. As we look toward the future, we must prioritize proactive measures to address these emerging threats and enhance global resilience against potential crises (Yang and Usman 2021). The time has come for concerted action integrating knowledge, resources, and commitment across borders to build a healthier world for all.

Table 12.1 Impact of climate change on crop yield percentage

S. N.	Crops	Yield rising %	Reason	Country	References
1	Corn	39	Slow warming	USA	Schlenker and Roberts (2009)
2	Soyabean	45	Rapid warming	USA	Schlenker and Roberts (2009)
3	Cotton	3	Mid GHG emission	California	Lee et al. (2011)
4	Sunflower	6	Lower GHG emission	California	Lee et al. (2011)
5	Wheat	9	Lower GHG emission	California	Lee et al. (2011)
6	Rice	7	Means of temperature	Multiple site in country	Zhao et al. (2017)
7	Maize	3	Means of temperature	Multiple sites of country	Zhao et al. (2017)
8	Rain feed corn	26	Means of temperature	Central Illinois	Cai et al. (2009)
9	Barley	9	Rising temperature	Europe	Ray et al. (2019)
10	Sugarcane	4	Rising temperature	Europe	Tao et al. (2006)
11	Sorghum	2	Rising temperature	Africa	Liu and Basso (2020)
12	Cassava	1.67	Rising temperature	Ocean	Kukal and Irmak (2018)

12.2.7.1 Heat-Related Illnesses

As temperatures rise, the incidence of heat-related illnesses, such as heatstroke, dehydration, and respiratory problems, increases, particularly in urban areas with high population densities. This trend highlights the urgent need for public health measures to address the risks associated with extreme heat, especially in densely populated regions where vulnerable populations may be at greater risk. Effective interventions could include increased access to cooling centres and public awareness campaigns about the signs and prevention of heat-related illnesses (Abbass et al. 2022).

12.2.7.2 Vector-Borne Diseases

Higher temperatures and altered weather conditions may extend the habitat of insects and the transmission of illness. Health conditions such as dengue fever, malaria, and the Lyme virus will probably expand their reach into fresh areas. This growth presents considerable difficulties for global health, as communities in these recently impacted regions may be deficient in immunity and resources to address these conditions. Consequently, the burden on healthcare systems could increase, necessitating proactive measures to mitigate the risks associated with these emerging threats (Santos et al. 2021). Temperature is essential for life in every environment. Therefore, an organism requires diverse temperatures to thrive on a planet. Precipitation, the second key element of CC, affects the infectious agent's movement and distribution. Numerous microorganisms are extinct due to global warming. This shifting climatic temperature may be causing species extinctions or promoting emerging species. Pathogenic microbes' responses to global warming illustrate this idea. A recent COVID-19 outbreak in China caused pneumonia and severe acute respiratory problems. Various animals, particularly bats and snakes, harbor multiple viruses that can subsequently transmit to humans. Climate change significantly affects the proliferation of vectors responsible for transmitting illnesses (Cui et al. 2021). The intricate relationship between climate change and animal vectors, such as bats and snakes, reveals a troubling scenario for public health and biodiversity. As global temperatures rise, these creatures, known to harbor a plethora of viruses, experience shifts in their habitats and behaviors that facilitate the proliferation of disease-carrying vectors (Ogden 2018). This environmental upheaval alters the dynamics of these populations and enhances the transmission potential of various illnesses to both wildlife and humans. The interplay between climate change and vector dynamics underscores the urgency for comprehensive research and proactive measures to mitigate its consequences, emphasizing that safeguarding our ecosystems is paramount in protecting public health against emerging infectious diseases.

12.2.7.3 Mental Health

The anxiety, trauma, and displacement caused by climate-induced disasters can have profound mental health consequences. People in vulnerable communities may experience depression, post-traumatic stress disorder (PTSD), and other mental health challenges. The lack of resources and support systems in these communities can exacerbate these mental health issues, further complicating recovery. As a result, addressing mental health in the context of climate change is crucial for fostering resilience and well-being. The profound impact of climate-induced disasters extends beyond physical devastation, deepening the emotional and psychological fabric of affected communities (Pal 2021). As anxiety and trauma take root, particularly among vulnerable populations, the prevalence of depression and PTSD becomes alarmingly evident. The scarcity of resources and support amplifies these mental health challenges, leaving many without the necessary tools to cope effectively. Therefore, prioritizing mental health interventions is not merely an auxiliary concern; it is essential for fostering resilience and ensuring the well-being of those enduring the relentless consequences of climate change. By recognizing and addressing these intertwined issues, we can begin to build stronger, more supportive communities capable of withstanding both environmental upheavals and their psychological toll (Ogden 2018; Abbass et al. 2022).

12.2.7.4 Antimicrobial Resistance

Antimicrobial resistance (AMR) is a growing worldwide health issue. Health experts worldwide are concerned about this phenomenon, which might undo practically all health advances. Microbes produce a significant amount of antibiotics, but they also develop resistance to them, potentially destabilizing both national and global economies. This is encouraged since AMR is not spreading in one area or nation. It thrives in all geographic regions (Salam et al. 2023). This epidemic hastened the post-antibiotic era, during which antibiotic-susceptible bacteria may resurrect endemics and pandemics with resistance. Unfortunately, this claim might increase the hazards of complex procedures, including chemotherapy, joint replacements, and organ transplantation (Su et al. 2018). The rise of medication resistance has made treating common diseases, including pneumonia, post-surgical infections, HIV/AIDS, TB, malaria, and others challenging and costly. For instance, antibiotic-resistant bacteria have the potential to spread quickly among individuals and across borders (Berendonk et al. 2015). Second- and third-generation antibiotics, like the well-known cephalosporin antibiotics, are more expensive, broad-spectrum, toxic, and require more extended treatment periods. This situation also showed that Southern microorganisms were more resistant. This scenario shows how CC-induced global warming may propagate antibiotic-resistant strains in the biological community, making it more expensive to create novel antibiotics.

12.2.8 Economic Sector

The climate significantly influences total productivity and economic expansion. Given its expanding global presence and impact on the global economy, climate change (CC) has become a significant concern for environmental policymakers at both local and global levels. The complex interplay between climate change and the global economy highlights the critical necessity for cooperative efforts among countries and decision-makers. Moving climate trends and catastrophic incidents destroy environments, affecting overall efficiency, hamper economic development, and pose risks to global earnings. This interrelationship underscores that climate change is not solely an ecological issue but a significant financial issue that requires urgent attention. By prioritizing sustainable practices and innovative solutions, we can mitigate these impacts, ensuring a resilient future where our planet and economies can thrive. The time to act is now; our collective response will shape the global economy's health for generations to come (Abbass et al. 2022). The growing occurrence of severe storms, rising ocean levels, and temperature extremes can damage infrastructure-roads, bridges, buildings, power grids, and communication networks, leading to costly repairs and economic disruptions. These impacts strain public resources and hinder economic growth as communities struggle to recover and adapt to ongoing environmental challenges. Consequently, resilient infrastructure becomes imperative to withstand future climate-related threats. Climate-induced disasters can force people from their homes, creating large-scale displacement. Migration pressures from areas hit by extreme weather events or sea-level rise could lead to political instability, especially in regions that lack the resources to handle large numbers of refugees. This influx of displaced individuals can strain local infrastructure and services, potentially igniting tensions between host communities and newcomers. Consequently, the instability may hinder effective governance and exacerbate existing social divides. Climate change has initially presented several beneficial impacts; however, the inevitable warming of the environment constitutes an adverse effect (Hassan et al. 2024).

An increase in temperature exceeding 3 °C yields predominantly adverse outcomes, while temperatures surpassing 70 °C can result in complete welfare loss. Global projections place the anticipated societal impact of carbon emissions at USD 29 per tonne of carbon in 2015, with an expected annual increase of 2%. Implementing climate change mitigation plans could benefit Solomon Island's fishery industry economically. Climate change will significantly impact agricultural markets, leading to a decrease of 0.25% in global GDP (Costinot et al. 2016). An estimated yearly decline of 0.2–1% in family welfare is expected if the climate conditions displayed for the 2080s had been realized. With a 1 °C increase in mean world temperatures, we anticipate that both market and non-market losses will increase in a quadratic manner, representing 1.19% of GDP. If we model future mitigation initiatives after past ones, we anticipate a 22% decline in world income by 2100, leading to an increased wealth disparity. The projection indicates a reduction in global economic expansion by 0.28% annually (Carleton and Hsiang 2016). Table 12.2 illustrates the economic advantages of various emerging agriculture technologies.

12.2.9 Social Inequality

Social inequality has an essential influence on climate change mitigation, which involves attempts to decrease GHGE and adaptation, which focuses on adjusting to and minimizing the effects of climate change. The repercussions arise from financial status, assets, and power disparities, which influence how various social groups encounter and react to climate change. These disparities can lead to unequal access to resources and opportunities for both mitigation and adaptation efforts, ultimately exacerbating the challenges faced by marginalized communities. Consequently, addressing social inequality is crucial for creating effective and equitable climate change strategies (Green and Healy 2022).

Table 12.2 Crop-related innovative technology, their efficiency, and benefits

S. N.	Crops	Innovate technology	Places	Efficiency	Benefits	References
1	Rice	Site-dependent nutrient modulation	India, Vietnam, Philippines	Rising nitrogen productivity	168 US$/ha, 34 US$/ha, 106 US$/ha	Pampolino et al. (2007)
2	Wheat	Laser land leveling	Pakistan	Save water and irrigation period	INR 23,250/ acre	Wagan et al. (2015)
3	Rice	Zero tillage	Pakistan	Save water and irrigation period		Latif et al. (2013)
4	Multiple crops	Crops tolerate adversity	Kenya	Enhance breeding livestock	80% greater income	Ogada et al. (2020)
5	Groundnut	Drought-tolerant variations	India	Enhance crop yield	Cost reduced 15%	Birthal et al. (2012)
6	Wheat	Zero tillage	India	Reduced cultivation cost	Higher-income	Tripathi et al. (2013)
7	Rice and wheat	Laser land leveling	India	Reduced electricity, irrigation period and rising crop yield percentage	US$ 142/ha/ year	Aryal et al. (2015a)
8	Wheat	Zero tillage	India	Reduced GHGs	US$ 97/ha	Aryal et al. (2015b)
9	Okra	Drip irrigation	India	Reduced electricity, irrigation period and rising crop yield percentage	INR 72,711/ acre	Narayanamoorthy and Devika (2017)
10	Eggplant	Drip irrigation	India	Lower fertilizer use, electricity bill, irrigation time	Higher-income	Narayanamoorthy et al. (2018)

12.2.9.1 Vulnerability of Poor Communities

Climate change disproportionately affects marginalized and poorer communities, who have fewer resources to adapt. This can exacerbate existing inequalities and social tensions. As these communities struggle to cope with the impacts of climate change, their limited access to resources can hinder their ability to recover and rebuild. Consequently, this situation can increase frustration and conflict, further widening the gap between different societal groups (Nguyen et al. 2023).

12.2.9.2 Climate Refugees

Rising sea levels, desertification, and other climate impacts are already displacing people, leading to "climate refugees"—individuals or communities who must move due to environmental changes. This can lead to conflicts over resources and migration-related issues. As these displaced populations seek new areas to inhabit, competition for limited resources can intensify, creating tension between newcomers and established communities. Consequently, this situation may exacerbate existing social and political conflicts, further complicating the challenges of climate change adaptation (Balsari et al. 2020).

12.2.9.3 Tipping Points

Climate change has the potential to initiate feedback chains that intensify the impacts of rising temperatures. The melting of ice sheets and glaciers results in a decrease in the Earth's albedo (reflection), which leads to an increase in the capture of daylight by the Earth's surface, consequently enhancing rising temperatures. The melting of glaciers results in the discharge of methane, a highly effective GHG, into the environment, thereby contributing to the acceleration of global warming (Bergmann et al. 2022).

12.2.10 Temperature Rising

Anthropogenic actions, which contribute to increased greenhouse gas emissions (GHGE), are primarily responsible for rising global temperatures, leading to a warming planet and subsequent climate change. This may influence weather variations, contribute to ice caps melting, lead to elevated ocean levels, and result in environmental alterations. A temperature increase in a specific area might be due to seasonal changes, heatwaves, or urbanization. Heatwaves, for example, are becoming more common and intense (Amoo and Fagbenle 2020).

12.2.10.1 Global Warming

Global temperatures have risen notably, with the highest levels recorded in recent decades. The increase in temperature is not consistent, as certain areas are experiencing more rapid warming than others. As the evidence mounts regarding the alarming rise in global temperatures, it becomes increasingly clear that we are witnessing an unprecedented climatic shift. The recent decades have marked a period of record-breaking heat, with consistent increases observed across various regions of the globe. This upward trend highlights the urgent need for comprehensive climate action and serves as a stark reminder of our responsibility to safeguard the environment for future generations. The implications of these rising temperatures extend beyond mere discomfort; they threaten ecosystems, disrupt weather patterns, and challenge our way of life. Collectively addressing this pressing issue is imperative to foster a sustainable future that restores and maintains the delicate balance of our planet (Amoo and Fagbenle 2020).

12.2.10.2 Heatwaves

Heatwaves are more frequent and severe, which may lead to health concerns, lower agricultural output, and put pressure on energy infrastructure. These impacts can exacerbate community vulnerabilities, potentially leading to increased healthcare costs and food insecurity. The strain on energy resources may also result in higher utility bills and increased demand for sustainable energy solutions. Heat waves' rising frequency and severity have far-reaching implications for communities, intertwining health concerns with economic challenges. As extreme temperatures become more common, they not only elevate the risk of heat-related illnesses but also significantly burden already strained healthcare systems, potentially driving up costs for individuals and families. With agricultural output declining due to unfavorable conditions, food insecurity looms large, further exacerbating the vulnerabilities faced by marginalized populations. Compounding these issues is the pressure on energy infrastructure, which leads to increased demand for cooling solutions and, subsequently, higher utility bills for consumers (Godde et al. 2021). Climate change and its impact in various fields are given in Fig. 12.1.

12.3 Mitigation and Adaptation Strategies

Climate change and mitigation highlight the initiatives to reduce or prevent GHGs that contribute to global warming and climate change. Unlike adaptation, which focuses on managing the inevitable effects of climate change, mitigation focuses on addressing its underlying causes by reducing the levels of greenhouse gases in the environment. Mitigation involves implementing strategies aimed at minimizing climate change's impact through reducing GHGs or by improving the mechanisms

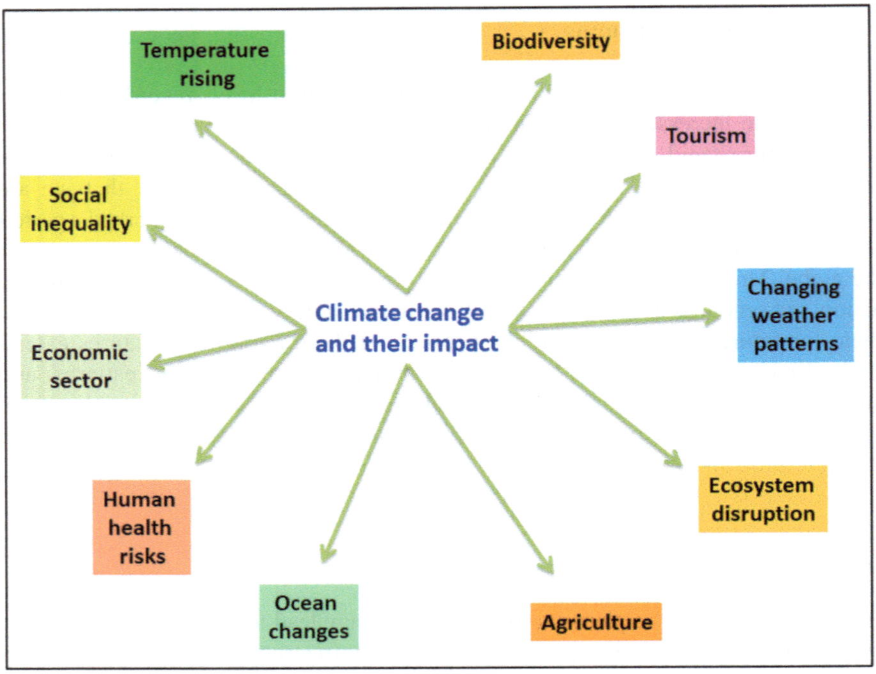

Fig. 12.1 Climate change and its impacts in various fields

that eliminate these gases from the environment, also known as "carbon sinks." These measures aim to stabilize the climate and prevent catastrophic changes in ecosystems, economies, and human well-being (Dwivedi et al. 2022). Addressing climate change involves mitigation (lowering GHGs) and adaptation (balancing the impacts of climate change).

- **Mitigation**: This encompasses the shift to renewable energy sources, enhancement of energy utilization, minimization of forest destruction, advancement of sustainable agriculture, and the development of CCSU techniques.
- **Adaptation**: This involves the development of more substantial infrastructure, enhancement of water conservation practices, the establishment of early-warning systems for severe weather occurrences, and developing policies that help vulnerable populations adapt to new conditions.

12.3.1 Strategies for Mitigation

Strategies for mitigation refer to deliberate actions and approaches taken to reduce, prevent, or counteract the negative impacts of a specific issue or threat. In the context of climate change, mitigation strategies focus on reducing greenhouse gas emissions and enhancing carbon sinks to limit global warming. These strategies aim to address the root causes of a problem to minimize or avoid potential harms, such as environmental damage, public health risks, or economic disruptions. Mitigation can apply to various fields, including climate, health, disaster management, and urban planning, and generally involves both proactive measures (like adopting cleaner technologies) and adaptive practices (like urban planning to reduce heat) (Abbass et al. 2022).

12.3.1.1 Reducing Greenhouse Gas Emissions

Minimizing GHGs requires adopting procedures, technological advances, and regulations to decrease the accidental discharge of gases that lead to the greenhouse effect and climate change. The main GHGs are carbon dioxide, methane, nitrous oxide, and fluorinated gases. The GHGs retain warmth within the planet's environment, increasing surface temperatures and playing a significant role in global warming (Abbass et al. 2022). To reduce GHG emissions, efforts focus on

- **Transition to renewable energy**: Transitioning from fossil fuels (coal, oil, and natural gas) to alternative forms of energy, including wind, solar, hydro, and geothermal. This represents the most straightforward approach to decreasing carbon dioxide emissions associated with generating energy (Battisti and Nylor 2009).
- **Energy efficiency**: Improving the efficiency of energy use in industry, transportation, buildings, and appliances can significantly lower emissions. This includes better insulation, energy-efficient appliances, and more efficient industrial processes (Diffenbaugh et al. 2017).
- **Electrification of transportation**: Transitioning from fossil-fuel-powered vehicles to electric vehicles (EVs), which can be powered by renewable energy, reducing emissions from one of the largest sectors contributing to global warming (Anwar et al. 2021).
- **Carbon capture and storage (CCS)**: The procedure involves the capture of CO_2 emissions from significant point sources, such as power plants and industrial operations, and their subsequent underground storage or alternative utilization methods, such as improved oil recovery or the production of synthetic gasoline (Hussain et al. 2020).

12.3.1.2 Reducing Land-Use Emissions

- **Reforestation and afforestation**: Forests are carbon sinks that absorb CO_2 from the atmosphere. Protecting existing forests, restoring degraded forests, and planting new forests can offset emissions and help mitigate climate change.
- **Sustainable agriculture**: Practices like agroforestry, no-till farming, and rotational grazing can reduce emissions from agriculture, which is responsible for a significant portion of global methane and nitrous oxide emissions.

- **Soil carbon sequestration**: Improving land management techniques to increase carbon storage in soils can help offset emissions. Practices include adding organic matter (e.g., compost) to the soil, which enhances its ability to store carbon.
- **Reducing deforestation**: Deforestation, particularly in tropical regions like the Amazon, is a major source of CO_2 emissions. Protecting forests from logging and land conversion is critical to mitigating climate change (Karkkainen et al. 2020).

12.3.1.3 Promoting Sustainable Consumption and Production

- **Circular economy**: Shifting to a circular economy that emphasizes reusing, recycling, and reducing waste helps minimize resource extraction and energy use, which are key drivers of emissions.
- **Sustainable food systems**: Changing diets, reducing food waste, and promoting plant-based food production can lower emissions from agriculture and food supply chains, especially livestock-related emissions, which produce methane.
- **Green buildings**: Designing and constructing buildings that use fewer resources, require less energy, and are energy-efficient throughout their lifecycle can significantly reduce emissions from the built environment (Usman and Makhdum 2021).

12.3.1.4 Adopting Low-Carbon Technologies

- **Clean hydrogen**: Green hydrogen, generated through renewable energy sources, represents an important replacement for fossil fuels, particularly in sectors that face challenges in electrification, such as heavy industry and transporting goods over long distances.
- **Nuclear energy**: While controversial, nuclear energy is a low-carbon energy source that some countries use as part of their mitigation strategy, given its potential to produce large amounts of electricity with minimal GHG emissions.
- **Geoengineering**: While still largely experimental, geoengineering approaches such as solar radiation management or ocean fertilization are being researched to potentially lower global temperatures or remove CO_2 from the atmosphere. However, these methods have significant risks and uncertainties (Vale et al. 2021).

12.3.1.5 Climate Policy and Global Cooperation

- **Carbon pricing**: Mechanisms such as carbon taxes or cap-and-trade systems incentivize organizations to lower their emissions. Implementing a carbon pricing mechanism encourages industries to transition toward more sustainable technologies and practices (Baum et al. 2022).
- **International agreements**: Global agreements like the Paris Agreement (2015) are essential in promoting collaboration among countries to restrict global warming to significantly below 2 °C (ideally 1.5 °C) above early industrial stages.

Countries establish emission reduction objectives, recognizing the necessity for collective international action to tackle climate change.

- **Green finance**: Financial markets are increasingly being directed to support sustainable projects through green bonds, climate financing, and ESG (Environmental, Social, and Governance) investment practices. Public and private investments in green technologies are crucial to scaling up mitigation efforts (Jahanzad et al. 2020).

12.3.2 Challenges to Climate Change Mitigation

Challenges to climate change mitigation are the various obstacles hindering efforts to lower GHGE and limit global warming. These challenges can be technical, economic, political, or social and may vary by region and sector (Usman and Makhdum 2021). Key challenges include.

12.3.2.1 Political and Economic Barriers

- **Lack of political will**: Climate change mitigation requires long-term planning and international cooperation, which can be hindered by short-term political cycles and conflicting national interests.
- **Economic costs**: Transitioning to a low-carbon economy can involve high upfront costs, and countries or businesses with limited resources may struggle to afford the necessary investments.
- **Resistance from vested interests**: Industries like fossil fuels, agriculture, and transportation may resist mitigation efforts because they would disrupt established business models and lead to job losses or higher costs (Jahanzad et al. 2020).

12.3.2.2 Technological and Infrastructure Challenges

- **Scaling up clean technologies**: While renewable energy technologies like solar and wind are advancing rapidly, their deployment still requires significant infrastructure development, including energy storage, grid modernization, and scaling up manufacturing.
- **Energy access in developing countries**: Many developing countries still rely heavily on fossil fuels for energy. Transitioning to cleaner energy sources without adequate infrastructure can be challenging and costly (Michel et al. 2021).

12.3.2.3 Global Cooperation and Equity

- **Developing versus developed countries**: Developing countries, which often have lower per capita emissions, argue that developed nations, historically contributing more to emissions, should bear a more significant share of the mitigation burden. Ensuring equitable action and financing is crucial.

- **Loss and damage**: Some nations may face irreversible damage due to climate change, even if mitigation efforts are successful. Addressing the loss of ecosystems, livelihoods, and infrastructure and compensating affected populations remains an ongoing challenge (Botzen et al. 2021).

12.3.3 Role of Global Cooperation

Climate change is a global problem, and effective mitigation requires coordinated action internationally. Efforts like the **Paris Agreement** involve all countries committing to reducing emissions, with developed nations providing financial and technological support to developing nations. Global cooperation is vital for the following:

- Setting and meeting emission reduction targets.
- Sharing technology and expertise for clean energy and sustainable practices.
- Ensuring that vulnerable countries are supported in mitigating and adapting to climate change (Waheed et al. 2021).

To combat CC, adaptation and mitigation are essential. Researchers describe climate change mitigation, whereas adaptation immediately affects it, like floods. Mitigation reduces greenhouse gas emissions, making it both a financial and environmental concern. Researchers are concerned with industrial and geographical adaptation and mitigation methods. Policy adaptation and mitigation primarily concentrate on farming, manufacturing, forest management, transportation, and land management. Both national and international focus are essential for effective adaptation and mitigation efforts. Economic and social progress needs adaptation to climate change, which has plagued the globe for decades. Develop worldwide policies and strategies to adapt and minimize CC (Abbass et al. 2022).

12.4 Climate Change and Adaptation

Climate change and adaptation involve adjusting procedures, mechanisms, and buildings to reduce damage and leverage possibilities in light of climate change's actual or anticipated effects. As global temperatures rise, adaptation strategies aim to reduce vulnerability and enhance adaptability among various populations, environments, and financial markets. Adaptation involves implementing measures to mitigate the negative impacts of climate change or to capitalize on future possibilities. While mitigation efforts (like reducing greenhouse gas emissions) are essential, adaptation is also necessary to deal with inevitable changes (Amoo and Fagbenle 2020). Adaptation strategies can be grouped into several categories:

12.4.1 Infrastructure Adaptation

Infrastructure adaptation refers to modifying, upgrading, or designing new infrastructure to withstand the impacts of climate change. This approach ensures that buildings, roads, transportation systems, utilities, and other critical infrastructure remain safe, functional, and resilient under changing environmental conditions (Ng 2021). Key components of infrastructure adaptation include:

- **Resilient Building Design**: Constructing buildings that can withstand extreme weather events (e.g., floods, hurricanes) using durable materials and elevated structures in flood-prone areas.
- **Flood and Storm Protection**: Implementing flood barriers, levees, seawalls, and stormwater management systems to protect urban and coastal areas from flooding and sea-level rise.
- **Climate-Resilient Transportation**: Adapting roads, bridges, railways, and airports to handle extreme temperatures, heavy rains, or storms to prevent disruptions in transportation networks (Bastidas-Arteaga et al. 2013).
- **Upgrading Utilities and Energy Systems**: Strengthening electricity grids, water systems, and waste facilities to reduce vulnerability to climate impacts, ensuring continuous service during extreme conditions.
- **Green Infrastructure**: Incorporating natural solutions, like green roofs, rain gardens, permeable pavements, and urban forests, to reduce heat, manage stormwater, and enhance urban resilience. Infrastructure adaptation is essential for protecting public safety, economic stability, and critical services, particularly as climate risks become more severe and frequent (Bobylev 2009).

12.4.2 Ecosystem-Based Adaptation

Ecosystem-based adaptation (EBA) is a strategy for addressing climate change that leverages ecological diversity and ecosystem services to assist individuals in adjusting to the negative impacts of climate change. Through ecosystem conservation, restoration, and sustainable management, EBA seeks to improve organic adaptability against adverse climate effects, including floods, droughts, and temperature fluctuations, while delivering advantages to both populations and the surroundings (Scarano 2017). Key aspects of ecosystem-based adaptation include:

- **Coastal Protection**: Restoring mangroves, coral reefs, and coastal marshes to protect shorelines from storm surges, erosion, and sea-level rise, reducing risks to coastal communities.
- **Water Management**: Protecting wetlands and watershed areas to improve water quality, reduce flood risks, and ensure a sustainable freshwater supply during droughts (Ferdowsi et al. 2024).

- **Forest Conservation and Reforestation**: Preserving forests or reforesting areas to prevent landslides, reduce soil erosion, and sequester carbon, which also helps regulate local temperatures (Ferdowsi et al. 2024).
- **Agroecological Practices**: Integrating natural elements like windbreaks, diverse crops, and soil conservation techniques into agricultural practices to improve crop resilience against climate variability (Ferdowsi et al. 2024).
- **Urban Green Spaces**: Creating parks, green roofs, and urban forests to reduce heat, absorb excess rainfall, and improve city air quality (Ferdowsi et al. 2024).

12.4.3 Agricultural Adaptation

Agricultural adaptation refers to adjusting farming practices, systems, and technologies to better cope with the impacts of climate change, such as increased temperatures, altered precipitation patterns, droughts, and extreme weather events. Agricultural adaptation aims to maintain food security, enhance crop resilience, and sustain farmers' livelihoods under changing environmental conditions (Sheikh et al. 2024). Key strategies in agricultural adaptation include:

- **Drought-Resistant Crops**: Developing and using crop varieties that can tolerate drought, heat, or salinity, enabling farmers to maintain yields in adverse conditions.
- **Improved Irrigation Techniques**: Using efficient irrigation systems, like drip irrigation or rainwater harvesting, to conserve water and ensure crops receive sufficient moisture during dry periods.
- **Soil Conservation**: Utilizing methods like no-till gardening, cover cropping, and the inclusion of organic matter to sustain soil fertility, enhance drainage, and minimize soil erosion.
- **Agroforestry**: Integrating trees and shrubs into farmland to provide shade, prevent erosion, improve soil health, and create microclimates that protect crops.
- **Diversified Cropping Systems**: Growing various crops to reduce the risk of crop failure and enhance resilience against pests, diseases, and climate variability.
- **Climate-Smart Pest and Disease Management**: Monitoring and controlling pests and diseases that may increase due to changing climate patterns using integrated pest management and resilient crop varieties. (Sheikh et al. 2024).

12.4.4 Urban Adaptation

Urban adaptation refers to modifying cities and metropolitan areas to enhance their resilience to the impacts of climate change, such as extreme heat, heavy rainfall, flooding, and sea-level rise. This approach involves planning, designing, and managing urban infrastructure, policies, and systems to protect public health, reduce vulnerability, and improve quality of life under changing climate conditions (Dharmarathne et al. 2024). Key strategies in urban adaptation include:

- **Heat mitigation**: Creating green spaces, such as parks, urban forests, and green roofs, to reduce urban heat and improve air quality. Using reflective and permeable materials for pavements and building surfaces to decrease heat absorption.
- **Flood control and stormwater management**: Building resilient drainage systems, flood barriers, and rainwater harvesting facilities to manage stormwater and reduce flood risks. Implementing green infrastructure like rain gardens, bioswales, and permeable pavements to absorb excess rainwater (Dharmarathne et al. 2024).
- **Resilient building design**: Constructing or retrofitting buildings to withstand extreme weather, using materials and designs that provide better insulation, reduce energy use, and maintain indoor comfort during extreme conditions (Dharmarathne et al. 2024).
- **Coastal protection**: Implementing seawalls, levees, and wetland restoration in coastal cities to protect against sea-level rise and storm surges.
- **Public health and emergency preparedness**: Developing systems to respond to climate-related health threats, such as heatwaves and vector-borne diseases, and establishing cooling centers and early warning systems to protect vulnerable populations (Rezvani et al. 2023).
- **Sustainable transportation**: Encouraging public transit, cycling, and walking infrastructure to reduce emissions and enhance urban mobility during extreme weather events (Rezvani et al. 2023).

12.4.5 Community-Based Adaptation:

Community-based adaptation (CBA) refers to the process of involving local communities in the design, implementation, and management of climate change adaptation strategies tailored to their specific needs, vulnerabilities, and capacities. This approach emphasizes local knowledge, participation, and decision-making, empowering communities to build resilience to the impacts of climate change while considering social, cultural, and economic contexts (Selje et al. 2024). Key components of community-based adaptation include:

- **Local Knowledge and Participation**: Actively involving community members in identifying climate risks, assessing vulnerabilities, and developing adaptation solutions. This often incorporates traditional knowledge alongside scientific data.
- **Capacity Building**: Strengthening the skills, resources, and awareness of community members and local institutions to enable them to manage climate risks and adaptation strategies effectively (Patnaik 2021).
- **Sustainable Livelihoods**: Supporting the development of livelihood strategies that are both climate-resilient and economically viable, such as promoting sustainable agriculture, fishing, or alternative income sources.

- **Inclusive and Equitable Approaches**: Ensuring that vulnerable groups, such as women, children, indigenous people, and marginalized communities, have a voice in the adaptation process and benefit from climate resilience efforts.
- **Local Resource Management**: Encouraging sustainable management of local natural resources, such as water, land, and forests, to enhance ecosystem resilience and protect communities from climate impacts.
- **Climate-Resilient Infrastructure**: Developing or improving local infrastructure, such as roads, water supply systems, and shelters that can withstand extreme weather events and climate-related hazards (Barrowman and Butler 2020).

12.4.6 Economic and Policy Adaptation

Economic and policy adaptation refers to the strategic adjustments in economic policies, frameworks, and practices designed to reduce vulnerability to climate change while promoting resilience and sustainable development. This type of adaptation focuses on integrating climate risks into economic planning, ensuring that policies support climate-resilient growth, and making economic systems more adaptive to changing environmental conditions (Eriksen et al. 2021). Key components of economic and policy adaptation include:

- **Insurance and risk financing**: Providing insurance schemes to protect vulnerable populations from climate-related losses.
- **Climate-resilient policies**: Developing policies that integrate climate risks into national and regional planning, ensuring that infrastructure, urban development, and agricultural practices are resilient to climate impacts.
- **Carbon pricing**: Execute strategies such as carbon taxes or cap-and-trade systems to encourage sustainable advancement while facilitating adaptation (Biagini et al. 2014).

Climate change is a global issue; adaptation must be approached through international collaboration. Climate financing mechanisms like the Green Climate Fund aim to support adaptation in developing countries, while programs like the Paris Agreement emphasize the need for a global effort to both mitigate and adapt to climate change. Adaptation is not just about minimizing damage; it's also about creating more resilient societies, economies, and ecosystems that can cope with the inevitable changes ahead. While mitigation is necessary to reduce the overall severity of climate change, adaptation helps us manage the risks and protect vulnerable populations from the worst impacts (Abbass et al. 2022). Climate change and their various adaptations are given in Fig. 12.2.

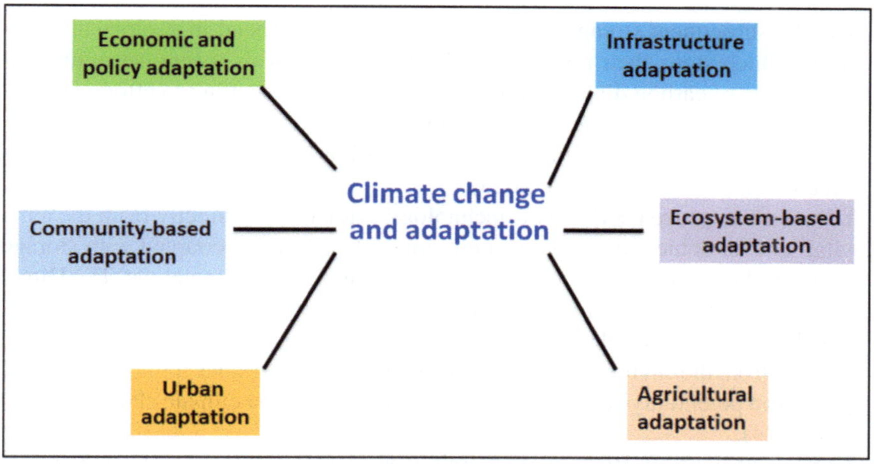

Fig. 12.2 Climate change and their various adaptations

12.5 Emerging Technology for Climate Change Mitigation and Adaptation

Emerging climate change mitigation and adaptation technology refers to innovative tools, techniques, and systems developed to reduce greenhouse gas emissions, slow global warming, and help communities adapt to climate change impacts. These technologies harness advancements in fields like renewable energy, data analytics, biotechnology, and engineering to create solutions that are more efficient, scalable, and accessible. Emerging technologies for climate change mitigation and adaptation are crucial to addressing the global challenges posed by climate change. These innovations span a wide range of sectors, from renewable energy to carbon capture, and from agricultural practices to climate forecasting. The emerging technologies for mitigation and adaptation are discussed step by step below.

12.5.1 Carbon Capture, Utilization, and Storage (CCUS)

CCUS is a technique suite aimed at capturing carbon dioxide emissions from sources like power plants and industrial facilities before they reach the atmosphere, and either utilizing the CO_2 for various purposes or securely storing it underground. CCUS is considered a crucial technology for mitigating climate change, especially for challenging sectors to decarbonize (Ma et al. 2022). Some CCUS are described the following to step wise:

- **Carbon capture and storage (CCS)**: CCS involves the collecting of carbon dioxide emissions at their source, such as power stations or industrial operations, and the subsequent storage of these emissions underground in geological structures, which stops their release into the atmosphere. This technique minimizes the release of carbon dioxide into the atmosphere, contributing to efforts in mitigating climate change. The carbon capture and storage (CCS) process helps create a more sustainable energy future by storing carbon dioxide in the subsurface in a secure manner (Bhavsar et al. 2023).

- **Direct air capture (DAC)**: This technology extracts CO_2 directly from the air, a vital tool for achieving "negative emissions." DAC can be coupled with storage or utilized to create synthetic fuels, chemicals, and building materials. The Direct Air Capture (DAC) system not only contributes to reducing greenhouse emissions but also supplies several businesses with vital resources. Direct removal of carbon dioxide from the environment accomplishes this. Given its dual applications, DAC presents a potentially valuable solution in the fight against climate change (Li and Yao 2024).
- **Carbon utilization**: Rather than merely storing captured carbon, it can be transformed into valuable products such as building materials (e.g., carbon concrete), plastics, or fuels, thereby decreasing dependence on fossil fuels. This strategy effectively reduces waste while developing sustainable alternatives that contribute to climate change mitigation. Transforming carbon emissions into valuable assets promotes a circular economy and enhances resilience to climate change (Anonymous 2021).

12.5.2 Renewable Energy Technologies

Energy sources and techniques refer to technologies that harness energy from quickly replenished sources such as sunlight, wind, water, geothermal heat, and organic materials. Implementing these technologies is crucial for minimizing GHGE, addressing climate change, and facilitating a transition to a cleaner and more sustainable energy future. The transition to renewable energy technologies signifies a crucial advancement in our shared pursuit of a cleaner and more sustainable future. Utilizing replenished sources allows us to generate energy while substantially decreasing GHGE, thus contributing significantly to the fight against climate change. This transition is not just an option but a critical strategy for safeguarding the environment's health and future generations. Adopting such creative approaches will enable us to develop an energy landscape that emphasizes environmental integrity while addressing the growing needs of society. As we progress, the dedication to renewable energy will foster a robust world where sustainability and dynamism coexist (Jaiswal et al. 2022). Key types of renewable energy technologies consist:

12.5.2.1 Solar Power

Solar power uses photovoltaic (PV) cells or solar thermal systems to convert sunlight directly into electricity or to heat fluids.

- **Photovoltaic (PV) Cells**: These cells capture sunlight and convert it into electricity through semiconductor materials.
- **Solar thermal** systems concentrate sunlight to generate heat, producing steam to drive electricity turbines or directly heat buildings and water (Maka and Alabid 2022).

12.5.2.2 Wind Power

Wind power employs turbines to capture wind energy and convert its kinetic energy into electricity. Utilizing wind energy via turbines signifies a significant advancement in sustainable energy solutions. The technique converts the kinetic energy of wind into electricity, offering a renewable power source while substantially contributing to reducing our carbon footprint. In our ongoing commitment to environmental conservation and the pursuit of alternatives to fossil fuels, wind energy emerges as a practical and effective solution. The continuous improvements in turbine design and efficiency significantly increase its potential, positioning wind power as a crucial element in the global shift toward cleaner energy sources. Adopting this innovative approach can facilitate a more sustainable future, emphasizing the significance of effectively using natural resources (Galdo et al. 2024).

12.5.2.3 Hydropower

Water flow, typically originating from rivers or dams, produces electricity through the operation of turbines. Utilizing the natural power of water flow is a crucial approach for electricity generation, especially through the use of rivers and dams. The technique entails the strategic management of water currents to drive turbines, converting kinetic energy into electrical energy. Utilizing this renewable resource allows us to leverage a sustainable energy source while reducing our dependence on fossil fuels, offering an environmentally responsible option for future generations. The incorporation of hydropower within our energy frameworks highlights the significance of creative approaches that harmonize with environmental responsibility while addressing the increasing need for electricity. With advanced exploration and enhancement of these technologies, the potential for water flow to play a significant role in our energy landscape remains robust and encouraging (Sarkar et al. 2023) Run-of-river systems utilize natural river flow without large dams, causing minimal disruption to water ecosystems. Dams store water, which can be released to generate electricity on demand, providing a reliable power source (Sarkar et al. 2023).

12.5.2.4 Geothermal Energy

The natural heat in the Earth's crust serves as a renewable energy source for geothermal energy. The decay of radioactive materials, the residual heat from Earth's formation, and tectonic movements generate this energy, ensuring a consistent and dependable power source. This dependable power source can be utilized for a range of usages, such as electricity generation and direct heating, establishing it as a significant alternative to fossil fuels. Consequently, geothermal energy plays a significant role in decreasing greenhouse gas emissions and fostering sustainability. Geothermal power plants extract geothermal fluids or steam to drive turbines for electricity generation. Geothermal heat pumps usage stable underground temperatures to heat and cool buildings (Spijkerboer et al. 2022).

12.5.2.5 Biomass and Bioenergy

Biomass and bioenergy involve the conversion of organic materials such as wood, crop waste, and algae into electricity, heat, or biofuels. The conversion of organic materials into electricity, heat, or biofuels via biomass and bioenergy processes signifies a crucial advancement in sustainable energy solutions. Utilizing readily available organic resources allows us to effectively mitigate waste and decrease our dependence on fossil fuels, which in turn contributes to a reduction in greenhouse gas emissions. This approach effectively addresses energy demands while promoting agricultural practices that enhance soil health and biodiversity. The optimization of these processes, the incorporation of biomass into our energy systems presents the potential for a cleaner, more resilient future—one in which renewable resources are essential in addressing global energy demands while promoting environmental responsibility (Ibitoye et al. 2023).

- **Direct combustion**: Burning biomass for heat or electricity.
- **Biogas production**: Using anaerobic digestion to convert organic matter into methane-rich biogas, which can be burned for energy.
- **Biofuels**: Producing ethanol and biodiesel from crops like corn and sugarcane to power vehicles.

12.5.2.6 Ocean Energy

Ocean Energy: Harnesses energy from oceanic movements, including tides, waves, and temperature variations. Utilizing ocean energy represents a significant opportunity in the pursuit of sustainable power solutions. Utilizing the natural forces of tides and waves, along with leveraging temperature variations, allows for the generation of electricity while effectively reducing our carbon footprint. The capacity of these renewable resources not only meets the increasing demand for clean energy but also contributes to alleviating the effects of climate change. With the advancement of technology and higher expenditures in infrastructure, ocean energy capture is set to play a crucial role in the world's energy picture. Adopting this creative plan will improve energy safety and promote sustainability, leading to a healthier and more adaptable future. Tidal Power uses tidal movements to turn turbines and generate electricity. Wave Power harnesses the energy of ocean waves to drive generators, and Ocean thermal energy conversion (OTEC) utilizes temperature disparities between warmer surface water and chilly ocean depths to generate electricity (Dey et al. 2022).

12.5.2.7 Advanced Solar Cells

Recent advancements in photovoltaic materials, including perovskite solar cells and tandem solar cells, enhance efficiency and reduce costs compared to conventional silicon-based solar cells. These advancements facilitate enhanced energy conversion and lower the overall costs associated with solar technology, thereby increasing the accessibility of renewable energy sources. Consequently, they are crucial in the ongoing shift toward sustainable energy solutions (Ibitoye et al. 2023).

12.5.2.8 Enhanced Geothermal Systems (EGS)

EGS facilitates the capture of geothermal energy from areas that were once deemed inappropriate for traditional geothermal generation of electricity. This innovation can significantly enhance the worldwide capacity for geothermal energy. These approaches deliver healthy, sustainable energy, minimize dependence on fossil fuels, and contribute to combating climate change by lowering GHGE. Utilizing geothermal energy from previously unexplored areas offers an enormous chance to transform the worldwide energy landscape. Utilizing this clean and sustainable power source allows us to significantly reduce our reliance on fossil fuels, thereby fostering a more robust energy future reference. The shift from electricity to geothermal energy facilitates the move toward more sustainable alternatives and significantly contributes to the fight against climate change by successfully reducing GHGE. With these potential energy reservoirs, we advanced toward establishing a sustainable world where clean energy is accessible and plentiful, securing a more sustainable environment for the future (Ang et al. 2022).

12.5.3 Energy Storage and Grid Technologies

Energy storage and grid technologies encompass a range of systems and innovations that store energy for later use and improve electricity grids' efficiency, reliability, and resilience. These technologies are vital for integrating renewable energy sources, like solar and wind, into the power grid, as they help balance supply and demand, manage peak loads, and provide backup power during outages (Amir et al. 2023).

12.5.3.1 Solid-State Batteries

Solid-state batteries offer enhanced safety and the potential for greater efficiency compared to conventional lithium-ion batteries. Solid electrolytes replace liquid ones, reducing the risk of fire and boosting energy absorption (Bates et al. 2022).

12.5.3.2 Grid-Scale Energy Storage

Large-scale energy storage technologies, such as flow batteries and pumped hydro storage, can store excess renewable energy generated during the day (from solar or wind) for use during periods of low generation, helping balance intermittent renewable sources with demand (Jafarizadeh et al. 2024).

12.5.3.3 Smart Grids

Incorporating digital technologies and artificial intelligence into energy grids facilitates real-time monitoring, enhances energy distribution efficiency, and improves the installation of renewable energy sources (Jafarizadeh et al. 2024).

12.5.4 Agriculture and Food Systems Innovation

Agriculture and food systems innovation comprises the creation of innovative techniques, technological advances, and strategies to enhance the environmental responsibility, effectiveness, and adaptability of food manufacturing, distribution, and consumption. These innovations aim to meet the growing global demand for food, address climate change impacts, reduce environmental degradation, and support the livelihoods of farmers and food producers (Çakmakçı et al. 2023).

12.5.4.1 Precision Agriculture

Drones, IoT sensors, and AI-driven analytics can optimize water use, fertilizers, and pesticides, reducing emissions and improving crop yields (Guareschi et al. 2023).

12.5.4.2 Vertical Farming and Indoor Agriculture

Growing crops in controlled indoor environments or vertically stacked structures uses less land and water and can be integrated with renewable energy sources like solar or wind (Guareschi et al. 2023).

12.5.4.3 Lab-Grown Meat and Plant-Based Proteins

These innovations aim to reduce the environmental impact of livestock farming, which is responsible for significant greenhouse gas emissions, land degradation, and water consumption (Çakmakçı et al. 2023).

12.5.5 Nature-Based Solutions and Environmental Restoration

Nature-based solutions (NbS) and **environmental restoration** leverage natural processes and ecosystems to address societal challenges, such as climate change, biodiversity loss, and water management. These solutions aim to restore, protect, and sustainably manage natural and modified ecosystems, promoting resilience and creating benefits for people and biodiversity (Waylen et al. 2024).

12.5.5.1 Reforestation and Afforestation

Large-scale planting of trees to sequester carbon and restore ecosystems is a crucial mitigation strategy. Advances in drone-assisted tree planting and genetically modified trees that grow faster are improving the efficiency of these efforts (Mohan et al. 2021).

12.5.5.2 Soil Carbon Sequestration

Enhanced agricultural techniques, including no-till farming, cover cropping, and agroforestry, can potentially augment the carbon storage capacity of the soil, effectively transforming agricultural lands into carbon sinks. Boosting soil health and improving biodiversity through these practices sequesters carbon while contributing to increased ecosystem resilience and productivity. The transition to sustainable agriculture is essential in addressing the harmful effects of climate change (Mohan et al. 2021).

12.5.5.3 Ocean-Based Solutions

Technologies such as ocean alkalinization, which increases the ocean's ability to absorb CO_2, and seaweed cultivation, which absorbs CO_2 and can be used in biofuels or as a food source, are emerging as ways to enhance carbon removal from the atmosphere (Bloomfield et al. 2020).

12.5.6 Climate Engineering and Geoengineering

Climate engineering and **geoengineering** refer to deliberate, large-scale interventions in the Earth's natural systems to counteract climate change. These methods aim to reduce greenhouse gases in the atmosphere or manage solar radiation to mitigate global warming. While they hold the potential to address climate impacts rapidly, they are often controversial due to risks, uncertainties, and ethical considerations (Proelss and Steenkamp 2023).

12.5.6.1 Solar Radiation Management (SRM)

SRM techniques injecting aerosols into the stratosphere to reflect sunlight can potentially lower global temperatures temporarily but also come with significant risks and uncertainties. Investigating solar radiation management (SRM) techniques, especially temporary temperature reduction via aerosol injection for sunlight reflection, initiates a nuanced discussion on climate intervention. This novel strategy is essential to acknowledge the considerable uncertainties and risks that accompany it. Possible changes in weather patterns, ecological disturbances, and geopolitical tensions pose significant challenges that require comprehensive analysis. At this critical juncture in our fight against climate change, it must carefully weigh the benefits of Sustainable Development Management (SRM) against its potential drawbacks. It presents the potential to alleviate specific impacts of global warming; however, it guides our actions with scientific knowledge, ethical principles, and a commitment to sustainable solutions for future generations (Honegger et al. 2021).

12.5.6.2 Ocean Fertilization

This technique entails the introduction of nutrients such as iron into the ocean to promote the growth of plankton, which can absorb CO_2 from the atmosphere. The ecological impact is currently under investigation. The implications of ocean fertilization on marine ecosystems remain inadequately comprehended, prompting apprehensions regarding unintended impacts on biodiversity and food webs. Researchers are currently examining the benefits and drawbacks of this carbon capture method (Kashyap et al. 2024).

12.5.6.3 Bioenergy with Carbon Capture and Storage (BECCS)

This technology:involves growing biomass (plants), converting it into energy, and capturing and storing the resulting CO_2 emissions, creating a net-negative carbon emissions cycle. It combines bioenergy (growing biomass to produce energy) with carbon capture and storage, sequestering CO_2 that the plants absorb from the atmosphere (Hait et al. 2024a, b, c, d, e, f, g; Honegger et al. 2021).

12.5.6.4 Direct Air Capture (DAC)

Direct Air Capture (DAC) utilizes chemical processes that remove CO_2 directly from the atmosphere, followed by underground storage or integration into various products. The potential of Direct Air Capture (DAC) presents a significant method for addressing climate change through the chemical extraction of carbon dioxide from the atmosphere. This advanced technology successfully decreases GHGE while providing sustainable methods for underground CO_2 storage, ensuring it does not re-enter the atmosphere. Furthermore, a range of products can employ captured carbon, fostering a circular economy that strengthens ecological and economic durability (Yafiee et al. 2024).

12.5.6.5 Stratospheric Aerosol Injection (SAI)

This method includes the introduction of reflective particles, such as sulfur aerosols, into the stratosphere to reflect sunlight. This method mimics the cooling effect that takes place after significant volcanic eruptions. Through the dispersion of these particles, SAI seeks to lower global temperatures and mitigate the impacts of climate change. Nonetheless, it brings to light concerns about possible ecological effects and ethical considerations (Laakso et al. 2022).

12.5.6.6 Marine Cloud Brightening

The technique employs sea salt aerosols in low-lying marine clouds to boost their reflectivity, aiding in rejuvenating the Earth's surface by reflecting more sunlight. The novel approach of Marine Cloud Brightening offers a promising strategy for tackling climate change by enhancing natural atmospheric functions. Spraying sea salt into low-lying clouds enhances their reflectivity, enabling them to reflect more sunlight and cool the Earth's surface. This approach provides a potential strategy for addressing rising temperatures and highlights the significance of utilizing existing natural mechanisms to tackle environmental challenges. As we proceed with exploring and implementing geoengineering solutions, Marine Cloud Brightening exemplifies human ingenuity in pursuing ecological equilibrium and sustainability (Horowitz et al. 2020).

12.5.6.7 Space-Based Reflectors

Space-based reflectors involve the strategic placement of mirrors or other reflective things in space to block a small fraction of arriving direct sunlight. This technique is primarily theoretical, facing significant technical and financial challenges. The idea of space-based reflectors offers an intriguing and intricate approach to tackling climate change through the strategic positioning of reflective items in orbit to obstruct a minor fraction of arriving direct sunlight. The theoretical advantages are persuasive, including the potential to cool the planet and alleviate specific impacts of global warming. However, considerable technical and financial challenges hinder the practical implementation of this concept. As researchers investigate creative approaches to utilize space technology for environmental applications, it is essential to carefully consider the challenges related to deploying and managing these reflectors. Space-based reflectors present a potential pathway for humanity to participate actively in climate management. However, their practicality will hinge on engineering advancements, solid funding strategies, and a strong dedication to sustainable practices on Earth (Chad et al. 2022).

12.5.6.8 Surface Albedo Modification

Applying white paint or reflective materials enhances the reflectivity (albedo) of surfaces, such as urban areas, deserts, or ice sheets, contributing to the cooling of specific regions. The technique of surface albedo alterations offers an exciting plan for addressing restricted warming impacts by increasing reflectivity. Reflective materials, such as particular paints or coatings, can alter different surfaces to efficiently redirect sunlight, leading to a notable temperature reduction in urban and rural environments. This method effectively meets the urgent requirement for temperature control while also playing a significant role in broader climate adaptation initiatives. As metropolitan areas expand and climate challenges become more pronounced, leveraging the potential of surface albedo modification may become an essential strategy in our efforts to create ecological settings that can flourish despite increasing global temperatures (Ziaeemehr et al. 2023).

12.5.7 Artificial Intelligence and Big Data for Climate Solutions

Artificial intelligence (AI) and big data for climate solutions involve using advanced computing techniques and vast datasets to monitor, predict, and address climate change. By analyzing complex climate data, AI and big data can improve climate models, optimize resource use, enhance resilience, and support decision-making for climate action. These technologies play a crucial role in adapting to and mitigating the effects of climate change (Chen et al. 2023). AI and big data detect emissions hotspots and reduce carbon footprints throughout supply chains, improving public and private sector decision-making. Using historical and real-time data, AI improves climate models to anticipate future climatic situations more accurately. Machine learning algorithms from large datasets can swiftly refine temperature, precipitation, severe events, and sea-level rise projections. AI-based Discount models warn of hurricanes, floods, droughts, and wildfires, enabling communities to plan and react. Satellites, sensors, and AI algorithms provide transparent and trustworthy greenhouse gas emissions data from industrial sites, urban regions, and ecosystems in real-time. Optimizing energy utilization using AI reduces emissions in numerous areas. Machine learning improves power grids, industrial operations, and building energy management, lowering energy waste and expenses. AI and big data allow intelligent networks to react to changes in wind and solar energy. AI balances grids and decreases fossil fuel use by anticipating supply and demand in real-time. Machine learning methods anticipate solar and wind energy availability, improving planning and storage to meet demand and avoid peak system stress (Chen et al. 2023).

12.5.8 Sustainable Transportation

Sustainable transportation refers to systems, modes, and practices that reduce environmental impact, promote social equity, and support economic efficiency, aiming to decrease greenhouse gas emissions, reduce air pollution, and lessen dependence on fossil fuels. Sustainable transportation prioritizes energy-efficient vehicles, alternative fuels, improved public transit, and infrastructure for active transportation, such as walking and cycling (Heidari et al. 2023). Electric cars, trucks, and buses are becoming more practical as the battery industry advances and countries build EV charging infrastructure. An automated electric car might minimize energy use and boost efficiency in transportation. Fuel cells might decarbonize heavy industries and long-haul transportation, which are difficult to electrify. Green hydrogen from renewable energy sources is becoming a clean fossil fuel replacement. Technologies with low emissions renewable energy sources like hybrids, biodiesel, and others reduce pollution Building dense, mixed-use districts near transportation hubs decreases long-distance automobile commuting and optimizes land usage. Smart traffic systems decrease fuel usage and emissions by using AI and sensors to maximize traffic flow, decrease congestion, and reduce idle time. Sustainability reduces ecological damage and promotes equitable mobility, making cities more sustainable, healthier, and adaptable. It is crucial to long-term climate and environmental objectives (Chen et al. 2023).

12.5.9 Water Management and Desalination

Water management and desalination are essential to ensure a reliable, sustainable water supply in response to growing demand, climate change, and regional water scarcity. Effective water management includes carefully using, distributing, and conserving water resources. At the same time, desalination provides a way to convert seawater and brackish water into fresh water, which is especially useful in arid regions and for coastal cities facing freshwater shortages (Pagliero et al. 2024). Recent developments in desalination technologies, such as reverse and forward osmosis, present practical solutions for supplying fresh water in arid regions. When integrated with alternative energy sources, desalination has the potential to deliver climate-resilient water supplies. Recent advancements in filtering and sterilization technologies are enhancing water reuse efficiency, facilitating a more sustainable approach to freshwater resource utilization across industrial, agricultural, and municipal systems. Integrated water resources management (IWRM) is a holistic approach that synchronizes the management of water, land, and associated resources to enhance economic and social welfare while preserving ecosystems (Hait et al. 2024a, b, c, d, e, f, g). The initiative encompasses collaboration across various sectors, including agriculture, industry, municipalities, and ecosystems. Sustainable extraction and recharge practices safeguard groundwater sources from overuse, ensuring their long-term availability. Monitoring aquifers and establishing groundwater usage limits are essential for preventing depletion and ensuring the maintenance of water quality (Rathore et al. 2023). Electrodialysis and reversal (ED/EDR) use electric fields to remove salt ions from water. This makes them helpful in desalinating brackish water with lower salt levels. These systems demand less energy than reverse osmosis for brackish water and are appropriate for inland desalination applications. Forward osmosis (FO) represents an innovative technology that utilizes a natural osmotic gradient to extract freshwater through a membrane, effectively leaving salts behind. The technology demonstrates the potential for desalination but remains predominantly in the experimental stage (Hait et al. 2024a, b, c, d, e, f, g). The various emerging technologies applied for climate change mitigation and adaptation are given in Fig. 12.3.

12.6 Upcoming Perspectives

Emerging technologies are rapidly transforming the climate change mitigation and adaptation landscape, offering potential solutions to reduce GHGE, enhance resilience, and foster sustainable development. There are significant perspectives on the future or upcoming of these technologies that are addressed below.

Fig. 12.3 Emerging technology for climate change mitigation and adaptation (a = carbon capture storage unit; b = solar panel; c = electric vehicle charger; d = battery grid; f = smart battery; g, h = fungicide and herbicide remover techniques; i = desalination plant; j = smart transportation; k = solar wind grid; l = Stratospheric aerosol injection for rain; m = smart spraying gadgets; h = acidification plant; o = smart gadget for agriculture; p = electric vehicle; q = novel techniques for direct air capture

12.6.1 Carbon Capture, Utilization, and Storage (CCUS)

CCUS technologies are expected to be critical in reducing emissions from industrial processes and power generation. These technologies can capture CO_2 emissions at the source and either store them underground or use them in various industrial applications. Emerging innovations are focused on making CCUS more cost-effective, scalable, and applicable to various industries. **Direct air capture (DAC)** technology extracts CO_2 directly from the atmosphere, and holds the potential to create "negative emissions." Innovations in energy-efficient capture materials and lower-cost, modular systems could accelerate the growth of DAC. **Bioenergy with carbon capture and storage (BECCS):** by combining bioenergy production with carbon capture, BECCS can help offset emissions from hard-to-abate sectors like aviation and cement production (Nguyen et al. 2023).

12.6.2 Renewable Energy Technologies

Continued advancements in renewable energy technologies such as solar, wind, and hydropower are central to achieving global decarbonization targets. Innovations in energy storage, grid integration, and offshore technologies are accelerating the transition to renewable energy. Next-generation solar cells (Perovskite solar cells), cheaper to manufacture and potentially more efficient than current silicon-based solar cells, could lead to widespread adoption of solar power, even in regions with limited access to conventional grids. Floating offshore wind and tidal energy technologies could unlock new renewable energy resources in previously inaccessible ocean regions. Integrating artificial intelligence and machine learning with energy management systems will allow for real-time optimization of renewable energy generation and distribution, minimizing waste and ensuring reliability (Yafiee et al. 2024).

12.6.3 Artificial Intelligence (AI) and Machine Learning

AI is becoming a cornerstone in climate change science, from improving energy efficiency to predicting climate patterns and optimizing natural resource use. Machine learning models are being deployed to enhance climate models, enabling better understanding and predicting extreme weather events. Predictive modeling for climate risks: AI-powered models can help predict the impacts of climate change with greater accuracy, assisting in disaster response, urban planning, and agricultural adaptation. AI can optimize energy systems, transportation networks, and supply chains, reducing emissions and waste. For example, AI-powered systems can optimize building heating/cooling, transportation routes, and even agricultural practices. AI-driven tools will revolutionize farming by using data from sensors and satellites to maximize water use, fertilizer application, and crop rotation, increasing productivity while minimizing environmental impact (Chen et al. 2023).

12.6.4 Climate-Resilient Infrastructure and Building Materials

As climate impacts intensify, building resilient infrastructure is critical to safeguarding communities. Advanced materials and construction techniques are being developed to help infrastructure withstand extreme weather events and changing environmental conditions. Self-healing concrete and innovative materials can repair their damage and respond dynamically to environmental conditions. This could extend the lifespan of infrastructure and reduce maintenance costs. Advances in climate-informed urban planning, including green roofs, urban cooling, and stormwater management, will help cities adapt to heat waves, flooding, and other climate impacts. Prefabricated and modular building techniques can improve the speed and cost of constructing climate-resilient housing, especially in disaster-prone regions (Bhavsar et al. 2023).

12.6.5 Geo-engineering

Geo-engineering approaches, including solar radiation management and ocean fertilization, are gaining attention as potential tools to counteract the effects of climate change. However, these technologies remain controversial due to their risks and uncertainties. Solar radiation management (SRM) technologies such as stratospheric aerosol injection could reflect sunlight away from the Earth's surface, lowering global temperatures. However, SRM raises concerns about potential side effects on regional weather patterns and ecosystems. Enhancing the ocean's ability to absorb CO_2 through controlled algae blooms could sequester significant amounts of carbon, though this approach requires further study to understand its environmental impacts. **Ocean fertilization** and **algae blooms** represent a controversial area of climate intervention, where nutrients are added to ocean waters to stimulate algae growth, which absorbs carbon dioxide from the atmosphere. The goal of this approach is to enhance carbon sequestration by promoting large algae blooms, which, upon dying, may sink to the ocean floor, potentially trapping carbon for long periods. While ocean fertilization could contribute to carbon removal, it has uncertain outcomes and ecological risks, influencing its future scope in research and potential deployment (Proelss and Steenkamp 2023).

12.6.6 Advanced Battery and Energy Storage Technologies

The widespread adoption of renewable energy requires reliable and cost-effective storage solutions to balance intermittent supply with demand. Advances in energy storage technologies will be crucial for decarbonizing energy systems. Solid-state batteries are expected to surpass current lithium-ion technology in energy density, safety, and longevity, making them ideal for large-scale energy storage and EVs. New storage technologies like pumped hydro, compressed air, and flow batteries are poised to become integral parts of the grid to store renewable energy for later use, providing stability to energy markets. **Grid-scale energy storage** is crucial for advancing climate change adaptation and mitigation by stabilizing renewable energy integration, supporting grid resilience, and enabling flexible, reliable power systems. As countries transition to low-carbon energy systems, grid-scale storage addresses the intermittency of renewables like wind and solar, making it foundational for clean energy expansion. It will also significantly decarbonize sectors like transportation and industry by enhancing energy availability and efficiency (Selje et al. 2024).

12.6.7 Sustainable Agriculture Technologies

Agriculture is a source of greenhouse gas emissions and a key sector impacted by climate change. New farming practices, food systems, and land management technologies are essential for mitigating emissions and adapting to changing climates. Vertical farming and aquaponic technologies enable food production in urban environments with reduced water use and land requirements. Using indoor farming systems can improve food security in climate-vulnerable regions. Regenerative agriculture technologies that support soil health (e.g., precision agriculture tools, cover cropping, and reduced tillage) can sequester carbon, improve resilience to droughts, and reduce emissions. The development of alternative proteins could reduce the environmental footprint of food production, especially livestock farming, which is a significant emitter of methane (Ng 2021).

12.6.8 Decentralized Energy Systems

The shift toward decentralized energy production, such as rooftop solar panels, microgrids, and community energy storage systems, can increase resilience, reduce reliance on centralized energy sources, and empower local communities. Blockchain and smart contracts could allow individuals to trade renewable energy directly, creating decentralized energy markets that increase energy access and equity. Microgrids and community energy systems maintain energy independence and ensure continued access to power during extreme weather events, which are becoming more frequent due to climate change (Dey et al. 2022).

12.6.9 Nature-Based Solutions (NBS)

Nature-based solutions, like reforestation, wetland restoration, and soil carbon sequestration, harness the natural environment to mitigate climate change and enhance resilience to its impacts. Reforestation and afforestation support large-scale reforestation efforts, including drones for planting trees and satellite monitoring to track forest health, which will be pivotal in sequestering carbon. **Reforestation** and **afforestation** are vital in combating climate change, enhancing biodiversity, and restoring degraded ecosystems. Technological advances are improving these efforts' scale, efficiency, and sustainability, making them more impactful for carbon sequestration, soil health, and ecosystem resilience. **Blue carbon initiatives** focus on protecting, restoring, and sustainably managing coastal and marine ecosystems—such as mangroves, salt marshes, and seagrass beds, to capture and store carbon. These ecosystems sequester carbon more efficiently than terrestrial forests, making blue carbon an increasingly important area for climate change mitigation and adaptation (Laakso et al. 2024). The future scope of blue carbon initiatives involves expanding their impact through scientific advances, innovative financing, and integration into climate policies. Restoration of coastal environments, including mangroves, sea grasses, and salt marshes, plays a crucial role in carbon sequestration and safeguarding coastal regions against increasing ocean levels and hurricanes. Emerging climate change mitigation and adaptation technologies offer immense potential to address the global climate crisis. However, their deployment will require overcoming technical, economic, and political challenges, including scaling up innovations, securing funding, and ensuring equitable access. Collaboration between governments, industries, and civil society will ensure these technologies reach their full potential, creating a sustainable and resilient future for all. Upcoming perspectives of emerging technologies for climate change, mitigation and adaptations are given in Fig. 12.4.

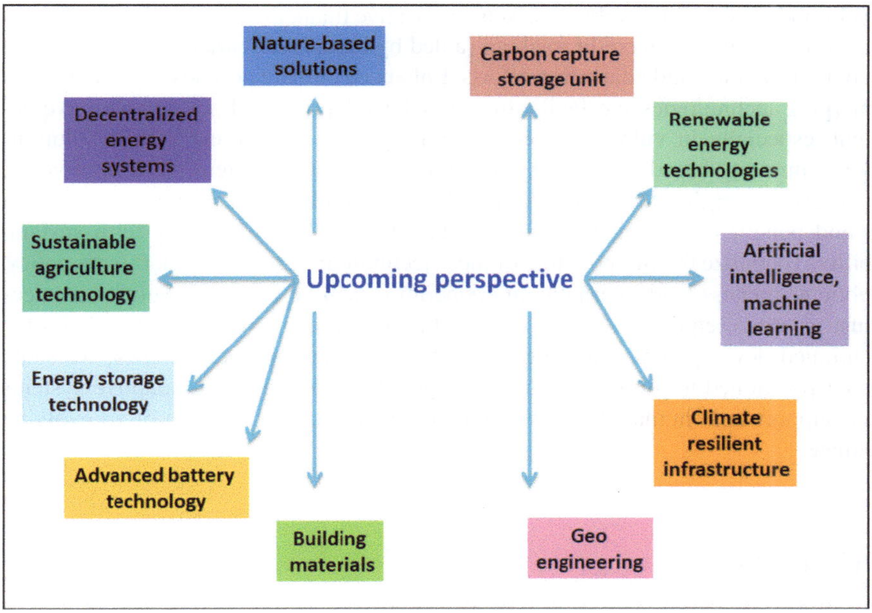

Fig. 12.4 Upcoming perspectives of emerging technologies for climate change, mitigation and adaptations

12.7 Conclusions

Emerging technologies for climate change mitigation and adaptation represent a transformative force in the global response to climate challenges. These technologies ranging from renewable energy and advanced energy storage to AI, big data, and nature-based solutions are reshaping our capacity to reduce emissions, protect ecosystems, and enhance resilience against climate impacts. Innovations like carbon capture and storage (CCS), advanced renewables, and sustainable transportation are advancing rapidly, enabling substantial cuts in greenhouse gas emissions across sectors. The integration of AI and data analytics optimizes energy efficiency, resource management, and emissions monitoring, pushing the boundaries of what is achievable. Technologies supporting water management, desalination, climate-resilient agriculture, and ecosystem restoration are vital for adapting to climate-induced stresses. Smart infrastructure, decentralized power grids, and precision agriculture empower communities to withstand and recover from climate events while preserving livelihoods. Expanding carbon markets, innovative financing models like blue bonds, and transparent carbon credit tracking (aided by blockchain) are driving investment in both mitigation and adaptation efforts. Enhanced climate policies that incorporate emerging technologies are facilitating broader adoption and sustainable development, especially in vulnerable regions. Engaging communities through solutions like community-based adaptation, urban greening, and ecosystem-based approaches is crucial. Emerging technologies provide communities with the tools to take part in and benefit from climate action, leading to locally driven, long-term sustainability. To realize the full potential of these technologies, robust support from policy, public awareness, and international cooperation will be essential. Future advances must prioritize equitable access, sustainability, and ecological integrity. With the continued development and integration of these emerging technologies, we stand better positioned to mitigate climate change and create resilient, adaptive systems that support both human and environmental well-being in the face of an evolving climate.

References

Abbass K, Qasim MZ, Song H (2022) A review of the global climate change impacts, adaptation, and sustainable mitigation measures. Environ Sci Pollut Res 29:42539–42559. https://doi.org/10.1007/s11356-022-19718-6

Amir M, Deshmukh RG, Khalid HM (2023) Energy storage technologies: an integrated survey of developments, global economical/environmental effects, optimal scheduling model, and sustainable adaption policies. J Energy Storage 72:108694. https://doi.org/10.1016/j.est.2023.108694

Amoo LM, Fagbenle RL (2020) Climate change in developing nations of the world. In: Fagbenle RO, Amoo OM, Aliu S (eds) Applications of heat, mass, and fluid boundary layers. Woodhead Publishing, pp 437–471. https://doi.org/10.1016/B978-0-12-817949-9.00023-2

Ang TZ, Salem M, Kamarol M (2022) A comprehensive study of renewable energy sources: classifications, challenges and suggestions. Energy Strategy Rev 43:100939. https://doi.org/10.1016/j.esr.2022.100939

Anonymous (2021) International Energy Agency, Net zero by 2050: a roadmap for the global energy sector. OECD Publishing.

Anwar A, Sinha A, Sharif A (2021) The nexus between urbanization, renewable energy consumption, financial development, and CO_2 emissions: evidence from selected Asian countries. Environ Dev Sustain. https://doi.org/10.1007/s10668-021-01716-2

Arshad MI, Iqbal MA, Shahbaz M (2018) Pakistan tourism industry and challenges: a review. Asia Pac J Tour Res 23(2):121–132

Aryal JP, Mehrotra MB, Jat ML (2015a) Impacts of laser land leveling in rice-wheat systems of the north-western Indo-Gangetic plains of India. Food Sec 7:725–738. https://doi.org/10.1007/s12571-015-0460-y

Aryal JP, Sapkota TB, Jat ML (2015b) On-farm economic and environmental impact of zero-tillage wheat: a case of north-west India. Exp Agric 51:1–16. https://doi.org/10.1017/S001447971400012X

Balsalobre-Lorente D, Ibánez-Luzón L, Usman M et al (2022) The environmental Kuznets curve, based on the economic complexity, and the pollution haven hypothesis in PIIGS countries. Renew Energy 185:1441–1455

Balsari S, Dresser C, Leaning J (2020) Climate change, migration, and civil strife. Curr Environ Health Rep 7:404–414. https://doi.org/10.1007/s40572-020-00291-4

Barrowman HM, Butler JRA (2020) Beyond challenges in community-based adaptation: critical insights from the human ecology framework. Hum Ecol Rev 26:73–94

Bastidas-Arteaga E, Schoefs F, Stewart MG (2013) Influence of global warming on durability of corroding RC structures: a probabilistic approach. Eng Struct 51:259–266

Bates AM, Preger Y, Torres-Castro L (2022) Are solid-state batteries safer than lithium-ion batteries? Joule 6(4):742–755. https://doi.org/10.1016/j.joule.2022.02.007

Battisti DS, Naylor RL (2009) Historical warnings of future food insecurity with unprecedented seasonal heat. Science 323(5911):240–244

Baul TK, McDonald M (2015) Integration of indigenous knowledge in addressing climate change. Indian J Tradit Knowl 1:20–27

Baum CM, Low S, Sovacool BK (2022) Between the sun and us: expert perceptions on the innovation, policy, and deep uncertainties of space-based solar geoengineering. Renew Sustain Energy Rev 158:112179. https://doi.org/10.1016/j.rser.2022.112179

Berendonk TU, Manaia CM, Merlin C (2015) Tackling antibiotic resistance: the environmental framework. Nat Rev Microbiol 13:310–317

Bergmann M, Collard F, Fabres J et al (2022) Plastic pollution in the Arctic. Nat Rev Earth Environ 3:323–337. https://doi.org/10.1038/s43017-022-00279-8

Betts RA, Alfieri L, Bradshaw C (2018) Changes in climate extremes, fresh water availability, and vulnerability to food insecurity projected at 1.5°C and 2°C global warming with a higher-resolution global climate model. Philos Trans A Math Phys Eng Sci 376:20160452. https://doi.org/10.1098/rsta.2016.0452.

Bhavsar A, Hingar D, Ostwal S (2023) The current scope and stand of carbon capture storage and utilization: a comprehensive review. Case Stud Chem Environ Eng 8:100368. https://doi.org/10.1016/j.cscee.2023.100368

Biagini B, Bierbaum R, Stults M (2014) A typology of adaptation actions: a global look at climate adaptation actions financed through the Global Environment Facility. Glob Environ Change 25:97–108. https://doi.org/10.1016/j.gloenvcha.2014.01.003

Birthal PS, Nigam AN, Narayanan AV (2012) Potential economic benefits from adoption of improved drought-tolerant groundnut in India. Agric Econ Res Rev 25:1–14

Bloomfield LS, McIntosh TL, Lambin EF (2020) Habitat fragmentation, livelihood behaviors, and contact between people and nonhuman primates in Africa. Landsc Ecol 35:985–1000. https://doi.org/10.1007/s10980-020-00995-w

Bobylev M (2009) Urban underground infrastructure and climate change: opportunities and threats. In: Proceedings of the fifth urban research symposium, Marseille, pp 28–30

Bolan S, Padhye LP, Jasemizad T, Govarthanan M, Karmegam N, Wijesekara H, Amarasiri D, Hou D, Zhou P, Biswal BK, Balasubramanian R, Wang H, Siddique KHM, Rinklebe J, Kirkham MB, Bolan N (2024) Impacts of climate change on the fate of contaminants through extreme weather events. Sci Total Environ 909:168388. https://doi.org/10.1016/j.scitotenv.2023.168388

Botzen W, Duijndam S, van Beukering P (2021) Lessons for climate policy from behavioral biases towards COVID-19 and climate change risks. World Dev 137:105214

Bujosa A, Riera A, Torres CM (2015) Valuing tourism demand attributes to guide climate change adaptation measures efficiently: the case of the Spanish domestic travel market. Tour Manag 47:233–239

Cai X, Wang D, Laurent R (2009) Impact of climate change on crop yield: a case study of rainfed corn in Central Illinois. J Appl Meteorol Climatol 48:1868–1881. https://doi.org/10.1175/2009JAMC1880.1

Çakmakçı R, Salık MA, Çakmakçı S (2023) Assessment and principles of environmentally sustainable food and agriculture systems. Agriculture 13(5):1073. https://doi.org/10.3390/agriculture13051073

Carleton TA, Hsiang SM (2016) Social and economic impacts of climate. Science 353:9837. https://doi.org/10.1126/science.aad9837

Chad MB, Sean L, Benjamin KS (2022) Between the sun and us: expert perceptions on the innovation, policy, and deep uncertainties of space-based solar geoengineering. Renew Sustain Energy Rev 158:112179. https://doi.org/10.1016/j.rser.2022.112179

Chen L, Chen Z, Zhang Y, Liu Y, Osman AI, Fargali M, Hua J, Al-Fatesh A, Ihara I, Roony DW, Yap PS (2023) Artificial intelligence-based solutions for climate change: a review. Environ Chem Lett 21:2525–2557. https://doi.org/10.1007/s10311-023-01617-y

Christiansen L, Olhoff A, Traerup S (2011) Technologies for adaptation: perspectives and practical experiences. United Nations Environment Programme Risø Centre on Energy, Climate and Sustainable Development. http://www.unep.org/pdf/TechnologiesAdaptation_PerspectivesExperiences.pdf

Costinot A, Donaldson D, Smith C (2016) Evolving comparative advantage and the impact of climate change in agricultural markets: evidence from 1.7 million fields around the world. J Pol Econ 124:20–25

Cruz RV, Harasawa H, Lal M, Wu S, Anokhin Y, Punsalmaa B, Honda Y, Jafari M, Li C, Huu Ninh N (2007) Asia. In: Parry ML, Canziani OF, Palutikof JP, van der Linden PJ, Hanson CE (eds) Climate change 2007: impacts, adaptation and vulnerability. Contribution of Working Group II to the Fourth Assessment Report of the Intergovernmental Panel on Climate Change. Cambridge University Press, Cambridge, UK

Cui W, Ouyang T, Qiu Y, Cui D (2021) Literature review of the implications of exercise rehabilitation strategies for SARS patients on the recovery of COVID-19 patients. Paper presented at Healthcare

Deryng D, Elliott J, Folberth C (2016) Regional disparities in the beneficial effects of rising CO_2 concentrations on crop water productivity. Nat Clim Change 6:786–790. https://doi.org/10.1038/nclimate2995

Dey S, Sreenivasulu A, Veerendra GTN (2022) Renewable energy present status and future potentials in India: an overview. Innov Green Dev 1(1):100006. https://doi.org/10.1016/j.igd.2022.100006

Dharmarathne G, Waduge AO, Bogahawaththa M (2024) Adapting cities to the surge: a comprehensive review of climate-induced urban flooding. Results Eng 22:102123. https://doi.org/10.1016/j.rineng.2024.102123

Diffenbaugh NS, Singh D, Mankin JS (2017) Quantifying the influence of global warming on unprecedented extreme climate events. Proc Natl Acad Sci USA 114:4881–4886

Dwivedi YK, Hughes L, Kar AK (2022) Climate change and COP26: are digital technologies and information management part of the problem or the solution? Int J Inf Manag 63:102456. https://doi.org/10.1016/j.ijinfomgt.2021.102456

Ebi KL, Vanos J, Baldwin JW (2021) Extreme weather and climate change: population health and health system implications. Annu Rev Public Health 42:293–315. https://doi.org/10.1146/annurev-publhealth-012420-105026

Egboduku WO, Egboduku T, Golohor OM, Imarhiagbe O, Ogwu MC (2024) Cassava as raw material for sustainable bioeconomy development. In: Ogwu MC, Izah SC, Cunha Alves AA, Babu SC (eds) Plant biology, sustainability and climate change, sustainable cassava. Academic, pp 57–73. https://doi.org/10.1016/B978-0-443-21747-0.00022-9

Eriksen S, Schipper ELF, Scoville-Simonds M (2021) Adaptation interventions and their effect on vulnerability in developing countries: help, hindrance, or irrelevance? World Dev 141:105383. https://doi.org/10.1016/j.worlddev.2020.105383

FAO (2018) The state of the world's forests 2018: forest pathways to sustainable development

Fargnoli M, Lombardi M (2020) Safety vision of agricultural tractors: an engineering perspective based on recent studies (2009–2019). Safety 6:1. https://doi.org/10.3390/safety6010001

Ferdowsi A, Piadeh F, Behzadian K (2024) Urban water infrastructure: a critical review on climate change impacts and adaptation strategies. Urban Climate 58:102132. https://doi.org/10.1016/j.uclim.2024.102132

Ferreira JJ, Fernandes CI, Ferreira FA (2020) Technology transfer, climate change mitigation, and environmental patent impact on sustainability and economic growth: a comparison of European countries. Technol Forecast Soc Change 150:119770

Furszyfer Del Rio DD, Sovacool BK, Griffiths S (2021) Culture, energy and climate sustainability, and smart home technologies: a mixed methods comparison of four countries. Energy Climate Change 2:100035. https://doi.org/10.1016/j.egycc.2021.100035

Galdo MIL, Cartelle Barros JJ, Orosa García JA, Rodríguez García JD, Couce Casanova A (2024) Introduction to wind turbines. In: Rahimpour MR (ed) Encyclopedia of renewable energy, sustainability and the environment, 1st edn. Elsevier, pp 331–342. https://doi.org/10.1016/B978-0-323-93940-9.00054-2

Godde CM, Mason-D'Croz D, Mayberry DE (2021) Impacts of climate change on the livestock food supply chain: a review of the evidence. Glob Food Sec 28:100488. https://doi.org/10.1016/j.gfs.2020.100488

Green F, Healy N (2022) How inequality fuels climate change: the climate case for a Green New Deal. One Earth 5:635–649. https://doi.org/10.1016/j.oneear.2022.05.005

Guareschi M, Mancini MC, Lottici C (2023) Strategies for the valorization of sustainable productions through an organic district model. Agroecol Sustain Food Syst 47:100–125. https://doi.org/10.1080/21683565.2022.2134270

Hait M, Kashyap NK, Bhardwaj AK (2024a) Emerging pollutants from urbanization and industrialization in the global south. In: Izah SC, Ogwu MC, Hamidifar H (eds) Biomonitoring of pollutants in the global south. Springer, Singapore. https://doi.org/10.1007/978-981-97-1658-6_2

Hait M, Kashyap NK, Bhardwaj AK (2024b) Emerging pollutants from urbanization and industrialization in the Global South. In: Izah SC, Ogwu MC, Hamidifar H (eds) Biomonitoring of pollutants in the global south. Springer, Singapore, pp 39–87. https://doi.org/10.1007/978-981-97-1658-6_2

Hait M, Kashyap NK, Giri AK (2024c) Synthesis of metal NPs using plant flowers and fruits. In: Bhardwaj AK, Srivastav AL, Rai S (eds) Biogenic wastes-enabled nanomaterial synthesis. Springer, Cham. https://doi.org/10.1007/978-3-031-59083-2_8

Hait M, Kashyap NK, Giri AK (2024d) Synthesis of metal NPs using plant flowers and fruits. In: Bhardwaj AK, Srivastav AL, Rai S (eds) Biogenic wastes-enabled nanomaterial synthesis. Springer, Cham, pp 195–237. https://doi.org/10.1007/978-3-031-59083-2_8

Hait M, Kashyap NK, Sahu P (2024e) Fabrication of metal NPs from plant root and tuber. In: Bhardwaj AK, Srivastav AL, Rai S (eds) Biogenic wastes-enabled nanomaterial synthesis. Springer, Cham, pp 29–71. https://doi.org/10.1007/978-3-031-59083-2

Hait M, Patel D, Izah SC (2024f) Molecular techniques and technologies in biomonitoring for environmental sustainability. In: Izah SC, Ogwu MC, Hamidifar H (eds) Biomonitoring of

pollutants in the global south. Springer, Singapore, pp 605–637. https://doi.org/10.1007/978-981-97-1658-6_17
Hait M, Sahu P, Biswas S, Izah SC (2024) Sustainable application of artificial intelligence in biomonitoring for environmental sustainability: challenges and prospects. In: Izah SC, Ogwu MC, Hamidifar H (eds) Biomonitoring of pollutants in the global south. Springer, Singapore, pp 747–778. https://doi.org/10.1007/978-981-97-1658-6_22
Hassan ZM, Langås EF, Sanfilippo F (2024) Exploring the synergies between collaborative robotics, digital twins, augmentation, and industry 5.0 for smart manufacturing: a state-of-the-art review. Rob Comput-Integr Manuf 89:102769. https://doi.org/10.1016/j.rcim.2024.102769
Heidari I, Toloie Eshlaghy A, Seyyed Hoseini SM (2023) Sustainable transportation: definitions, dimensions, and indicators—case study of importance-performance analysis for the city of Tehran. Heliyon 9:e20457. https://doi.org/10.1016/j.heliyon.2023.e20457
Hoegh-Guldberg O, Poloczanska ES, Skirving W (2017) Coral reef ecosystems under climate change and ocean acidification. Front Mar Sci 4:158. https://doi.org/10.3389/fmars.2017.00158
Honegger M, Michaelowa A, Pan J (2021) Potential implications of solar radiation modification for achievement of the sustainable development goals. Mitig Adapt Strateg Glob Change 26:21. https://doi.org/10.1007/s11027-021-09958-1
Horowitz HM, Holmes C, Wright A, Sherwen T, Wang X, Evans M, Huang J, Jaeglé L, Chen Q, Zhai S, Alexander B (2020) Effects of sea salt aerosol emissions for marine cloud brightening on atmospheric chemistry: implications for radiative forcing. Geophys Res Lett 47(4):e2019GL085838. https://doi.org/10.1029/2019GL085838
Hussain M, Butt AR, Uzma F (2020) A comprehensive review of climate change impacts, adaptation, and mitigation on environmental and natural calamities in Pakistan. Environ Monit Assess 192:48
Ibitoye SE, Mahamood RM, Jen TC (2023) An overview of biomass solid fuels: biomass sources, processing methods, and morphological and microstructural properties. J Bioresour Bioprod 8(4):333–360. https://doi.org/10.1016/j.jobab.2023.09.005
Imarhiagbe O, Okafor AC, Ikponmwosa BO, Ogwu MC (2023) Sustainable agricultural pest control strategies to boost food and socioecological security: the allelopathic strategy. In: Ogwu MC, Chibueze Izah S (eds) One health implications of agrochemicals and their sustainable alternatives. Sustainable development and biodiversity. Springer, Singapore, pp 637–657. https://doi.org/10.1007/978-981-99-3439-3_23
Intergovernmental Panel on Climate Change (IPCC) (2012) Managing the risks of extreme events and disasters to advance climate change adaptation. In: Field CB, Barros V, Stocker TF, Qin D, Dokken DJ, Ebi KL, Mastrandrea MD, Mach KJ, Plattner GK, Allen SK, Tignor M, Midgley PM (eds) A special report of Working Groups I and II of the Intergovernmental Panel on Climate Change. Cambridge University Press, Cambridge, UK and New York, USA
Intergovernmental Panel on Climate Change (IPCC) (2014) Asia. In: Stocker TF, Qin D, Plattner GK, Tignor M, Allen SK, Boschung J, Nauels A, Xia Y, Bex V, Midgley PM (eds) Contribution of Working Group II to the fifth assessment report of the Intergovernmental Panel on Climate Change. Cambridge University Press, Cambridge, UK and New York, USA. http://ipcc-wg2.gov/AR5/images/uploads/WGIIAR5-Chap24_FGDall.pdf
IPCC (2013) Summary for policymakers. Clim Chang Phys Sci Basis Contrib Work Gr I Fifth Assess Rep
IPCC (2014) Climate change 2014: synthesis report. In: Pachauri RK, Meyer LA (eds) Contribution of Working Groups I, II, and III to the fifth assessment report of the intergovernmental panel on climate change. IPCC, Geneva, p 151
Ishikawa-Ishiwata Y, Furuya J (2022) Economic evaluation and climate change adaptation measures for rice production in Vietnam using a supply and demand model: special emphasis on the Mekong River Delta region in Vietnam. Interlocal adaptations to climate change in East and Southeast Asia. Springer, Cham, pp 45–53
Izah SC, Ogwu MC, Akram M (2023) Herbal medicine phytochemistry: applications and trends. Springer, Cham, Switzerland. https://doi.org/10.1007/978-3-031-21973-3

Jacob DE, Nelson IU, Oluwafemi OJ, Izah SC, Ogwu MC (2024) Environmental stewardship: safeguarding biodiversity in protected landscapes and recreational parks using biosecurity. In: Izah SC, Ogwu MC, Hamidifar H (eds) Biomonitoring of pollutants in the global south. Springer, Singapore. https://doi.org/10.1007/978-981-97-1658-6_23

Jafarizadeh H, Yamini E, Zolfaghari SM (2024) Navigating challenges in large-scale renewable energy storage: barriers, solutions, and innovations. Energy Rep 12:2179–2192. https://doi.org/10.1016/j.egyr.2024.08.019

Jahanzad E, Holtz BA, Zuber CA (2020) Orchard recycling improves climate change adaptation and mitigation potential of almond production systems. PLoS ONE 15

Jaiswal KK, Chowdhury CR, Yadav D (2022) Renewable and sustainable clean energy development and impact on social, economic, and environmental health. Energy Nexus 7:100118. https://doi.org/10.1016/j.nexus.2022.100118

Karkkainen L, Lehtonen H, Helin J (2020) Evaluation of policy instruments for supporting greenhouse gas mitigation efforts in agricultural and urban land use. Land Use Policy 99:104991

Kashyap NK, Hait M, Bhardwaj AK (2024) Planktons as a sustainable biomonitoring tool of aquatic ecosystem. In: Izah SC, Ogwu MC, Hamidifar H (eds) Biomonitoring of pollutants in the global south. Springer, Singapore, pp 275–319. https://doi.org/10.1007/978-981-97-1658-6_8

Kaur M, Hemavathi B, Shalini KV, Kashyap NK (2023) Biomedical signal processing and healthcare applications: innovations and challenges. J Adv Zool 44:5279–5286. http://jazindia.com/index.php/jaz/article/view/2714

Kosoe EA, Achana GTW, Ogwu MC (2023) Meta-evaluation of the one health implication on food systems of agrochemical use. In: Ogwu MC, Chibueze Izah S (eds) One health implications of agrochemicals and their sustainable alternatives. Sustainable development and biodiversity. Springer, Singapore, pp 387–409. https://doi.org/10.1007/978-981-99-3439-3_14

Kukal MS, Irmak S (2018) Climate-driven crop yield and yield variability and climate change impacts on the U.S. great plains agricultural production. Sci Rep 8:3450. https://doi.org/10.1038/s41598-018-21848-2

Laakso A, Niemeier U, Visioni D, Tilmes S, Kokkola H (2022) Dependency of the impacts of geoengineering on the stratospheric sulfur injection strategy—Part 1: Intercomparison of modal and sectional aerosol modules. Atmos Chem Phys 22:93–118. https://doi.org/10.5194/acp-22-93-2022

Latif A, Shakir AS, Rashid MU (2013) Appraisal of economic impact of zero tillage, laser land leveling, and bed-furrow intervention in Punjab, Pakistan. Pak J Eng Appl Sci 13:65–81

Lee J, Gryze SD, Six J (2011) Effect of climate change on field crop production in California's central valley. Clim Change 109–S353. https://doi.org/10.1007/s10584-011-0305-4

Lemery J, Knowlton K, Sorensen C (2021) Global climate change and human health: from science to practice. Wiley

Li G, Yao J (2024) Direct air capture (DAC) for achieving net-zero CO_2 emissions: advances, applications, and challenges. Eng 5:1298–1336. https://doi.org/10.3390/eng5030069

Licardo JT, Domjan M, Orehovački T (2024) Intelligent robotics—a systematic review of emerging technologies and trends. Electron 13(3):542. https://doi.org/10.3390/electronics13030542

Liu L, Basso B (2020) Impacts of climate variability and adaptation strategies on crop yields and soil organic carbon in US Midwest. PLoS ONE 15. https://doi.org/10.1371/journal.pone.0225433

Long SP, Ainsworth EA, Leakey AD, Nosberger J, Ort DR (2006a) Food for thought: lower-than-expected crop yield stimulation with rising CO_2 concentrations. Science 312(5782):1918–1921

Long SP, Ainsworth EA, Leakey ADB (2006b) Food for thought: lower-than-expected crop yield stimulation with rising CO_2 concentrations. Science 312:1918–1921. https://doi.org/10.1126/science.1114722

Ma J, Li L, Wang H (2022) Carbon capture and storage: history and the road ahead. Engineering 14:33–43. https://doi.org/10.1016/j.eng.2021.11.024

Madan P, Jagadish SVK, Craufurd PQ et al (2012) Effect of elevated CO_2 and high temperature on seed-set and grain quality of rice. J Exp Bot 63:3843–3852. https://doi.org/10.1093/jxb/ers077

Maka AOM, Alabid JM (2022) Solar energy technology and its roles in sustainable development. Clean Energy 6:476–483. https://doi.org/10.1093/ce/zkac023
Malhi GS, Kaur M, Kaushik P (2021) Impact of climate change on agriculture and its mitigation strategies: a review. Sustainability 13(3):1318. https://doi.org/10.3390/su13031318
Manes S, Costello JM, Beckett H, Debnath A, Devenish-Nelson E, Kerry-Anne G, Jenkins R, Khan TM, Kiessling W, Krause C, Maharaj SS, Midgley GF, Price J, Talukdar G, Vale MM (2021) Endemism increases species' climate change risk in areas of global biodiversity importance. Biological Conservation 257:109070. https://doi.org/10.1016/j.biocon.2021.109070
Meierrieks D (2021) Weather shocks, climate change and human health. World Dev 138:105228. https://doi.org/10.1016/j.worlddev.2020.105228
Michel D, Eriksson M, Klimes M (2021) Climate change and (in)security in transboundary river basins. In: Edward Elgar Publishing (ed) Handbook of security and the environment
Mimura N (2013) Sea-level rise caused by climate change and its implications for society. Proc Jpn Acad Ser B Phys Biol Sci 89(7):281–301. https://doi.org/10.2183/pjab.89.281
Mohan M, Richardson G, Gopan G (2021) UAV-supported forest regeneration: current trends, challenges and implications. Remote Sens 13(13):2596. https://doi.org/10.3390/rs13132596
Narayanamoorthy A, Bhattarai M, Joshi P (2018) An assessment of the economic impact of drip irrigation in vegetable production in India. Agric Econ Res Rev 31:105–112
Narayanamoorthy A, Devika N (2017) Economic and resource impacts of drip method of irrigation on okra cultivation: an analysis of field survey data. J Land Rural Stud 6:15–33. https://doi.org/10.1177/2321024917731840
Ng WS (2021) Impact of climate change on infrastructure. In: Leal Filho W, Azul AM, Brandli L (eds) Industry, innovation and infrastructure. Encyclopedia of the UN sustainable development goals. Springer, Cham. https://doi.org/10.1007/978-3-319-95873-6_10
Nguyen TT, Grote U, Neubacher F (2023) Security risks from climate change and environmental degradation: implications for sustainable land use transformation in the global south. Curr Opin Environ Sustain 63:101322. https://doi.org/10.1016/j.cosust.2023.101322
Obahiagbon EG, Ogwu MC (2023) Consumer perception and demand for sustainable herbal medicine products and market. In: Izah SC, Ogwu MC, Akram M (eds) Herbal medicine phytochemistry. Reference series in phytochemistry. Springer, Cham. https://doi.org/10.1007/978-3-031-21973-3_65-1
Obahiagbon EG, Ogwu MC (2024) The nexus of business, sustainability, and herbal medicine. In: Izah SC, Ogwu MC, Akram M (eds) Herbal medicine phytochemistry. Reference series in phytochemistry. Springer, Cham. https://doi.org/10.1007/978-3-031-21973-3_67-1
Ogada MJ, Rao EJO, Radeny M (2020) Climate-smart agriculture, household income, and asset accumulation among smallholder farmers in Nyando basin of Kenya. World Dev Perspect 18:100203. https://doi.org/10.1016/j.wdp.2020.100203
Ogden LE (2018) Climate change, pathogens, and people: the challenges of monitoring a moving target. Bioscience 68:733–739
Ogwu MC (2019) Towards sustainable development in Africa: the challenge of urbanization and climate change adaptation. In: Cobbinah PB, Addaney M (eds) The geography of climate change adaptation in urban Africa. Springer, Switzerland, pp 29–55. https://doi.org/10.1007/978-3-030-04873-0_2
Ogwu MC, Izah SC (2023) One health implications of agrochemicals and their sustainable alternatives (First Edition). In: Ramawat KG (ed) Sustainable development and biodiversity series, vol 34. Springer, Singapore, 826p. Hardcover. ISBN: 978-981-99-3438-6. https://doi.org/10.1007/978-981-99-3439-3
Ogwu MC, Osawaru ME, Ahana CM (2014) Challenges in conserving and utilizing plant genetic resources (PGR). Int J Genet Mol Biol 6(2):16–22. https://doi.org/10.5897/IJGMB2013.0083
Ortiz-Bobea A, Ault TR, Carrillo CM, Chambers RG, Lobell DB (2021) Anthropogenic climate change has slowed global agricultural productivity growth. Nat Clim Chang 11:306–312. https://doi.org/10.1038/s41558-021-01000-1

Pagliero L, McIntyre N, Aitken D, Bolz P, Jamett N, Pérez-Murillo G, Rivero F, Herrera-León S, Ordens CM, Campos L, García G, Cisternas LA (2024) Sustainable integration of desalinated seawater into regional water supply networks using a participatory modelling framework. Environ Sci Policy 155:103714. https://doi.org/10.1016/j.envsci.2024.103714
Pal JK (2021) Visualizing the knowledge outburst in global research on COVID-19. Scientometrics 126:4173–4193
Pampolino MF, Manguiat IJ, Ramanathan S (2007) Environmental impact and economic benefits of site-specific nutrient management (SSNM) in irrigated rice systems. Agric Syst 93:1–24. https://doi.org/10.1016/j.agsy.2006.04.002
Patel D, Chaturwedi AK, Kashyap NK (2024) Quality control and standardization of nanomaterials. In: Thakur A, Kumar A (eds) Sustainability, safety, and applications of nanomaterials-based corrosion inhibitors. IGI Global, USA, pp 297–325. https://doi.org/10.4018/979-8-3693-7640-9.ch012
Patnaik H (2021) Gender and participation in community-based adaptation: evidence from the decentralized climate funds project in Senegal. World Dev 142:105448
Precht WF, Aronson RB (2004) Climate flickers and range shifts of reef corals. Front Ecol Environ 2:307–314. https://doi.org/10.1890/1540-9295(2004)002[0307:CFARSO]2.0.CO;2
Proelss A, Steenkamp RC (2023) Geoengineering: methods, associated risks and international liability. In: Gailhofer P, Krebs D, Proelss A (eds) Corporate liability for transboundary environmental harm. Springer, Cham. https://doi.org/10.1007/978-3-031-13264-3_9
Qi D, Chen L, Chen B (2017) Increase in acidifying water in the western Arctic Ocean. Nat Clim Change 7:195–199. https://doi.org/10.1038/nclimate3228
Rathore RK, Kashyap NK, Annamareddy SHK, Hait M, Roymahapatra G, Vaishnav MM (2023) Appraisal of physicochemical qualities of grazing land soil in Kartala block, Chhattisgarh, India. ES Food Agrofor 11:856. https://doi.org/10.30919/esfaf856
Ray DK, West PC, Clark M (2019) Climate change has likely already affected global food production. PLoS ONE 14. https://doi.org/10.1371/journal.pone.0217148
Rehman A, Ma H, Ahmad M (2021) Towards environmental sustainability: devolving the influence of carbon dioxide emission to population growth, climate change, forestry, livestock and crops production in Pakistan. Ecol Indic 125:107460. https://doi.org/10.1016/j.ecolind.2021.107460
Rezvani SMHS, de Almeida NM, Falcão MJ (2023) Climate adaptation measures for enhancing urban resilience. Buildings 13:2163. https://doi.org/10.3390/buildings13092163
Richard G, Hait M, Ibrahim M (2024) Emerging chemical air contaminants and their associated human health effects. In: The handbook of environmental chemistry. Springer, Berlin, Heidelberg, pp 1–24. https://doi.org/10.1007/698_2024_1130
Ricke KL, Orr JC, Schneider K (2013) Risks to coral reefs from ocean carbonate chemistry changes in recent earth system model projections. Environ Res Lett 8:034003. https://doi.org/10.1088/1748-9326/8/3/034003
Robinson A, Lehmann J, Barriopedro D, Rahmstorf S, Coumou D (2021) Increasing heat and rainfall extremes now far outside the historical climate. Clim Atmos Sci 4:45. https://doi.org/10.1038/s41612-021-00202-w
Salam MA, Al-Amin MY, Salam MT (2023) Antimicrobial resistance: a growing serious threat for global public health. Healthcare (Basel) 11:1946. https://doi.org/10.3390/healthcare11131946
Santos WS, Gurgel-Gonçalves R, Garcez LM (2021) Deforestation effects on Attalea palms and their resident Rhodnius, vectors of Chagas disease, in eastern Amazonia. PLoS ONE 16
Sarkar S, Nag AK, Kumar R (2023) An overview of small-scale hydropower and its recent development. In: Jeguirim M, Dutournié P (eds) Advances in renewable energy technologies, renewable energy production and distribution, vol 2. Academic Press, 249–314. https://doi.org/10.1016/B978-0-443-18439-0.00009-4.
Saunders MI, Leon JX, Callaghan DP (2014) Interdependency of tropical marine ecosystems in response to climate change. Nat Clim Change 4:724–729. https://doi.org/10.1038/nclimate2274

Scarano FR (2017) Ecosystem-based adaptation to climate change: concept, scalability, and a role for conservation science. Perspect Ecol Conserv 15:65–73. https://doi.org/10.1016/j.pecon.2017.05.003

Schlenker W, Roberts MJ (2009) Nonlinear temperature effects indicate severe damages to U.S. crop yields under climate change. Proc Natl Acad Sci U S A 106:15594–15598. https://doi.org/10.1073/pnas.0906865106

Scott D (2021) Sustainable tourism and the grand challenge of climate change. Sustainability 13(4):1966

Selje T, Schmid LA, Heinz B (2024) Community-based adaptation to climate change: core issues and implications for practical implementations. Climate 12:155. https://doi.org/10.3390/cli12100155

Sheikh ZA, Ashraf S, Weesakul S (2024) Impact of climate change on farmers and adaptation strategies in Rangsit, Thailand. Environ Chall 15:100902. https://doi.org/10.1016/j.envc.2024.100902

Spijkerboer RC, Turhan E, Roos A (2022) Out of steam? A social science and humanities research agenda for geothermal energy. Energy Res Soc Sci 92:102801. https://doi.org/10.1016/j.erss.2022.102801

Strengers Y, Nicholls L (2017) Convenience and energy consumption in the smart home of the future: Industry visions from Australia and beyond. Energy Res Soc Sci 32:86–93. https://doi.org/10.1016/j.erss.2017.02.008

Su H-C, Liu Y-S, Pan C-G (2018) Persistence of antibiotic resistance genes and bacterial community changes in drinking water treatment systems: from drinking water source to tap water. Sci Total Environ 616:453–461

Sugg MM, Ryan S, Spurlock T, Duncan S, Hege A, Ogwu MC, Tyson J, Shay E, Renwick K, Hendren C, Roy M, Runkle JD (2023) Climate change and health in rural mountain environments: summary of a workshop on knowledge gaps, barriers, and opportunities for action. GeoJournal. https://doi.org/10.1007/s10708-023-10916-4

Tao F, Yokozawa M, Xu Y (2006) Climate changes and trends in phenology and yields of field crops in China, 1981–2000. Agric for Meteorol 138:82–92. https://doi.org/10.1016/j.agrformet.2006.03.014

Tripathi RS, Raju R, Thimmappa K (2013) Impacts of zero tillage on economics of wheat production in Haryana. Agric Econ Res Rev 26:101–108

Uddling J, Broberg MC, Feng Z (2018) Crop quality under rising atmospheric CO_2. Curr Opin Plant Biol 45:262–267. https://doi.org/10.1016/j.pbi.2018.06.001

Usman M, Makhdum MSA (2021) What abates ecological footprint in BRICS-T region? Exploring the influence of renewable energy, non-renewable energy, agriculture, forest area, and financial development. Renew Energy 179:12–28

Vale MM, Arias PA, Ortega G (2021) Climate change and biodiversity in the Atlantic Forest: best climatic models, predicted changes and impacts, and adaptation options. In: The Atlantic Forest. Springer, pp 253–267

Wagan SA, Memon QUA, Wagan TA (2015) Economic analysis of laser land leveling technology, water use efficiency, and crop productivity of wheat crop in Sindh, Pakistan. J Environ Earth Sci 5:21–25

Waheed A, Fischer TB, Khan MI (2021) Climate change policy coherence across policies, plans, and strategies in Pakistan—implications for the China-Pakistan Economic Corridor Plan. Environ Manag 67:793–810

Watts N, Adger WN, Agnolucci P (2015) Health and climate change: policy responses to protect public health. Lancet 386:1861–1914

Waylen KA, Wilkinson ME, Blackstock KL (2024) Nature-based solutions and restoration are intertwined but not identical: highlighting implications for societies and ecosystems. Nat-Based Solut 5:100116. https://doi.org/10.1016/j.nbsj.2024.100116

World Health Organization (2021) COP26 special report on climate change and health: the health argument for climate action. World Health Organization, Geneva

Yafiee OA, Mumtaz F, Kumari P, Karanikolos GN, Decarlis A, Dumée LF (2024) Direct air capture (DAC) vs. direct ocean capture (DOC)—a perspective on scale-up demonstrations and environmental relevance to sustain decarbonization. Chem Eng J 497:154421. https://doi.org/10.1016/j.cej.2024.154421

Yamano H, Sugihara K, Nomura K (2011) Rapid poleward range expansion of tropical reef corals in response to rising sea surface temperatures. Geophys Res Lett 38. https://doi.org/10.1029/2010gl046474

Yang B, Usman M (2021) Do industrialization, economic growth XE "Growth, economic vs. ecological", and globalization processes influence the ecological footprint and healthcare expenditures? Fresh insights based on the STIRPAT model for countries with the highest healthcare expenditures. Sust Prod Cons 28:893–910

Yao T, Huang J, Xu B, Wang Y, Chen X, Liu J, Duan Q, Wu G, Wang W (2023) Climate change adaptation and mitigation strategies for building a Green Silk Road. Bull Chin Acad Sci 38(9):Article 5. https://doi.org/10.16418/j.issn.1000-3045.20230715001

Zafar MH, Langas EF, Sanfilippo F (2024) Exploring the synergies between collaborative robotics, digital twins, augmentation, and industry 5.0 for smart manufacturing: a state-of-the-art review. Robot Comput Integr Manuf 89:102769. https://doi.org/10.1016/j.rcim.2024.102769

Zhao C, Liu B, Piao S (2017) Temperature increase reduces global yields of major crops in four independent estimates. Proc Natl Acad Sci U S A 114:9326–9331. https://doi.org/10.1073/pnas.1701762114

Ziaeemehr B, Jandaghian Z, Ge H, Lacasse M, Moore T (2023) Increasing solar reflectivity of building envelope materials to mitigate urban heat islands: state-of-the-art review. Build 13(11):2868. https://doi.org/10.3390/buildings13112868

Chapter 13
Energy Efficiency and Smart Grid Technologies for Environmental Sustainability

Moses Fayiah and Mohamed Koiva Kallon

Abstract This chapter examines the principles, applications, and challenges of energy efficiency and smart grid technologies in promoting environmental sustainability. With escalating global energy demands, these innovations emerge as critical strategies to mitigate the environmental impacts of energy production and consumption. Energy efficiency, often termed the "first fuel," minimizes energy waste while enhancing service quality. Smart grids, characterized by advanced communication systems and real-time monitoring, optimize energy distribution and integrate renewable energy sources seamlessly into traditional grids. Together, they foster sustainable energy practices, reduce greenhouse gas emissions, and support the global transition to carbon neutrality. Technological advancements, such as energy-efficient appliances, smart meters, and IoT-enabled systems, are highlighted for their significant contributions to energy conservation. Metrics and standards like ISO 50001 and Energy Star certification are discussed as tools to measure and benchmark energy efficiency efforts. In parallel, the chapter examines smart grid technologies, including demand response systems, distributed energy resources, and blockchain-enabled energy transactions. These technologies enhance grid resilience, enable real-time energy management, and facilitate the adoption of renewable energy, reducing dependency on fossil fuels. Case studies from smart cities and international initiatives illustrate successful applications and scalable models for integrating these technologies. Despite their potential, the chapter acknowledges barriers to adoption, including economic constraints, technical challenges, and cybersecurity concerns. It underscores the need for robust policy frameworks and international collaboration to overcome these obstacles. By addressing these challenges, energy efficiency and smart grid technologies can pave the way for a sustainable, resilient, and low-carbon energy future.

M. Fayiah (✉)
Department of Forestry, School of Natural Resources Management, Njala University, Njala, Sierra Leone
e-mail: mfayiah@njala.edu.sl

M. K. Kallon
School of Grassland Science, Beijing Forestry University, Beijing, China
e-mail: mohamedkoivakallon@bjfu.edu.cn

M. C. Ogwu and S. Chibueze Izah (eds.), *Evaluating Environmental Processes and Technologies*, Environmental Science and Engineering,
https://doi.org/10.1007/978-3-031-85327-2_13

Keywords Energy efficiency · Smart grids · Environmental sustainability · Renewable energy · Carbon emissions · Energy storage · Smart cities · Carbon neutrality

13.1 Introduction

As global energy demands continue to rise, the need for sustainable energy management strategies has become increasingly urgent. Energy efficiency and smart grid technologies have emerged as pivotal tools in mitigating environmental impacts while addressing the challenges of resource depletion and climate change. These approaches aim to optimize energy consumption, reduce greenhouse gas (GHG) emissions, and modernize traditional energy systems to align with environmental sustainability goals. Energy efficiency is often regarded as the "first fuel" of a sustainable energy future. It encompasses various strategies and technologies designed to minimize energy waste while maintaining or improving service quality (IEA 2020). From advanced appliances and industrial processes to building management systems, energy efficiency interventions are critical in reducing energy intensity and lowering carbon emissions. For instance, improvements in energy-efficient technologies have led to significant reductions in residential and industrial energy use, showcasing their potential to drive global energy transitions (Ürge-Vorsatz et al. 2012).

The significance of energy efficiency in achieving environmental sustainability cannot be overstated. Energy efficiency is crucial in reducing greenhouse gas emissions, minimizing resource depletion, and promoting sustainable development. By optimizing energy use, we can significantly lower the environmental impact associated with energy production and consumption. For instance, energy-efficient buildings and technologies can reduce energy demand, thereby decreasing reliance on fossil fuels and mitigating climate change effects (Dabaieh 2013; Kluczek 2018). Furthermore, integrating renewable energy sources into the energy mix is essential for achieving long-term sustainability goals, as it reduces the carbon footprint associated with traditional energy systems (Lu et al. 2017; Vlasov et al. 2021). Smart grid technologies represent a transformative approach to energy management that enhances energy efficiency and supports environmental sustainability. The smart grid incorporates advanced communication and information technologies to facilitate two-way communication between utilities and consumers, enabling real-time monitoring and management of energy resources (Hussain et al. 2018; Law et al. 2014). This intelligent infrastructure allows for integrating distributed energy resources, such as solar and wind power, into the grid, thereby enhancing the reliability and efficiency of energy distribution (Kim et al. 2019; Güngör et al. 2011). Moreover, smart grids enable demand-side management strategies that empower consumers to optimize their energy use, further contributing to sustainability efforts (Zunnurain et al. 2018; Erol-Kantarci and Mouftah 2015).

Smart grid technologies complement energy efficiency by enabling dynamic, decentralized, and efficient energy management. Unlike traditional grids, smart grids integrate advanced sensors, communication systems, and automation tools to facilitate real-time monitoring, demand response, and renewable energy integration (Fang et al. 2012). The Internet of Things (IoT), artificial intelligence (AI), and blockchain have further enhanced the capabilities of smart grids, allowing predictive maintenance, decentralized energy transactions, and improved system reliability (Morstyn et al. 2018; Amabie et al. 2024). By integrating distributed energy resources such as solar panels and battery storage, smart grids reduce reliance on fossil fuels and enhance grid resilience, contributing to sustainable development. Despite their transformative potential, adopting energy efficiency measures and smart grid technologies face several barriers. Economic constraints, technical limitations, and cybersecurity concerns remain critical challenges (He et al. 2018). Moreover, effective implementation requires robust policy frameworks, international collaboration, and public awareness to bridge the gap between technological advancements and practical deployment (IEA 2021). Successful case studies, such as the European Union's energy efficiency directives and smart grid programs in the United States, provide valuable insights into scalable and replicable strategies. The smart grid's innovative design is rooted in its ability to facilitate efficient energy generation, transmission, and consumption. It employs various communication networks, including Wide Area Networks (WANs), Neighbourhood Area Networks (NANs), and Home Area Networks (HANs), to coordinate energy resources effectively (Asuhaimi et al. 2018). This interconnectedness improves the power grid's reliability and enhances the efficiency of energy management systems (Brahmia et al. 2020; Lampropoulos et al. 2017). By leveraging advanced technologies such as smart meters and demand response systems, the smart grid can significantly reduce energy waste and improve the economic viability of renewable energy sources (Sarikaya and Partal 2022; Gercek et al. 2019).

The aim of this chapter is to examine the critical role of energy efficiency and smart grid technologies in achieving environmental sustainability. It seeks to explore the foundational principles of energy efficiency, highlighting its importance in reducing carbon emissions and optimizing resource use while introducing key metrics and standards for assessment. The chapter overviews smart grid technologies, detailing their evolution, components, and transformative potential in modernizing energy systems. By discussing technological advancements such as energy-efficient appliances, IoT integration, AI-driven predictive maintenance, and blockchain for decentralized energy management, the chapter emphasizes innovative solutions to energy challenges. Additionally, it addresses economic, technical, and cybersecurity barriers to adoption, offering real-world case studies and successful regional implementations as models for scalability. Through an analysis of global policy frameworks and emerging trends, the chapter aims to align energy efficiency and smart grid strategies with broader sustainability goals, paving the way for a carbon–neutral future.

13.2 Energy Efficiency: Concepts and Principles

Energy efficiency is defined as the ability to use less energy to provide the same level of energy service. This concept encompasses various applications, from industrial processes and commercial buildings to residential energy use and transportation systems (Kluczek 2018; Moussa 2023). The scope of energy efficiency extends beyond mere energy savings; it includes optimizing energy systems, integrating renewable energy sources, and implementing advanced technologies that enhance energy performance (Bisello 2023; Moussa 2023). As urbanization and energy consumption continue to rise, the importance of energy efficiency becomes increasingly critical in addressing global energy challenges and promoting sustainable development (Fichera et al. 2016; Igaliyeva et al. 2020; Ogwu 2019a, b). Energy efficiency measures can be categorized into technological improvements, behavioral changes, and policy initiatives. Technological advancements, such as energy-efficient appliances, smart meters, and building insulation, significantly reduce energy consumption (Picot 2023; Linnenberg et al. 2013). Behavioral changes, such as promoting energy conservation practices among consumers, are equally important for achieving energy efficiency goals (Igaliyeva et al. 2020). Furthermore, policy initiatives, including energy efficiency standards and incentives, are essential for fostering an environment conducive to energy-efficient practices (Moussa 2023; Igaliyeva et al. 2020).

To effectively measure energy efficiency, various metrics and standards have been developed. Key performance indicators (KPIs) are commonly used to assess energy efficiency initiatives' effectiveness and track progress toward sustainability goals (Table 13.1; Guarino et al. 2021; Barni et al. 2022). These metrics can include energy use intensity (EUI), which measures energy consumption per unit of output or floor area, and the overall energy savings achieved through specific interventions (Picot 2023; Guarino et al. 2021). Standards such as ISO 50001 provide a framework for organizations to establish energy management systems that promote continuous improvement in energy performance (Moussa 2023; Picot 2023). The Energy Star program also offers certification for energy-efficient products and buildings, helping consumers identify and choose energy-saving options (Kluczek 2018; Bisello 2023). These standards and metrics ensure transparency and accountability in energy efficiency efforts, allowing stakeholders to evaluate their initiatives' impact effectively.

Energy efficiency is pivotal in reducing carbon emissions and mitigating climate change. Decreasing energy consumption, and energy efficiency measures can directly contribute to lower greenhouse gas emissions associated with energy production, particularly from fossil fuel sources (Brien et al. 2010; Picot 2023). For instance, implementing energy-efficient technologies in buildings can significantly reduce energy demand, thereby reducing the need for energy generation from carbon-intensive sources (Dabaieh 2013; Igaliyeva et al. 2020). Moreover, energy efficiency is often viewed as a cost-effective strategy for achieving climate goals. According to the International Energy Agency (IEA), improving energy efficiency could account

Table 13.1 Key metrics and standards for assessing energy efficiency across sectors

Metric/standard	Definition	Application area	Relevance to energy efficiency
Energy intensity (EI)	Ratio of energy consumption to economic output (e.g., GDP)	National/sectoral performance	Indicates energy efficiency improvements at macroeconomic levels
Energy performance index (EPI)	Measurement of energy use per unit of output or service	Buildings, industries, and appliances	Evaluates efficiency in operational energy consumption
Energy efficiency ratio (EER)	Ratio of cooling capacity (BTUs) to energy consumed (watts)	HVAC systems and appliances	Quantifies the efficiency of cooling and heating systems
Building energy efficiency ratio (BEER)	Energy consumption per square meter of floor area	Commercial and residential buildings	Helps compare energy efficiency across buildings of different sizes and types
Energy Star Certification	Performance-based certification for energy-efficient products and buildings	Appliances, electronics, lighting, and buildings	Identifies and promotes the most energy-efficient technologies and practices
Leadership in energy and environmental design (LEED)	Green building certification for design, construction, and operations	Buildings and infrastructure	Encourages sustainable and energy-efficient building practices
ISO 50001: Energy Management Systems	International standard for optimizing energy use and reducing emissions	Industries and organizations	Provides a structured framework for improving energy efficiency
Specific energy consumption (SEC)	Energy consumed per unit of production or output	Manufacturing and industrial processes	Assesses energy efficiency within industrial sectors
Thermal efficiency	Ratio of useful energy output to energy input in thermal systems	Power plants, engines, and industrial heating	Measures the efficiency of energy conversion processes
Load factor	Ratio of actual energy consumed to the maximum possible energy consumption	Power generation and distribution systems	Indicates efficiency in energy utilization over a given period
Coefficient of performance (COP)	Ratio of useful heating or cooling provided to energy input	Heat pumps, refrigerators, and air conditioners	Measures the energy efficiency of thermal systems
Carbon intensity	Amount of CO_2 emissions per unit of energy consumed	Energy production and consumption systems	Highlights the environmental impact of energy use and incentivizes efficiency

(continued)

Table 13.1 (continued)

Metric/standard	Definition	Application area	Relevance to energy efficiency
Kilowatt-hour per square meter (kWh/m^2)	Energy consumption per square meter	Buildings and facilities	Helps benchmark energy performance in the building sector
Energy return on investment (EROI)	Ratio of energy output to energy input during energy production	Renewable and non-renewable energy generation	Evaluates the net energy benefit of energy sources and technologies

for a substantial portion of the emissions reductions needed to meet global climate targets (Moussa 2023; Igaliyeva et al. 2020). This is particularly relevant in sectors such as transportation and industry, where energy efficiency improvements can lead to significant emissions reductions without compromising economic growth (Kluczek 2018; Moussa 2023).

13.3 Technological Innovations in Energy Efficiency

Developing advanced energy-efficient appliances (Table 13.2) is a cornerstone of modern energy efficiency initiatives. These appliances are designed to consume less energy while maintaining or enhancing performance levels, thereby contributing to significant reductions in energy consumption and greenhouse gas emissions. For instance, innovations in refrigeration technology, such as improved compressor designs and enhanced insulation materials, have led to substantial energy savings in household refrigerators (Belman-Flores et al. 2015; Ibrik and Sulaiman 2019). The introduction of energy labeling systems has also played a crucial role in guiding consumers toward more efficient choices, as evidenced by studies indicating that informed consumers are more likely to purchase energy-efficient products (Li et al. 2019; Olatunde 2024). Moreover, integrating smart technologies into household appliances has further amplified energy efficiency. Smart appliances can optimize energy use by adjusting their operation based on real-time energy prices and the availability of renewable energy sources (Lucía et al. 2015; Mishima et al. 2014). For example, induction heating technologies have emerged as a highly efficient method for cooking, offering faster heating times and reduced energy waste compared to traditional methods (Lucía et al. 2015). These advancements improve household energy efficiency and contribute to broader sustainability goals by reducing overall energy demand. Building Management Systems (BMS) are critical for optimizing energy use in commercial and industrial buildings. These systems utilize advanced sensors, controls, and software to monitor and manage various building operations, including heating, ventilation, air conditioning (HVAC), lighting, and security systems (Ružičić et al. 2021; Гусев et al. 2018). By integrating these systems, BMS can significantly enhance energy efficiency through automated controls that adjust energy consumption based on occupancy patterns and environmental conditions (Гусев et al. 2018).

Research has shown that implementing BMS can lead to energy savings of up to 30% in commercial buildings (Ružičić et al. 2021; Гусев et al. 2018). For instance, BMS can optimize HVAC operations by adjusting temperatures based on real-time occupancy data, thereby reducing unnecessary energy use during unoccupied periods. Additionally, BMS can facilitate the integration of renewable energy sources, such as solar panels, by managing energy storage and consumption dynamically (Ružičić et al. 2021). This capability enhances energy efficiency and supports the transition to a more sustainable energy system. In the industrial sector, technological innovations drive significant improvements in energy efficiency across manufacturing processes. Adopting advanced manufacturing technologies, such as automation, robotics, and data analytics, has enabled industries to optimize their energy use and reduce waste (Liu et al. 2023; Yanfen et al. 2022). For example, the implementation of energy management systems in manufacturing plants allows for real-time monitoring of energy consumption, enabling operators to identify inefficiencies and implement corrective measures (Гусев et al. 2018). Furthermore, innovations in process design, such as using energy-efficient machinery and waste heat recovery systems, have enhanced energy efficiency in industrial operations (Liu et al. 2023; Yanfen et al. 2022). Research indicates that industries embracing these technologies can achieve energy savings of 20% or more energy savings, significantly reducing their carbon footprint (Liu et al. 2023). Additionally, coupling technological innovation with industrial structure upgrading is crucial for enhancing green economic efficiency, as it promotes the adoption of cleaner production techniques and sustainable practices (Xu 2023).

13.4 Smart Grid Technologies and Their Applications for Environmental Sustainability

Smart grid technologies (Table 13.3) represent a significant evolution in the management and distribution of electricity, integrating advanced technologies to enhance power systems' efficiency, reliability, and sustainability. A smart grid is defined as an electricity supply network that uses digital communication technology to detect and react to local changes in usage, allowing for a two-way flow of electricity and information between utilities and consumers (Si et al. 2023). Key features of smart grids include advanced metering infrastructure (AMI), real-time monitoring and control, integration of renewable energy sources, and enhanced demand response capabilities (Zakareya 2018). These features collectively contribute to improved energy management, reduced operational costs, and enhanced environmental sustainability by facilitating the integration of cleaner energy sources and optimizing energy consumption patterns.

Table 13.2 Examples of advanced energy-efficient appliances

Appliance	Type	Key energy-efficient characteristics	Environmental benefits
Energy-efficient refrigerators	Refrigerators designed to minimize energy consumption	High-efficiency compressors, advanced insulation, inverter technology	Reduces electricity usage, lowering carbon footprint and energy bills
LED lighting	Lighting systems using light-emitting diodes for illumination	Low energy consumption, long lifespan, minimal heat emission	Decreases electricity use and reduces waste with longer-lasting bulbs
Smart thermostats	Programmable thermostats with AI and IoT integration for climate control	Real-time energy monitoring, automatic scheduling, remote control	Optimizes energy use, reduces HVAC-related energy consumption, and lowers emissions
Energy star washing machines	Washing machines meeting Energy Star efficiency criteria	Reduced water usage, optimized spin cycles, high-efficiency motors	Conserves water and electricity, reducing strain on natural resources
High-efficiency HVAC systems	Heating, ventilation, and air conditioning systems with advanced efficiency	Variable speed motors, smart controls, advanced heat exchangers	Lowers energy consumption, reduces greenhouse gas emissions, and improves air quality
Induction cooktops	Cooking appliances using magnetic induction to directly heat cookware	Faster heating, minimal energy loss, precise temperature control	Saves energy compared to conventional electric or gas stoves, and reduces indoor air pollution
Energy-efficient water heaters	Water heaters using advanced technologies like heat pumps or solar thermal systems	High energy factor (EF), insulated tanks, on-demand water heating	Decreases energy consumption and supports renewable energy adoption
Smart dishwashers	Dishwashers designed for minimal water and energy usage	Energy-saving cycles, soil-sensing technology, delay start options	Reduces water and electricity usage while maintaining performance
Advanced ceiling fans	Ceiling fans designed with high-efficiency motors and aerodynamic blades	Energy-efficient DC motors, smart controls, and adjustable speed settings	Reduces HVAC reliance by improving air circulation, lowering overall energy use
Efficient air purifiers	Air purifiers designed for energy efficiency while maintaining indoor air quality	Low-energy fans, smart sensors for air quality monitoring, and auto shut-off	Reduces power consumption and enhances indoor air quality with minimal environmental impact

(continued)

Table 13.2 (continued)

Appliance	Type	Key energy-efficient characteristics	Environmental benefits
Solar-powered outdoor lights	Lighting powered by solar panels, storing energy for nighttime illumination	Renewable energy source, automatic on/off sensors, and durable materials	Eliminates energy costs, reduces reliance on the grid, and uses clean energy
High-efficiency televisions	TVs incorporating energy-saving technologies like OLED or QLED displays	Automatic brightness adjustment, energy-saving modes, and efficient backlighting systems	Lowers electricity use while delivering high-quality visual performance
Energy-saving computers	Laptops and desktops optimized for reduced power consumption	Low-power processors, energy-efficient power supplies, and sleep modes	Conserves energy during operation and standby, supporting sustainable IT practices

The transition from traditional power grids to smart grids has been driven by the need for more efficient and sustainable energy systems. Traditional grids are characterized by a unidirectional flow of electricity from centralized power plants to consumers, with limited communication and control capabilities (Vlasov et al. 2021). This model is increasingly inadequate due to rising energy demand, aging infrastructure, and the urgent need to reduce greenhouse gas emissions. The evolution to smart grids involves incorporating digital technologies, enabling real-time data collection and analysis, which enhances grid reliability and facilitates the integration of distributed energy resources (DERs) such as solar and wind (Liu et al. 2014). This shift improves operational efficiency and empowers consumers to participate actively in energy management through demand response programs and smart appliances. A smart grid comprises several interconnected components that work together to optimize energy distribution and consumption (Fig. 13.1). Key components include smart meters, which provide real-time data on energy usage; advanced sensors that monitor grid conditions; and automated control systems that manage energy flow and respond to changes in demand (Güngör et al. 2011). Additionally, communication networks facilitate the exchange of information between utilities and consumers, enabling better decision-making and coordination of energy resources (Lyulyov et al. 2021). The integration of these components allows for enhanced grid resilience, improved energy efficiency, and the ability to incorporate renewable energy sources seamlessly into the existing infrastructure. One of the most significant advantages of smart grids is their ability to integrate renewable energy sources into the electricity supply system. As the share of renewable energy in the energy mix increases, smart grids enable the management of intermittent energy generation from sources such as solar and wind (Ullah et al. 2020). Advanced forecasting tools and energy management systems help utilities balance supply and demand, ensuring grid stability while maximizing the use of renewable resources (Si et al. 2023). This

Table 13.3 Some smart grid technologies and their applications

Smart grid technology	Description	Application area	Contribution to environmental sustainability Environmental sustainability "
Advanced metering infrastructure (AMI)	Digital systems for real-time energy consumption monitoring and communication between consumers and utilities	Residential, commercial, and industrial buildings	Promotes energy efficiency by enabling demand-side management and reducing energy waste
Distributed energy resources (DERs)	Integration of small-scale, decentralized energy sources like solar, wind, and storage systems	Renewable energy systems, local energy grids	Reduces dependency on fossil fuels, enhances grid reliability, and minimizes greenhouse gas emissions
Demand response (DR)	Systems that adjust energy demand based on supply availability and price signals	Industries, buildings, and households	Balances energy supply and demand, reducing peak load and minimizing the need for additional fossil-fuel-based power generation
Smart sensors and IoT devices	Devices that provide real-time data on grid performance, energy usage, and environmental conditions	Smart homes, industries, and grid infrastructure	Enables optimized energy consumption, predictive maintenance, and environmental monitoring, reducing emissions and energy loss
Energy storage systems	Technologies like batteries and thermal storage for storing excess energy from renewable sources	Renewable energy systems, microgrids	Stabilizes grid operations, reduces energy wastage, and supports the integration of intermittent renewable energy
Advanced grid automation	Automated control systems for managing grid performance and ensuring reliable energy distribution	Power transmission and distribution networks	Increases grid efficiency and reliability, reducing energy losses and the carbon footprint of energy distribution

(continued)

Table 13.3 (continued)

Smart grid technology	Description	Application area	Contribution to environmental sustainability Environmental sustainability "
Electric vehicle (EV) integration	Systems that incorporate EVs into the grid as both energy consumers and potential storage units	Transportation and energy grids	Reduces transportation emissions and supports renewable energy adoption through vehicle-to-grid (V2G) integration
Renewable energy integration systems	Platforms for seamlessly integrating renewable energy sources into existing grid infrastructure	Solar farms, wind farms, and grid infrastructure	Accelerates the transition to low-carbon energy systems and reduces dependency on non-renewable energy sources
Microgrid technology	Localized energy grids that can operate independently or in conjunction with the main grid	Communities, campuses, and industrial zones	It enhances energy resilience, reduces transmission losses, and supports using renewable energy locally
Blockchain in energy	Decentralized ledger systems for managing energy transactions and peer-to-peer energy trading	Energy markets and grid operations	Enhances transparency, efficiency, and traceability in energy trading, fostering sustainable energy practices

integration not only reduces reliance on fossil fuels but also contributes to lower carbon emissions, aligning with global sustainability goals (Widergren et al. 2011).

Demand response (DR) management is a critical feature of smart grids that enhances energy efficiency and sustainability. DR programs incentivize consumers to adjust their energy usage during peak demand periods, alleviating grid stress and reducing the need for additional power generation from fossil fuel sources (Rottondi et al. 2015). Utilities can implement dynamic pricing models that encourage consumers to shift their energy consumption to off-peak times by employing real-time data and advanced analytics. This leads to more balanced load profiles and reduces operational costs ("Smart Assimilation of Renewable Based Microgrids", 2019). This proactive approach enhances grid reliability and supports the integration of renewable energy sources by aligning demand with variable supply. Energy storage solutions play a vital role in the functionality of smart grids, enabling the effective integration of renewable energy sources and enhancing grid stability. Technologies such as batteries, pumped hydro storage, and flywheels allow for storing excess energy generated during periods of low demand or high renewable output, which can then

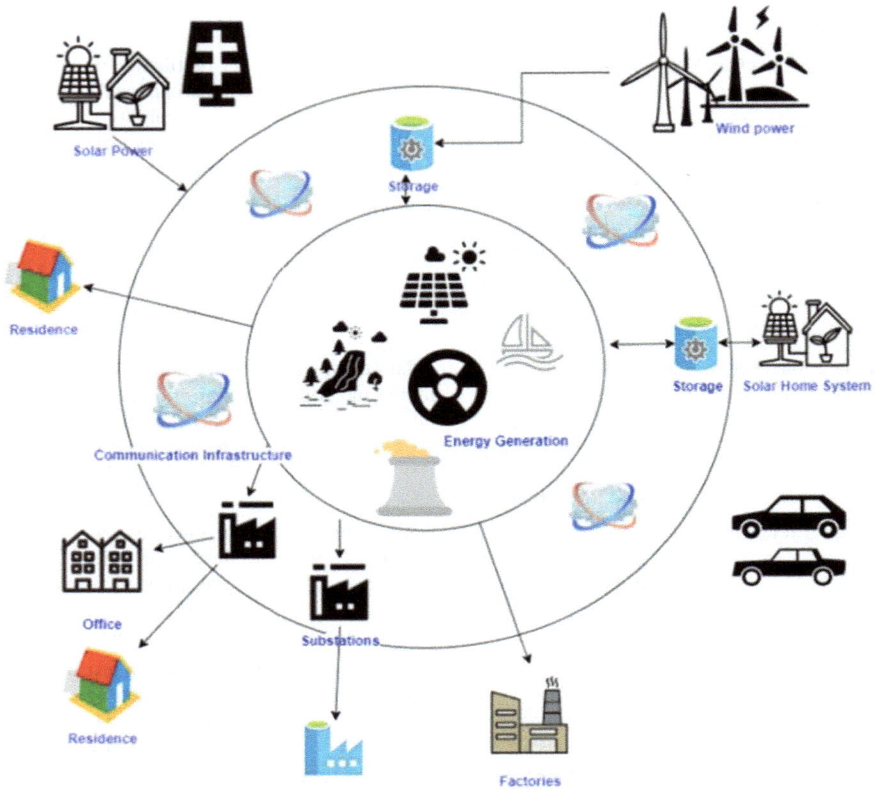

Fig. 13.1 Basic component and structure of a smart grid. *Source* Thilakarathne et al. (2021)

be released during peak demand (Durairaj et al. 2022). This capability helps to mitigate the intermittency of renewable energy sources, ensuring a reliable power supply while reducing greenhouse gas emissions associated with fossil fuel-based generation (McDonald et al. 2013). The environmental benefits of energy storage extend beyond emissions reductions; they also contribute to improved grid resilience and the ability to support electric vehicle (EV) charging infrastructure, further promoting sustainable transportation solutions. Integrating EVs into smart grids represents a transformative opportunity for enhancing sustainability in the energy sector. EVs can serve as both energy consumers and storage units, enabling vehicle-to-grid (V2G) technologies that allow for bidirectional energy flow (Dimitriou and Karame 2013). This capability supports grid stability by providing additional energy storage and encourages using renewable energy sources for charging, thereby reducing reliance on fossil fuels. Furthermore, the widespread adoption of EVs can significantly reduce greenhouse gas emissions from the transportation sector, aligning with global efforts to combat climate change and promote sustainable urban development.

13.5 Role of Advanced Technologies in Smart Grids

The Internet of Things (IoT) revolutionizes energy management within smart grids by enabling seamless communication between devices, systems, and users. IoT technologies facilitate real-time data collection and analysis, allowing for enhanced monitoring and control of energy consumption across various sectors (Bandyopadhyay and Sen 2011; Hurtado et al. 2014). For instance, smart meters equipped with IoT capabilities provide consumers and utilities with detailed insights into energy usage patterns, enabling more informed decision-making regarding energy consumption and conservation strategies (Liu et al. 2019; Anvari-Moghaddam et al. 2015). Moreover, IoT applications extend to smart homes and buildings, where interconnected devices can optimize energy use by adjusting settings based on occupancy, weather conditions, and energy prices (Anvari-Moghaddam et al. 2015). This integration enhances energy efficiency and empowers consumers to participate actively in demand response programs, which can lead to significant cost savings and reduced peak demand on the grid (Ma et al. 2015). The ability of IoT to facilitate two-way communication between utilities and consumers is pivotal in creating a more responsive and resilient energy system (Bandyopadhyay and Sen 2011).

Artificial Intelligence (AI) and machine learning are increasingly used in smart grids for predictive maintenance and operational optimization. These technologies analyze vast amounts of data collected from various grid components to identify patterns and predict potential failures before they occur (Chauhan 2024; Anvari-Moghaddam et al. 2017). By leveraging AI algorithms, utilities can optimize maintenance schedules, reduce downtime, and extend the lifespan of critical infrastructure, ultimately leading to enhanced reliability and efficiency of the power supply (Chauhan 2024; Rocha et al. 2015). For example, AI-driven energy management systems can forecast energy demand and supply fluctuations, enabling utilities to make informed decisions about resource allocation and grid management (Chauhan 2024; Rocha et al. 2015). Additionally, machine learning models can optimize the operation of distributed energy resources (DERs), such as solar panels and battery storage systems, by predicting their performance based on historical data and environmental conditions (Anvari-Moghaddam et al. 2017; Lee et al. 2014). This proactive approach not only improves operational efficiency but also contributes to the overall sustainability of the energy system by minimizing waste and maximizing the use of renewable resources.

Blockchain technology is emerging as a transformative force in the energy sector, particularly in facilitating decentralized energy transactions. By providing a secure and transparent platform for peer-to-peer energy trading, blockchain enables consumers to buy and sell excess energy generated from renewable sources, such as solar panels, directly with one another ("Application of BIM in Public Building Energy Management System", 2023; Li et al. 2017). This decentralized approach not only empowers consumers but also enhances the resilience of the energy grid by diversifying energy sources and reducing reliance on centralized power plants (Li et al. 2017; Anvari-Moghaddam et al. 2015). Furthermore, blockchain can streamline the management of energy certificates and carbon credits, ensuring that transactions are recorded accurately and transparently ("Application of BIM in Public Building Energy Management System", 2023; Anvari-Moghaddam et al. 2015). This capability is particularly important in promoting renewable energy adoption and supporting sustainability initiatives. As the energy landscape evolves, integrating blockchain technology into smart grids will play a crucial role in fostering innovation, enhancing energy security, and promoting environmental sustainability (Anvari-Moghaddam et al. 2015; Huang 2014).

13.6 Case Studies in Energy Efficiency and Smart Grids

Smart cities are at the forefront of promoting energy sustainability by integrating advanced technologies and innovative urban planning. These cities leverage smart grid technologies to optimize energy consumption, reduce waste, and enhance their residents' overall quality of life. The concept of smart cities encompasses a holistic approach to urban development, where energy efficiency is a key component of sustainability initiatives (Toli and Murtagh 2020; Saker 2023). For instance, smart cities utilize IoT devices to monitor energy usage in real-time, enabling better demand-side management and reducing peak load pressures on the grid (Haque 2024). Moreover, smart cities often integrate renewable energy sources, such as solar panels and wind turbines, into the smart grid framework. This integration reduces reliance on fossil fuels and promotes local energy production, thereby enhancing energy security and resilience (Anjimoon 2024; Avwioroko 2024; Nizzy and Ogwu 2024; Ogwu et al. 2024). The role of smart cities in promoting energy sustainability is further exemplified by their emphasis on community engagement and education, fostering a culture of energy conservation among residents (Toli and Murtagh 2020).

Several regions worldwide have successfully implemented smart grid technologies, showcasing the potential benefits of these systems. For example, in the United States, the city of San Diego has made significant strides in integrating smart grid technologies to enhance energy efficiency and sustainability. The city has deployed advanced metering infrastructure (AMI) and demand response programs, resulting in a 20% reduction in peak energy demand (Kushawaha 2024). San Diego's commitment to renewable energy integration has also led to establishing microgrids that support local energy resilience and sustainability (Anjimoon 2024). Another notable success story is found in Europe, where countries like Germany and Denmark have pioneered smart grid initiatives. Germany's Energiewende (energy transition) policy has facilitated the integration of renewable energy sources into the grid, supported by smart grid technologies that enhance grid stability and efficiency (Haque 2024; Hussain et al. 2018). Denmark, known for its wind energy production, has implemented smart grid solutions allowing real-time monitoring and management of energy resources, significantly reducing carbon emissions and promoting energy sustainability (Janjić et al. 2016; Abdulwahid and Ateeq 2019).

Energy efficiency programs in both industrial and residential sectors play a crucial role in maximizing the benefits of smart grids. In the industrial sector, companies increasingly adopt smart manufacturing practices that leverage IoT and AI technologies to optimize energy use and reduce waste (Hsu 2023). For instance, smart energy management systems enable manufacturers to monitor energy consumption patterns and implement predictive maintenance strategies, resulting in substantial energy savings and improved operational efficiency (Ezeigweneme 2024). In the residential sector, energy efficiency programs often focus on educating consumers about energy-saving practices and providing incentives for adopting energy-efficient appliances and technologies. Smart meters and home energy management systems empower homeowners to track their energy usage in real-time, allowing for better decision-making regarding energy consumption (Hussain et al. 2018; Saleem et al. 2022). Programs that promote demand response participation further enhance energy efficiency by incentivizing consumers to reduce or shift their energy usage during peak demand periods, thereby alleviating stress on the grid (Hussain et al. 2018; Kushawaha 2024).

13.7 Challenges and Barriers to Implementing Energy Efficiency and Smart Grid Technologies Globally

13.7.1 Economic and Policy Challenges in Adopting Smart Grids

The transition to smart grids is fraught with economic and policy challenges that can impede their successful implementation. One of the primary economic barriers is the high initial investment required to deploy smart grid technologies. These costs include upgrading existing infrastructure, installing advanced metering systems, and integrating renewable energy sources (Dantas et al. 2018). Furthermore, traditional business models in the electricity sector often do not align with the incentives necessary for promoting smart grid technologies. For instance, fixed uniform pricing policies can deter utilities from investing in dynamic pricing models that are essential for realizing the full benefits of smart grids (Dantas et al. 2018). Additionally, the regulatory framework governing the electricity sector can pose significant challenges. In many regions, outdated regulations do not accommodate the innovative business models and technologies associated with smart grids, leading to a lack of clarity and uncertainty for investors (Brown et al. 2018). Policymakers must, therefore, create supportive regulatory environments encouraging investment in smart grid technologies, including incentives for utilities and consumers to adopt these systems (Brown et al. 2018).

13.7.2 Technical Limitations and Infrastructure Gaps

Technical limitations and infrastructure gaps represent another significant barrier to implementing smart grids. Many existing power grids are aging and lack the necessary infrastructure to support the advanced technologies required for smart grid operations (Shuqair 2023; Cowan 2013). This includes outdated communication systems, insufficient data management capabilities, and inadequate cybersecurity measures (Bari et al. 2014). Moreover, integrating distributed energy resources (DERs) such as solar panels and wind turbines into the existing grid poses technical challenges related to grid stability and reliability (Guo et al. 2015). The intermittent nature of renewable energy sources requires advanced forecasting and management tools to ensure a balanced supply and demand, which many current systems are not equipped to handle (Bari et al. 2014). Addressing these technical limitations requires significant investment in modernizing infrastructure and developing new technologies that can seamlessly integrate renewable energy into the grid (Shuqair 2023; Cowan 2013).

13.7.3 Cybersecurity and Data Privacy Concerns

Cybersecurity and privacy concerns are paramount as smart grids rely heavily on digital communication and data exchange. The interconnected nature of smart grid systems makes them vulnerable to cyberattacks, which can disrupt power supply, compromise sensitive data, and undermine consumer trust (Qureshi et al. 2021). The potential for data breaches and unauthorized access to personal information raises significant privacy concerns for consumers, who may be hesitant to adopt smart grid technologies due to fears of surveillance and data misuse (Qureshi et al. 2021). To mitigate these risks, robust cybersecurity measures must be implemented to protect smart grid infrastructure from cyber threats. This includes adopting advanced encryption techniques, intrusion detection systems, and regular security audits (Qureshi et al. 2021). Additionally, there is a need for clear policies and regulations governing data privacy and security in smart grid operations to ensure that consumer rights are protected (Qureshi et al. 2021).

13.8 Policy and Regulatory Frameworks for Energy Efficiency and Smart Grid Technologies

The global landscape for energy efficiency and smart grid development is shaped by various policies and regulatory frameworks aimed at promoting sustainable energy practices. One notable example is the Energy Independence and Security Act of 2007 in the United States, which mandates the development of a framework for smart grid interoperability (Greer et al. 2014). This legislation emphasizes the need for coordinated efforts among various stakeholders, including utilities, technology providers, and regulatory bodies, to facilitate the integration of smart grid technologies and enhance energy efficiency (Greer et al. 2014; Gopstein et al. 2020). Internationally, organizations such as the International Energy Agency (IEA) and the United Nations have established guidelines and frameworks to support adopting smart grid technologies. The IEA's reports outline strategies for improving energy efficiency across sectors, emphasizing the role of smart grids in facilitating the transition to a low-carbon economy (Lammers and Heldeweg 2016). Similarly, the United Nations' Sustainable Development Goals (SDGs) include energy efficiency and sustainable energy systems targets, encouraging countries to adopt policies that promote smart grid development (Liu et al. 2023). Some policies and regulatory frameworks for energy efficiency and smart grid technologies are presented in Table 13.4.

The European Union has implemented the Clean Energy for All Europeans package, which includes directives to enhance energy efficiency and promote the deployment of smart grids across member states (Zaheeruddin and Manas 2015). This comprehensive policy framework encourages investments in smart grid technologies, supports the integration of renewable energy sources, and fosters innovation in energy management practices (Zaheeruddin and Manas 2015). Governments and international organizations play a pivotal role in promoting the adoption of energy efficiency and smart grid technologies through various initiatives and funding programs. For instance, the U.S. government has invested significantly in smart grid modernization, with a historic allocation of $3.4 billion in grants to support deploying advanced metering infrastructure and other smart grid technologies (Güngör et al. 2011). This investment enhances grid reliability and encourages utilities to adopt innovative practices that improve energy efficiency (Güngör et al. 2011).

International organizations, such as the World Bank and the Asian Development Bank, have also recognized the importance of smart grids in achieving sustainable energy goals. These organizations provide financial assistance and technical expertise to developing countries to help them implement smart grid projects that enhance energy access and efficiency (Gopstein et al. 2020). For example, the World Bank's Energy Sector Management Assistance Program (ESMAP) supports initiatives that promote smart grid technologies in low- and middle-income countries, facilitating the transition to more sustainable energy systems (Gopstein et al. 2020). Furthermore, collaborative efforts among governments, industry stakeholders, and research institutions are essential for fostering innovation in smart grid technologies. Initiatives such as the Smart Grid Interoperability Panel (SGIP) in the U.S. bring together diverse stakeholders to develop interoperability standards and best practices for smart grid implementation (Greer et al. 2014). This collaborative approach ensures that the benefits of smart grid technologies are maximized while addressing the challenges associated with their deployment (Greer et al. 2014).

Table 13.4 Energy policy framework, application and impact

Policy/framework	Focus	Application area	Impacts
Energy Policy Act (EPA)	U.S. legislation promoting energy conservation and efficiency	Energy utilities, industries, and households	Encourages energy efficiency through incentives, standards, and research funding
EU Energy Efficiency Directive	European Union directive to reduce energy use and enhance efficiency	Buildings, industries, and energy systems	Sets binding targets for energy savings and promotes smart grid development
Renewable Energy Portfolio Standards (RPS)	Mandates specific levels of renewable energy integration in power generation	Utilities and energy producers	Promotes grid modernization and integration of renewable energy into the grid
ISO 50001: Energy Management Systems	International standard for continuous improvement in energy management	Industries, organizations, and large facilities	Provides a systematic framework to enhance energy performance and support smart grid adoption
Smart Grid Investment Grant (SGIG)	U.S. government initiative funding smart grid infrastructure	Utilities, smart grid developers	Accelerates the deployment of smart meters, grid automation, and demand response technologies
Clean Energy Standard (CES)	Policy requiring a certain percentage of electricity to come from low-carbon sources	Energy generation and utilities	Drives energy efficiency by fostering the adoption of cleaner energy sources and smart grid technologies
Building Energy Codes	Regulations governing energy efficiency standards for new and existing buildings	Residential and commercial buildings	Encourages the adoption of energy-efficient technologies and appliances, supporting integration with smart grid systems
Net Metering Policies	Regulations allowing consumers to sell excess renewable energy back to the grid	Solar and renewable energy users	Promotes energy efficiency and grid interactivity by incentivizing renewable energy adoption
Energy Efficiency Resource Standards (EERS)	Mandates energy savings targets for utilities and other entities	Power utilities and energy providers	Encourages utilities to implement programs that improve energy efficiency in homes and businesses
Carbon Pricing and Emission Trading Schemes (ETS)	Market-based mechanisms to reduce carbon emissions	Industries and energy producers	Incentivizes energy efficiency by making carbon-intensive energy production more costly

(continued)

Table 13.4 (continued)

Policy/framework	Focus	Application area	Impacts
Federal Energy Regulatory Commission (FERC) Orders 745 and 2222	U.S. regulations supporting demand response and distributed energy resources	Power markets and distributed energy systems	Enhances grid efficiency by integrating distributed energy resources and demand-side energy management
Green Building Certification Programs	Standards like LEED and BREEAM for sustainable building design and operation	Construction and real estate sectors	Encourages energy-efficient practices in building construction and operation, supporting smart grid energy monitoring
National Smart Grid Policy (NSGP)	Policies focused on modernizing electricity networks	National energy grids	Provides strategic direction for smart grid development and integration with renewable energy sources
Energy Performance Certificates (EPCs)	Certification system assessing building energy performance	Residential and commercial buildings	Encourages property owners to improve energy efficiency through transparent performance metrics
Demand-Side Management (DSM) Programs	Initiatives to reduce or shift energy consumption through consumer incentives	Utilities, industries, and households	Supports energy efficiency by integrating load-shifting technologies and optimizing grid operations

13.9 Conclusion

Energy efficiency is a multifaceted concept encompassing a wide range of practices and technologies to reduce energy consumption while maintaining service levels. By employing key metrics and standards to measure efficiency, stakeholders can effectively track progress and implement strategies contributing to environmental sustainability. Ultimately, energy efficiency is critical in the global effort to reduce carbon emissions and combat climate change. Technological innovations in energy-efficient appliances, building management systems, and industrial processes are pivotal in advancing energy efficiency. Smart grid technologies offer a comprehensive framework for enhancing environmental sustainability through improved energy management, integration of renewable resources, and innovative demand response strategies. As the energy landscape continues to evolve, adopting smart grid solutions will be essential for achieving a sustainable and resilient energy future. Advanced technologies such as IoT, AI, and blockchain are integral to developing and operating smart grids. These technologies enhance energy management, optimize operational efficiency, and facilitate decentralized energy transactions, ultimately contributing to a more sustainable and resilient energy future. As the demand for cleaner and more efficient energy solutions grows, the continued integration of these advanced technologies will be essential for achieving environmental sustainability goals. These innovations contribute to significant energy savings and are vital in reducing carbon emissions and promoting environmental sustainability. As the demand for energy continues to rise, the ongoing development and implementation of these technologies will be essential for achieving a sustainable energy future. While smart grids offer significant potential for enhancing energy efficiency and sustainability, their implementation is hindered by a range of economic, technical, and cybersecurity challenges. Addressing these barriers will require coordinated efforts from policymakers, utilities, and technology providers to create an enabling environment for the successful adoption of smart grid technologies. The policy and regulatory frameworks supporting energy efficiency and smart grid development are critical for facilitating the transition to sustainable energy systems. Through coordinated efforts at the national and international levels, governments and organizations can promote the adoption of innovative technologies that enhance energy efficiency, reduce carbon emissions, and contribute to a more resilient energy future.

References

Abdulwahid A, Ateeq A (2019) Innovative differential protection scheme for microgrids based on RC current sensor.https://doi.org/10.5772/intechopen.85473
Amabie T, Izah SC, Ogwu MC, Hait M (2024) Harmonizing tradition and technology: the synergy of artificial intelligence in traditional medicine. In: Izah SC, Ogwu MC, Akram M (eds) Herbal medicine phytochemistry. Reference series in phytochemistry. Springer, Cham. https://doi.org/10.1007/978-3-031-21973-3_70-1
Anjimoon S (2024) Secure and sustainable energy distribution through blockchain technology in smart grids. E3s Web Conf 505:02002. https://doi.org/10.1051/e3sconf/202450502002
Anvari-Moghaddam A, Guerrero J, Vasquez J, Monsef H, Rahimi-Kian A (2017) Efficient energy management for a grid-tied residential microgrid. IET Gener Transm Distrib 11(11):2752–2761. https://doi.org/10.1049/iet-gtd.2016.1129
Anvari-Moghaddam A, Monsef H, Rahimi-Kian A, Guerrero J, Vasquez J (2015) Optimized energy management of a single-house residential micro-grid with automated demand response, pp 1–6. https://doi.org/10.1109/ptc.2015.7232243
Asuhaimi F, Bu S, Imran M (2018) Joint resource allocation and power control in heterogeneous cellular networks for smart grids, pp 1–6. https://doi.org/10.1109/glocom.2018.8647668
Avwioroko A (2024) Smart grid integration of solar and biomass energy sources. Eur J Comput Sci Inf Technol 12(3):1–14. https://doi.org/10.37745/ejcsit.2013/vol12n3114
Bandyopadhyay D, Sen J (2011) Internet of things: applications and challenges in technology and standardization. Wirel Pers Commun 58(1):49–69. https://doi.org/10.1007/s11277-011-0288-5
Bari A, Jiang J, Saad W, Jaekel A (2014) Challenges in the smart grid applications: an overview. Int J Distrib Sens Netw 10(2):974682. https://doi.org/10.1155/2014/974682
Barni A, Capuzzimati C, Fontana A, Pirotta M, Hänninen S, Räikkönen M, Uusitalo T (2022) Design of a lifecycle-oriented environmental and economic indicators framework for the mechanical manufacturing industry. Sustainability 14(5):2602. https://doi.org/10.3390/su14052602
Belman-Flores J, Barroso-Maldonado J, Rodríguez-Muñoz A, Camacho-Vázquez G (2015) Enhancements in domestic refrigeration, approaching a sustainable refrigerator—a review. Renew Sustain Energy Rev 51:955–968. https://doi.org/10.1016/j.rser.2015.07.003
Bisello A (2023) Multicriteria spatial economic decision support systems to support positive energy districts: a literature review, pp 15–31. https://doi.org/10.1007/978-3-031-39206-1_2

Brahmia I, Wang J, Oliveira L, Xu H (2020) Hierarchical smart energy management strategy based on cooperative distributed economic model predictive control for multi-microgrids systems

Brien S, Lorenzetti D, Lewis S, Kennedy J, Ghali W (2010) Overview of a formal scoping review on health system report cards. Implement Sci 5(1). https://doi.org/10.1186/1748-5908-5-2

Brown M, Zhou S, Ahmadi M (2018) Smart grid governance: an international review of evolving policy issues and innovations. Wiley Interdiscip Rev Energy Environ 7(5). https://doi.org/10.1002/wene.290

Chauhan B (2024) AI-driven energy management systems for smart buildings. PST 48(1):322–337. https://doi.org/10.52783/pst.280

Cowan K (2013) A new roadmapping technique for creatively managing the emerging smart grid. Creativ Innov Manag 22(1):67–83. https://doi.org/10.1111/caim.12017

Dabaieh M (2013) Energy efficient design strategies for contemporary vernacular buildings in Egypt, pp 599–604. https://doi.org/10.1201/b15685-104

Dantas G, Castro N, Dias L, Antunes C, Vardiero P, Brandão R, Zamboni L (2018) Public policies for smart grids in Brazil. Renew Sustain Energy Rev 92:501–512. https://doi.org/10.1016/j.rser.2018.04.077

Dimitriou T, Karame G (2013) Privacy-friendly tasking and trading of energy in smart grids. https://doi.org/10.1145/2480362.2480488

Durairaj D, Wróblewski Ł, Sheela A, Hariharasudan A, Urbański M (2022) Random forest based power sustainability and cost optimization in smart grid. https://doi.org/10.20944/preprints202201.0138.v1

Erol-Kantarci M, Mouftah H (2015) Energy-efficient information and communication infrastructures in the smart grid: a survey on interactions and open issues. IEEE Commun Surv Tutor 17(1):179–197. https://doi.org/10.1109/comst.2014.2341600

Ezeigweneme C (2024) Smart grids in industrial paradigms: a review of progress, benefits, and maintenance implications: analyzing the role of smart grids in predictive maintenance and the integration of renewable energy sources, along with their overall impact on the industri. Eng Sci Technol J 5(1):1–20. https://doi.org/10.51594/estj.v5i1.719

Fang X, Misra S, Xue G, Yang D (2012) Smart grid—the new and improved power grid: a survey. IEEE Commun Surv Tutor 14(4):944–980

Fichera A, Frasca M, Volpe R (2016) On energy distribution in cities: a model based on complex networks. Int J Heat Technol 34(4):611–615. https://doi.org/10.18280/ijht.340409

Gercek C, Schram W, Lampropoulos I, Sark W, Reinders A (2019) A comparison of households' energy balance in residential smart grid pilots in the Netherlands. Appl Sci 9(15):2993. https://doi.org/10.3390/app9152993

Gopstein A, Nguyen C, Byrnett D, Worthington K, Villarreal C (2020) Framework and roadmap for smart grid interoperability standards regional roundtables summary report. https://doi.org/10.6028/nist.ir.8284

Greer C, Wollman D, Prochaska D, Boynton P, Mazer J, Nguyen C, Bushby S (2014) NIST framework and roadmap for smart grid interoperability standards, release 3.0. https://doi.org/10.6028/nist.sp.1108r3

Guarino F, Bisello A, Frieden D, Bastos J, Brunetti A, Cellura M, Zhang X (2021) State of the art on sustainability assessment of positive energy districts: methodologies, indicators and future perspectives, pp 479–492. https://doi.org/10.1007/978-981-16-6269-0_40

Güngör V, Sahin D, Koçak T, Ergüt S, Buccella C, Cecati C, Hancke G (2011) Smart grid technologies: communication technologies and standards. IEEE Trans Ind Informatics 7(4):529–539. https://doi.org/10.1109/tii.2011.2166794

Guo C, Bond C, Narayanan A (2015) The adoption of new smart-grid technologies: incentives, outcomes, and opportunities. https://doi.org/10.7249/rr717

Haque A (2024) Technical and economic challenges and future prospects of a smart grid—a case study. Control Syst Optim Lett 1(3):186–193. https://doi.org/10.59247/csol.v1i3.57

He X, Yan Z, Xu W (2018) Cybersecurity risks and challenges in smart grids. IEEE Access 6:23479–23487

Hsu C (2023) A survey on recent applications of artificial intelligence and optimization for smart grids in smart manufacturing. Energies 16(22):7660. https://doi.org/10.3390/en16227660
Huang G (2014) A hybrid technique to improve measurement accuracy and reliability in ac systems. HVAC&R Res 20(5):581–591. https://doi.org/10.1080/10789669.2014.920223
Hurtado L, Nguyen P, Kling W (2014) Agent-based control for building energy management in the smart grid framework. https://doi.org/10.1109/isgteurope.2014.7028937
Hussain H, Javaid N, Iqbal S, Hasan Q, Aurangzeb K, Alhussein M (2018) An efficient demand side management system with a new optimized home energy management controller in smart grid. Energies 11(1):190. https://doi.org/10.3390/en11010190
Ibrik I, Sulaiman T (2019) Impact of implementation energy efficient appliances in Palestinian market. Open J Energy Efficiency 08(04):149–165. https://doi.org/10.4236/ojee.2019.84009
Igaliyeva L, Niyazbekova S, Serikova M, Kenzhegaliyeva Z, Mussirov G, Zueva A, Maisigova L (2020) Towards environmental security via energy efficiency: a case study. J Entrep Sustain Issues 7(4):3488–3499. https://doi.org/10.9770/jesi.2020.7.4(61)
International Energy Agency (IEA) (2020) Energy efficiency 2020. IEA Publications, Paris
International Energy Agency (IEA) (2021). Smart grids: deploying smarter electricity networks. Retrieved from https://www.iea.org
Janjić A, Savić S, Janaćković G, Stanković M, Velimirović L (2016) Multi-criteria assessment of the smart grid efficiency using the fuzzy analytic hierarchy process. Facta Univ Ser Electron Energ 29(4):631–646. https://doi.org/10.2298/fuee1604631j
Kim S, Lee T, Kim S, Park L, Park S (2019) Security issues on smart grid and blockchain-based secure smart energy management system. MATEC Web Conf 260:01001. https://doi.org/10.1051/matecconf/201926001001
Kluczek A (2018) An environmental sustainability assessment of energy efficiency for production systems. DEStech Trans Eng Technol Res (ICPR). https://doi.org/10.12783/dtetr/icpr2017/17693
Kushawaha V (2024) Enhancing energy efficiency: advances in smart grid optimization. Int J Innov Res Eng Manag 11(2):100–105. https://doi.org/10.55524/ijirem.2024.11.2.20
Lammers I, Heldeweg M (2016) Smart design rules for smart grids: analysing local smart grid development through an empirico-legal institutional lens. Energy Sustain Soc 6(1). https://doi.org/10.1186/s13705-016-0102-z
Lampropoulos I, Broek M, Sark W, Hoofd E, Hommes K (2017) Enabling flexibility from demand-side resources through aggregator companies, pp 333–353. https://doi.org/10.1007/978-3-319-54558-5_16
Law Y, Pota H, Jin J, Man Z, Palaniswami M (2014) Control and communication techniques for the smart grid: an energy efficiency perspective. IFAC Proc Vol 47(3):987–998. https://doi.org/10.3182/20140824-6-za-1003.01736
Lee J, Kim N, Kim H (2014) A testing technique of microgrid ems using the hardware-in-the loop simulation (HILS) system. Int J Energy Inf Commun 5(2):53–60. https://doi.org/10.14257/ijeic.2014.5.2.04
Li G, Li W, Jin Z, Wang W (2019) Influence of environmental concern and knowledge on households' willingness to purchase energy-efficient appliances: a case study in Shanxi, China. Sustainability 11(4):1073. https://doi.org/10.3390/su11041073
Li W, Logenthiran T, Phan V, Woo W (2017) Housing development building management system (HDBMS) for optimized electricity bills. Trans Environ Electr Eng 2(2):64. https://doi.org/10.22149/teee.v2i2.113
Linnenberg T, Mueller A, Christiansen L, Seitz C, Fay A (2013) Ontoenergy—a lightweight ontology for supporting energy-efficiency tasks—enabling generic evaluation of energy efficiency in the engineering phase of automated manufacturing plants. https://doi.org/10.5220/0004622903370344
Liu G, Li D, Zhang J, Jia M (2014) High efficiency wide range bidirectional dc/dc converter for OBCM application, pp 1434–1438. https://doi.org/10.1109/peac.2014.7038075
Liu L, Liu S, Yang Y, Gong X, Zhao Y, Jin R, Pan J (2023) How do different types of environmental

Liu Z, Wu Q, Shahidehpour M, Li C, Huang S, Wei W (2019) Transactive real-time electric vehicle charging management for commercial buildings with PV on-site generation. IEEE Trans Smart Grid 10(5):4939–4950. https://doi.org/10.1109/tsg.2018.2871171

Lu Z, Sun C, Cheng J, Yang L, Li Y, Wen X (2017) SDN-enabled communication network framework for energy internet. J Comput Netw Commun 2017:1–13. https://doi.org/10.1155/2017/8213854

Lucía Ó, Sarnago H, Burdío J (2015) Soft-stop optimal trajectory control for improved performance of the series-resonant multiinverter for domestic induction heating applications. IEEE Trans Ind Electron 62(10):6251–6259. https://doi.org/10.1109/tie.2015.2417132

Lyulyov O, Vakulenko I, Pimonenko T, Kwiliński A, Dźwigoł H, Dźwigoł-Barosz M (2021) Comprehensive assessment of smart grids: is there a universal approach? Energies 14(12):3497. https://doi.org/10.3390/en14123497

Ma J, Chen H, Song L, Li Y (2015) Residential load scheduling in smart grid: a cost efficiency perspective. IEEE Trans Smart Grid 7:771–784. https://doi.org/10.1109/tsg.2015.2419818

Manas M (2015) Analysis of design of technologies, tariff structures, and regulatory policies for sustainable growth of the smart grid. Energy Technol Policy 2(1):28–38. https://doi.org/10.1080/23317000.2015.1015083

McDonald M, Dilley L, Barber M, Mahler R, Adam J, Wolcott M, Ford A (2013) Small-scale energy storage for integration of renewable supplies with smart grids. https://doi.org/10.2495/esus130031

Mishima T, Takami C, Nakaoka M (2014) A new current phasor-controlled ZVS twin half-bridge high-frequency resonant inverter for induction heating. IEEE Trans Ind Electron 61(5):2531–2545. https://doi.org/10.1109/tie.2013.2274420

Morstyn T, Farrell N, Darby SJ, McCulloch MD (2018) Using peer-to-peer energy-trading platforms to incentivize prosumers to form federated power plants. Nat Energy 3(2):94–101

Moussa J (2023) Sustainable energy investment and environmental development. J Poverty Invest Dev 8(3). https://doi.org/10.47604/jpid.2095

Nizzy AM, Ogwu MC (2024) Valorization of cassava processing by-products into biofuel for a sustainable environment. In: Ogwu MC, Izah SC, Cunha Alves AA, Babu SC (eds) Plant biology, sustainability and climate change, sustainable cassava. Academic, pp 291–309. https://doi.org/10.1016/B978-0-443-21747-0.00010-2

Ogwu MC (2019) Towards sustainable development in Africa: the challenge of urbanization and climate change adaptation. In: Cobbinah PB, Addaney M (eds) The geography of climate change adaptation in urban Africa. Springer, Switzerland, pp 29–55. https://doi.org/10.1007/978-3-030-04873-0_2

Ogwu MC (2019) Understanding the composition of food waste: an "-omics" approach to food waste management. In: Gunjal AP, Waghmode MS, Patil NN, Bhatt P (eds) Global initiatives for waste reduction and cutting food loss. IGI Global, Pennsylvania, USA, pp 212–236. https://doi.org/10.4018/978-1-5225-7706-5.ch011

Ogwu MC, Odozi IP, Ahonsi OC, Uleanya KO, Odozi EB (2024) Organic acid production from cassava. In: Ogwu MC, Izah SC, Cunha Alves AA, Babu SC (eds) Sustainable cassava. Academic, pp 395–418. https://doi.org/10.1016/B978-0-443-21747-0.00009-6

Olatunde T (2024) Reviewing the impact of energy-efficient appliances on household consumption. Int J Sci Technol Res Arch 6(2):001–011. https://doi.org/10.53771/ijstra.2024.6.2.0038

Picot J (2023) Definition and implementation of an energy metering plans for energy efficiency. Eng Technol J 08(04). https://doi.org/10.47191/etj/v8i4.06

Qureshi K, Islam M, Jeon G (2021) A trust evaluation model for secure data aggregation in smart grids infrastructures for smart cities. J Ambient Intell Smart Environ 1–18. https://doi.org/10.3233/ais-210602

Rocha P, Siddiqui A, Städler M (2015) Improving energy efficiency via smart building energy management systems: a comparison with policy measures. Energy Build 88:203–213. https://doi.org/10.1016/j.enbuild.2014.11.077

Rottondi C, Duchon M, Koss D, Verticale G, Schätz B (2015) An energy management system for a smart office environment. https://doi.org/10.1109/netsys.2015.7089084

Ružičić M, Miletić M, Dobrota M (2021) Does a national innovation system encourage sustainability? Lessons from the construction industry in Serbia. Sustainability 13(7):3591. https://doi.org/10.3390/su13073591

Saker A (2023) Nurturing sustainable urban space: integrating smart city innovations and earthship design principles for eco-friendly futures. Int J Archit Eng Urban Res 6(2):376–400. https://doi.org/10.21608/ijaeur.2024.265873.1066

Saleem M, Usman M, Usman M, Politis C (2022) Design, deployment and performance evaluation of an IoT based smart energy management system for demand side management in smart grid. IEEE Access 10:15261–15278. https://doi.org/10.1109/access.2022.3147484

Sarikaya T, Partal S (2022) A case study of load scheduling for home energy management with integrated renewable energy. Int J Eng Manag Res 12(5):27–34. https://doi.org/10.31033/ijemr.12.5.4

Shuqair Y (2023) Towards a sustainable energy future—the case for smart grids in Jordan. Ecol Eng Environ Technol 24(8):82–96. https://doi.org/10.12912/27197050/171569

Si S, Maheswari D, Anitha M, Indira Priyadharshini R (2023) Optimizing renewable energy management in smart grids using machine learning. E3s Web Conf 387:02006. https://doi.org/10.1051/e3sconf/202338702006

Thilakarathne NN, Samarasinghe R, Kagita MK, Lanka S, Ahmad H (2021) Smart grid: a survey of architectural elements, machine learning and deep learning applications and future directions. J Intell Syst Internet Things (JISIoT) 3(1):32–42

Toli A, Murtagh N (2020) The concept of sustainability in smart city definitions. Front Built Environ 6. https://doi.org/10.3389/fbuil.2020.00077

Ullah K, Ali S, Khan T, Khan I, Jan S, Shah I, Hafeez G (2020) An optimal energy optimization strategy for smart grid integrated with renewable energy sources and demand response programs. Energies 13(21):5718. https://doi.org/10.3390/en13215718

Ürge-Vorsatz D et al (2012) Best practices in energy efficiency. Annu Rev Environ Resour 37(1):1–31

Vlasov A, Adamova A, Selivanov K (2021) Development of smart grid technologies: organizational and communication aspects. E3s Web Conf 250:08001. https://doi.org/10.1051/e3sconf/202125008001

Widergren S, Paget M, Secrest T, Balducci P, Orrell A, Bloyd C (2011) Using smart grids to enhance use of energy-efficiency and renewable-energy technologies. https://doi.org/10.2172/1013938

Xu S (2023) Impact of manufacturing agglomeration on the green innovation efficiency—spatial effect based on china's provincial panel data. Int J Environ Res Public Health 20(5):4238. https://doi.org/10.3390/ijerph20054238

Yanfen Z, Huang M, Xiang W, Lu L, Lu Y, Gao J, Cheng Y (2022) The impact of high-tech industry development on energy efficiency and its influencing mechanisms. Front Environ Sci 10. https://doi.org/10.3389/fenvs.2022.962627

Zakareya O (2018) Vulnerability and forensics associated with the smart grid: cyber attacks. Int J Comput Appl 181(22):32–38. https://doi.org/10.5120/ijca2018917952

Zunnurain I, Maruf M, Rahman M, Shafiullah G (2018) Implementation of advanced demand side management for microgrid incorporating demand response and home energy management system. Infrastructures 3(4):50. https://doi.org/10.3390/infrastructures3040050

Гусев B, Naumkina T, Voytolovskiy N, Minaeva E, Zemskova O (2018) Methods for assessing the efficiency of innovation activity of enterprises of the electric power industry. MATEC Web Conf 170:01098. https://doi.org/10.1051/matecconf/201817001098

Chapter 14
Sustainable Urban Planning and Environmental Management

Enoch Akwasi Kosoe and Matthew Chidozie Ogwu

Abstract Sustainable urban planning and environmental management are critical to addressing the growing challenges of rapid urbanization, climate change, and resource depletion. This chapter examines the principles, strategies, and innovations necessary for fostering resilient and livable cities while ensuring ecological integrity and social equity. Key principles such as smart growth, mixed-use development, and green space preservation are examined for their role in promoting sustainable urban environments. Integrating advanced technologies and renewable energy systems is highlighted as pivotal in optimizing resource use and enhancing urban resilience. Furthermore, the chapter emphasizes the importance of participatory approaches, stakeholder engagement, and inclusive governance to align urban development with community needs and sustainability goals. It delves into environmental management strategies, focusing on pollution control, resource efficiency, and ecosystem service integration to mitigate urban ecological challenges. Case studies of successful urban sustainability initiatives—such as Copenhagen's carbon-neutral infrastructure, Singapore's "Garden City" approach, and Curitiba's (Brazil) innovative transport systems—demonstrate the transformative potential of localized solutions and adaptive policies. However, economic constraints, social inequities, and technological barriers underscore the need for interdisciplinary and collaborative frameworks. By synthesizing global best practices, the research outlines a roadmap for achieving sustainable urban futures through the interplay of technology, policy, and community-driven approaches. It concludes with recommendations for policymakers, urban planners, and researchers to advance sustainable urban planning and

E. A. Kosoe (✉)
Department of Environment and Resource Studies, Faculty of Integrated Development Studies, Simon Diedong Dombo University of Business and Integrated Development Studies, Wa, Upper West Region, Ghana
e-mail: ekosoe@ubids.edu.gh

M. C. Ogwu
Goodnight Family Department of Sustainable Development, Appalachian State University, Boone, NC, USA
e-mail: ogwumc@appstate.edu

M. C. Ogwu and S. Chibueze Izah (eds.), *Evaluating Environmental Processes and Technologies*, Environmental Science and Engineering,
https://doi.org/10.1007/978-3-031-85327-2_14

environmental management in the face of unprecedented urbanization and ecological pressures.

Keywords Sustainable urban planning · Environmental management · Green infrastructure · Urban resilience · Smart growth · Mixed-use development · Climate change · Participatory governance

14.1 Introduction

Urban areas are increasingly recognized as critical sites for implementing sustainability initiatives due to their significant environmental footprints and the concentration of human activities. Sustainable urban planning is essential for fostering resilient communities that can adapt to environmental challenges while enhancing the quality of life for residents. Sustainable urban planning is paramount in addressing the multifaceted challenges posed by urbanization, including environmental degradation, social inequity, and economic instability. As urban areas expand, they often lead to urban sprawl, exacerbating habitat fragmentation and straining public infrastructure (Wu et al. 2017). Effective spatial planning can mitigate these issues by promoting compact, mixed-use developments that reduce automobile reliance and enhance public transport accessibility (Fossheim and Andersen 2017). Furthermore, sustainable urban planning integrates ecological considerations into the development process, ensuring that urban growth does not come at the expense of natural resources (Coutinho et al. 2018). The principles of sustainable urban design, such as those outlined in the LEED-ND framework, emphasize resource efficiency and creating livable spaces, which are crucial for fostering community well-being (Shawabkeh 2018).

Moreover, urban planning must prioritize the involvement of local stakeholders to align development projects with community needs and expectations. Engaging residents in the planning process enhances the relevance of urban interventions and fosters a sense of ownership and responsibility toward the urban environment (Azhimov and Manukhina 2023). This participatory approach is vital for creating sustainable urban landscapes that reflect the cultural and social dynamics of the community (Meyer and Auriacombe 2019). The objectives of environmental management in urban contexts are multifaceted, encompassing the preservation of natural resources, the reduction of pollution, and the enhancement of urban resilience. Effective environmental management seeks to balance economic growth with ecological integrity, ensuring that urban development does not compromise the health of ecosystems (Salati et al. 2022). For instance, strategic environmental assessments (SEA) and environmental impact assessments (EIA) are critical tools that guide urban planners in identifying potential ecological risks and mitigating negative impacts associated with urban projects (Coutinho et al. 2018).

Additionally, urban environmental management aims to improve air and water quality, which is essential for public health. Studies have shown that urbanization often increases pollution levels, necessitating robust management strategies to monitor and control emissions (Gao et al. 2019). Implementing green infrastructure, such as urban forests and green roofs, can significantly enhance urban biodiversity while providing essential ecosystem services (Ogwu et al. 2016a, b; Wei et al. 2018). Such initiatives contribute to environmental sustainability and improve the aesthetic and recreational value of urban spaces (Dou et al. 2014). The interconnections between urban planning and sustainability are evident in the frameworks and policies that guide urban development. Sustainable urban planning is not merely an aesthetic endeavor but a comprehensive approach that integrates social, economic, and environmental dimensions (Tang and Lee 2016). For instance, the New Urbanism movement emphasizes the creation of walkable neighborhoods that promote social interaction and reduce environmental impacts through mixed-use development ("Values of the New Urbanist Neighborhood Planning Based on the Concept of Sustainability", 2023). This approach aligns with sustainability goals by fostering community cohesion and reducing reliance on fossil fuels. Furthermore, adopting sustainability indicators in urban planning facilitates assessing progress towards sustainability goals. These indicators can encompass a range of factors, including energy consumption, waste generation, and social equity metrics, providing a holistic view of urban sustainability (Salati et al. 2022). By utilizing such indicators, urban planners can make informed decisions that align with sustainability objectives, ultimately leading to more resilient urban environments.

This chapter explores the principles, strategies, and innovations in sustainable urban planning and environmental management, focusing on their role in addressing the challenges posed by rapid urbanization and climate change. It seeks to provide a comprehensive understanding of how urban planning can integrate environmental sustainability through smart growth, resource efficiency, and green infrastructure. The chapter emphasizes the importance of preserving urban biodiversity, managing natural resources, and minimizing pollution to create resilient and livable cities. It also examines the role of advanced technologies, such as GIS, IoT, and renewable energy systems, in enhancing urban sustainability. By presenting case studies and best practices, the chapter highlights successful approaches to sustainable urban development while addressing the economic, social, and policy-related barriers to implementation. Ultimately, it aspires to guide policymakers, urban planners, and researchers in fostering equitable, resilient, and sustainable urban environments.

14.2 Key Principles of Sustainable Urban Planning

Sustainable urban planning is predicated on several key principles that aim to create livable, resilient, and environmentally friendly urban environments. Three critical principles for sustainable urban planning and development are smart Growth and

Compact Cities, Mixed-Use Development and Land-Use Optimization, and the Preservation of Green Spaces and Biodiversity.

14.2.1 Smart Growth and Compact Cities

Smart growth is a planning approach that emphasizes sustainable urban development through compact, efficient land use. This principle advocates for higher-density development, which can reduce urban sprawl and its associated environmental impacts. Compact cities facilitate the efficient use of resources, promote public transportation, and enhance walkability, thereby reducing reliance on automobiles (Seong et al. 2021). Research indicates that urban areas with smart growth principles can significantly reduce greenhouse gas emissions and energy consumption (Dur and Yiğitcanlar 2014). The concept of compact cities is closely linked to the New Urbanism movement, which promotes the design of pedestrian-friendly neighbourhoods integrated with public transport systems (Huang and Hsieh 2014). Studies have shown that higher densities and mixed land uses in urban areas encourage walking and cycling, thus fostering healthier lifestyles (Suek 2024). Furthermore, compact urban forms can enhance social interactions and community engagement, contributing to residents' sense of place and belonging (Almansoub et al. 2022). However, implementing smart growth strategies must consider local contexts and community needs to avoid potential gentrification and displacement of existing residents (Kumar 2024). Therefore, while compact cities offer numerous benefits, planners must balance growth with social equity and community well-being.

Figure 14.1 presents a holistic framework for achieving urban sustainability by integrating three key dimensions: social, economic, and environmental. It emphasizes the interconnectedness of population, land, and industry, highlighting "optimal matching" to maximize resource efficiency while minimizing environmental damage (Yang et al. 2024). Strategies such as smart growth (compact development, greenbelt preservation, and mixed land uses) and smart shrinkage (downsizing, demolishing unused infrastructure, and enhancing public services) address the challenges of urban sprawl and urban shrinkage. The framework promotes compact and intensive development through sustainable urban planning that balances equity, economic productivity, and environmental preservation (Yang et al. 2024). By prioritizing resource efficiency and green infrastructure, Fig. 14.4 underscores the importance of inclusive, adaptable, and resilient approaches to urban management, supporting long-term environmental sustainability and livable cities.

14.2.2 Mixed-Use Development and Land-Use Optimization

Mixed-use development is a cornerstone of sustainable urban planning, integrating residential, commercial, and recreational spaces within proximity. This approach

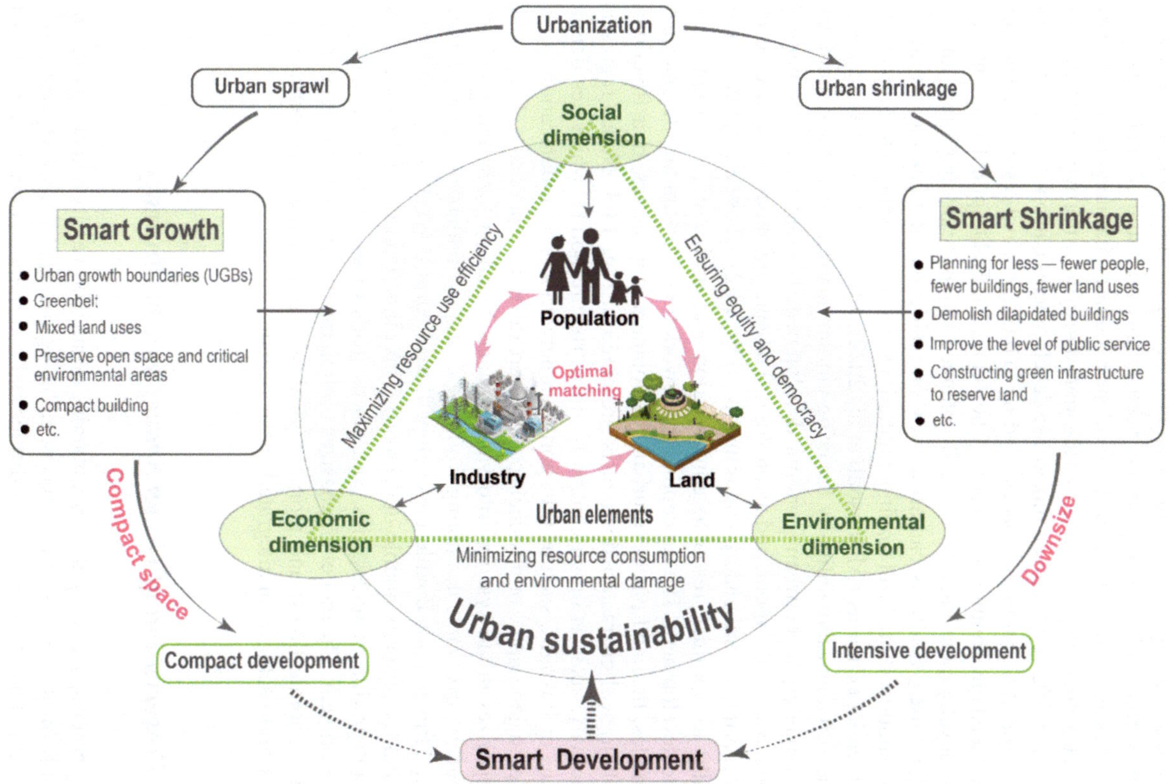

Fig. 14.1 Smart development theoretical framework for urban sustainability. *Source* Yang et al. (2024)

enhances urban areas' functionality and promotes social interaction and economic vitality (Yang et al. 2021). Research has consistently shown that mixed land use can lead to increased walking and reduced vehicle trips, lowering traffic congestion and pollution (Seong et al. 2021; Huang and Hsieh 2014). The relationship between land use mix and walking behavior has been extensively studied, particularly in high-density cities (Seong et al. 2021). Evidence suggests that neighborhoods with diverse land use encourage residents to walk more and cycle, improving public health outcomes (Suek 2024). Moreover, mixed-use developments can optimize land use by maximizing the utility of urban spaces, leading to more efficient resource consumption and reduced environmental impacts (Chen et al. 2022). Despite the advantages, mixed-use development must be carefully planned to ensure compatibility among different land uses. Conflicts can arise between residential and commercial activities, necessitating thoughtful zoning regulations and community engagement to address potential concerns (Almansoub et al. 2022). Effective mixed-use planning can enhance urban sustainability by fostering vibrant, interconnected communities.

There is an interplay between economic development, key resources, and enabling factors such as capital, labour, and technological advancements in mixed-use development scenarios in land undergoing transition (Fig. 14.2). These elements interact dynamically to optimize resource configuration and utilization, contributing to sustainable development. The conceptual framework for understanding the interplay between nodes, ties, and places in urban development emphasizes their functional and morphological characteristics and is presented in Fig. 14.3. Nodes represent hubs of activity, such as urban centres or infrastructure points, while ties denote connections or linkages, such as transportation networks or communication flows. Places refer to geographic or physical spaces where development occurs. The framework categorizes urban systems into various states based on their functional and morphological development, ranging from dispersed and underdeveloped configurations (e.g., functionally double-undeveloped) to integrated systems (e.g., morphologically and functionally double-developed). This classification highlights how the balanced integration of nodes, ties, and places leads to sustainable and cohesive urban development. The model emphasizes optimizing functional ties (e.g., connectivity and economic flows) and morphological ties (e.g., spatial arrangement and infrastructure) to achieve an integrated urban system.

14.2.3 Preservation of Green Spaces and Biodiversity

Preserving green spaces is vital for maintaining urban biodiversity and enhancing the quality of life in cities. Urban green spaces, including parks, gardens, and green roofs, provide essential habitats for various species and contribute to ecological resilience (Kumar 2024). Studies have shown that urban areas with abundant green spaces support higher levels of biodiversity, enhancing ecosystem services such as air purification, temperature regulation, and stormwater management (Beninde et al. 2015; Stagoll et al. 2012). Moreover, green spaces are crucial in promoting mental health

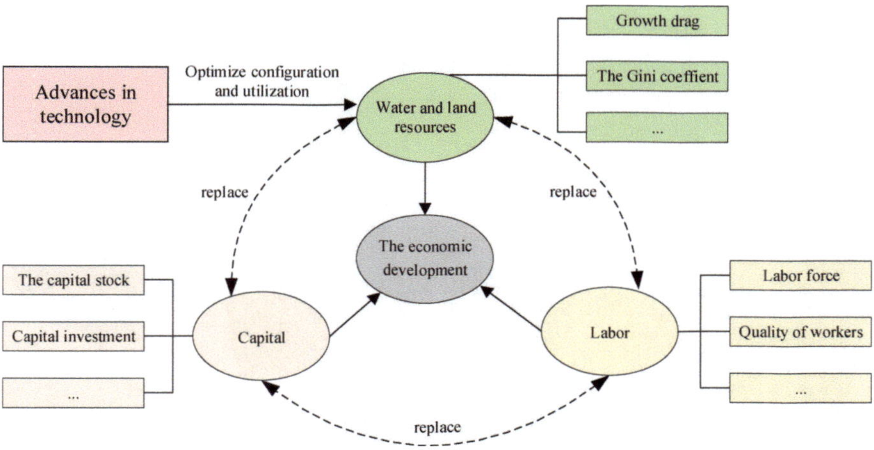

Fig. 14.2 Interplay between economic development, key resources, and enabling factors such as capital, labour, and technological advancements in mixed-used development scenario in land under transition. *Source* Pan et al. (2021)

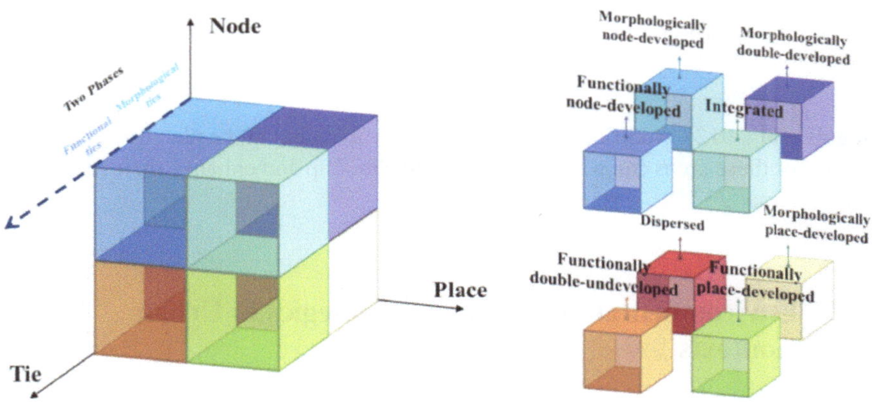

Fig. 14.3 A framework for understanding the interplay between nodes, ties, and places in urban development. *Source* Berawi et al. (2020)

and well-being among urban residents. Access to nature has been linked to reduced stress levels and improved physical health outcomes (Daniels et al. 2020). Therefore, integrating green infrastructure into urban planning is essential for creating sustainable and livable cities (Kumar 2024). However, urbanization often leads to the loss of natural habitats and green spaces, which can threaten local biodiversity (Park et al. 2020). Urban planners must prioritize preserving and enhancing green spaces within city designs to counteract this trend. This includes implementing policies that protect existing natural areas and promote the creation of new green spaces (Kumar 2024;

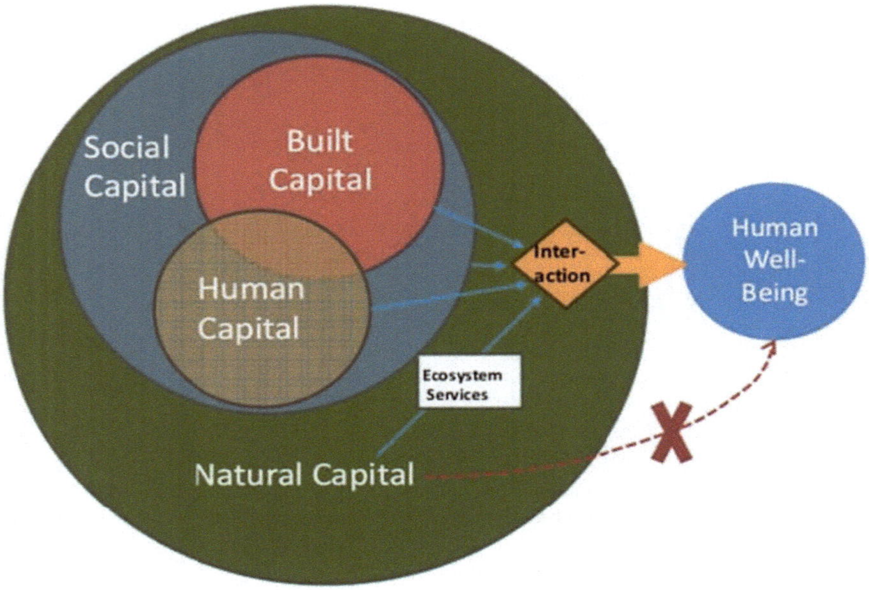

Fig. 14.4 Interdependence of natural, human, social, and built capital supports human well-being. *Source* Fistola (2023)

Stagoll et al. 2012). Additionally, community involvement in the planning and maintenance of these areas can foster a sense of stewardship and enhance their ecological value (Kumar 2024).

14.3 Environmental Management Strategies in Urban Contexts

Urban environments face numerous challenges, including pollution, resource depletion, and the degradation of natural ecosystems. Effective environmental management strategies are essential to address these issues and promote sustainable urban development. Urban ecosystems provide a range of services that are vital for human well-being and environmental sustainability. These services include air and water purification, climate regulation, and recreational opportunities, collectively enhancing the quality of life in urban areas (Ogwu 2019; Dushkova and Haase 2020). The concept of natural capital refers to the world's stocks of natural assets, including geology, soil, air, water, and all living things, which provide ecosystem services essential for human survival (Fig. 14.4; Ncube and Arthur 2021). The interdependence of natural, human, social, and built capital supports human well-being (Fig. 14.4). Natural capital forms the foundation, providing essential ecosystem services that underpin the other forms of capital. Human, social, and built capital interact dynamically, leveraging natural resources to enhance quality of life and societal development. However, Fig. 14.4 highlights a critical warning: disrupting natural capital's ecosystem services can break the connection, jeopardizing human well-being. This underscores the need for sustainable management of natural resources to maintain the balance and ensure long-term societal prosperity.

Research indicates that integrating ecosystem services into urban planning can lead to more resilient and sustainable cities. For instance, Dushkova and Haase (2020) highlight how nature-based solutions can effectively mitigate climate change impacts, regulate water flows, and improve urban biodiversity. Similarly, Guerry et al. (2015) emphasize recognizing the interdependence of ecosystems and human well-being, advocating for policies that restore and sustainably manage natural capital. Moreover, integrating ecosystem services into urban governance frameworks can enhance decision-making processes. For example, the UK National Planning Policy Framework encourages planning systems to recognize the benefits of natural capital and ecosystem services, promoting sustainable urban development (Ncube and Arthur 2021). This approach supports biodiversity, fosters community engagement, and enhances the aesthetic value of urban spaces (Pearson 2021).

Pollution control is a critical aspect of environmental management in urban contexts. Urban areas are significant sources of air, water, and soil pollution, which pose serious health risks to residents and degrade the environment (Thomson and Newman 2018). Effective pollution control strategies involve implementing regulations, promoting cleaner technologies, and enhancing public awareness about pollution sources and impacts. Resource efficiency is closely linked to pollution control, as it focuses on minimizing waste and optimizing the use of resources in urban environments. Schebek and Lützkendorf (2022) propose a methodological framework for assessing resource efficiency in urban neighborhoods, emphasizing the need for indicators to guide urban planning decisions. Adopting resource-efficient practices can help cities reduce their ecological footprints and enhance sustainability. Furthermore, innovative approaches such as urban metabolism studies can provide valuable insights into city resource flows and waste generation. Peña et al. (2022) argue that integrating spatially explicit data into urban planning can facilitate the development of resource-efficient open space networks, ultimately contributing to sustainable urban environments. This holistic approach to resource management addresses pollution and promotes economic efficiency and social equity.

Water management is critical to sustainable urban planning, particularly in increasing urbanization and climate change (Ogwu et al. 2024; Ogwu and Kosoe 2024). Sustainable Urban Drainage Systems (SUDS) are designed to manage stormwater runoff to mimic natural drainage processes, thereby reducing flooding and improving water quality. These systems incorporate green infrastructure elements, such as permeable pavements, green roofs, and bioretention areas, which enhance urban resilience and biodiversity (Ghofrani et al. 2020). Integrating blue-green infrastructure in urban planning can significantly improve water management outcomes. Ncube and Arthur (2021) highlight the importance of combining blue (water) and green (vegetation) infrastructure to enhance natural and human-derived capital in urban drainage planning. This approach addresses water management challenges and provides additional benefits, such as improved air quality and enhanced recreational opportunities (Iyiola et al. 2023, 2024). Moreover, effective water management strategies must consider the interconnectedness of urban water systems with other urban functions. Ahern et al. (2014) propose a framework for incorporating ecosystem services into adaptive urban planning and design, emphasizing the need for innovative solutions addressing water management and urban resilience. By adopting a comprehensive approach to water management, cities can enhance their sustainability and adaptability to climate change.

14.4 Technological Innovations Supporting Urban Sustainability

Technological innovations such as those outlined in Table 14.1 are pivotal in advancing urban sustainability by enhancing environmental management, optimizing resource use, and fostering community engagement. Smart cities leverage advanced technologies to improve urban living conditions and enhance environmental management. They utilize data-driven approaches to optimize resource allocation, reduce waste, and improve service delivery (Bibri 2019). Bibri (2019) emphasizes that smart, sustainable cities are characterized by their ability to balance economic, social, and environmental goals by integrating innovative technologies. This integration allows for real-time monitoring of urban systems, enabling cities to respond effectively to environmental challenges such as air pollution and traffic congestion. Moreover, smart cities foster stakeholder collaboration, including government agencies, private sector actors, and citizens. Meijer and Thaens (2016) propose a sociotechnical framework for analyzing smart city projects, highlighting the importance of collaborative governance in achieving sustainable urban outcomes. This collaborative approach enhances the legitimacy of smart city initiatives and encourages citizen participation in environmental management (Deakin 2014). However, the implementation of smart city technologies must be approached critically. Yiğitcanlar (2020) argues that many smart city initiatives focus primarily on efficiency improvements without addressing broader social and environmental issues. Therefore, a balanced approach integrating technology, policy, and community engagement is essential for realizing smart cities' full potential in promoting urban sustainability.

Urban data analytics and GIS support sustainable urban planning and management. These technologies enable cities to collect, analyze, and visualize data related to various urban systems, facilitating informed decision-making (Schinkel et al. 2022). For instance, GIS applications can help urban planners assess land use patterns, identify areas of environmental concern, and optimize infrastructure development (Kencono 2023). Integrating big data analytics into urban management allows identifying trends and patterns that can inform sustainability initiatives. Bibri (2019) notes that using big data in smart cities can enhance the understanding of urban dynamics

Table 14.1 Technological innovations supporting urban sustainability

Technology	Description	Applications	Benefits
Smart sensors and IoT	Devices that collect and transmit real-time data on urban systems and infrastructure	Smart cities, traffic management, water systems	Enables data-driven decision-making, improves efficiency, and reduces resource wastage
GIS and remote sensing	Tools for mapping and analyzing spatial data for urban planning and resource management	Land use planning, disaster management	Optimizes land use, enhances disaster preparedness, and monitors environmental changes
Renewable energy systems	Solar panels, wind turbines, and other technologies for generating sustainable energy	Urban power supply, residential buildings	Reduces reliance on fossil fuels, decreases greenhouse gas emissions, and lowers energy costs
Green infrastructure	Nature-based solutions integrated into urban design, such as green roofs and rain gardens	Stormwater management, urban cooling	Mitigates urban heat islands improves water quality, and enhances biodiversity
Energy storage systems	Technologies like batteries and thermal storage to store renewable energy for later use	Microgrids, electric vehicles	Stabilizes energy supply, supports renewable energy integration, and enhances energy resilience
Smart grids	Intelligent electricity networks that optimize energy distribution and demand	Urban energy systems	Improves energy efficiency, facilitates renewable energy integration, and reduces outages
Electric vehicles (EVs)	Electrified transport systems that reduce reliance on fossil fuels	Public transportation, private vehicles	Cuts transportation emissions, improves air quality, and integrates with renewable energy systems
AI and machine learning	Algorithms that optimize urban operations through predictive analytics	Traffic flow, waste management, energy systems	Increases efficiency, reduces operational costs and supports proactive urban management

(continued)

Table 14.1 (continued)

Technology	Description	Applications	Benefits
Waste-to-energy technologies	Systems that convert urban waste into usable energy or materials	Urban waste management, energy production	Reduces landfill waste, generates energy, and supports circular economy principles
Blockchain technology	Decentralized systems for managing energy transactions and urban services	Peer-to-peer energy trading, water distribution	Enhances transparency, improves service efficiency, and fosters community-based resource sharing

and improve the planning and governance of urban systems. Harnessing data from various sources, including sensors and social media, can help cities develop more responsive and adaptive management strategies (Parjanen et al. 2018). Furthermore, urban data analytics can support the evaluation of sustainability indicators, enabling cities to track progress toward sustainability goals. Schinkel et al. (2022) propose a set of indicators to assess the performance of urban technologies and infrastructures in contributing to sustainable development. This data-driven approach enhances transparency and fosters accountability in urban governance.

Integrating renewable energy sources and green infrastructure is crucial for achieving urban sustainability. Renewable energy technologies, such as solar panels and wind turbines, can significantly reduce greenhouse gas emissions and dependence on fossil fuels (Angelidou et al. 2017). Smart cities often incorporate renewable energy solutions into their energy systems, promoting energy efficiency and resilience (D'Auria et al. 2018). Green infrastructure, including parks, green roofs, and permeable surfaces, enhances urban ecosystems and mitigates environmental impacts. It provides essential services like stormwater management, air quality improvement, and habitat creation (Bifulco et al. 2016). Combining renewable energy and green infrastructure can lead to synergistic benefits, enhancing environmental quality and urban livability (Nuutinen and Lappalainen 2022). Moreover, adopting innovative technologies in urban green infrastructure can facilitate monitoring and managing ecosystem services. For example, IoT devices can be used to track environmental conditions and optimize the performance of green infrastructure systems (Acharyya 2023). Integrating technology into green infrastructure enhances its effectiveness and promotes community engagement in sustainability efforts (Li 2023).

14.5 Some Case Studies and Best Practices in Sustainable Urban Planning and Environmental Management

Sustainable urban planning and environmental management are increasingly critical as cities face rapid urbanization, climate change, and resource depletion challenges. Numerous cities worldwide have adopted innovative strategies to promote sustainability in urban planning. For instance, Copenhagen has emerged as a leader in sustainable urban development through its ambitious climate action plan, which aims to become carbon neutral by 2025. The city has invested heavily in cycling infrastructure, public transportation, and renewable energy sources, demonstrating the effectiveness of integrated urban planning Wiengarten et al. (2016). The success of Copenhagen's initiatives illustrates the importance of comprehensive policy frameworks that align environmental goals with urban development objectives. Another notable example is Singapore, which has implemented a "Garden City" vision that integrates green spaces into urban environments. The city-state has prioritized vertical gardens, green roofs, and urban parks, enhancing biodiversity while improving residents' quality of life. Singapore's approach to sustainable urban planning emphasizes the need for innovative design solutions that harmonize urbanization with ecological preservation (Nie et al. 2022).

Additionally, the city of Curitiba in Brazil is often cited as a model for sustainable urban transport. Its Bus Rapid Transit (BRT) system has significantly reduced traffic congestion and pollution while providing efficient public transportation options (Shekfa and Ahmed 2022). Curitiba's experience underscores the importance of investing in sustainable transport infrastructure as a means to enhance urban mobility and reduce environmental impacts. Regional adaptations to environmental management practices are essential for addressing local challenges and leveraging unique opportunities. In the United Arab Emirates, urban densification has been explored as a strategy for sustainable development. Research have also highlighted the impacts of urban densification on social sustainability, emphasizing the need for policies that promote inclusive urban environments while managing the challenges associated with rapid growth (Enoguanbhor et al. 2019). This case illustrates how regional contexts shape the implementation of sustainable urban planning practices. In Nigeria, the integration of GIS and remotely sensed data has been pivotal in land use planning. Enoguanbhor et al. (2019) demonstrate how these technologies can support sustainable land management by identifying discrepancies between planned land use and actual urban expansion (Tung et al. 2014). This approach highlights the importance of data-driven decision-making in addressing urbanization challenges in the Global South. Furthermore, the role of environmental regulation in fostering regional innovation has been explored in China. Nie et al. (2022) argue that effective environmental regulations can stimulate green innovation and enhance regional competitiveness, demonstrating how policy frameworks can drive sustainable development. This case emphasizes the need for adaptive regulatory approaches that encourage innovation while addressing environmental concerns.

Successful urban sustainability initiatives provide valuable lessons for cities seeking to enhance their environmental management practices. One key lesson is the importance of stakeholder engagement in the planning process. Effective communication and collaboration among government agencies, private sector actors, and community members are crucial for building consensus and ensuring that sustainability initiatives reflect local needs and priorities (Burki and Dahlstrøm 2017). Additionally, integrating green infrastructure into urban planning has proven to be a successful strategy for enhancing environmental resilience. Steenkamp et al. highlight the role of urban agriculture in addressing food security risks, advocating for its incorporation into urban planning practices (Steenkamp et al. 2021). This approach improves food access, fosters community engagement, and enhances urban biodiversity. Moreover, adopting environmental management systems (EMS) has improved organizational performance in various sectors. Tung et al. (2014) demonstrate that organizations implementing EMS can achieve better environmental outcomes while enhancing operational efficiency. This finding underscores the potential benefits of adopting systematic approaches to environmental management in urban contexts.

14.6 Challenges and Future Directions in Sustainable Urban Planning and Environmental Management

Sustainable urban planning and environmental management face numerous challenges that must be addressed to create resilient and livable cities. Urbanization is a significant driver of change in cities, leading to increased population density and the expansion of urban areas. This rapid growth places immense pressure on infrastructure, resources, and the environment. He et al. (2019) highlight that urbanization can exacerbate climate change vulnerability, particularly in densely populated areas where the capacity to adapt is often limited. The challenges associated with urbanization include inadequate housing, transportation congestion, and increased demand for public services, which can strain local governments and resources (Table 14.2). Moreover, urban sprawl, characterized by the uncontrolled expansion of urban areas, can lead to habitat destruction and increased greenhouse gas emissions (Zhang and Ramos 2022). As cities grow, they often encroach on natural ecosystems, resulting in biodiversity loss and diminished ecosystem services (He et al. 2015). Therefore, effective urban planning must address these challenges by promoting compact, mixed-use developments that optimize land use and reduce environmental impacts.

Climate change poses significant risks to urban areas, including increased temperatures, extreme weather events, and rising sea levels. Mishra and Lettenmaier (2011) note that urban areas are particularly vulnerable to climate change due to the urban heat island effect, which exacerbates heat stress and increases energy demand for cooling. As cities continue to grow, the impacts of climate change will likely intensify, necessitating adaptive strategies to mitigate these effects. Urban flooding is another critical concern, as climate change is expected to increase the frequency

Table 14.2 Challenges in sustainable urban planning for environmental management

Challenge	Related issues	Implications
Urbanization pressure	Rapid population growth leads to unplanned expansion of urban areas	Increases resource consumption, contributes to urban sprawl, and strains infrastructure and services
Climate change impacts	Rising temperatures, extreme weather, and sea level rise affect urban areas	Exacerbates risks to infrastructure, disrupts ecosystems, and increases the vulnerability of urban populations
Resource scarcity	Limited availability of natural resources like water, land, and energy	Compromises sustainability goals, increases competition for resources, and impacts economic development
Inefficient land use	Poor zoning and land management practices lead to fragmented urban growth	Resulting in habitat loss, reduced agricultural land, and inefficient infrastructure use
Economic constraints	Lack of funding for implementing sustainable urban projects	Limits investments in green infrastructure, renewable energy, and sustainable transportation systems
Social inequity	Unequal access to urban services and amenities among different social groups	Creates disparities in living standards and reduces inclusivity in urban development
Governance and policy gaps	Weak or inconsistent regulations for sustainable planning and development	Leads to uncoordinated efforts, delayed projects, and a lack of accountability
Technological barriers	Limited access to advanced technologies for urban management	Hinders the implementation of smart solutions for energy, water, and waste management
Community engagement challenges	Low levels of public participation in urban planning processes	Reduces the effectiveness of sustainable projects and increases resistance to change
Waste management issues	Inefficient handling and disposal of urban waste	Causes environmental pollution, health hazards, and loss of resources that could be recycled or reused
Transportation emissions	High reliance on private vehicles and lack of sustainable transit options	Contributes to air pollution, greenhouse gas emissions, and traffic congestion
Data and monitoring limitations	Insufficient data for planning and evaluating sustainability initiatives	Restricts evidence-based decision-making and tracking of progress toward sustainability goals

and intensity of heavy rainfall events. Studies indicate that urbanization increases surface runoff and decreases infiltration, leading to heightened flood risks (Molavi et al. 2011; Blair and Sanger 2016). Effective stormwater management strategies, such as green infrastructure and sustainable urban drainage systems, enhance urban resilience to climate change (Kang et al. 2016). Furthermore, the interplay between urbanization and climate change is complex, as urban expansion can exacerbate climate vulnerabilities while offering mitigation opportunities through sustainable design and planning (Argüeso et al. 2015). Therefore, urban planners must adopt integrated approaches that consider both climate adaptation and mitigation in their strategies.

Implementing sustainable urban planning practices often faces economic and social constraints. Limited financial resources can hinder investments in infrastructure and services necessary for sustainability initiatives. Ghahramanpouri et al. (2013) emphasize that social sustainability is frequently overlooked in urban planning despite its critical role in fostering community engagement and equity. Addressing social disparities and ensuring that all community members benefit from sustainability initiatives is essential for achieving long-term success. Moreover, the lack of data and information can impede effective decision-making in urban planning. Many cities, particularly in developing countries, struggle with inadequate data on urban dynamics, environmental conditions, and community needs (Hardoy and Ruete 2013). This knowledge gap can lead to poorly informed policies and practices that fail to address the unique challenges urban populations face.

Emerging trends in sustainable urban design focus on creating resilient, adaptable, and inclusive urban environments. Innovations such as biophilic design, which integrates nature into urban spaces, can enhance well-being and promote biodiversity (Haqi 2016). Smart technologies and data analytics can also optimize urban systems and improve resource efficiency, contributing to more sustainable urban futures (Anttiroiko 2012). The integration of circular economy principles into urban management is gaining traction to reduce waste and promote resource efficiency. Cities can minimize their environmental impact and enhance resilience by designing urban systems that prioritize recycling, reuse, and sustainable consumption (Chapman et al. 2019). This approach encourages stakeholder collaboration, fostering a sustainability culture that extends beyond traditional urban planning practices. Creating resilient and equitable urban futures requires a holistic approach that addresses the interconnected challenges of climate change, social inequality, and economic development. Collaborative governance models that engage diverse stakeholders in decision-making processes can enhance community resilience and ensure that sustainability initiatives reflect local needs (Pasquini et al. 2021). Furthermore, integrating social equity considerations into urban planning can help address disparities and promote inclusive growth (Lundgren and Kjellström 2013).

14.7 Policy and Governance for Sustainable Urban Development

Effective policy and governance frameworks are essential for promoting sustainable urban development. Urban planning policies and regulations serve as the backbone for sustainable urban development. They establish the framework for urban growth and development, guiding land use, infrastructure investment, and environmental protection. Navarra and Milio (2015) highlight that comprehensive governance frameworks incorporating sustainability principles are crucial for effective urban planning. These frameworks should be informed by interdisciplinary approaches addressing the complex social, economic, and environmental interactions (Navarra and Milio 2015). Moreover, effective urban planning policies must be adaptable to changing circumstances, including demographic shifts and environmental challenges. For instance, the Sustainable Flood Risk Appraisal (SFRA) framework developed for the Pearl River Delta integrates local sensitivity and specific criteria to address flood risks while promoting sustainability (Chan et al. 2013). This case illustrates how localized policies can enhance resilience to climate-related hazards while ensuring sustainable urban development. In addition, the integration of sustainability principles into urban planning regulations can facilitate the transition to smart sustainable cities. Ablyazov and Vishnivetskaya (2021) emphasize the importance of aligning financing methods with urban development goals to ensure that projects contribute to the overall sustainability of the urban environment. This alignment is essential for fostering long-term investments in infrastructure that support sustainable urban growth.

Multistakeholder collaboration and public participation are critical components of effective urban governance. Engaging diverse stakeholders, including government agencies, private sector actors, and community members, fosters a sense of ownership and accountability in urban planning processes. Hergül and Göker (2021) argue that transforming urban planning requires not only changes in physical infrastructure but also a shift in governance mechanisms to promote sustainability. Public participation in urban planning processes enhances transparency and ensures that the needs and preferences of local communities are considered. This participatory approach can lead to more equitable and inclusive urban development outcomes. For example, the involvement of local communities in the planning of green spaces can enhance their ecological value and promote social cohesion (Lundgren and Kjellström 2013). Furthermore, effective collaboration among stakeholders can facilitate the sharing of knowledge and resources, leading to more innovative solutions to urban challenges. The integration of digital finance and green finance into urban governance frameworks can enhance cities' capacity to fund sustainable initiatives and promote environmental sustainability (Li et al. 2022; Zakari and Khan 2022).

Financing mechanisms play a crucial role in supporting sustainable urban development initiatives. Traditional funding sources often fall short of meeting the growing infrastructure needs of rapidly urbanizing areas. Therefore, innovative financing approaches are necessary to mobilize resources for sustainable urban projects. For instance, land-based financing strategies can leverage increased urban land values to fund infrastructure development (Gandhi and Phatak 2016). This approach is particularly relevant in metropolitan areas experiencing significant population growth. Green finance has emerged as a vital tool for promoting sustainability in urban development. It encompasses a range of financial instruments aimed at supporting environmentally friendly projects. Tian and Hou (2022) demonstrate how green finance can enhance industrial land use efficiency, thereby contributing to sustainable urban development. Prioritizing investments in green technologies and infrastructure can also help cities reduce their carbon footprints and promote resource efficiency. Public–private partnerships (PPPs) offer a promising avenue for financing urban sustainability initiatives. Leveraging private sector expertise and investment to implement innovative projects will contribute to address pressing urban challenges. Hu et al. (2022) emphasizes the importance of coordinated governance among regional governments to optimize the impact of green finance on sustainable development (Hu et al. 2022).

14.8 Conclusion

Sustainable urban planning and environmental management are intrinsically linked through their shared goals of promoting ecological integrity, enhancing quality of life, and fostering community resilience. As urban areas continue to evolve, integrating sustainability principles into planning processes will be crucial for addressing the pressing challenges of the twenty-first century. The key principles of sustainable urban planning—Smart Growth and Compact Cities, Mixed-Use Development and Land-Use Optimization, and the Preservation of Green Spaces and Biodiversity—are interconnected and essential for fostering resilient urban environments. By adhering to these principles, urban planners can create cities that are not only sustainable but also vibrant and inclusive. Environmental management strategies in urban contexts are essential for promoting sustainable development and enhancing the quality of life for urban residents. By focusing on urban ecosystem services and natural capital, pollution control and resource efficiency, and water management through sustainable urban drainage systems, cities can address pressing environmental challenges and foster resilient urban environments. Technological innovations are essential for supporting urban sustainability through the development of smart cities, the application of urban data analytics and GIS, and the integration of renewable energy and green infrastructure. By leveraging these technologies, cities can enhance their environmental management practices, optimize resource use, and foster community participation, ultimately leading to more sustainable urban futures. Sustainable urban planning and environmental management face multifaceted challenges, requiring innovative solutions and collaborative approaches. By addressing the pressures of urbanization, the impacts of climate change, and the economic and social constraints in implementation, cities can pave the way for resilient and equitable urban futures. There is also a need to learn from global examples, regional adaptations, and successful sustainability initiatives so that cities can develop effective strategies to address the pressing challenges of urbanization and environmental degradation. Effective policy and governance frameworks are essential for promoting sustainable urban development. Urban planning policies and regulations provide the necessary structure for guiding growth, while multistakeholder collaboration and public participation ensure that diverse perspectives are considered in decision-making processes. Innovative financing mechanisms, including green and land-based financing, are crucial for mobilizing resources and supporting sustainable urban initiatives. By addressing these key areas, cities can pave the way for a more sustainable and resilient urban future.

References

Ablyazov T, Vishnivetskaya A (2021) Financing the human environment development projects based on the concept of a smart sustainable city. E3s Web Conf 295:03002. https://doi.org/10.1051/e3sconf/202129503002

Acharyya S (2023) Revolutionizing natural language understanding with prompt engineering: a comprehensive study. Int Res J Innov Eng Technol 07(10):692–695. https://doi.org/10.47001/irjiet/2023.710091

Ahern J, Cilliers S, Niemelä J (2014) The concept of ecosystem services in adaptive urban planning and design: a framework for supporting innovation. Landsc Urban Plan 125:254–259. https://doi.org/10.1016/j.landurbplan.2014.01.020

Almansoub Y, Zhong M, Raza A, Muhammad S, Dahou A, Al-qaness M (2022) Exploring the effects of transportation supply on mixed land use at the parcel level. Land 11(6):797. https://doi.org/10.3390/land11060797

Angelidou M, Psaltoglou A, Komninos N, Kakderi C, Tsarchopoulos P, Panori A (2017) Enhancing sustainable urban development through smart city applications. J Sci Technol Policy Manag 9(2):146–169. https://doi.org/10.1108/jstpm-05-2017-0016

Anttiroiko A (2012) Urban planning 2.0. Int J E Plan Res 1(1):16–30. https://doi.org/10.4018/ijepr.2012010103

Argüeso D, Evans J, Pitman A, Luca A (2015) Effects of city expansion on heat stress under climate change conditions. PLoS One 10(2):e0117066. https://doi.org/10.1371/journal.pone.0117066

Azhimov T, Manukhina L (2023) Increasing the environmental potential of territories through the implementation of redevelopment projects. E3s Web Conf 403:05013. https://doi.org/10.1051/e3sconf/202340305013

Beninde J, Veith M, Hochkirch A (2015) Biodiversity in cities needs space: a meta-analysis of factors determining intra-urban biodiversity variation. Ecol Lett 18(6):581–592. https://doi.org/10.1111/ele.12427

Berawi MA, Saroji G, Iskandar FA, Ibrahim BE, Miraj P, Sari M (2020) Optimizing land use allocation of transit-oriented development (TOD) to generate maximum ridership. Sustainability 12(9):3798. https://doi.org/10.3390/su12093798

Bibri S (2019) The anatomy of the data-driven smart sustainable city: instrumentation, datafication, computerization and related applications. J Big Data 6(1). https://doi.org/10.1186/s40537-019-0221-4

Bifulco F, Tregua M, Amitrano C, D'Auria A (2016) ICT and sustainability in smart cities management. Int J Public Sect Manag 29(2):132–147. https://doi.org/10.1108/ijpsm-07-2015-0132

Blair A, Sanger D (2016) Climate change and watershed hydrology—heavier precipitation influence on stormwater runoff. Geosciences 6(3):34. https://doi.org/10.3390/geosciences6030034

Burki U, Dahlstrøm R (2017) Mediating effects of green innovations on interfirm cooperation. Australas Mark J (AMJ) 25(2):149–156. https://doi.org/10.1016/j.ausmj.2017.05.001

Chan F, Mitchell G, Cheng X, Adekola O, McDonald A (2013) Developing a sustainable flood risk appraisal (SFRA) framework for the Pearl River Delta. Environ Urban ASIA 4(2):301–323. https://doi.org/10.1177/0975425313510770

Chapman S, Thatcher M, Salazar Á, Watson J, McAlpine C (2019) The impact of climate change and urban growth on urban climate and heat stress in a subtropical city. Int J Climatol 39(6):3013–3030. https://doi.org/10.1002/joc.5998

Chen H, Su K, Peng L, Bi G, Zhou L, Yang Q (2022) Mixed land use levels in rural settlements and their influencing factors: a case study of Pingba Village in Chongqing, China. Int J Environ Res Public Health 19(10):5845. https://doi.org/10.3390/ijerph19105845

Coutinho P, Silva N, Fulgêncio C, Cañas A, Alcobia S (2018) Environmental management practices for the sustainable development of a new city in Mozambique. Braz J Oper Prod Manag 15(1):112–126. https://doi.org/10.14488/bjopm.2018.v15.n1.a10

D'Auria A, Tregua M, Martos M (2018) Modern conceptions of cities as smart and sustainable and their commonalities. Sustainability 10(8):2642. https://doi.org/10.3390/su10082642

Daniels B, Jedamski J, Ottermanns R, Roß-Nickoll M (2020) A "plan bee" for cities: pollinator diversity and plant-pollinator interactions in urban green spaces. PLoS One 15(7):e0235492. https://doi.org/10.1371/journal.pone.0235492

Deakin M (2014) Smart cities: the state-of-the-art and governance challenges. Triple Helix J 1(1). https://doi.org/10.1186/s40604-014-0007-9

Dou X, Wang J, He Y (2014) Urban green planning and sustainable development: a case of China. Appl Mech Mater 522–524:1622–1626. https://doi.org/10.4028/www.scientific.net/amm.522-524.1622

Dur F, Yiğitcanlar T (2014) Assessing land-use and transport integration via a spatial composite indexing model. Int J Environ Sci Technol 12(3):803–816. https://doi.org/10.1007/s13762-013-0476-9

Dushkova D, Haase D (2020) Not simply green: nature-based solutions as a concept and practical approach for sustainability studies and planning agendas in cities. Land 9(1):19. https://doi.org/10.3390/land9010019

Enoguanbhor E, Gollnow F, Nielsen J, Lakes T, Walker B (2019) Land cover change in the Abuja city-region, Nigeria: integrating GIS and remotely sensed data to support land use planning. Sustainability 11(5):1313. https://doi.org/10.3390/su11051313

Fistola R (2023) Ecosystem services for the city as a complex system: a methodological proposal. Sustainability 15(12):9318. https://doi.org/10.3390/su15129318

Fossheim K, Andersen J (2017) Plan for sustainable urban logistics—comparing between Scandinavian and UK practices. Eur Transp Res Rev 9(4). https://doi.org/10.1007/s12544-017-0270-8

Gandhi S, Phatak V (2016) Land-based financing in metropolitan cities in India: the case of Hyderabad and Mumbai. Urbanisation 1(1):31–52. https://doi.org/10.1177/2455747116644106

Gao H, Yuan P, Liu R, Han L, Song Y (2019) Nationwide assessment of urban surface water environment status in China. E3s Web Conf 81:01003. https://doi.org/10.1051/e3sconf/201981 01003

Ghahramanpouri A, Lamit H, Sedaghatnia S (2013) Urban social sustainability trends in the research literature. Asian Soc Sci 9(4). https://doi.org/10.5539/ass.v9n4p185

Ghofrani Z, Sposito V, Faggian R (2020) Maximising the value of natural capital in a changing climate through the integration of blue-green infrastructure. J Sustain Dev Energy Water Environ Syst 8(1):213–234. https://doi.org/10.13044/j.sdewes.d7.0279

Guerry AD, Polasky S, Lubchenco J, Chaplin-Kramer R, Daily GC, Griffin R, Ruckelshaus M, Bateman IJ, Duraiappah A, Elmqvist T, Feldman MW (2015) Natural capital and ecosystem services informing decisions: from promise to practice. Proc Natl Acad Sci 112(24):7348–7355. https://doi.org/10.1073/pnas.1503751112

Haqi F (2016) Sustainable urban development and social sustainability in the urban context. Emara Indones J Arch 2(1):21–26. https://doi.org/10.29080/eija.v2i1.15

Hardoy J, Ruete R (2013) Incorporating climate change adaptation into planning for a liveable city in Rosario, Argentina. Environ Urban 25(2):339–360. https://doi.org/10.1177/095624781349 3232

He C, Zhao Y, Huang Q, Zhang Q, Zhang D (2015) Alternative future analysis for assessing the potential impact of climate change on urban landscape dynamics. Sci Total Environ 532:48–60. https://doi.org/10.1016/j.scitotenv.2015.05.103

He C, Zhou L, Ma W, Wang Y (2019) Spatial assessment of urban climate change vulnerability during different urbanization phases. Sustainability 11(8):2406. https://doi.org/10.3390/su11082406

Hergül Ö, Göker P (2021) Eko kentleri insan doğa etkileşimi içerisinde sürdürülebilir bir bakış açısıyla değerlendirmek. Eur J Sci Technol. https://doi.org/10.31590/ejosat.822062

Hu Y, Dai X, Zhao L (2022) Digital finance, environmental regulation, and green technology innovation: an empirical study of 278 cities in China. Sustainability 14(14):8652. https://doi.org/10.3390/su14148652

Huang S, Hsieh H (2014) The study of the relationship between accessibility and mixed land use in Tainan, Taiwan. Int J Environ Sci Dev 5(4):352–356. https://doi.org/10.7763/ijesd.2014.v5.508

Iyiola AO, Afolabi OA, Alimi SK, Akingba OO, Izah SC, Ogwu MC (2024) Climate change and water crisis in the global south. In: Izah SC, Ogwu MC, Loukas A, Hamidifar H (eds) Water crises and sustainable management in the global south. Springer, Singapore. https://doi.org/10.1007/978-981-97-4966-9_4

Iyiola AO, Izah SC, Morya S, Akinsorotan A, Ogwu MC (2023) Owalla reservoir in south-western Nigeria: assessment of fish distribution, biological diversity, and water quality index. Indonesian Journal of Agricultural Research 6(02):106–125. https://doi.org/10.32734/injar.v6i2.12207

Kang N, Kim S, Kim Y, Noh H, Hong S, Kim H (2016) Urban drainage system improvement for climate change adaptation. Water 8(7):268. https://doi.org/10.3390/w8070268

Kencono D (2023) The use of CCTV model analysis for a smart environment in Semarang City. E3s Web Conf 440:07006. https://doi.org/10.1051/e3sconf/202344007006

Kumar S (2024) The impact of urban planning on the creation and preservation of green spaces for biodiversity in metropolitan areas. Int J Res Publ Rev 5(1):4654–4658. https://doi.org/10.55248/gengpi.5.0124.0339

Li F (2023) Structural diffusion model and urban green innovation efficiency—a hybrid study based on DEA-SBM, NCA, and FSQCA. Sustainability 15(17):12705. https://doi.org/10.3390/su151712705

Li G, Zhang R, Feng S, Wang Y (2022) Digital finance and sustainable development: evidence from environmental inequality in China. Bus Strateg Environ 31(7):3574–3594. https://doi.org/10.1002/bse.3105

Lundgren K, Kjellström T (2013) Sustainability challenges from climate change and air conditioning use in urban areas. Sustainability 5(7):3116–3128. https://doi.org/10.3390/su5073116

Meijer A, Thaens M (2016) Urban technological innovation: developing and testing a sociotechnical framework for studying smart city projects. Urban Aff Rev 54(2):363–387. https://doi.org/10.1177/1078087416670274

Meyer N, Auriacombe C (2019) Good urban governance and city resilience: an Afrocentric approach to sustainable development. Sustainability 11(19):5514. https://doi.org/10.3390/su11195514

Mishra V, Lettenmaier D (2011) Climatic trends in major U.S. urban areas, 1950–2009. Geophys Res Lett 38(16). https://doi.org/10.1029/2011gl048255

Molavi S, Muttil N, Tran H (2011) Urban flood modelling and climate change: a Melbourne area case study. https://doi.org/10.36334/modsim.2011.i6.molavi

Navarra D, Milio S (2015) Sustainable urban development, governance and policy: a comparative overview of EU policies and projects, 393–415. https://doi.org/10.1007/978-3-319-12784-2_18

Ncube S, Arthur S (2021) Influence of blue-green and grey infrastructure combinations on natural and human-derived capital in urban drainage planning. Sustainability 13(5):2571. https://doi.org/10.3390/su13052571

Nie X, Wu J, Li L, Huang C, Li W, Wei Z (2022) Booster or stumbling block? The role of environmental regulation in the coupling path of regional innovation under the Porter hypothesis. Sustainability 14(5):2876. https://doi.org/10.3390/su14052876

Nuutinen M, Lappalainen I (2022) Services for sustainable future lifestyle in smart blocks—empowering people to make a difference. IOP Conf Ser Earth Environ Sci 1122(1):012013. https://doi.org/10.1088/1755-1315/1122/1/012013

Ogwu MC, Kosoe EA (2024) Place of cultural diversity in sustainable water resource management in Ghana. In: Izah SC, Ogwu MC, Loukas A, Hamidifar H (eds) Water crises and sustainable management in the global south. Springer, Singapore. https://doi.org/10.1007/978-981-97-4966-9_14

Ogwu MC, Chibuogwu AC, Osawaru ME (2016a) Trees in the service of man: University of Benin flora as a case study. UNILAG J Med Sci Technol 4(2):71–87

Ogwu MC, Osawaru ME, Obayuwana KO (2016b) Diversity and abundance of tree species in the University of Benin, Benin City, Nigeria. Appl Trop Agric 21(3):46–54

Ogwu MC (2019) Towards sustainable development in Africa: the challenge of urbanization and climate change adaptation. In: Cobbinah PB, Addaney M (eds) The geography of climate change adaptation in Urban Africa. Springer Nature, Switzerland, pp 29–55. https://doi.org/10.1007/978-3-030-04873-0_2

Ogwu MC, Pamela BO, Dare AF, Ovuru KF, Iyiola AO (2024) Traditional and conventional water treatment methods: a sustainable approach. In: Izah SC, Ogwu MC, Loukas A, Hamidifar H (eds) Water crises and sustainable management in the global south. Springer, Singapore. https://doi.org/10.1007/978-981-97-4966-9_15

Pan Y, Ma L, Tang H, Wu Y, Yang Z (2021) Land use transitions under rapid urbanization in Chengdu-Chongqing Region: a perspective of coupling water and land resources. Land 10(8):812. https://doi.org/10.3390/land10080812

Parjanen S, Hyypiä M, Martikainen S, Hennala L (2018) Elements of socially sustainable innovation processes in Finnish urban development. Sustain Dev 27(3):281–288. https://doi.org/10.1002/sd.1886

Park Y, Kim C, Choi M, Do Y (2020) A comparative study of carabid beetles in green spaces and former natural habitats. Diversity 12(12):479. https://doi.org/10.3390/d12120479

Pasquini L, Petrik D, Nyamakura B, Strachan K, Spires M, Shackleton S, Ziervogel G (2021) Effective collaborative climate change governance in urban areas. https://doi.org/10.4337/9781800883666.00030

Pearson D (2021) Lifestyle properties, ecosystem services, and biodiversity protection in Peri-Urban Aotearoa-New Zealand: a case study from Peri-Urban Palmerston North. Land 10(12):1345. https://doi.org/10.3390/land10121345

Peña D, Perrotti D, Mohareb E (2022) Advancing urban metabolism studies through GIS data: resource flows, open space networks, and vulnerable communities in Mexico City. J Ind Ecol 26(4):1333–1349. https://doi.org/10.1111/jiec.13261

Salati M, Bragança L, Mateus R (2022) Sustainability assessment on an urban scale: context, challenges, and most relevant indicators. Appl Syst Innov 5(2):41. https://doi.org/10.3390/asi5020041

Schebek L, Lützkendorf T (2022) Assessing resource efficiency of city neighbourhoods: a methodological framework for structuring and practical application of indicators in urban planning. Sustainability 14(13):7951. https://doi.org/10.3390/su14137951

Schinkel U, Becker N, Trapp M, Speck M (2022) Assessing the contribution of innovative technologies to sustainable development for planning and decision-making processes: a set of indicators to describe the performance of sustainable urban infrastructures (ISI). Sustainability 14(4):1966. https://doi.org/10.3390/su14041966

Seong E, Lee N, Choi C (2021) Relationship between land use mix and walking choice in high-density cities: a review of walking in Seoul, South Korea. Sustainability 13(2):810. https://doi.org/10.3390/su13020810

Shawabkeh R (2018) New urban regeneration for the City of Amman using sustainable urban design principles: Al-Abdali area as a case study. Int J Appl Eng Res 13(19):14345. https://doi.org/10.37622/ijaer/13.19.2018.14345-14358

Shekfa L, Ahmed K (2022) The other side of the urban densification 'coin': impacts on social sustainability in redeveloped urban sprawled communities in united arab emirates. Futur Cities Environ 8(1). https://doi.org/10.5334/fce.155

Stagoll K, Lindenmayer D, Knight E, Fischer J, Manning A (2012) Large trees are keystone structures in urban parks. Conserv Lett 5(2):115–122. https://doi.org/10.1111/j.1755-263x.2011.00216.x

Steenkamp J, Cilliers E, Cilliers S, Lategan L (2021) Food for thought: addressing urban food security risks through urban agriculture. Sustainability 13(3):1267. https://doi.org/10.3390/su13031267

Suek J (2024) Enhancing workers' quality of life: why land use mix should be increased? (case study: Semarang City). IoP Conf Ser Earth Environ Sci 1318(1):012019. https://doi.org/10.1088/1755-1315/1318/1/012019

Tang H, Lee Y (2016) The making of sustainable urban development: a synthesis framework. Sustainability 8(5):492. https://doi.org/10.3390/su8050492

Thomson G, Newman P (2018) Cities and the anthropocene: urban governance for the new era of regenerative cities. Urban Stud 57(7):1502–1519. https://doi.org/10.1177/0042098018779769

Tian F, Hou S (2022) The impact of green finance on industrial land use efficiency: evidence from 279 cities in China. Sustainability 14(10):6184. https://doi.org/10.3390/su14106184

Tung A, Baird K, Schoch H (2014) The association between the adoption of an environmental management system with organisational environmental performance. Australas J Environ Manag 21(3):281–296. https://doi.org/10.1080/14486563.2014.932721

Wei J, Qian J, Yu T, Feng H, Ou W (2018) Evaluating spatial priority of urban green infrastructure for urban sustainability in areas of rapid urbanization: a case study of Pukou in China. Sustainability 10(2):327. https://doi.org/10.3390/su10020327

Wiengarten F, Humphreys P, Onofrei G, Fynes B (2016) The adoption of multiple certification standards: perceived performance implications of quality, environmental and health & safety certifications. Prod Plan Control 28(2):131–141. https://doi.org/10.1080/09537287.2016.1239847

Wu Y, Fan P, Li B, Ouyang Z, Liu Y, You H (2017) The effectiveness of planning control on urban growth: evidence from Hangzhou, China. Sustainability 9(5):855. https://doi.org/10.3390/su9050855

Yang H, Fu M, Wang L, Tang F (2021) Mixed land use evaluation and its impact on housing prices in Beijing based on multi-source big data. Land 10(10):1103. https://doi.org/10.3390/land10101103

Yang Y, Dong Z, Zhou B-B, Liu Y (2024) Smart growth and smart shrinkage: a comparative review for advancing urban sustainability. Land 13(5):660. https://doi.org/10.3390/land13050660

Yiğitcanlar T (2020) Smart city beyond efficiency: technology–policy–community at play for sustainable urban futures. Hous Policy Debate 31(1):88–92. https://doi.org/10.1080/10511482.2020.1846885

Zakari A, Khan I (2022) The introduction of green finance: a curse or a benefit to environmental sustainability? Energy Res Lett 3(3). https://doi.org/10.46557/001c.29977

Zhang X, Ramos B (2022) Urban sprawl and warming—research on the evolution of the urban sprawl of Chinese municipalities and its relationship with climate warming in the past three decades. Int Arch Photogramm Remote Sens Spat Inf Sci XLIII-B4-2022:209–215. https://doi.org/10.5194/isprs-archives-xliii-b4-2022-209-2022

Chapter 15
Environmental Economics: Valuing Ecosystem Services and Green Technologies

Matthew Chidozie Ogwu and Enoch Akwasi Kosoe

Abstract Environmental economics integrates ecological and economic perspectives to address sustainability challenges, focusing on valuing ecosystem services and advancing green technologies. This chapter delves into the theoretical underpinnings of environmental economics, emphasizing its role in assessing ecosystem services' economic value and informing sustainable policy decisions. It highlights ecosystem services' classification and valuation methods, including contingent valuation, cost-benefit analysis, and market-based approaches, supported by real-world case studies. Additionally, the chapter explores economic instruments for environmental management, such as carbon pricing, emission trading schemes, and subsidies for green technologies, showcasing their potential to align economic growth with ecological preservation. The discussion extends to green technologies as pivotal drivers of economic sustainability. Examples include renewable energy, energy-efficient systems, and sustainable agricultural practices, all contributing to job creation, reduced operational costs, and resilience against climate change. Cost-benefit analyses underscore the long-term economic advantages of green innovations despite higher initial investments. Financing mechanisms like green bonds and public-private partnerships are also explored as enablers for adopting sustainable practices. Challenges such as equity, access, and limitations in valuation techniques are critically analyzed, emphasizing the need for integrating ecological considerations into economic frameworks. The chapter concludes with future directions, advocating for natural capital

M. C. Ogwu
Goodnight Family Department of Sustainable Development, Appalachian State University, Boone, NC, USA
e-mail: ogwumc@appstate.edu

E. A. Kosoe (✉)
Department of Environment and Resource Studies, Faculty of Integrated Development Studies, Simon Diedong Dombo University of Business and Integrated Development Studies, Wa, Upper West Region, Ghana
e-mail: ekosoe@ubids.edu.gh

M. C. Ogwu and S. Chibueze Izah (eds.), *Evaluating Environmental Processes and Technologies*, Environmental Science and Engineering,
https://doi.org/10.1007/978-3-031-85327-2_15

accounting, interdisciplinary approaches, and innovative green financing. By synthesizing economic and ecological insights, the chapter aims to provide actionable strategies for achieving sustainable development while addressing pressing environmental and economic issues.

Keywords Environmental economics · Ecosystem services · Green technologies · Carbon pricing · Sustainable development · Economic instruments · Natural capital accounting · Cost-benefit analysis · Green financing

15.1 Introduction

Environmental economics is a subfield of economics that focuses on the relationship between economic systems and the environment. It seeks to understand how economic activities impact the environment and how environmental policies can be designed to promote sustainable development. The scope of ecological economics encompasses the valuation of ecosystem services, the analysis of trade-offs between economic development and environmental conservation, and the evaluation of green technologies to reduce ecological footprints. Environmental economics studies how economic activities affect the environment and how economic principles can be applied to environmental issues. It involves assessing environmental costs and benefits, valuing natural resources, and designing policies that promote sustainable resource management. The discipline emphasizes the importance of integrating ecological considerations into economic decision-making, recognizing that natural capital is essential for long-term economic viability (Obst et al. 2020). The scope of environmental economics extends to various areas, including analyzing market failures related to environmental degradation, implementing regulatory frameworks, and exploring innovative financial instruments for environmental protection. For instance, the System of Environmental-Economic Accounting (SEEA) has been developed to provide a comprehensive framework for measuring the stocks and flows of natural capital, thereby supporting informed policymaking (Obst et al. 2020). Furthermore, integrating environmental management accounting with corporate strategies is crucial for enhancing environmental and economic performance (Christine et al. 2019).

Economics plays a pivotal role in addressing environmental challenges by providing tools and frameworks for understanding the trade-offs involved in resource allocation. The relationship between economic growth and ecological sustainability is complex. In contrast, economic development often leads to increased resource consumption and environmental degradation; it can also provide the means for investment in cleaner technologies and sustainable practices (Wang et al. 2011). For example, empirical research indicates that investments in environmental protection can yield long-term economic benefits, as seen in the case of Shanghai, where environmental investments have been linked to economic growth (Wang et al. 2011).

Moreover, the concept of trade-offs is central to environmental economics. Policymakers must often navigate the competing demands of economic development and ecological conservation. This is particularly evident when resource extraction and agricultural land use conflict with biodiversity preservation and ecosystem services (Pagani-Núñez et al. 2022; Zheng et al. 2019). Recognizing these trade-offs allows for developing strategies to balance economic objectives with ecological integrity, thereby fostering sustainable development (Fidelman et al. 2012).

The trade-offs between development and conservation are a critical aspect of environmental economics. As Fletcher (2012) highlighted, neoliberal conservation approaches often exacerbate inequalities, as the competition for scarce resources can lead to conflicts between environmental protection and poverty alleviation. This dynamic is particularly pronounced in regions where marginalized communities rely on natural resources for their livelihoods, creating a tension between conservation efforts and economic necessity (Fidelman et al. 2012; Gill et al. 2019). Marine conservation, for instance, the allocation of resource rights and the capacity of local communities to benefit from conservation initiatives are key determinants of the success of these programs (Gill et al. 2019; Iyiola et al. 2023, 2024a, b, c). Effective governance that acknowledges and addresses these trade-offs is essential for achieving both conservation goals and socio-economic development (Fidelman et al. 2012; Brooks 2016). Furthermore, integrating community-based conservation approaches has shown promise in reconciling these competing interests by involving local stakeholders in decision-making processes (Brooks 2016; Fortnam 2023).

This chapter explores the role of environmental economics in valuing ecosystem services and assessing the economic feasibility of green technologies, to promote sustainable development. It seeks to equip readers with the knowledge and tools to analyze natural resources' monetary and non-monetary value, evaluate the effectiveness of market-based instruments and policy incentives, and critically assess the cost-benefit dynamics of adopting sustainable technologies. The chapter will contribute to understanding how economic principles can guide environmental management, conservation strategies, and the transition to a green economy while addressing equity, access, and ecological preservation challenges.

15.2 Valuation of Ecosystem Services

Ecosystem service valuation has emerged as a critical area of study within environmental economics. It provides a framework for understanding the economic benefits derived from natural ecosystems. This valuation is essential for informing policy decisions, guiding resource management, and promoting sustainable development. The following sections will explore the types of ecosystem services, methods for their valuation, and relevant case studies that illustrate these concepts.

Table 15.1 Types of ecosystem services

Type of service	Associated benefits	Examples
Provisioning services	Tangible goods are provided by ecosystems that directly benefit humans	Food (crops, fish), fresh water, timber, medicinal plants
Regulating services	Benefits obtained from the regulation of ecosystem processes contribute to environmental stability	Climate regulation, air quality maintenance, water purification, erosion control, and pollination
Cultural services	Non-material benefits are derived from ecosystems that enhance human well-being and cultural enrichment	Recreation, spiritual significance, aesthetic appreciation, educational value
Supporting services	Fundamental ecosystem processes are necessary to produce all other services	Soil formation, nutrient cycling, primary production, habitat provision

15.2.1 Types of Ecosystem Services

Ecosystem services are typically categorized into four main types: provisioning, regulating, cultural, and supporting services (Table 15.1). Provisioning services refer to the tangible products obtained from ecosystems, such as food, fresh water, timber, and fibre (Pandeya et al. 2016). Regulating services encompass the benefits of ecosystem processes that regulate environmental conditions, including climate regulation, water purification, and pollination (Sharma et al. 2021). Cultural services provide non-material benefits, such as recreational opportunities, aesthetic enjoyment, and spiritual enrichment (Arias-Arévalo et al. 2018). Finally, supporting services are necessary to produce all other ecosystem services, including soil formation, nutrient cycling, and primary production (Sharma et al. 2021). The classification of ecosystem services is crucial for understanding their multifaceted roles in human well-being and ecological health. For instance, Yang et al. (2022a, b) highlight the importance of grassland ecosystems in providing both provisioning and regulating services, emphasizing their role in maintaining biodiversity and supporting livelihoods in the Yellow River Basin. This classification not only aids in assessing ecosystem service values but also informs conservation strategies by identifying which services are most at risk from human activities (Pandeya et al. 2016).

15.2.2 Methods for Valuing Ecosystem Services

Valuing ecosystem services involves various methodologies, each with its strengths and limitations. Figure 15.1 provides a framework for understanding ecosystem services valuation, distinguishing between utilitarian and non-utilitarian approaches. The utilitarian approach emphasizes tangible and intangible benefits derived from

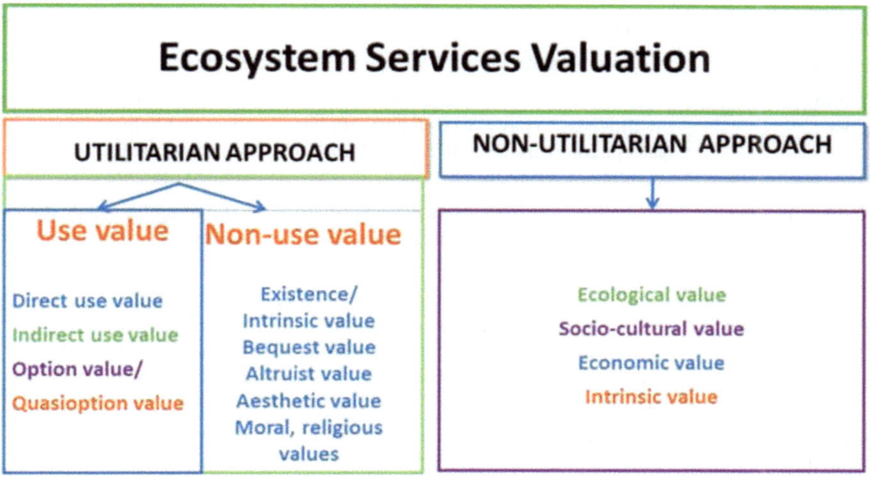

Fig. 15.1 Ecosystem valuation approaches based on utilization approaches. *Source* Baciu et al. (2021)

ecosystems, categorized into use values (e.g., direct benefits like resources, indirect benefits like ecosystem functions and future options) and non-use values (e.g., intrinsic, bequest, aesthetic, and moral values associated with ecosystem preservation). On the other hand, the non-utilitarian approach focuses on holistic aspects such as ecological, socio-cultural, economic, and intrinsic values, reflecting the broader significance of ecosystems beyond human use. Together, these approaches highlight the diverse perspectives necessary for valuing ecosystem services in environmental decision-making.

In addition to the utilitarian approach to valuing ecosystem services, the primary methods include contingent valuation, cost-benefit analysis, and market-based approaches (Table 15.2).

15.2.2.1 Contingent Valuation

Contingent valuation (CV) is a survey-based economic technique used to estimate the value of non-market resources by asking individuals their willingness to pay (WTP) for specific ecosystem services (Pandeya et al. 2016). This method is particularly useful for valuing services that do not have a direct market price, such as clean air or biodiversity conservation. For example, Guo (2023) discusses the application of CV in assessing marine ecosystem services in China, highlighting its potential to inform policy decisions regarding marine conservation. However, CV is often criticized for its reliance on hypothetical scenarios, which may lead to biases in respondents' answers (Pandeya et al. 2016).

Table 15.2 Methods for valuing ecosystem services

Method	Estimation method	Examples of application
Contingent valuation (CV)	It uses surveys to estimate individuals' willingness to pay (WTP) for benefits or accept compensation (WTA) for losses of ecosystem services	Valuing clean water sources, biodiversity conservation efforts, or scenic landscapes
Hedonic pricing	Examines how the presence or quality of ecosystem services affects market prices, such as property values or wages	Estimating the value of air quality, noise reduction, or proximity to parks and green spaces
Travel cost method (TCM)	It uses the travel expenses incurred by individuals visiting a site to estimate the value of recreational and cultural ecosystem services	Valuing tourism and recreation at national parks, beaches, or wildlife reserves
Replacement cost method	Estimates the costs of replacing natural ecosystem services with artificial or engineered alternatives	Valuing wetlands for flood protection or forests for their role in carbon sequestration
Market-based approaches	It relies on the market prices of goods and services directly derived from ecosystems	Pricing timber, fishery products, or crops cultivated using natural ecosystem functions
Cost-benefit analysis (CBA)	Weighs the costs of conservation, restoration, or environmental protection against the quantifiable benefits provided by ecosystem services	Assessing the economic impact of wetland restoration, forest conservation, or pollution abatement projects
Benefit transfer method	Applies valuation data from similar ecosystems or contexts to a new, unstudied area or situation	Estimating the value of biodiversity conservation in a new forest using data from comparable areas
Production function approach	This linkage directly connects ecosystem services to their contribution to economic or production processes, quantifying their value in supporting human activity	Valuing pollination services for agricultural productivity or soil fertility for crop yields

15.2.2.2 Cost-Benefit Analysis

Cost-benefit analysis (CBA) is another widely used method for valuing ecosystem services. It involves comparing the total expected costs of a project or policy against its total expected benefits, expressed in monetary terms. This approach systematically evaluates trade-offs associated with different management options (Pandeya et al. 2016). For instance, Sharma et al. (2021) emphasize the importance of integrating ecological and economic perspectives in CBA to ensure that ecosystem services are adequately represented in decision-making processes. However, CBA can be limited by difficulties in quantifying certain ecosystem services and the potential for double counting (Fu et al. 2010).

15.2.2.3 Market-Based Approaches

Market-based approaches to ecosystem service valuation involve using market mechanisms to assign economic value to ecosystem services. This includes establishing markets for ecosystem services, such as carbon trading or payments for ecosystem services (PES) schemes (Pandeya et al. 2016). These approaches aim to create financial incentives for conservation and sustainable management practices. For example, Milon and Álvarez (2019) discuss the role of PES in coastal resource management, highlighting its effectiveness in promoting conservation while providing economic benefits to local communities. However, market-based approaches may also face challenges related to equity and access, particularly for marginalized communities (Pandeya et al. 2016).

Numerous case studies illustrate the application of ecosystem service valuation methods in real-world contexts. For instance, the study by Keyue et al. (2019) examines the impact of land use changes on ecosystem service values in the Shangzhou District, demonstrating how shifts in land management practices can significantly alter the value of ecosystem services. This case highlights the dynamic nature of ecosystem service valuation and the need for ongoing assessment as environmental conditions change. Another notable example is the valuation of marine ecosystem services in South Africa, as discussed by Musetsho et al. (2022). The study utilized the Costing Nature platform to assess the economic value of ecosystem services in a critical biodiversity area, providing insights into the spatial distribution of benefits and informing conservation planning (Musetsho et al. 2022). Such case studies underscore the importance of context-specific approaches to ecosystem service valuation, as local conditions and stakeholder needs can greatly influence the outcomes of valuation efforts (Pandeya et al. 2016).

15.3 Economic Instruments for Environmental Management

Economic instruments for environmental management are essential tools that governments and organizations utilize to address environmental challenges while promoting sustainable development. These instruments can take various forms, including market-based tools, conservation incentives, and green technologies subsidies. Market-based tools are designed to harness market forces to achieve environmental objectives. They include carbon pricing, emission trading schemes, and pollution taxes, which aim to internalize the external costs of environmental degradation.

15.3.1 Carbon Pricing

Carbon pricing is a market-based approach that assigns a cost to carbon emissions, thereby incentivizing reductions in greenhouse gas emissions. This can be implemented through carbon taxes or cap-and-trade systems. The effectiveness of carbon pricing in reducing emissions has been widely studied, with evidence suggesting that it can significantly reduce carbon outputs when appropriately designed (Zhu et al. (2021; Santis and Lasinio 2016). For instance, implementing carbon pricing in various jurisdictions has demonstrated its potential to stimulate green innovation and transition towards low-carbon economies (He 2023). However, challenges remain in ensuring that carbon prices are set at levels that effectively drive behavioural change among businesses and consumers (Yang et al. 2022a, b).

15.3.2 Emission Trading Schemes

Emission trading schemes (ETS) allow companies to buy and sell allowances for emissions, creating a market for pollution rights. This flexibility enables firms to find the most cost-effective ways to reduce emissions. The European Union Emission Trading System (EU ETS) is one of the most prominent examples, which has been credited with reducing emissions in the power sector (Lin and Zhao 2023). However, the effectiveness of ETS can be influenced by factors such as the initial allocation of allowances and market volatility, which can undermine the intended environmental benefits (Yin et al. 2022). Studies have shown that while ETS can lead to emissions reductions, their success often hinges on robust regulatory frameworks and market stability (Chen et al. 2022).

15.3.3 Pollution Taxes

Pollution taxes impose a fee on the emission of pollutants, thereby providing a financial incentive for firms to reduce their emissions. This approach is effective in various contexts, as it encourages companies to innovate and adopt cleaner technologies (Guo and Фy 2021). For example, research indicates that pollution taxes can improve environmental performance in manufacturing sectors by promoting investments in cleaner production methods (Liu et al. 2022). However, implementing pollution taxes must consider equity implications, as they can disproportionately affect lower-income households unless mitigated by compensatory measures (Pan et al. 2022).

Incentives for conservation are crucial for promoting sustainable practices and protecting natural resources. One prominent mechanism is Payment for Ecosystem Services (PES) programs. PES programs provide financial incentives to landowners and resource managers for maintaining or enhancing ecosystem services. These programs have gained traction globally as a means to align economic interests with environmental conservation (Alesina and Passarelli 2014). For instance, PES has been effectively used in forest conservation efforts, where landowners are compensated for preserving forested areas that provide critical ecosystem services such as carbon sequestration and biodiversity conservation (Wang et al. 2022a, b). The success of PES programs often depends on clear definitions of ecosystem services, effective monitoring, and the engagement of local communities (Lim and Prakash 2014). However, challenges such as ensuring equitable benefit distribution and addressing potential market failures remain significant (Sikdar and Lapan 2011).

Subsidies and tax breaks for green technologies are essential tools for promoting and adopting environmentally friendly practices and innovations. These financial incentives can lower the cost barriers associated with transitioning to sustainable technologies. Subsidies for renewable energy technologies, such as solar and wind power, have been shown to increase their adoption rates (Yamazaki 2022) significantly. Research indicates that such subsidies can lead to substantial reductions in greenhouse gas emissions while fostering economic growth in the renewable energy sector (Wang et al. 2022a, b). Similarly, tax breaks for energy-efficient appliances and electric vehicles have effectively encouraged consumers to make greener choices (Yi et al. 2019). However, the design of these subsidies must be carefully considered to avoid unintended consequences, such as market distortions or dependency on government support.

15.4 Green Technologies and Economic Sustainability

Integrating green technologies into economic systems is increasingly essential for sustainable development. Green technologies, particularly renewable energy sources and sustainable practices, offer numerous economic benefits, contribute to environmental protection, and foster resilience against climate change (Table 15.3). The transition to renewable energy sources such as solar, wind, hydroelectric, and geothermal power presents significant economic advantages. One primary benefit is reducing greenhouse gas emissions, which is crucial for mitigating climate change. By decreasing reliance on fossil fuels, renewable energy can help stabilize energy prices and reduce the economic volatility associated with fluctuating oil and gas markets Eze (2023). This stability is particularly important for developing economies that are vulnerable to external shocks in energy prices.

Moreover, the renewable energy sector has proven to be a substantial job creator. According to the International Renewable Energy Agency (IRENA), the renewable energy sector employed over 11 million people globally in 2018, with projections indicating continued growth as investments in green technologies increase (Chou 2023). This job creation spans various sectors, including manufacturing, installation, and maintenance of renewable energy systems, contributing to local economies and enhancing energy security (Azeakpono and Lloyd 2020). In addition to job creation,

Table 15.3 Economic instruments used for environmental management

Instrument	Implementation strategies	Examples of application	Advantages	Challenges
Carbon pricing	Charges a fee per unit of greenhouse gas emissions to encourage a reduction in carbon output	Carbon taxes on industries, fees for emissions, or cap-and-trade programs	Provides a direct incentive to reduce emissions and generates revenue for green initiatives	Difficult to set an appropriate carbon price; potential resistance from industries and consumers
Emission trading schemes (ETS)	Allows organizations to trade emission allowances under a capped limit to reduce overall emissions	The European Union Emissions Trading System or regional carbon markets	Creates flexibility in meeting emission reduction targets; incentivizes innovation in reducing emissions	Requires effective regulation to prevent manipulation and risk of creating regional disparities
Pollution taxes	Imposes taxes on pollutants to incentivize reducing waste and adopting cleaner technologies	Taxes on sulfur dioxide emissions, plastic production, or industrial wastewater discharges	Encouraging polluters to adopt cleaner technologies can reduce pollution levels significantly	It can increase costs for businesses and consumers and requires careful monitoring to prevent evasion
Subsidies for green technologies	Financial incentives are provided to support adopting renewable energy, energy efficiency, or sustainable practices	Subsidies for solar panels, wind farms, or energy-efficient appliances	Accelerates the adoption of clean technologies; lowers upfront costs for individuals and businesses	Risk of over-reliance on subsidies; may lead to market distortions if poorly implemented
Payment for ecosystem services (PES)	Compensates landowners or communities for maintaining or enhancing ecosystem services	Programs pay farmers to preserve forests, wetlands, or biodiversity	Provides financial support for conservation; promotes sustainable land management	Requires accurate valuation of ecosystem services; may exclude marginalized communities
Resource pricing and royalties	Implements fees or royalties on resource extraction to reflect environmental costs	Fees for water usage, mining royalties, or charges on logging activities	Ensures that the costs of resource use are reflected in prices; generates revenue for conservation	This may lead to higher costs for end users; and requires enforcement to prevent illegal exploitation

(continued)

Table 15.3 (continued)

Instrument	Implementation strategies	Examples of application	Advantages	Challenges
Tradable resource quotas	Allocates fixed quotas for resource use or harvesting, which can be traded in markets	Fisheries quotas for sustainable harvests or tradable water rights in arid regions	Encourages sustainable resource management; provides economic incentives for efficient use	This may result in overharvesting if quotas are not properly set; and requires strong regulatory frameworks
Green bonds and sustainable financing	Funds are raised through bonds dedicated to environmental projects and sustainability efforts	Financing for renewable energy projects, afforestation initiatives, or clean transportation	Attracts investment for green projects; promotes long-term sustainability	Requires transparency and accountability to ensure funds are used effectively
Deposit-refund systems	Requires a refundable deposit to encourage recycling or proper disposal of waste products	Refund systems for plastic bottles, aluminium cans, or electronic waste	Increases recycling rates; reduces waste in landfills and the environment	Requires infrastructure to collect and refund deposits; may not cover all waste streams
Eco-labelling and certification	Labels and certifications indicating environmentally friendly products to encourage sustainable consumption	Certifications like Fair Trade, Forest Stewardship Council (FSC), or Energy Star	Educates consumers about sustainable choices; creates market demand for eco-friendly products	Can be costly for producers to obtain certification; potential for greenwashing if standards are weak
Tax breaks for environmental actions	Reductions in taxes to incentivize environmentally beneficial activities	Tax credits for electric vehicles, renewable energy installations, or reforestation efforts	Reduces financial barriers to sustainable practices; encourages investment in green initiatives	May reduce government revenue; and require proper targeting to avoid benefiting non-environmental projects
User fees	Charges are imposed on the use of environmental resources or services to discourage overuse	Park entrance fees, water usage fees, or tolls for roads in environmentally sensitive areas	Generates revenue for conservation; promotes sustainable use of resources	May restrict access for low-income groups; and requires enforcement and proper allocation of collected fees

renewable energy technologies can significantly save healthcare costs associated with air pollution. The World Health Organization (WHO) estimates that air pollution contributes to millions of premature deaths annually, with substantial economic costs related to healthcare and lost productivity (Inglesi-Lotz and Doğan 2018; Izah et al. 2023). Transitioning to cleaner energy sources can alleviate these burdens, resulting in healthier populations and reduced healthcare expenditures. Furthermore, sustainable practices, such as energy efficiency improvements and waste reduction, can enhance productivity and reduce business operational costs. For instance, companies that invest in energy-efficient technologies often experience lower energy bills and improved competitiveness (Singh et al. 2019). Sustainable agricultural practices can also increase yields and reduce input costs, contributing to rural food security and economic resilience (Li et al. 2023; Kosoe et al 2023; Egboduku et al. 2024).

Conducting a cost-benefit analysis (CBA) of green technologies compared to traditional technologies is crucial for understanding their economic viability. CBA involves quantifying the costs and benefits associated with each technology, allowing policymakers and stakeholders to make informed decisions. In many cases, the initial capital costs of renewable energy technologies can be higher than those of traditional fossil fuel systems. However, the long-term benefits often outweigh these initial investments. For example, a study by the International Energy Agency (IEA) found that the levelized cost of electricity (LCOE) for solar and wind energy has decreased significantly over the past decade, making them increasingly competitive with coal and natural gas (Hanssen et al. 2018). Additionally, the operational costs of renewable energy systems are generally lower, as they do not require fuel purchases and have lower maintenance costs (Lian et al. 2018). When considering environmental benefits, the CBA for green technologies often includes the value of ecosystem services, such as carbon sequestration and improved air quality. These benefits are typically not accounted for in traditional energy systems, leading to underestimating the true costs of fossil fuel consumption (Inglesi-Lotz and Doğan 2018). For instance, the social cost of carbon, which estimates the economic damages associated with a metric ton of carbon dioxide emissions, can significantly alter the cost-benefit landscape in favour of renewable technologies (Iqbal 2023). Moreover, the CBA should also consider the potential for job creation and economic development associated with green technologies. Investments in renewable energy infrastructure can stimulate local economies, create jobs, and enhance energy independence, which are critical factors for long-term economic sustainability (Jia et al. 2023).

The successful adoption of green technologies often hinges on the availability of financing mechanisms that can support initial investments and reduce financial risks. Several innovative financing options have emerged to facilitate the transition to renewable energy and sustainable practices. Green bonds are fixed-income financial instruments specifically earmarked for funding projects that have positive environmental impacts, such as renewable energy installations and energy efficiency improvements. The market for green bonds has grown exponentially, with issuances reaching record levels in recent years (Shahbaz et al. 2020). These bonds allow investors to support sustainable projects while earning a return on their investment. The appeal of green bonds lies in their ability to attract a diverse range of investors, including institutional investors who are increasingly prioritizing environmental, social, and governance (ESG) criteria in their investment decisions (Çevik et al. 2020). Furthermore, green bonds can help lower the cost of capital for renewable energy projects, making them more financially viable and attractive to developers (Callaghan and Williams 2014).

Public-private partnerships (PPPs) are another effective financing mechanism for adopting green technology. These collaborations between government entities and private companies can leverage public resources and expertise to implement large-scale renewable energy projects. PPPs can facilitate the sharing of risks and rewards, making it easier for private investors to commit capital to green initiatives (Filho et al. 2022). For example, many countries have successfully implemented PPPs for solar and wind energy projects, increasing capacity and reducing costs. These partnerships can also foster innovation by combining the strengths of both sectors, leading to the development of cutting-edge technologies and solutions (Guðlaugsson et al. 2021). Sustainable investment funds focus on investing in companies and projects prioritizing environmental sustainability and social responsibility. These funds can provide essential capital for green technology projects, helping to bridge the financing gap that often exists in the renewable energy sector (Yang et al. 2012). The growth of sustainable investment funds reflects a broader trend toward responsible investing, with investors increasingly seeking to align their portfolios with their values. This shift has led to a surge in demand for green investments, prompting financial institutions to develop new products and services that cater to this market (Sun et al. 2010) (Table 15.4).

15.5 Challenges in Economic Environmental Valuation

Economic environmental valuation is critical for understanding the value of natural resources and ecosystem services. However, several challenges complicate the effective application of valuation techniques. Equity and access are fundamental concerns in the economic valuation of environmental resources. The valuation process often reflects specific demographic groups' preferences and willingness to pay, which can lead to inequitable outcomes. For instance, marginalized communities may have limited representation in valuation studies, resulting in their ecosystem service needs being undervalued or overlooked (Giménez and Mas 2020). This inequity can exacerbate existing social disparities, as decisions based on biased valuations may prioritize the interests of wealthier populations over those of disadvantaged groups (Jílková et al. 2010). Moreover, the concept of "access" in environmental valuation extends beyond economic factors to include social and cultural dimensions. Access to natural resources is often tied to cultural practices and community identities, which are not easily quantifiable in monetary terms (Matos et al. 2012). For example, indigenous communities may derive significant cultural value from ecosystems that cannot be captured through conventional economic valuation methods. This disconnect can lead to conflicts over resource management and conservation strategies, as decisions made

Table 15.4 Green technologies useful for driving economic sustainability

Technology	Description	Applications	Economic benefits	Challenges
Renewable energy	Harnesses natural energy sources like solar, wind, hydro, and geothermal to generate electricity	Solar panels, wind turbines, hydropower dams, geothermal power plants	Reduces energy costs in the long term; creates jobs in clean energy sectors; reduces dependence on fossil fuels	High initial investment costs; intermittency issues (e.g., solar and wind); requires supportive infrastructure
Energy efficiency technologies	Reduces energy consumption through advanced systems and practices	LED lighting, energy-efficient appliances, smart grid systems, and green buildings	Lowers operational costs; reduces greenhouse gas emissions; improves resource efficiency	Initial retrofitting costs require public awareness and skilled labour
Electric vehicles (EVs)	Uses electricity as a cleaner alternative to gasoline or diesel for transportation	Electric cars, buses, scooters, and charging infrastructure	Reduces fuel costs, decreases air pollution, and stimulates battery and vehicle technology innovation	Limited charging infrastructure; high battery cost, and environmental concerns about battery disposal
Green building technologies	Incorporates sustainable materials, energy-efficient systems, and water-saving designs	Passive cooling designs, solar-powered heating, rainwater harvesting, and green roofs	Reduces energy and water costs; increases property value; improves occupant health and productivity	Higher upfront costs; requires technical expertise; limited adoption in certain regions
Sustainable agriculture	Employs methods to enhance productivity while preserving the environment	Precision farming, drip irrigation, vertical farming, organic farming	Increases yield per acre; reduces water and fertilizer costs; preserves soil health and biodiversity	Access to technology may be limited for smallholder farmers and requires investment in training
Waste-to-energy technologies	Converts waste materials into energy, such as biogas or electricity	Anaerobic digesters, incineration with energy recovery, and landfill gas capture systems	Reduces landfill waste; generates additional income streams; creates renewable energy	High initial infrastructure costs; air pollution concerns with some methods (e.g., incineration)

(continued)

Table 15.4 (continued)

Technology	Description	Applications	Economic benefits	Challenges
Water purification and recycling	Utilizes advanced techniques to clean and recycle water for reuse	Reverse osmosis, UV treatment, desalination, and greywater recycling	Ensures a sustainable water supply; reduces water costs; supports agriculture and industry in water-scarce areas	High maintenance costs; energy-intensive processes; may require government subsidies
Carbon capture and storage (CCS)	Captures CO2 emissions from industrial processes and stores them underground to prevent release	Industrial plants, power stations, and bioenergy with carbon capture systems	Mitigates climate change; enables continued use of fossil fuels during transition periods	Expensive implementation; requires safe and secure storage sites; limited public acceptance
Bioenergy technologies	Converts organic materials into biofuels and bioproducts	Ethanol and biodiesel production, algae-based fuels, and biomass power plants	Creates renewable energy; supports rural economies through biomass supply chains; reduces carbon footprint	Competition with food crops for resources; requires investment in research and scaling
Circular economy innovations	Focuses on reusing, repairing, and recycling materials to minimize waste	Upcycling, industrial symbiosis, and recycling infrastructure	Reduces material costs; minimizes waste disposal costs; promotes job creation in recycling industries	Requires cultural shifts in consumption habits; limited infrastructure in some areas
Green hydrogen technology	Produces hydrogen from renewable energy sources for use in energy storage or fuel cells	Fuel cells for transport, industrial processes, and renewable energy storage solutions	Offers a clean fuel alternative; reduces industrial emissions; supports energy storage for renewables	High production costs; requires significant infrastructure investments; efficiency challenges
Sustainable forestry practices	Combines sustainable harvesting with forest conservation and management	Agroforestry, selective logging, and forest certification programs (e.g., FSC certification)	Preserves biodiversity; generates income from sustainably managed forests; supports carbon sequestration	Monitoring and enforcement of practices can be difficult; with potential conflicts with local communities

(continued)

Table 15.4 (continued)

Technology	Description	Applications	Economic benefits	Challenges
Smart grid technologies	Uses digital communication to optimize energy distribution and consumption	Real-time energy monitoring, demand-response systems, and energy storage integration	It improves energy efficiency, reduces transmission losses, and enables the integration of renewable energy sources	High deployment and maintenance costs; cybersecurity risks; requires regulatory and policy support
Biodegradable materials	Develops materials that break down naturally, reducing environmental pollution	Bioplastics, compostable packaging, and biodegradable textiles	Reduces landfill waste; decreases reliance on fossil fuel-based plastics; supports sustainable industries	Limited availability: higher production costs compared to conventional plastics
Desalination technologies	Removes salt and impurities from seawater to produce fresh water	Reverse osmosis desalination plants and solar desalination systems	Provides a sustainable water source in arid regions; supports agriculture and industry	Energy-intensive process; environmental concerns about brine disposal

without considering these cultural values may alienate local populations (Kapsalis et al. 2019). Additionally, the distribution of benefits derived from environmental resources can be uneven. Wealthier individuals or corporations may disproportionately benefit from ecosystem services, while poorer communities bear the brunt of environmental degradation ("Study on the Economic Valuation of Water Resources and the Allocation of Rights and Interests in Water Resources", 2023; Ogwu and Kosoe 2024; Ogwu et al. 2024). This disparity raises ethical questions about the fairness of environmental policies that rely on economic valuation, as they may perpetuate existing inequalities.

Current valuation techniques, while useful, have significant limitations that can undermine their effectiveness. One major limitation is the reliance on monetary metrics to quantify ecosystem services. Many valuation methods, such as contingent valuation and hedonic pricing, attempt to assign a dollar value to non-market goods, which can be problematic (Giménez and Mas 2020). Critics argue that reducing complex ecological and social values to monetary terms oversimplifies nature's intrinsic worth and can lead to misguided policy decisions (Ali and Khalidi 2020). Furthermore, existing valuation techniques often struggle to capture the full range of ecosystem services and their interdependencies. For example, traditional valuation methods may focus on direct-use values, such as timber or water, while neglecting non-use values, such as biodiversity and cultural significance (Jílková et al. 2010). This narrow focus can result in an incomplete understanding of ecosystems' ecological and social functions, leading to suboptimal management strategies. Another challenge is the temporal aspect of environmental valuation. Ecosystem services often provide benefits over long time horizons, making it difficult to assess their present value (Muhammad et al. 2023). Discounting future benefits can lead to the undervaluation of long-term ecological health, which is crucial for sustainable development. This temporal disconnect can skew decision-making processes in favour of short-term economic gains at the expense of long-term ecological integrity (Favré 2021).

Moreover, the uncertainty inherent in ecological systems poses additional challenges for valuation. Ecosystems are complex and dynamic, and predicting their responses to human activities can be uncertain (Wasis et al. 2018). This unpredictability complicates the application of valuation techniques, as it can lead to inaccurate assessments of ecosystem service values and their potential changes over time.

The challenge of balancing economic growth with ecological preservation is a central theme in environmental valuation. Economic development often relies on exploiting natural resources, which can lead to ecological degradation and loss of ecosystem services (Ghisellini et al. 2016). This tension raises critical questions about the sustainability of current economic models and the need for alternative approaches that prioritize ecological health. One approach to achieving this balance is sustainable development, which seeks to integrate economic, social, and environmental objectives. Sustainable development emphasizes the importance of maintaining ecological integrity while promoting economic growth, recognizing that long-term prosperity is contingent upon healthy ecosystems (Kapsalis et al. 2019). However, implementing sustainable development principles in practice can be challenging, as it requires a fundamental shift in how economic success is defined and measured. Additionally, integrating ecological considerations into economic decision-making processes is essential for balancing growth and preservation. This integration can be facilitated by adopting innovative policy frameworks that prioritize ecosystem service values in economic planning (Matos et al. 2012). For example, incorporating natural capital accounting into national economic indicators can help policymakers recognize the economic importance of ecosystems and make more informed decisions supporting economic growth and ecological health (Muhammad et al. 2023). Furthermore, fostering collaboration between stakeholders, including government agencies, businesses, and local communities, is crucial for achieving sustainable outcomes. Engaging diverse perspectives in the valuation process can help ensure that the needs and values of all stakeholders are considered, leading to more equitable and effective resource management strategies ("Study on the Economic Valuation of Water Resources and the Allocation of Rights and Interests in Water Resources", 2023). Collaborative approaches can also facilitate the development of adaptive management practices that respond to changing ecological conditions and societal needs.

15.6 Future Directions in Environmental Economics

As the global community increasingly recognizes the urgent need for sustainable development, future directions in environmental economics must focus on integrating ecological considerations into economic frameworks. Integrating natural capital into national accounting systems represents a significant shift in how economies assess their wealth and sustainability. Traditional economic indicators, such as Gross Domestic Product (GDP), often overlook the depletion of natural resources and the degradation of ecosystems. This oversight can lead to policies prioritizing short-term economic gains at the expense of long-term ecological health (Bisultanova 2023). Natural capital accounting aims to provide a more comprehensive view of economic performance by including the value of ecosystem services and natural resources in national accounts. This approach aligns with the System of Environmental-Economic Accounting (SEEA), which offers a framework for measuring the stocks and flows of natural capital (Ferreira et al. 2020). By incorporating natural capital into national accounting, policymakers can better understand the trade-offs between economic growth and environmental sustainability, leading to more informed decision-making.

Countries such as the United Arab Emirates and the Netherlands have begun implementing natural capital accounting frameworks, demonstrating the feasibility and benefits of this approach (Adedoyin 2021). These initiatives enhance transparency and encourage investments in sustainable practices by highlighting the economic value of preserving natural resources. As more nations adopt natural capital accounting, it could pave the way for a global shift towards sustainable economic models prioritizing ecological integrity and economic development. Innovations in green financing are crucial for mobilizing the capital necessary to support the transition to a sustainable economy. Green financing encompasses a range of financial instruments and mechanisms designed to fund environmentally friendly projects and initiatives. Green bonds and climate finance mechanisms have gained prominence. Green bonds are fixed-income securities specifically earmarked for financing projects with positive environmental impacts, such as renewable energy, energy efficiency, and sustainable agriculture. The market for green bonds has expanded rapidly, with issuances reaching record levels in recent years (Fernandes et al. 2021). This growth reflects a broader trend among investors who are increasingly prioritizing environmental, social, and governance (ESG) criteria in their investment decisions. Green bonds offer several advantages, including attracting a diverse range of investors and lowering the cost of capital for green projects. Green bonds enhance transparency and accountability in the financing process by providing a clear framework for assessing the environmental impact of funded projects (Masud et al. 2023). Furthermore, the development of standardized frameworks for green bond issuance, such as the Green Bond Principles, has helped to establish credibility in the market and facilitate the growth of this financing mechanism.

Climate finance mechanisms play a vital role in supporting mitigation and adaptation efforts in the face of climate change. These mechanisms include funding from various sources, such as governments, international organizations, and private sector entities, to facilitate the transition to low-carbon economies (Ge et al. 2018). Climate finance is essential for implementing climate policies effectively, as it enables countries to reduce greenhouse gas emissions and enhance resilience to climate impacts. Innovative climate finance mechanisms, such as results-based financing and blended finance, are emerging as effective tools for mobilizing resources for climate action. Results-based financing links funding to achieving specific environmental outcomes, incentivizing project developers to deliver measurable results (Zhang 2023). Conversely, blended finance combines public and private capital to reduce investment risks and attract additional funding for climate-related projects (Strezov et al. 2016). These innovative approaches can enhance the effectiveness of climate finance and ensure that resources are allocated efficiently to address pressing environmental challenges.

Advancing interdisciplinary approaches to economic policies is essential for addressing the complex challenges of environmental sustainability. Environmental issues often intersect with various fields, including economics, sociology, political science, and ecology. By fostering collaboration among these disciplines, policymakers can develop more holistic and effective strategies for sustainable development. Interdisciplinary approaches can enhance understanding of sustainability's social, economic, and environmental dimensions. For instance, integrating insights from behavioural economics can help policymakers design interventions that encourage sustainable consumer behaviour ("The Influence of International Experience of Green Technology Innovation Policy on Regional Residents' Income Difference", 2021). Similarly, incorporating ecological knowledge into economic models can improve the assessment of ecosystem services and inform resource management decisions (Scott 2023). Furthermore, interdisciplinary collaboration can facilitate the development of innovative solutions to environmental challenges. For example, partnerships between scientists, policymakers, and business leaders can lead to co-creating sustainable technologies and practices that address economic and ecological goals (Nugroho et al. 2023). By breaking down silos and fostering dialogue among diverse stakeholders, interdisciplinary approaches can drive the transition to a more sustainable and resilient economy. Future directions in environmental economics must prioritize the integration of natural capital into national accounting systems, innovations in green financing, and interdisciplinary approaches to economic policies. By embracing these strategies, policymakers can create a more sustainable economic framework that values ecological integrity and promotes long-term prosperity.

15.7 Conclusion

Environmental economics provides a robust framework for understanding the intricate relationships between economic activities and ecological health. By valuing ecosystem services and recognizing the trade-offs inherent in development and conservation, policymakers can craft strategies that promote sustainable growth while safeguarding the environment for future generations. Evaluating ecosystem services is a vital component of environmental economics that informs policy and management decisions. By understanding the types of ecosystem services, employing various valuation methods, and examining relevant case studies, stakeholders can better appreciate the economic significance of ecosystems and work towards sustainable development. Economic instruments for environmental management, including market-based tools, incentives for conservation, and subsidies for green technologies, play a vital role in addressing environmental challenges. Their effectiveness largely depends on their design and implementation, as well as the broader regulatory and economic context in which they operate. Continued research and evaluation are necessary to optimize these tools and ensure they contribute effectively to sustainable development. The economic sustainability of green technologies is supported by numerous benefits, including job creation, reduced healthcare costs, and long-term savings. Cost-benefit analyses consistently show that while initial investments may be higher, the long-term advantages of renewable energy and sustainable practices far outweigh the costs. Financing mechanisms such as green bonds, public-private partnerships, and sustainable investment funds are crucial in facilitating the transition to a more sustainable economy. As the world continues to grapple with climate change and resource depletion challenges, adopting green technologies will be vital for achieving economic sustainability and environmental resilience. The challenges in economic environmental valuation are multifaceted, encompassing issues of equity and access, limitations of current valuation techniques, and the need to balance economic growth with ecological preservation. Addressing these challenges requires a comprehensive understanding of the complexities of ecosystems and the values they provide, as well as a commitment to integrating ecological considerations into economic decision-making processes. By fostering collaboration and adopting innovative policy frameworks, creating a more sustainable future that values economic prosperity and ecological integrity is possible.

References

Adedoyin F (2021) Global sustainability in the presence of green technology transfer, 1–11. https://doi.org/10.1007/978-3-030-38948-2_114-1

Alesina A, Passarelli F (2014) Regulation versus taxation. J Public Econ 110:147–156. https://doi.org/10.1016/j.jpubeco.2013.09.001

Ali S, Khalidi M (2020) Valuation of equity securities, private firms, and startups. Does Value Co Creat Impacts Cust Loyal Repurchase Intention? 16(1):125–140. https://doi.org/10.46745/ilma.jbs.2020.16.01.09

Arias-Arévalo P, Gómez-Baggethun E, Martín-López B, Pérez-Rincón M (2018) Widening the evaluative space for ecosystem services: a taxonomy of plural values and valuation methods. Environ Values 27(1):29–53. https://doi.org/10.3197/096327118x15144698637513

Azeakpono E, Lloyd A (2020) Renewable energy consumption and economic growth in Nigeria: any causal relationship? Bus Manag Rev 11(01). https://doi.org/10.24052/bmr/v11nu01/art-08

Baciu GE, Dobrotă CE, Apostol EN (2021) Valuing forest ecosystem services. Why is an integrative approach needed? Forests 12(6):677. https://doi.org/10.3390/f12060677

Bisultanova A (2023) Sustainable financial services as a tool for sustainable economic development. E3s Web Conf 458:05004. https://doi.org/10.1051/e3sconf/202345805004

Brooks J (2016) Design features and project age contribute to joint success in social, ecological, and economic outcomes of community-based conservation projects. Conserv Lett 10(1):23–32. https://doi.org/10.1111/conl.12231

Callaghan G, Williams D (2014) Teddy bears and tigers: how renewable energy can revitalise local communities. Local Econ J Local Econ Policy Unit 29(6–7):657–674. https://doi.org/10.1177/0269094214551254

Çevik E, Yıldırım D, Dibooğlu S (2020) Renewable and non-renewable energy consumption and economic growth in US: a Markov-switching var analysis. Energy Environ 32(3):519–541. https://doi.org/10.1177/0958305x20944035

Chen Z, Niu X, Gao X, Chen H (2022) How does environmental regulation affect green innovation? a perspective from the heterogeneity in environmental regulations and pollutants. Front Energy Res 10:885525. https://doi.org/10.3389/fenrg.2022.885525

Chou C (2023) Renewable energy integration for sustainable economic growth: insights and challenges via bibliometric analysis. Sustainability 15(20):15030. https://doi.org/10.3390/su152015030

Christine D, Yadiati W, Afiah N, Fitrijanti T (2019) The relationship of environmental management accounting, environmental strategy and managerial commitment with environmental performance and economic performance. Int J Energy Econ Policy 9(5):458–464. https://doi.org/10.32479/ijeep.8284

Egboduku WO, Egboduku T, Golohor OM, Imarhiagbe O, Ogwu MC (2024) Cassava as raw material for sustainable bioeconomy development. In: Ogwu MC, Izah SC, Cunha Alves AA, Babu SC (eds) Plant biology, sustainability and climate change, sustainable cassava. Academic Press, pp 57–73. https://doi.org/10.1016/B978-0-443-21747-0.00022-9

Eze V (2023) A comparative analysis of renewable energy policies and its impact on economic growth: a review. Int J Educ Sci Technol Eng 6(2):41–46. https://doi.org/10.36079/lamintang.ijeste-0602.555

Favré M (2021) Are households willing to finance the cost of individual water supply? a Case study in central Tunisia. Water Econ Policy 07(04). https://doi.org/10.1142/s2382624x21500168

Fernandes C, Veiga P, Ferreira J, Hughes M (2021) Green growth versus economic growth: do sustainable technology transfer and innovations lead to an imperfect choice? Bus Strateg Environ 30(4):2021–2037. https://doi.org/10.1002/bse.2730

Ferreira J, Fernandes C, Ferreira F (2020) Technology transfer, climate change mitigation, and environmental patent impact on sustainability and economic growth: a comparison of European countries. Technol Forecast Soc Chang 150:119770. https://doi.org/10.1016/j.techfore.2019.119770

Fidelman P, Evans L, Fabinyi M, Foale S, Cinner J, Rosen F (2012) Governing large-scale marine commons: contextual challenges in the coral triangle. Mar Policy 36(1):42–53. https://doi.org/10.1016/j.marpol.2011.03.007

Filho A, Araújo M, Salvador H, Ferreira P, Saba H (2022) Renewable sources to promote well-being in poor regions of Brazil. Front Phys 10:1048721. https://doi.org/10.3389/fphy.2022.1048721

Fletcher R (2012) Using the master's tools? neoliberal conservation and the evasion of inequality. Dev Chang 43(1):295–317. https://doi.org/10.1111/j.1467-7660.2011.01751.x

Fortnam M (2023) Practitioner approaches to trade-off decision-making in marine conservation development. People Nat 5(5):1636–1648. https://doi.org/10.1002/pan3.10530

Fu B, Su C, Wei Y, Willett I, Lü Y, Li G (2010) Double counting in ecosystem services valuation: causes and countermeasures. Ecol Res 26(1):1–14. https://doi.org/10.1007/s11284-010-0766-3

Ge J, Polhill G, Craig T, Liu N (2018) From oil wealth to green growth—an empirical agent-based model of recession, migration and sustainable urban transition. Environ Model Softw 107:119–140. https://doi.org/10.1016/j.envsoft.2018.05.017

Ghisellini P, Cialani C, Ulgiati S (2016) A review on circular economy: the expected transition to a balanced interplay of environmental and economic systems. J Clean Prod 114:11–32. https://doi.org/10.1016/j.jclepro.2015.09.007

Gill D, Cheng S, Glew L, Aigner E, Bennett N, Mascia M (2019) Social synergies, tradeoffs, and equity in marine conservation impacts. Annu Rev Environ Resour 44(1):347–372. https://doi.org/10.1146/annurev-environ-110718-032344

Giménez F, Mas C (2020) The valuation of recreational use of wetlands and the impact of the economic crisis. Int J Environ Res Public Health 17(9):3228. https://doi.org/10.3390/ijerph17093228

Guðlaugsson B, Ahmed T, Dawood H, Ogwumike C, Dawood N (2021) Application of cost benefits analysis for the implementation of renewable energy and smart solution technologies: a case study of integrity project, 15. https://doi.org/10.3390/environsciproc2021011015

Guo J (2023) Multilevel meta-analysis of stated preference values of marine ecosystem services in China. Mar Dev 1(1). https://doi.org/10.1007/s44312-023-00004-z

Guo X, Фу Л (2021) Can environmental regulations promote greenhouse gas abatement in OECD countries? command-and-control vs. market-based policies. Sustainability 13(12):6913. https://doi.org/10.3390/su13126913

Hanssen F, May R, Dijk J, Rød J (2018) Spatial multi-criteria decision analysis tool suite for consensus-based siting of renewable energy structures. JEAPM 20(03):1840003. https://doi.org/10.1142/s1464333218400033

He J (2023) The impact of environmental regulation on regional economic growth: a case study of the Yangtze River economic belt, china. PLoS One 18(9):e0290607. https://doi.org/10.1371/journal.pone.0290607

Inglesi-Lotz R, Doğan E (2018) The role of renewable versus non-renewable energy to the level of CO_2 emissions a panel analysis of Saharan Africa's βig 10 electricity generators. Renew Energy 123:36–43. https://doi.org/10.1016/j.renene.2018.02.041

Iqbal M (2023) Understanding the potential of hydrogen as a game-changing energy source. Chem Eng Technol 46(12):2577–2582. https://doi.org/10.1002/ceat.202300051

Iyiola AO, Izah SC, Morya S, Akinsorotan A, Ogwu MC (2023) Owalla reservoir in South-Western Nigeria: assessment of fish distribution, biological diversity, and water quality index. Indones J Agric Res 6(02):106–125. https://doi.org/10.32734/injar.v6i2.12207

Iyiola AO, Afolabi OA, Alimi SK, Akingba OO, Izah SC, Ogwu MC (2024a) Climate change and water crisis in the global south. In: Izah SC, Ogwu MC, Loukas A, Hamidifar H (eds) Water crises and sustainable management in the global south. Springer, Singapore. https://doi.org/10.1007/978-981-97-4966-9_4

Iyiola AO, Kolawole AS, Setufe SB, Bilikoni J, Ofori E, Ogwu MC (2024b) Fish as a sustainable biomonitoring tool in aquatic environments. In: Izah SC, Ogwu MC, Hamidifar H (eds) Biomonitoring of pollutants in the global south. Springer, Singapore. https://doi.org/10.1007/978-981-97-1658-6_12

Iyiola AO, Setufe SB, Ofori E, Bilikoni J, Ogwu MC (2024c) Biomarkers for the detection of pollutants from the water environment. In: Izah SC, Ogwu MC, Hamidifar H (eds) Biomonitoring of pollutants in the global south. Springer, Singapore. https://doi.org/10.1007/978-981-97-1658-6_16

Izah SC, Ogwu MC, Akram M (2023) Herbal medicine phytochemistry: applications and trends. Springer Cham, Switzerland. https://doi.org/10.1007/978-3-031-21973-3

Jia H, Fan S, Miao X (2023) The impact of renewable energy consumption on economic growth: evidence from countries along the belt and road. Sustainability 15(11):8644. https://doi.org/10.3390/su15118644

Jílková J, Hollander R, Kochmann L, Slavík J, Slavíková L (2010) Economic valuation of environmental resources and its use in local policy decision-making: a comparative Czech–German border study. J Comp Policy Anal Res Pract 12(3):299–309. https://doi.org/10.1080/13876981003714610

Kapsalis V, Kyriakopoulos G, Aravossis K (2019) Investigation of ecosystem services and circular economy interactions under an inter-organizational framework. Energies 12(9):1734. https://doi.org/10.3390/en12091734

Keyue Y, Fei L, Yang H, Yi-ming W (2019) The influence of land use change on ecosystem service value in Shangzhou district. Int J Environ Res Public Health 16(8):1321. https://doi.org/10.3390/ijerph16081321

Kosoe EA, Achana GTW, Ogwu MC (2023) Meta-evaluation of the one health implication on food systems of agrochemical use. In: Ogwu MC, Chibueze Izah S (eds) One health implications of agrochemicals and their sustainable alternatives. Sustainable development and biodiversity. Springer, Singapore, pp. 387–409. https://doi.org/10.1007/978-981-99-3439-3_14

Li R, Wen H, Huang X, Liu Y (2023) Renewable energy: a curse or blessing—international evidence. Sustainability 15(14):11103. https://doi.org/10.3390/su151411103

Lian J, Sun Y, Kalsi K, Widergren S, Wu D, Ren H (2018) Transactive system: part ii: analysis of two pilot transactive systems using foundational theory and metrics. https://doi.org/10.2172/1422303

Lim S, Prakash A (2014) Voluntary regulations and innovation: the case of ISO 14001. Public Adm Rev 74(2):233–244. https://doi.org/10.1111/puar.12189

Lin D, Zhao Y (2023) The impact of environmental regulations on enterprises' green innovation: the mediating effect of managers' environmental awareness. Sustainability 15(14):10906. https://doi.org/10.3390/su151410906

Liu L, Li M, Gong X, Pan J, Jin R, Zhang Y (2022) Influence mechanism of different environmental regulations on carbon emission efficiency. Int J Environ Res Public Health 19(20):13385. https://doi.org/10.3390/ijerph192013385

Masud M, Sahara J, Kabir H (2023) A relationship between climate finance and climate risk: evidence from the South Asian region. Climate 11(6):119. https://doi.org/10.3390/cli11060119

Matos A, Cabo P, Ribeiro M, Fernandes A (2012) Economic valuation of environmental goods and services, 139–170. https://doi.org/10.4324/9780203096246-17

Milon J, Álvarez S (2019) Coastal resources economics and ecosystem valuation. Water 11(11):2206. https://doi.org/10.3390/w11112206

Muhammad A, Idris M, Ishaq A, Umar U (2023) Using laplace series and partial integration in valuing environmental assets and estimating green GDP. J Environ Sci Econ 2(1):55–60. https://doi.org/10.56556/jescae.v2i1.477

Musetsho K, Chitakira M, Ramoelo A (2022) Ecosystem service valuation for a critical biodiversity area: the case of the Mphaphuli community, South Africa. Land 11(10):1696. https://doi.org/10.3390/land11101696

Nugroho L, Melati S, Wahyuliana I, Pawestri A, Kurniawan L (2023) The legal policy of implementation of green economy in the tourism sector to realize sustainable tourism and environment. IOP Conf Ser Earth Environ Sci 1181(1):012018. https://doi.org/10.1088/1755-1315/1181/1/012018

Obst C, Alfieri A, Kroese B (2020) Advancing environmental-economic accounting in the context of the system of economic statistics. Stat J IAOS 36(3):629–645. https://doi.org/10.3233/sji-200707

Ogwu MC, Kosoe EA (2024) Place of cultural diversity in sustainable water resource management in Ghana. In: Izah SC, Ogwu MC, Loukas A, Hamidifar H (eds) Water crises and sustainable management in the global south. Springer, Singapore. https://doi.org/10.1007/978-981-97-4966-9_14

Ogwu MC, Pamela BO, Dare AF, Ovuru KF, Iyiola AO (2024) Traditional and conventional water treatment methods: a sustainable approach. In: Izah SC, Ogwu MC, Loukas A, Hamidifar H (eds) Water crises and sustainable management in the global south. Springer, Singapore. https://doi.org/10.1007/978-981-97-4966-9_15

Pagani-Núñez, E., Yang, X., Yan, M., He, J., Jiang, Z., & Jiang, H. (2022). Trade-offs between economic development and biodiversity conservation on a tropical island. Conservation Biology, 36(5). https://doi.org/10.1111/cobi.13912

Pan J, Li M, Zhao Y, Gong X, Jin R, Zhang Y, Liu L (2022) Does environmental regulation improve carbon emission efficiency? Inspection of panel data from inter-provincial provinces in China. Sustainability 14(16):10448. https://doi.org/10.3390/su141610448

Pandeya B, Buytaert W, Zulkafli Z, Karpouzoglou T, Mao F, Hannah D (2016) A comparative analysis of ecosystem services valuation approaches for application at the local scale and in data-scarce regions. Ecosyst Serv 22:250–259. https://doi.org/10.1016/j.ecoser.2016.10.015

Santis R, Lasinio C (2016) Environmental policies, innovation and productivity in the EU. Glob Econ J 16(4):615–635. https://doi.org/10.1515/gej-2015-0060

Scott C (2023) Climate finance and its role in climate policy. J Clim Policy 2(1):54–66. https://doi.org/10.47941/jcp.1550

Shahbaz M, Raghutla C, Chittedi K, Jiao Z, Vo X (2020) The effect of renewable energy consumption on economic growth: evidence from the renewable energy country attractive index. Energy 207:118162. https://doi.org/10.1016/j.energy.2020.118162

Sharma S, Hussain S, Singh A (2021) Integrated approaches of ecology and economy for sustainable development with special emphasis on ecosystem services: a review. J Sci Res 65(03):89–110. https://doi.org/10.37398/jsr.2021.650312

Sikdar S, Lapan H (2011) Carbon leakage: the role of sequential policy setting. Environ Dev Econ 17(1):91–104. https://doi.org/10.1017/s1355770x1100026x

Singh N, Nyuur R, Richmond B (2019) Renewable energy development as a driver of economic growth: evidence from multivariate panel data analysis. Sustainability 11(8):2418. https://doi.org/10.3390/su11082418

Strezov V, Evans A, Evans T (2016) Assessment of the economic, social and environmental dimensions of the indicators for sustainable development. Sustain Dev 25(3):242–253. https://doi.org/10.1002/sd.1649

Sun T, Li B, Song X, Hou Y, Guo H (2010) Factor analysis of ecosystem services value in Huangfuchuan basin, 2405–2409. https://doi.org/10.1109/bmei.2010.5639683

Wang L, Yang Y, Wang L, Guo J (2011) Empirical research on the relationship between investment in environmental protection and economic growing of Shanghai. Adv Mater Res 361–363:1066–1070. https://doi.org/10.4028/www.scientific.net/amr.361-363.1066

Wang M, He Y, Zhou J, Ren K (2022a) Evaluating the effect of Chinese environmental regulation on corporate sustainability performance: the mediating role of green technology innovation. Int J Environ Res Public Health 19(11):6882. https://doi.org/10.3390/ijerph19116882

Wang Z, Wang N, Hu X, Wang H (2022b) Threshold effects of environmental regulation types on green investment by heavily polluting enterprises. Environ Sci Eur 34(1). https://doi.org/10.1186/s12302-022-00606-2

Wasis B, Saharjo B, Kusumadewi F, Utami N, Putra M (2018) Analysis of economic valuation of environmental damage due to sand mine in Gumulung Tonggoh, Cirebon District, West Java Province, Indonesia. Arch Agric Environ Sci 3(4):360–366. https://doi.org/10.26832/24566632.2018.030405

Yamazaki A (2022) Environmental taxes and productivity: lessons from Canadian manufacturing. J Public Econ 205:104560. https://doi.org/10.1016/j.jpubeco.2021.104560

Yang J, Wu W, Jiang X, Xing F (2012) Damage assessment of marine ecosystem service function loss caused by oil spill. Adv Mater Res 573–574:319–324. https://doi.org/10.4028/www.scientific.net/amr.573-574.319

Yang J, Xie B, Wang S, Zhang D, Liu C, Mak-Mensah E (2022a) Temporal and spatial characteristics of grassland ecosystem service value and its topographic gradient effect in the Yellow River Basin. PLoS One 17(12):e0278211. https://doi.org/10.1371/journal.pone.0278211

Yang Y, Peng Z, Tang D (2022b) The impact of heterogeneous environmental regulations on carbon neutrality in China: new evidence based on the spatial measurement model. Energy Environ 35(3):1456–1478. https://doi.org/10.1177/0958305x221140578

Yi M, Fang X, Wen L, Guang F, Yao Z (2019) The heterogeneous effects of different environmental policy instruments on green technology innovation. Int J Environ Res Public Health 16(23):4660. https://doi.org/10.3390/ijerph16234660

Yin K, Liu L, Gu H (2022) Green paradox or forced emission reduction—the dual effects of environmental regulation on carbon emissions. Int J Environ Res Public Health 19(17):11058. https://doi.org/10.3390/ijerph191711058

Zhang D (2023) Environmental sustainability and global value chains: challenges and opportunities for green supply chain management. Front Bus Econ Manag 10(2):155–160. https://doi.org/10.54097/fbem.v10i2.10898

Zheng H, Wang L, Peng W, Zhang C, Li C, Robinson BE, Wu X, Kong L, Li R, Xiao Y, Xu W (2019) Realizing the values of natural capital for inclusive, sustainable development: informing China's new ecological development strategy. Proc Natl Acad Sci 116(17):8623–8628. https://doi.org/10.1073/pnas.1819501116

Zhu Y, Sun Z, Zhang S, Wang X (2021) Economic policy uncertainty, environmental regulation, and green innovation—an empirical study based on Chinese high-tech enterprises. Int J Environ Res Public Health 18(18):9503. https://doi.org/10.3390/ijerph18189503

Chapter 16
Policy and Regulatory Frameworks for Sustainable Environmental Practices

Matthew Chidozie Ogwu, Soumia El Malahi, and Sylvester Chibueze Izah

Abstract This chapter underscores the pivotal role of robust policies and regulations in addressing global environmental challenges such as climate change, biodiversity loss, and resource depletion. It explores the design, implementation, and impact of international agreements, national legislation, and governance models in promoting sustainability. Emphasis is placed on frameworks like the Sustainable Development Goals and the Paris Agreement, which provide guidelines for coordinated action on ecological issues. Case studies, including Costa Rica's Payment for Ecosystem Services and the European Union's Water Framework Directive, highlight successful strategies for policy integration and stakeholder engagement. The chapter also evaluates diverse governance models, such as centralized and decentralized frameworks, participatory approaches, and integrating technology and data for environmental management. It also examines economic instruments like market-based mechanisms, subsidies, and Payment for Ecosystem Services, highlighting their potential and challenges. Legal frameworks and corporate accountability measures are reviewed, emphasizing the need for compliance, enforcement, and equity to achieve sustainable outcomes. Despite advancements, significant barriers remain, including political resistance, limited institutional capacity, and social opposition.

M. C. Ogwu
Goodnight Family Department of Sustainable Development, Appalachian State University, Boone, NC, USA
e-mail: ogwumc@appstate.edu

S. El Malahi (✉)
Laboratory of Microbial Biotechnologies and Plant Protection, Faculty of Science, IBN ZOHR University, Agadir, Morocco
e-mail: soumia.elmalahi98@gmail.com

Laboratory of Biotechnology, Faculty of Science, IBN ZOHR University, Agadir, Morocco

S. C. Izah
Department of Community Medicine, Faculty of Clinical Sciences, Bayelsa Medical University, Yenagoa, Bayelsa, Nigeria
e-mail: Sylvester.izah@bmu.edu.ng

Department of Microbiology, Faculty of Science, Bayelsa Medical University, Yenagoa, Bayelsa, Nigeria

M. C. Ogwu and S. Chibueze Izah (eds.), *Evaluating Environmental Processes and Technologies*, Environmental Science and Engineering,
https://doi.org/10.1007/978-3-031-85327-2_16

Future directions advocate for adaptive, technology-driven, and inclusive approaches to policy-making. The chapter calls for a multi-level governance strategy that bridges global agreements with local actions, fostering collaboration and ensuring equitable resource management. By aligning sustainability goals with innovative governance and economic instruments, this chapter provides a roadmap for achieving long-term environmental resilience.

Keywords Sustainable development goals (SDGs) · Governance models · Payment for ecosystem services · Market-based mechanisms · Environmental legislation · Corporate accountability

16.1 Introduction

The significance of policy and regulatory frameworks in fostering sustainable environmental practices cannot be overstated. These frameworks form the foundation for implementing strategies that mitigate ecological degradation while promoting economic growth and social equity. As global challenges such as climate change, biodiversity loss, and resource depletion intensify, the need for robust policies becomes increasingly critical. However, implementing sustainable practices faces numerous challenges, including political resistance, insufficient funding, and limited public awareness and engagement (Andreopoulou and Koliouska 2018; Fairbrass et al. 2023; Kazak and Świąder 2018). This chapter explores various policy tools, governance models, and regulatory frameworks that can effectively support sustainability initiatives, addressing their potential and limitations. Effective policy frameworks are indispensable for guiding the transition towards sustainability across diverse sectors. For instance, Andreopoulou and Koliouska (2018) emphasize that entrepreneurs must adopt modern environmental policies and sustainable marketing strategies to enhance their sustainability efforts. Furthemore, Fairbrass et al. (2023) critique the limitations of current monitoring frameworks, which often prioritize policy responses over measurable environmental outcomes, advocating for a more balanced approach to evaluating policy effectiveness.

These insights reveal a critical gap in the existing frameworks: the need to bridge the divide between policy design and tangible ecological benefits. Addressing this gap is essential to ensure that policies respond to environmental challenges and deliver long-term sustainable outcomes. Moreover, the Sustainable Development Goals (SDGs) provide a comprehensive framework for addressing global sustainability challenges. However, as Bengtsson et al. (2018)noted, the SDGs lack clear guidance on achieving sustainable consumption and production patterns, which are pivotal for long-term sustainability. This highlights the need for policymakers to develop more actionable strategies that align with the SDGs' overarching goals while addressing local contexts and specific challenges. Implementing sustainable environmental practices is fraught with numerous challenges that vary by region and sector. For example, Kazak and Świąder (2018) discuss the necessity of new investments in sustainable energy development strategies, which often require substantial financial resources and stakeholder engagement. Similarly, a study on the environmental sustainability assessment of healthcare facilities underscores the importance of aligning new frameworks with existing regulatory standards to ensure successful implementation (Moldovan 2023). These examples illustrate that the transition to sustainable practices can be impeded without adequate resources and alignment with existing policies. Furthermore, the role of public procurement as a policy tool is increasingly recognized in promoting sustainability. Uwaoma (2023) argues that green taxation and renewable energy investments can significantly reduce pollution and promote sustainable practices, but these must be carefully managed to maximize their effectiveness. Manika (2020) echoes this sentiment, emphasizing integrating social and environmental considerations into public procurement processes to achieve broader sustainability goals.

Governance models play a crucial role in shaping the effectiveness of sustainability policies. Mickwitz et al. (2021) propose a theory-based evaluation framework that integrates various theoretical perspectives to assess policies aimed at promoting sustainability transitions. This approach can help identify best practices and areas for improvement in existing policies. Additionally, Dedeurwaerdere (2013) highlights the need for transdisciplinary approaches in sustainability science, which can facilitate incremental institutional changes necessary for effective policy implementation. Moreover, the integration of technology into sustainability practices is gaining traction. Parmentola et al. (2021) explore how blockchain technology can enhance environmental sustainability by improving transparency and accountability in supply chains. This technological integration presents new opportunities for policymakers to leverage innovative solutions in their regulatory frameworks.

This chapter explores the role of policy and regulatory frameworks in advancing sustainable environmental practices by examining their design, implementation, and impact on global, national, and local levels. It seeks to analyze international agreements, national legislation, and governance models that support sustainability while highlighting the challenges and opportunities in integrating sustainability into various sectors. Through case studies and examples, it provides insights into successful frameworks and future directions for developing inclusive, participatory, and innovative policies that address environmental justice, ensure equitable resource management, and foster a global transition toward sustainability.

16.2 Global Perspectives on Environmental Policy

Environmental policies have evolved significantly over the past decades, shaped by international conventions, regional agreements, and national strategies to address critical environmental challenges. Some international agreements and treaties on ecological resources are presented in Table 16.1. Globally, frameworks such as the United Nations Sustainable Development Goals (SDGs) and the Paris Agreement on Climate Change have set ambitious targets for sustainability, emphasizing the need for coordinated efforts to combat climate change, conserve biodiversity, and promote resource efficiency (UN 2015; IPCC 2021). These agreements highlight the importance of global collaboration and the role of science-based policy in achieving environmental goals. Scholars have extensively examined the relationship between global policy frameworks and national-level implementation. For instance, studies reveal disparities in adopting and enforcing ecological policies between developed and developing countries. Developed nations often have the institutional capacity and financial resources to implement rigorous environmental regulations, while developing countries face challenges such as weak governance, lack of funding, and socioeconomic pressures (Sachs et al. 2019; Biermann 2020). These disparities underscore the need for capacity-building programs and international support mechanisms to bridge the gap. Emerging approaches, such as circular economy models and nature-based solutions, have gained prominence in global policy discussions, reflecting a shift toward integrative and innovative practices (Steffen et al. 2015). Policies promoting these approaches aim to balance economic development with ecological sustainability, addressing systemic issues such as overconsumption, waste generation, and ecosystem degradation. Despite progress, the literature highlights ongoing challenges in global environmental governance, including insufficient enforcement mechanisms, conflicting interests among stakeholders, and the need for greater inclusivity in policymaking processes. Scholars emphasize the importance of participatory approaches that engage local communities, Indigenous peoples, and other marginalized groups to ensure equity and effectiveness in policy outcomes (Ostrom 2009; Escobar 2018; Akudugu and Ogwu 2024). Overall, the global perspective on environmental policy underscores the interconnectedness of environmental issues and the critical need for a multi-level governance approach that bridges global agreements with local actions. As ecological challenges grow in complexity, the role of adaptive, science-driven, and equity-focused policies becomes increasingly vital.

16.2.1 International Agreements and Treaties on Environmental Resources

International agreements and treaties are pivotal in shaping global environmental policy and fostering cooperation among nations. The Paris Agreement, adopted in 2015, represents a landmark commitment by nearly all countries to combat climate change by limiting global warming to well below 2 degrees Celsius above pre-industrial levels, with efforts to limit the temperature increase to 1.5 degrees Celsius (Table 16.1; Falkner 2016; Pramudianto 2019). This agreement marked a significant shift from the Kyoto Protocol, which imposed binding targets primarily on developed nations, to a more inclusive framework that allows all countries to set their own Nationally Determined Contributions (NDCs) based on their national circumstances (Dagnet et al. 2021).

Table 16.1 Some international agreements and treaties on environmental resources

Agreement/treaty	Year adopted	Focal area	Major outcomes
United Nations Framework Convention on Climate Change (UNFCCC)	1992	Climate change mitigation and adaptation	Established the framework for international climate agreements, including the Kyoto Protocol and the Paris Agreement
Kyoto Protocol	1997	Greenhouse gas emissions reduction	Legally binding targets for developed countries to reduce greenhouse gas emissions
Paris Agreement	2015	Global temperature rise limit to 1.5 °C	Commitment to nationally determined contributions (NDCs) and financial support for climate action
Convention on Biological Diversity (CBD)	1992	Biodiversity conservation	Framework for protecting ecosystems, species, and genetic diversity; protocols on biosafety and access to resources
Ramsar Convention on Wetlands	1971	Wetland conservation and sustainable use	Designated over 2400 wetlands of international importance ("Ramsar Sites")
Stockholm Convention on Persistent Organic Pollutants	2001	Elimination of harmful chemicals	Banned or restricted the use of harmful persistent organic pollutants (POPs)
Basel Convention	1989	Hazardous waste management	Regulation of transboundary movements and disposal of hazardous wastes
Montreal Protocol	1987	Ozone layer protection	Phased out the production and consumption of ozone-depleting substances (ODS)
United Nations Convention to Combat Desertification (UNCCD)	1994	Land degradation and desertification	Promoted sustainable land management practices to combat desertification and mitigate drought
CITES (Convention on International Trade in Endangered Species of Wild Fauna and Flora)	1973	Wildlife trade regulation	Established protections for endangered species by regulating international trade
MARPOL Convention	1973	Marine pollution prevention	Prevented pollution from ships through regulations on oil spills, garbage disposal, and emissions
International Treaty on Plant Genetic Resources for Food and Agriculture	2001	Conservation of genetic resources	Framework for access to genetic resources and benefit-sharing for agriculture
Antarctic Treaty System	1959	Preservation of Antarctica	Banned military activity, protected the environment, and promoted scientific cooperation in Antarctica

The Convention on Biological Diversity (CBD) and the Kyoto Protocol are notable examples of international treaties addressing environmental issues. Established in 1992, the CBD seeks to conserve biological diversity, promote sustainable use of its components, and ensure fair sharing of benefits arising from genetic resources (Seddon et al. 2018). Similarly, the Kyoto Protocol, which came into force in 2005, set legally binding emission reduction targets for developed countries. However, its effectiveness has been questioned due to low compliance rates and the withdrawal of key nations (Nordhaus 2018). Despite the ambitious goals set by these agreements, challenges remain in achieving compliance and enforcement across nations. Roelfsema et al. (2020) highlight a significant gap between policy commitments and actual implementation, noting that even if NDCs were fully realized, the gap between current policies and the Paris goals would only be reduced by a third. Furthermore, the effectiveness of these treaties is often undermined by political and economic factors, including national interests that may conflict with global environmental objectives (Falkner 2016; Dagnet et al. 2021). The lack of binding enforcement mechanisms and the reliance on voluntary compliance further complicate the landscape of international environmental governance (Höhne et al. 2016).

16.2.2 Role of Multilateral Organizations

Multilateral organizations, such as the United Nations Environment Programme (UNEP) and the World Bank, play a critical role in facilitating international cooperation and supporting sustainable practices. UNEP is instrumental in coordinating global environmental efforts, providing a platform for dialogue, and promoting the integration of environmental considerations into national policies (Oppong 2023). The organization also plays a key role in capacity-building initiatives, helping countries develop the necessary frameworks and tools to implement their commitments under international agreements (Dagnet et al. 2021). The World Bank, through its financial support and technical assistance, supports projects aimed at enhancing environmental sustainability and addressing climate change. For instance, it has established various climate finance mechanisms to assist developing countries in transitioning to low-carbon economies (Michaelowa et al. 2020; Kissinger 2020). These initiatives enable countries to meet their NDCs and achieve SDGs (Zheng et al. 2021). However, the effectiveness of these funding mechanisms is often contingent on recipient countries' political will and ability to implement sustainable practices effectively (Silva et al. 2020). Moreover, the role of multilateral development banks in climate finance has gained increasing attention. They are seen as essential players in mobilizing resources for climate action. Yet, questions remain regarding the efficiency of fund allocation and the impact of these investments on achieving environmental goals (Michaelowa et al. 2020). As highlighted by Dagnet et al. (2021), leveraging insights from past experiences can enhance the effectiveness of climate finance initiatives and ensure that they align with the objectives of the Paris Agreement (Dagnet et al. 2021).

16.3 National Policy Frameworks for Environmental Sustainability

Table 16.2 provides examples of national policy frameworks various countries adopt to advance environmental sustainability. These frameworks represent diverse approaches to addressing critical environmental challenges such as climate change, biodiversity loss, pollution, and resource management. Each policy framework is uniquely tailored to address the respective countries' ecological, social, and economic conditions, demonstrating their commitment to sustainable development. Moreover, these policies emphasize the integration of environmental considerations into national governance, highlighting efforts to align environmental priorities with broader development goals.

16.3.1 Environmental Legislation

Environmental legislation is a cornerstone of national policy frameworks to promote sustainability and protect natural resources. Countries worldwide have enacted diverse and impactful laws to address environmental challenges, with notable examples including the Clean Air Act in the United States and the Environmental Protection Act in India. The Clean Air Act, passed in 1963 and significantly amended in subsequent years, establishes comprehensive federal and state regulations to control air pollution from stationary and mobile sources (Wilson et al. 2011). It has been instrumental in reducing emissions of harmful pollutants, thereby improving air quality and public health (Wilson et al. 2011). Similarly, the Environmental Protection Act of 1986 in India provides a framework for protecting and improving the environment, emphasizing the need for sustainable development (Walsh-Führing 2018).

Key components of effective environmental legislation include well-defined objectives, enforceable standards, and mechanisms for public participation and accountability. Effective laws should also incorporate adaptive management practices that allow for flexibility in response to new scientific findings and changing environmental conditions (Berry and Portney 2013). For instance, the European Union's Water Framework Directive exemplifies a comprehensive approach that integrates various aspects of water management, prioritizing stakeholder engagement and cross-sectoral coordination (Green 2017). Furthermore, the effectiveness of environmental legislation often hinges on the political will and capacity of regulatory agencies to enforce compliance (Switzer 2017).

Table 16.2 Examples of national policy frameworks for environmental sustainability

Country	Policy framework	Year enacted	Key focus areas	Major outcomes
United States	National Environmental Policy Act (NEPA)	1969	Environmental protection, impact assessment	Mandated Environmental Impact Statements (EIS) for federal projects, promoting transparency
Germany	Renewable Energy Sources Act (EEG)	2000	Renewable energy promotion	Achieved a significant increase in renewable energy use, particularly solar and wind
India	National Action Plan on Climate Change (NAPCC)	2008	Climate change mitigation and adaptation	Developed eight national missions, including solar energy, energy efficiency, and water management
China	Environmental Protection Law	2015	Pollution control and environmental governance	Strengthened penalties for polluters and improved environmental monitoring systems
South Africa	National Environmental Management Act (NEMA)	1998	Sustainable development and environmental rights	Promoted integrated environmental management and public participation
Australia	Environment Protection and Biodiversity Conservation Act	1999	Biodiversity and natural heritage protection	Established a framework for the protection of biodiversity and world heritage sites
Brazil	Forest Code	1965 (Revised 2012)	Deforestation control and forest conservation	Landowners are required to maintain a percentage of their land in forested areas
Canada	Federal Sustainable Development Strategy (FSDS)	2008	Sustainable development and green economy	Integrated sustainability goals into federal policies and programs
Japan	Basic Environment Law	1993	Pollution control, resource recycling	Established a foundation for Japan's circular economy initiatives
Kenya	Environmental Management and Coordination Act	1999	Environmental governance	Created a framework for environmental policy integration and management
United Kingdom	Climate Change Act	2008	Greenhouse gas reduction, renewable energy	Legally binding targets for reducing carbon emissions by 2050

(continued)

Table 16.2 (continued)

Country	Policy framework	Year enacted	Key focus areas	Major outcomes
Sweden	Environmental Code	1998	Sustainable resource management	Consolidated environmental laws to ensure sustainable use of natural resources
Norway	Nature Diversity Act	2009	Biodiversity conservation	Ensured sustainable use and protection of ecosystems and biodiversity
Ghana	Environmental Protection Agency Act	1994	Pollution control, environmental quality	Established the EPA to enforce regulations and promote environmental sustainability

16.3.2 Policy Integration Across Sectors

Incorporating sustainability into various sectors such as agriculture, industry, energy, and urban development is essential for achieving comprehensive environmental goals. Policy integration ensures that ecological considerations are systematically embedded in all aspects of governance rather than being treated as isolated issues. For example, the integration of sustainable agriculture practices can be seen in the European Union's Common Agricultural Policy, which promotes environmentally friendly farming practices and supports rural development (Konisky and Teodoro 2015). Case studies of successful cross-sectoral policy integration highlight the potential for creating synergistic benefits. In Sweden, the government has implemented a cohesive strategy that links energy, transport, and urban planning policies to achieve climate neutrality by 2045. This approach has significantly reduced greenhouse gas emissions while fostering economic growth and social equity (Walsh-Führing 2018). Similarly, South Korea's Green Growth Strategy has successfully integrated environmental sustainability into national development plans, leading to advancements in renewable energy and green technology (Grano 2014). Moreover, the role of non-governmental organizations (NGOs) in advocating for policy integration cannot be overlooked. NGOs often serve as key stakeholders in policymaking, raising awareness of environmental issues, and promoting sustainable practices across sectors (Azis 2022). Their involvement can enhance public engagement and ensure diverse perspectives are incorporated into developing environmental policies.

16.3.3 *Role of Political Will*

Political will is a crucial determinant of the effectiveness of environmental policies. Leadership and political commitment play pivotal roles in influencing the implementation of sustainability initiatives. Countries with progressive ecological policies often exhibit strong political leadership that places environmental issues at the forefront and fosters stakeholder collaboration. For instance, Costa Rica's commitment to environmental sustainability is reflected in its reforestation and biodiversity conservation policies, supported by a government that actively promotes green initiatives (Ba et al. 2022). Conversely, political ideology can significantly impact the design and enforcement of environmental policies. Research by Tawiah (2022) suggests that political ideologies shape the urgency with which governments address ecological sustainability, with more progressive parties typically advocating for stronger environmental regulations. In contrast, conservative administrations may prioritize economic growth over environmental protection, often weakening the enforcement of existing laws (Switzer 2017). The case of the United States under President Trump exemplifies this dynamic, where significant rollbacks of environmental regulations occurred, reflecting a realignment of political priorities (Jotzo et al. 2018; Bomberg 2021).

16.4 Economic Instruments in Environmental Regulation

Environmental challenges such as climate change, biodiversity loss, and pollution require innovative and effective policy tools to achieve sustainable outcomes. Among these tools, economic instruments have emerged as pivotal mechanisms to incentivize environmentally friendly practices while addressing market failures. These instruments leverage financial incentives and disincentives to encourage behavior that aligns with environmental goals. Figure 16.1 highlights three key categories of economic instruments: Market-Based Mechanisms, Incentives and Subsidies, and Payment for Ecosystem Services (PES). Each category demonstrates its unique contributions to environmental protection and sustainability alongside their associated challenges. This visualization emphasizes their roles in achieving greenhouse gas reductions, promoting renewable energy, and safeguarding ecosystem services, offering insights into their benefits and limitations.

16.4.1 *Market-Based Mechanisms*

Market-based mechanisms (MBMs) are increasingly recognized as practical tools for environmental regulation, particularly in the context of carbon pricing, emission trading schemes (ETS), and pollution taxes. Carbon pricing, which includes

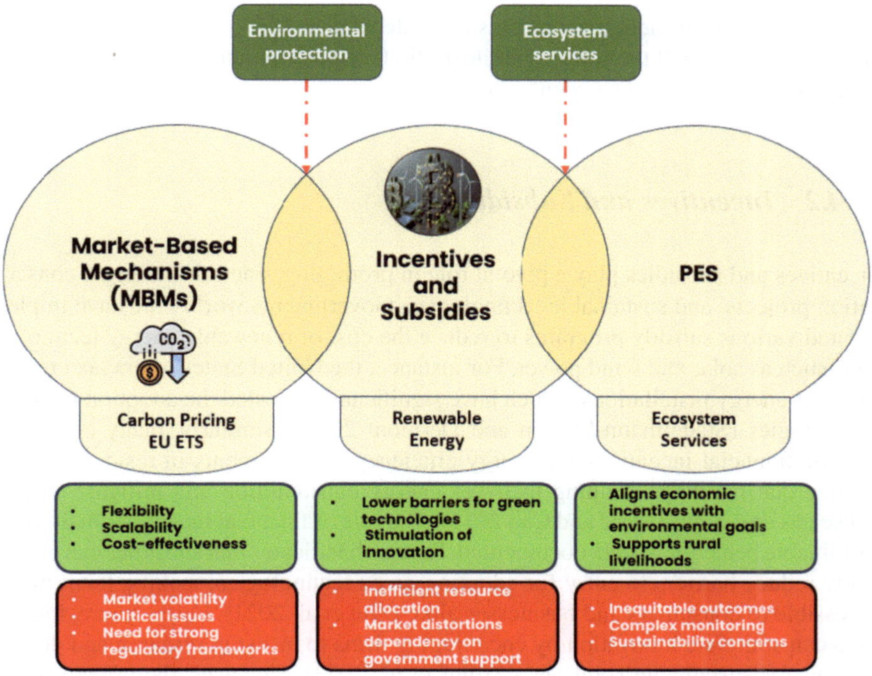

Fig. 16.1 Economic instruments for environmental protection and ecosystem services

both carbon taxes and cap-and-trade systems, aims to account for the external costs of carbon emissions, thereby incentivizing reductions in greenhouse gas emissions (Castiglione et al. 2014; Chistyakova et al. 2019). The European Union Emission Trading System (EU ETS) is a prominent example of an ETS, allowing companies to buy and sell emission allowances, creating a financial incentive to reduce emissions (Li et al. 2022). The advantages of market-based approaches include their flexibility, scalability, and potential for cost-effectiveness. By allowing firms to choose how to reduce emissions—whether through technological innovation or purchasing allowances—MBMs can lead to more efficient outcomes than traditional command-and-control regulations (Blackman et al. 2018). Furthermore, the double dividend hypothesis suggests that environmental taxes can yield both environmental and economic benefits, as they can generate revenue that can be reinvested in sustainable initiatives (Castiglione et al. 2014; Davydenko 2023). However, there are notable limitations to market-based mechanisms. One significant challenge is the potential for market manipulation and volatility, which can undermine the effectiveness of ETS (Choi and Cho 2021). Additionally, the implementation of carbon pricing can be politically contentious, as it may disproportionately affect low-income households unless accompanied by targeted compensatory measures (Zhou 2023). Moreover, the effectiveness of these mechanisms often depends on the robustness of the regulatory framework and the capacity of institutions to monitor and enforce compliance (Blackman et al. 2018; Davydenko 2023).

16.4.2 Incentives and Subsidies

Incentives and subsidies play a pivotal role in promoting renewable energy, conservation projects, and sustainable technologies. Governments worldwide have implemented various subsidy programs to reduce the cost of renewable energy technologies, such as solar and wind power. For instance, the United States offers tax credits for solar energy installations, which have significantly boosted the adoption of solar technologies (Stadelmann-Steffen and Dermont 2018). Similarly, many countries provide financial incentives for energy efficiency improvements in residential and commercial buildings, helping to lower energy consumption and mitigate greenhouse gas emissions (Ward and Cao 2012). The role of financial incentives in driving sustainable behavior is well-documented. Research indicates that subsidies can effectively reduce barriers to entry for adopting green technologies, making them more accessible to consumers and businesses alike (Zhu et al. 2021). Furthermore, incentives can stimulate innovation by encouraging firms to invest in research and development for sustainable solutions (Arthur et al. 2023). However, the effectiveness of subsidies can be hindered by issues such as inefficient resource allocation and the potential for creating dependency on government support (Guo and Фу 2021). Critics argue that poorly designed subsidy programs may lead to market distortions and unintended consequences, such as over-reliance on specific technologies without fostering broader sustainability practices (Sjögren 2010). Therefore, it is essential for policymakers to carefully assess and refine the design and implementation of subsidy programs to ensure they achieve desired environmental outcomes while avoiding negative side effects.

16.4.3 Payment for Ecosystem Services

Payment for Ecosystem Services (PES) is an innovative approach that compensates landowners and resource managers for maintaining or enhancing ecosystem services, such as carbon sequestration, water purification, and biodiversity conservation. PES programs have gained traction globally, with examples ranging from forest conservation initiatives in Costa Rica to urban green infrastructure projects in various cities (Angelis et al. 2019; Ejelöv et al. 2022). These programs aim to align economic incentives with environmental stewardship, encouraging landholders to adopt sustainable practices that benefit society. The framework of PES typically involves a formal contractual agreement where beneficiaries of ecosystem services (e.g., governments, NGOs, or private companies) provide financial compensation to landowners who manage their land in ways that preserve or enhance these services (Le et al. 2023). This approach not only promotes conservation but also supports rural livelihoods by providing alternative income sources for communities dependent on natural resources (Aiwansoba et al. 2019; Yang et al. 2022; Meijaard et al. 2023). However, challenges in scaling PES programs and ensuring fair and equitable benefit distribution remain significant. One major concern is the potential for inequitable outcomes, where wealthier landowners may disproportionately benefit from PES schemes, marginalizing vulnerable communities who may lack access to compensation (Pan 2023). Additionally, establishing effective monitoring and verification systems to ensure compliance and assess the impact of PES initiatives is often resource-intensive and logistically complex (Le et al. 2023). Moreover, the long-term sustainability of PES programs depends on the stability of funding sources and the political will to maintain commitments to ecosystem service payments (Willett et al. 2013). Therefore, while PES represents a promising tool for promoting environmental sustainability, its success hinges on careful design and implementation to ensure that it achieves both equitable and effective outcomes.

16.5 Governance Models for Sustainability

16.5.1 Centralized Versus Decentralized Models

Governance models for sustainability can be broadly categorized into centralized and decentralized approaches, each with its own advantages and disadvantages. Centralized governance typically involves a top-down approach where decision-making authority is concentrated in a single entity, often at the national or federal level. This model can lead to the efficient implementation of policies and regulations, as it allows for uniform standards and streamlined decision-making processes (Wang et al. 2022). For example, countries like France have successfully implemented national environmental policies through centralized governance, significantly increasing renewable energy adoption and reducing emissions (Okot 2023). However, centralized models may also face challenges such as bureaucratic inertia and a lack of responsiveness to local needs. Critics argue that this approach can overlook regional differences and the specific environmental challenges local communities face (Ma et al. 2018). In contrast, decentralized governance empowers local authorities and communities to take charge of environmental management, fostering greater participation and responsiveness to local conditions. For instance, the decentralized governance model in Costa Rica has enabled local communities to engage in reforestation and conservation efforts, leading to notable improvements in biodiversity and ecosystem health (Pospelova et al. 2023). Despite its advantages, decentralized governance can also present challenges, such as inconsistencies in policy implementation and potential conflicts between local and national priorities. The effectiveness of decentralized models often hinges on the capacity of local governments and communities to manage resources effectively (Ogwu 2019, 2023; Jebur 2022). Therefore, a hybrid approach that combines elements of centralized and decentralized governance may provide a more balanced solution, offering both the efficiency of centralized decision-making and the inclusiveness of local engagement in sustainability efforts.

16.5.2 Participatory and Inclusive Governance

Participatory and inclusive governance is essential for fostering sustainable development, as it engages communities, non-governmental organizations (NGOs), and the private sector in policy development and implementation. This approach recognizes the importance of diverse stakeholder perspectives and promotes collaboration among various actors in the governance process (García-Pérez and Mesa 2020). Engaging communities in decision-making can lead to more effective and equitable outcomes, as local knowledge and experiences are integrated into policy design. Examples of co-management and community-led initiatives illustrate the potential of participatory governance. In the Philippines, the Integrated Coastal Resource Management program has successfully involved local communities in managing coastal resources, resulting in improved fisheries management and enhanced livelihoods (Yuan 2021). Similarly, in Canada, Indigenous-led conservation initiatives have demonstrated the effectiveness of community-driven approaches in preserving biodiversity and traditional ecological knowledge (Li 2022). However, challenges remain in ensuring meaningful participation and inclusivity in governance processes. Power imbalances between stakeholders can undermine the effectiveness of participatory governance, as marginalized groups may struggle to have their voices heard (Birkel and Hartmann 2019). Additionally, the capacity of communities to engage in governance processes varies significantly, necessitating targeted resources and support to empower local actors (Copie et al. 2013). Therefore, fostering an inclusive governance framework requires ongoing efforts to build capacity, address power dynamics, and ensure that all stakeholders have a meaningful role in decision-making.

16.5.3 Role of Technology and Data

The role of technology and data in governance for sustainability is increasingly recognized as a critical factor in enhancing the effectiveness of environmental regulations. Digital tools, such as Geographic Information Systems (GIS) and remote sensing technologies, enable more accurate monitoring and enforcement of environmental policies (Memić et al. 2022). These technologies facilitate data collection and analysis, enabling better decision-making and resource management. Applications of big data, artificial intelligence (AI), and the Internet of Things (IoT) are transforming the landscape of environmental governance. For instance, IoT devices can provide real-time air and water quality data, enabling authorities to respond swiftly to ecological issues (Tan et al. 2014). In smart cities, IoT applications are being utilized to optimize energy consumption, reduce waste, and enhance urban planning efforts (Roşca et al. 2021). Moreover, AI algorithms can analyze vast amounts of environmental data to identify trends and predict future outcomes, aiding policymakers in developing proactive strategies for sustainability (Pei et al. 2021). Despite the potential benefits of technology in governance, challenges related to data privacy, security, and accessibility must be addressed. Ensuring that data is collected and used ethically is paramount, as is the need for equitable access to technology among different communities (Qin et al. 2022). Furthermore, the reliance on technology should not overshadow the importance of human judgment and local knowledge in governance processes (Routray et al. 2019). Therefore, integrating technology into governance models for sustainability requires a careful balance between leveraging data-driven insights and maintaining a human-centered approach to decision-making.

To better understand the advantages and challenges associated with various governance models for sustainability, the following matrix Chart presented in Fig. 16.2 summarizes the key features, strengths, and challenges of four prominent approaches: Centralized Governance, Decentralized Governance, Participatory and Inclusive Governance, and the Role of Technology and Data in environmental governance. This matrix highlights how each model's characteristics contribute to or hinder effective policy implementation and community engagement. It also showcases examples from around the world to illustrate how these governance models are applied in practice. The matrix serves as a valuable tool for comparing these models and provides insights into their potential benefits and limitations in the context of sustainable development.

16.6 Legal Frameworks for Evaluation of Sustainable Environmental Practices

Legal frameworks like those presented in Table 16.3 are critical in shaping and enforcing sustainable environmental practices worldwide. These frameworks provide the foundation for regulating ecological impacts, conserving natural resources, and promoting sustainable development. The following table highlights key legal frameworks from various countries and regions, showcasing their focus areas, provisions, and notable impacts. By examining these examples, we gain insights into how diverse legal instruments contribute to achieving environmental sustainability goals and addressing global challenges.

Environmental justice and equity are critical components of legal frameworks promoting sustainable practices. These frameworks address disparities in environmental impacts and access to resources, ensuring that environmental hazards do not disproportionately burden marginalized communities. Ecological justice emphasizes the fair distribution of environmental benefits and burdens, advocating for the inclusion of affected communities in decision-making processes Hörisch et al.

Governance	Key Features	Strengths	Examples
Centralized Governance	Top-down, uniform policies	Efficient implementation	France's national renewable energy policy
		Consistency and uniformity	
		Clear accountability	
		Ignores local needs	
		Bureaucratic inertia	
Decentralized Governance	Local authorities empowered	Responsive to local conditions	Costa Rica's reforestation programs
		Community participation	
		Policy inconsistencies	
		Conflicts with national policies	
Participatory and Inclusive Governance	Stakeholder engagement, collaboration	Inclusive decision-making	Canada's Indigenous-led conservation
		Integration of local knowledge	
		Power imbalances	
		Varied community capacity	
Role of Technology and Data	Use of AI, IoT, and big data for monitoring and policy	Real-time monitoring	Smart cities, IoT-based environmental monitoring
		Predictive insights	
		Data privacy	
		Unequal access to technology	

Fig. 16.2 Comparison of governance models for sustainability: key features, strengths, and examples

(2017). Numerous legal frameworks have been established to promote environmental justice. For instance, the United States' National Environmental Policy Act (NEPA) mandates that federal agencies consider the environmental effects of their proposed actions and involve the public in the decision-making process (Pavlovskaia 2014). Similarly, the Environmental Justice for All Act aims to strengthen protections for low-income and minority communities by requiring federal agencies to assess the cumulative impacts of their actions on these populations (Quesada et al. 2018). Internationally, frameworks such as the Aarhus Convention emphasize the rights of individuals to access information, participate in decision-making, and seek justice in environmental matters (Easter et al. 2020). These frameworks are essential for promoting equity and ensuring all communities have a voice in environmental governance. However, challenges remain in effectively implementing these legal frameworks, particularly in addressing systemic inequalities and providing meaningful participation from marginalized groups (Khan et al. 2023). Compliance and enforcement mechanisms are vital for ensuring adherence to environmental regulations and achieving sustainability goals. Effective legal frameworks must include robust mechanisms for monitoring compliance and enforcing rules to deter violations and promote accountability (Athapaththu and Karunasena 2018). Various approaches have been employed to enhance compliance, including inspections, reporting requirements, and penalties for non-compliance (Rowe et al. 2016). In developed countries, regulatory agencies often have the resources and capacity to enforce environmental laws

Table 16.3 Some legal frameworks for evaluating sustainable environmental practices

Legal framework	Country/region	Year established	Focus areas	Key provisions	Notable impact
National Environmental Policy Act (NEPA)	United States	1969	Environmental impact assessment, federal agency accountability	Requires federal agencies to evaluate the environmental impacts of actions/projects	Enhanced integration of environmental considerations into federal decision-making
Environmental Protection and Management Act (EPMA)	Singapore	1999	Pollution control, environmental health	Comprehensive guidelines for waste management, pollution control, and environmental monitoring	Improved urban environmental quality and public health outcomes
Environmental Impact Assessment Decree	Nigeria	1992	Environmental impact assessments, sustainable development	Mandates EIA for public and private projects likely to impact the environment	Increased awareness and monitoring of the environmental impacts of major projects
Environment Protection and Biodiversity Conservation Act (EPBC)	Australia	1999	Biodiversity, heritage conservation, environmental protection	Provides legal framework for environmental protection and biodiversity conservation	Strengthened habitat protection and sustainable use of natural resources
Kyoto Protocol Implementation Act	Canada	2007	Greenhouse gas reduction, climate change	Outlines Canada's obligations to meet GHG reduction targets under Kyoto Protocol	Contributed to national climate change strategies
Forest Rights Act	India	2006	Forest management, Indigenous rights	Recognizes forest resource rights of Indigenous and tribal communities	Empowered local communities in forest conservation efforts
European Union Water Framework Directive	European Union	2000	Integrated water resource management	Sets objectives for achieving good qualitative and quantitative status of water bodies	Improved water quality and integrated management of transboundary waters

(continued)

Table 16.3 (continued)

Legal framework	Country/region	Year established	Focus areas	Key provisions	Notable impact
The Climate Change Act	United Kingdom	2008	Climate change mitigation, carbon budgeting	Legally binding carbon reduction targets and establishment of the Climate Change Committee	Promoted leadership in climate action and carbon reduction strategies
Environmental Management and Coordination Act (EMCA)	Kenya	1999	Environmental governance, pollution control, natural resources	Establishes framework for coordinated environmental management	Strengthened environmental regulation and awareness in Kenya
Zero Waste Masterplan	South Korea	2010	Waste management, recycling	Sets ambitious targets for waste reduction and recycling initiatives	Increased national recycling rates and waste management efficiency

effectively. For example, the U.S. Environmental Protection Agency (EPA) employs various compliance tools, including inspections and enforcement actions, to ensure adherence to environmental regulations (Coker 2023). However, challenges persist, such as political pressures and budget constraints that can limit enforcement efforts (Vaka et al. 2020). In contrast, developing countries often face significant challenges in compliance and enforcement due to limited resources, capacity, and institutional frameworks (Izah and Ogwu 2023; Ogwu et al. 2022, 2023). For instance, in many African nations, weak regulatory institutions and corruption hinder effective enforcement of environmental laws (Ershov 2023). This disparity highlights the need for international cooperation and capacity-building initiatives to strengthen compliance mechanisms in developing countries (Mohamed 2023). Moreover, integrating technology in compliance monitoring, such as remote sensing and data analytics, offers promising solutions for enhancing enforcement capabilities in developed and developing contexts (Flores 2023). However, successfully implementing these technologies requires adequate infrastructure, training, and support for local institutions (Vorontsova et al. 2019).

Corporate accountability is increasingly recognized as a crucial element of legal frameworks for sustainability. Legal obligations for corporations to adopt sustainable practices are essential for promoting responsible business conduct and minimizing environmental impacts. Various regulatory frameworks, such as the European Union's Non-Financial Reporting Directive, require large companies to disclose information on their environmental, social, and governance (ESG) performance (Kock and Min 2013). This transparency is intended to hold corporations accountable for their sustainability practices and encourage responsible behavior. Green certifications and reporting frameworks, such as ISO 14001 and the Global Reporting Initiative (GRI), provide standards for companies to demonstrate their commitment to sustainability (Buana 2023). These frameworks enhance corporate accountability and foster a culture of sustainability within organizations. For example, companies that achieve ISO 14001 certification are recognized for their effective environmental management systems, which can lead to improved operational efficiency and reduced environmental impacts (Poto and Murray 2023). However, challenges remain in ensuring that corporate accountability mechanisms are effective and enforceable. Critics argue that many reporting frameworks lack rigor and may allow companies to engage in "greenwashing," where they present a misleading image of their environmental performance (Friede et al. 2015). Additionally, the lack of standardized metrics for measuring sustainability performance complicates comparisons across companies and sectors (Kostruba 2022). To address these challenges, a growing call is for more stringent regulations and standardized reporting requirements that ensure transparency and accountability in corporate sustainability practices (Boros and Fogarassy 2019). Legal scholars emphasize the importance of integrating sustainability into corporate governance frameworks to align business practices with broader societal goals (Tan et al. 2011).

16.7 Case Studies of Successful Policy Frameworks for Environmental Resource Evaluation and Sustainability

Implementing effective, sustainable policies is crucial for addressing environmental challenges and promoting resource conservation. The EU's Water Framework Directive (WFD) is a comprehensive legal framework aimed at protecting and enhancing the quality of water resources across Europe. Adopted in 2000, the WFD requires member states to achieve "good status" for all water bodies by 2027. The directive emphasizes integrated water management, public participation, and the establishment of river basin management plans (Anderson and Renaud 2021). The WFD has successfully promoted public participation in water management decisions, increasing transparency and accountability (Moraes et al. 2022). Engaging local communities and stakeholders in the planning process has proven essential for the acceptance and effectiveness of water management strategies. The directive's focus on integrated management across different sectors and levels of governance has facilitated coordinated action and resource sharing among various stakeholders (Werts and Brewer 2014).

Costa Rica's Payment for Ecosystem Services (PES) program, established in 1997, incentivizes landowners to conserve and restore forest ecosystems by providing financial compensation for their ecosystem services, such as carbon sequestration and biodiversity conservation. This innovative approach has led to significant reforestation and improved ecosystem health across the country (Suwandayani et al. 2021). The program's success is partly attributed to the involvement of local communities in decision-making processes, ensuring that the needs and knowledge of those directly affected by conservation efforts are considered (Politis et al. 2014). The Great Lakes Water Quality Agreement (GLWQA) is a bilateral agreement between the United States and Canada aimed at restoring and protecting the water quality of the Great Lakes. The agreement emphasizes collaboration between federal, state, provincial, and local governments and public participation in decision-making (Collins et al. 2019). The GLWQA exemplifies the importance of collaborative governance in addressing transboundary environmental issues. By fostering cooperation among various levels of government and stakeholders, the agreement has facilitated coordinated action to address pollution and habitat degradation (Saunders et al. 2017). The agreement incorporates adaptive management principles, allowing for adjustments based on monitoring and evaluation of environmental conditions. This flexibility has been crucial for responding to emerging challenges, such as invasive species and climate change impacts (Bailey et al. 2011).

Successful policy frameworks often prioritize stakeholder engagement as a fundamental component of their design and implementation. Engaging communities, NGOs, and the private sector enhances the legitimacy of policies and ensures that diverse perspectives are considered. For instance, the participatory approaches used in the WFD and Costa Rica's PES program can serve as models for other regions seeking to implement sustainable policies (Kotchen and Segerson 2018). Integrating scientific knowledge into policy-making processes is essential for developing effective environmental policies. The GLWQA's reliance on scientific assessments to inform decision-making demonstrates the value of evidence-based approaches. Policymakers should prioritize collaboration with researchers and scientists to ensure that policies are grounded in the best available evidence (Zanti and Thomas 2021).Adaptive management practices allow for flexibility and responsiveness in policy implementation. By incorporating monitoring and evaluation mechanisms, policymakers can assess the effectiveness of their strategies and make necessary adjustments. This approach has been successfully employed in the GLWQA and can be replicated in other contexts facing dynamic environmental challenges (Kotchen and Segerson 2018). Economic incentives, such as PES programs, can effectively motivate conservation efforts and promote sustainable practices. Policymakers should explore the potential for similar incentive-based approaches in their regions, tailoring them to local contexts and resource availability (Haddaway et al. 2017).

16.8 Challenges in Policy Implementation for Environmental Resource Evaluation and Future Directions

Various challenges, including limited funding, inadequate technical expertise, conflicting stakeholder interests, and weak enforcement mechanisms, often hinder the effective implementation of environmental resource evaluation policies. Addressing these barriers is essential to successfully adopting sustainable practices and achieving long-term environmental goals.

16.8.1 Political and Economic Barriers

One of the most significant challenges in implementing sustainable policies is the resistance from vested interests and industries that environmental regulations may adversely affect. For example, industries reliant on fossil fuels often lobby against policies to reduce carbon emissions, arguing that such measures threaten jobs and economic growth Nielsen et al. (2016). This resistance can manifest in various forms, including political lobbying, misinformation campaigns, and legal challenges to proposed regulations (Aguboshim et al. 2021). The interplay between short-term economic growth and long-term sustainability presents a complex dilemma for policymakers. While immediate economic benefits can be politically advantageous, they often conflict with the need for sustainable practices that may require upfront investments and long-term planning (Ljovkina et al. 2019). For instance, the agricultural sector may resist regulations to reduce pesticide use due to concerns about short-term productivity losses, despite the long-term benefits of sustainable farming practices (Xu et al. 2022).

The challenge of balancing short-term economic growth with long-term sustainability is evident in many policy frameworks. Policymakers often face pressure to prioritize economic development over environmental protection, leading to compromises that undermine sustainability goals (Ljovkina et al. 2019). For example, in developing countries, rapid industrialization can lead to significant environmental degradation, as seen in the deforestation associated with agricultural expansion in Brazil (Wang et al. 2021). Innovative policy approaches that align economic incentives with sustainability goals are essential to address these challenges. For example, integrating green finance mechanisms can provide the necessary funding for sustainable projects while promoting economic growth (Williams 2023).

16.8.2 Institutional Capacity

Environmental policy effectiveness heavily depends on the institutional enforcement capacity. Many countries, particularly in the developing world, struggle with weak regulatory institutions that lack the resources and expertise to implement and enforce environmental laws (Kanger et al. 2020). This can lead to widespread noncompliance and a lack of accountability for ecological violations (Bardosh et al. 2017). For instance, in Nigeria, the lack of institutional capacity has hindered the enforcement of environmental regulations, resulting in significant oil spills and environmental degradation in the Niger Delta region (Richard et al. 2023; "Pursuing higher environmental goals for agriculture in an interconnected world", 2023). Similarly, in many parts of Southeast Asia, inadequate enforcement of logging regulations has led to rampant deforestation and biodiversity loss (Ishak and Thiruchelvam 2023; Meijaard et al. 2023). Building institutional capacity requires investment in training, resources, and infrastructure. Collaborative approaches that involve partnerships between governmental agencies, NGOs, and local communities can enhance enforcement capabilities and promote accountability (Ali 2023).

16.8.3 Social and Cultural Resistance

Social and cultural resistance to environmental policies can pose significant barriers to implementation. Communities may be skeptical of new regulations, particularly if they perceive them as threats to their livelihoods or cultural practices (Fagerberg 2018). For example, indigenous communities may resist conservation policies restricting access to traditional lands and resources, leading to conflicts between conservation goals and indigenous rights (Li 2020; Ogwu and Osawaru 2022). Engaging communities in the policy-making process and ensuring their voices are heard is essential to overcome this resistance. Participatory approaches incorporating local knowledge and values can foster greater acceptance of environmental policies (Hamann et al. 2018). For instance, community-led conservation initiatives in Costa Rica have successfully integrated local perspectives into conservation strategies, resulting in improved outcomes for both biodiversity and local livelihoods (Oskam et al. 2018).

16.8.4 Future Directions in Policy Development

Innovative policy-making approaches are crucial for addressing the complex challenges of sustainability. Policymakers should explore the potential of adaptive management frameworks that allow for flexibility and responsiveness to changing environmental conditions (Томашук and Baldynyuk 2023). Additionally, integrating technology and data analytics can enhance decision-making processes and improve the effectiveness of environmental policies (Sun and Yu 2023). Emerging global initiatives, such as green finance and circular economy policies, offer promising avenues for advancing sustainability. Green finance mechanisms can mobilize investment for sustainable projects, while circular economy policies promote resource efficiency and waste reduction (Adenle et al. 2015). These initiatives can help bridge the gap between economic growth and environmental sustainability, providing a pathway for more sustainable development. The future of ecological policy requires a vision for an integrated and inclusive global framework that addresses the interconnectedness of ecological, social, and economic challenges. Such a framework should prioritize collaboration among nations, stakeholders, and communities, fostering a shared commitment to sustainability (Baquedano et al. 2022).

16.9 Conclusion

Developing and implementing effective policy and regulatory frameworks are paramount for advancing sustainable environmental practices. While significant challenges remain, integrating innovative governance models, public procurement strategies, and technological advancements can pave the way for more effective sustainability initiatives. Future research and policy efforts must focus on bridging the gaps identified in current frameworks, ensuring they are adaptable, actionable, and aligned with local and global sustainability goals. International agreements and multilateral organizations are fundamental to shaping global environmental policy. While significant progress has been made through treaties like the Paris Agreement and the CBD, challenges in compliance and enforcement persist. The role of multilateral organizations in providing funding and capacity-building initiatives is crucial for supporting countries in implementing sustainable practices.

Future efforts must focus on enhancing the effectiveness of these frameworks and ensuring that they are adaptable to the evolving global environmental landscape. In the same way, national policy frameworks for ecological sustainability are multifaceted and require robust environmental legislation, cross-sectoral policy integration, and strong political will. The interplay between these elements determines the effectiveness of sustainability initiatives and the ability of nations to address pressing environmental challenges. Economic instruments in environmental regulation, including market-based mechanisms, incentives, subsidies, and PES, offer diverse approaches to addressing environmental challenges. While each instrument has its advantages and limitations, their effectiveness often hinges on the political context, institutional capacity, and careful design of policies. As countries strive to meet their sustainability goals, integrating these economic instruments into comprehensive environmental strategies will be essential for fostering a more sustainable future.

Governance models for sustainability encompass a range of approaches, each with its strengths and weaknesses. Centralized and decentralized models offer different pathways for policy implementation, while participatory governance emphasizes the importance of engaging diverse stakeholders in decision-making. Integrating technology and data further enhances the effectiveness of governance efforts, providing valuable insights for environmental management. As countries strive to achieve sustainability goals, a comprehensive and adaptive governance framework incorporating these elements will be essential for addressing the complex challenges of environmental degradation and climate change. Legal frameworks for sustainable practices encompass a range of elements, including environmental justice, compliance and enforcement mechanisms, and corporate accountability. While significant progress has been made in establishing these frameworks, challenges remain in their implementation and effectiveness. Addressing disparities in environmental impacts, enhancing compliance and enforcement, and ensuring corporate accountability are essential for achieving sustainability goals. As the global community grapples with pressing ecological challenges, robust legal frameworks will be critical in guiding sustainable practices and promoting equity.

References

Adenle A, Azadi H, Arbiol J (2015) Global assessment of technological innovation for climate change adaptation and mitigation in developing world. J Environ Manage 161:261–275. https://doi.org/10.1016/j.jenvman.2015.05.040

Aguboshim F, Onwuka I, Obiokafor I, Oboti N (2021) Factors challenging research and innovation in tertiary education in Nigeria. World J Adv Res Rev 10(1):224–229. https://doi.org/10.30574/wjarr.2021.10.1.0167

Aiwansoba RO, Ogwu MC, Osawaru ME (2019) Assessing the relatedness of *Abelmoschus* accessions using morphological characters. J Trop Biol Conserv 16:193–207

Akudugu MA, Ogwu MC (2024) Sustainable development policies and interventions: a bibliometric analysis of the contributions of the academic community. J Clean Prod 434:139919. https://doi.org/10.1016/j.jclepro.2023.139919
Ali A (2023) Information sharing: the missing ingredient in science, technology and innovation policy of pakistan-agriculture perspective. https://doi.org/10.31223/x5wx0s
Anderson C, Renaud F (2021) A review of public acceptance of nature-based solutions: the 'why', 'when', and 'how' of success for disaster risk reduction measures. Ambio 50(8):1552–1573. https://doi.org/10.1007/s13280-021-01502-4
Andreopoulou Z, Koliouska C (2018) Benchmarking internet promotion of renewable energy enterprises: is sustainability present? Sustainability 10(11):4187. https://doi.org/10.3390/su10114187
Angelis E, Giacomo M, Vannoni D (2019) Climate change and economic growth: the role of environmental policy stringency. Sustainability 11(8):2273. https://doi.org/10.3390/su11082273
Arthur E, Gyamfi S, Gerstlberger W, Stejskal J, Prokop V (2023) Towards circular economy: unveiling heterogeneous effects of government policy stringency, environmentally related innovation, and human capital within oecd countries. Sustainability 15(6):4959. https://doi.org/10.3390/su15064959
Athapaththu K, Karunasena G (2018) Framework for sustainable construction practices in Sri Lanka. Built Environ Proj Asset Manag 8(1):51–63. https://doi.org/10.1108/bepam-11-2016-0060
Azis A (2022) The role of non-governmental organizations (ngo's) in shaping environmental policies. https://doi.org/10.2991/assehr.k.220102.064
Ba Y, Schwaeble K, Birkland T (2022) The united states in chinese environmental policy narratives: is there a trump effect? Rev Policy Res 39(6):708–729. https://doi.org/10.1111/ropr.12503
Bailey S, Deneau M, Jean L, Wiley C, Leung B, MacIsaac H (2011) Evaluating efficacy of an environmental policy to prevent biological invasions. Environ Sci Technol 45(7):2554–2561. https://doi.org/10.1021/es102655j
Baquedano F, Jelliffe J, Beckman J, Ivanic M, Zereyesus Y, Johnson M (2022) Food security implications for low- and middle-income countries under agricultural input reduction: the case of the european union's farm to fork and biodiversity strategies. Appl Econ Perspect Policy 44(4):1942–1954. https://doi.org/10.1002/aepp.13236
Bardosh KL, Scoones JC, Grace D, Kalema-Zikusoka G, Jones KE, de Balogh K, Waltner-Toews D, Bett B, Welburn SC, Mumford E, Dzingirai V (2017) Engaging research with policy and action: what are the challenges of responding to zoonotic disease in Africa? Philos Trans R Soc B Biol Sci 372(1725):20160172. https://doi.org/10.1098/rstb.2016.0172
Bengtsson M, Alfredsson E, Cohen M, Lorek S, Schroeder P (2018) Transforming systems of consumption and production for achieving the sustainable development goals: moving beyond efficiency. Sustain Sci 13(6):1533–1547. https://doi.org/10.1007/s11625-018-0582-1
Berry J, Portney K (2013) Sustainability and interest group participation in city politics. Sustainability 5(5):2077–2097. https://doi.org/10.3390/su5052077
Biermann F (2020) Earth system governance: world politics in the Anthropocene. MIT Press
Birkel H, Hartmann E (2019) Impact of IoT challenges and risks for SCM. Supply Chain Manag Int J 24(1):39–61. https://doi.org/10.1108/scm-03-2018-0142
Blackman A, Li Z, Liu A (2018) Efficacy of command-and-control and market-based environmental regulation in developing countries. Annu Rev Resour Econ 10(1):381–404. https://doi.org/10.1146/annurev-resource-100517-023144
Bomberg E (2021) The environmental legacy of president trump. Policy Studies 42(5–6):628–645. https://doi.org/10.1080/01442872.2021.1922660
Boros A, Fogarassy C (2019) Relationship between corporate sustainability and compliance with state-owned enterprises in central-europe: a case study from hungary. Sustainability 11(20):5653. https://doi.org/10.3390/su11205653

Buana A (2023) The role of customary law in natural resource management: a comparative study between indonesia and australia. Gold Rat Mapp Idea Lit Format 3(2):167–186. https://doi.org/10.52970/grmilf.v3i2.400

Castiglione C, Infante D, Smirnova J (2014) Is there any evidence on the existence of an environmental taxation Kuznets curve? the case of european countries under their rule of law enforcement. Sustainability 6(10):7242–7262. https://doi.org/10.3390/su6107242

Chistyakova G, Rolgayzer A, Bondareva E, Schlee I (2019) International practice of environmental challenges regulation. E3s Web Conf 105:02024. https://doi.org/10.1051/e3sconf/201910502024

Choi G, Cho S (2021) Effect of market versus nonmarket environmental policy stringency on knowledge sourcing behavior for green technology: evidence from oecd countries. Rev Int Bus Strat 32(1):72–93. https://doi.org/10.1108/ribs-09-2020-0113

Coker J (2023) The role of legal practitioners in mitigating corporate risks in nigeria: a comprehensive review of existing literature on the strategies and approaches adopted by legal practitioners in nigeria to mitigate corporate risks. Financ Account Res J 5(10):309–332. https://doi.org/10.51594/farj.v5i10.579

Collins A, Coughlin D, Randall N (2019) Engaging environmental policymakers with systematic reviews: challenges, solutions and lessons learned. Environ Evid 8(1). https://doi.org/10.1186/s13750-018-0144-0

Copie A, Fortiş T, Munteanu V, Negru V (2013) From cloud governance to IoT governance. https://doi.org/10.1109/waina.2013.169

Dagnet Y, Simeonova K, Cogswell N, Holt M, Viña T, Warszawski N (2021) Toward more effective implementation of the paris agreement: learning from 30 years of experience. WRIPUB. https://doi.org/10.46830/wriwp.20.00069

Davydenko N (2023) Fiscal regulation policy and financial incentives impact on the environmental management. IOP Conf Ser Earth Environ Sci 1269(1):012010. https://doi.org/10.1088/1755-1315/1269/1/012010

Dedeurwaerdere T (2013) Transdisciplinary sustainability science at higher education institutions: science policy tools for incremental institutional change. Sustainability 5(9):3783–3801. https://doi.org/10.3390/su5093783

Easter S, Ceulemans K, Kelly D (2020) Bridging research-practice tensions: exploring day-to-day engaged scholarship investigating sustainable development challenges. Eur Manag Rev 18(2):9–23. https://doi.org/10.1111/emre.12443

Ejelöv E, Harring N, Hansla A, Jagers S, Nilsson A (2022) Push, pull, or inform—an empirical taxonomy of environmental policy support in Sweden. J Publ Policy 42(3):529–552. https://doi.org/10.1017/s0143814x21000271

Ershov D (2023) Legal framework for sustainable development and current global challenges. Sustain Social Dev 1(1):1–12. https://doi.org/10.54517/ssd.v1i1.2196

Escobar A (2018) Designs for the pluriverse: radical interdependence, autonomy, and the making of worlds. Duke University Press

Fagerberg J (2018) Mobilizing innovation for sustainability transitions: a comment on transformative innovation policy. Res Policy 47(9):1568–1576. https://doi.org/10.1016/j.respol.2018.08.012

Fairbrass A, O'Sullivan A, Campbell J, Ekins P (2023) The SDG monitoring framework provides limited evidence that environmental policies are delivering multiple ecological and social benefits. https://doi.org/10.22541/essoar.167527462.20087822/v1

Falkner R (2016) The paris agreement and the new logic of international climate politics. Int Aff 92(5):1107–1125. https://doi.org/10.1111/1468-2346.12708

Flores G (2023) Fostering awareness of integrating education for sustainable development in teacher education institutions across the Caraga Region, Philippines. Int J Membr Sci Technol 10(3):1176–1183. https://doi.org/10.15379/ijmst.v10i3.1700

Friede G, Busch T, Bassen A (2015) ESG and financial performance: aggregated evidence from more than 2000 empirical studies. J Sustain Financ Invest 5(4):210–233. https://doi.org/10.1080/20430795.2015.1118917

García-Pérez A, Mesa J (2020) Sustainable engineering and internet of things (IoT): trends and perspectives. https://doi.org/10.18687/laccei2020.1.1.109

Grano S (2014) Change and continuities: Taiwan's post-2008 environmental policies. J Curr Chin Aff 43(3):129–159. https://doi.org/10.1177/186810261404300306

Green J (2017) Transnational delegation in global environmental governance: when do non-state actors govern? Regul Gov 12(2):263–276. https://doi.org/10.1111/rego.12141

Guo X, Фу Л (2021) Can environmental regulations promote greenhouse gas abatement in oecd countries? command-and-control vs. market-based policies. Sustainability 13(12):6913. https://doi.org/10.3390/su13126913

Haddaway NR, Kohl C, Rebelo da Silva N, Schiemann J, Spök A, Stewart R, Sweet JB, Wilhelm R (2017) A framework for stakeholder engagement during systematic reviews and maps in environmental management. Environ Evid 6(1):1–14. https://doi.org/10.1186/s13750-017-0089-8

Hamann M, Berry K, Chaigneau T, Curry T, Heilmayr R, Henriksson PJ, Hentati-Sundberg J, Jina A, Lindkvist E, Lopez-Maldonado Y, Nieminen E (2018) Inequality and the biosphere. Annu Rev Environ Resour 43(1):61–83. https://doi.org/10.1146/annurev-environ-102017-025949

Höhne N, Kuramochi T, Warnecke C, Röser F, Fekete H, Hagemann M, Day T, Tewari R, Kurdziel M, Sterl S, Gonzales S (2016) The paris agreement: resolving the inconsistency between global goals and national contributions. Clim Policy 17(1):16–32. https://doi.org/10.1080/14693062.2016.1218320

Hörisch J, Burritt R, Christ K, Schaltegger S (2017) Legal systems, internationalization and corporate sustainability: an empirical analysis of the influence of national and international authorities. Corp Gov 17(5):861–875. https://doi.org/10.1108/cg-08-2016-0169

Intergovernmental Panel on Climate Change (IPCC) (2021) Climate change 2021: the physical science basis. Cambridge University Press. Retrieved from https://www.ipcc.ch

Ishak N, Thiruchelvam V (2023) Sustainable innovations in Malaysia's public procurement: strategic policy initiatives and coherences. Int J Innov Sci 16(2):338–372. https://doi.org/10.1108/ijis-08-2022-0144

Izah SC, Ogwu MC (2023) Sustainable utilization and conservation of Africa's biological resources and environment. In: Ramawat KG (ed) Sustainable development and biodiversity series, 1 edn, vol 32. Springer Nature, Singapore, 691 p. https://doi.org/10.1007/978-981-19-6974-4. ISBN 978-981-19-6973-7

Jebur, T. (2022). Greening the internet of things: a comprehensive review of sustainable IoT solutions from an educational perspective. Indones J Educ Res Technol 3(3):247–256. https://doi.org/10.17509/ijert.v3i3.57428

Jotzo F, Depledge J, Winkler H (2018) US and international climate policy under President Trump. Clim Policy 18(7):813–817. https://doi.org/10.1080/14693062.2018.1490051

Kanger L, Sovacool B, Noorkõiv M (2020) Six policy intervention points for sustainability transitions: a conceptual framework and a systematic literature review. Res Policy 49(7):104072. https://doi.org/10.1016/j.respol.2020.104072

Kazak J, Świąder M (2018) SOLIS—a novel decision support tool for the assessment of solar radiation in ArcGIS. Energies 11(8):2105. https://doi.org/10.3390/en11082105

Khan N, Ali H, Alam H (2023) Addressing sustainability challenges as part of director's duty in Malaysia. Int J Law Manag 65(6):538–559. https://doi.org/10.1108/ijlma-03-2023-0044

Kissinger G (2020) Policy responses to direct and underlying drivers of deforestation: examining rubber and coffee in the central highlands of Vietnam. Forests 11(7):733. https://doi.org/10.3390/f11070733

Kock C, Min B (2013) Legal origins, corporate governance and environmental outcomes. Acad Manag Proc 2013(1):15932. https://doi.org/10.5465/ambpp.2013.15932abstract

Konisky D, Teodoro M (2015) When governments regulate governments. Am J Polit Sci 60(3):559–574. https://doi.org/10.1111/ajps.12221

Kostruba A (2022) Legal aspects of transnational scale corporations' activity in terms of sustainable development. Rivista Di Studi Sulla Sostenibilita 2:49–61. https://doi.org/10.3280/riss2021-002005

Kotchen M, Segerson K (2018) On the use of group performance and rights for environmental protection and resource management. Proc Natl Acad Sci 116(12):5285–5292. https://doi.org/10.1073/pnas.1802881115

Le D, Ren F, Li Y (2023) The effect of energy use rights trading policy on environmental performance: evidence from Chinese 262 cities. Int J Environ Res Public Health 20(4):3570. https://doi.org/10.3390/ijerph20043570

Le TT, Vodden K, Wu J, Bullock R, Sabau G (2023) Payments for ecosystem services programs: a global review of contributions towards sustainability. Heliyon 10(1): e22361. https://doi.org/10.1016/j.heliyon.2023.e22361c

Li L (2020) The governance of low-carbon transitions in a multilevel perspective framework: how does the concept of 'system transformation' work? Energy Res J 11(1):45–53. https://doi.org/10.3844/erjsp.2020.45.53

Li L, Jin X, Li Y, Chen H, Wang Y (2022) The impact of carbon trading policy on regional ecological risk: synergy between market-based environmental policy and government intervention. Front Environ Sci 10:1010522. https://doi.org/10.3389/fenvs.2022.1010522

Li M (2022). The effects of heterogeneous environmental regulation on the effectiveness of corporate green investment in china. https://doi.org/10.21203/rs.3.rs-2212028/v1

Ljovkina A, Dusseault D, Zaharova O, Klochkov Y (2019) Managing innovation resources in accordance with sustainable development ethics: typological analysis. Resources 8(2):82. https://doi.org/10.3390/resources8020082

Ma Y, Wu C, Kang P, Hong C, Jiang C (2018) Internet of things applications in public safety management: a survey. Library Hi Tech 38(1):133–144. https://doi.org/10.1108/lht-12-2017-0275

Manika S (2020) Encouraging sustainability of European smart cities through green and sustainable public procurement. Int J Adv Res 8(5):668–675. https://doi.org/10.21474/ijar01/10973

Meijaard E, Unus N, Ariffin T, Dennis RA, Ancrenaz M, Wich S, Wunder S, Goh CS, Sherman J, Ogwu MC, Refisch J, Ledgard J, Sheil D, Hockings KJ (2023) Apes and agriculture. Front Conserv Sci 4. https://doi.org/10.3389/fcosc.2023.1225911

Memić B, Džubur A, Avdagić-Golub E (2022) Green IoT: sustainability environment and technologies. Sci Eng Technol 2(1), 24–29. https://doi.org/10.54327/set2022/v2.i1.25

Michaelowa K, Michaelowa A, Reinsberg B, Shishlov I (2020) Do multilateral development bank trust funds allocate climate finance efficiently? Sustainability 12(14):5529. https://doi.org/10.3390/su12145529

Mickwitz P, Neij L, Johansson M, Benner M, Sandin S (2021) A theory-based approach to evaluations intended to inform transitions toward sustainability. Evaluation 27(3):281–306. https://doi.org/10.1177/1356389021997855

Mohamed Y (2023) Climate change policy: the critical role of legal scholarship in shaping sustainable futures. ISSLP 2(1), 1–3. https://doi.org/10.61838/kman.isslp.2.1.1

Moldovan F (2023) The environmental sustainability assessment of an orthopedics emergency hospital supported by a new innovative framework. Sustainability 15(18):13402. https://doi.org/10.3390/su151813402

Moraes R, Reguero B, Mazarrasa I, Ricker M, Juanes J (2022) Nature-based solutions in coastal and estuarine areas of Europe. Front Environ Sci 10. https://doi.org/10.3389/fenvs.2022.829526

Nielsen K, Reisch L, Thøgersen J (2016) Sustainable user innovation from a policy perspective: a systematic literature review. J Clean Prod 133:65–77. https://doi.org/10.1016/j.jclepro.2016.05.092

Nordhaus W (2018) Projections and uncertainties about climate change in an era of minimal climate policies. Am Econ J Econ Pol 10(3):333–360. https://doi.org/10.1257/pol.20170046

Ogwu MC (2019) Towards sustainable development in Africa: the challenge of urbanization and climate change adaptation. In: Cobbinah PB, Addaney M (eds) The geography of climate change adaptation in urban Africa. Springer Nature, Switzerland, pp 29–55. https://doi.org/10.1007/978-3-030-04873-0_2
Ogwu MC (2023) Local food crops in Africa: sustainable utilization, threats, and traditional storage strategies. In: Izah SC, Ogwu MC (eds) Sustainable utilization and conservation of Africa's biological resources and environment. Sustainable development and biodiversity, vol 888. Springer, Singapore, pp 353–374. https://doi.org/10.1007/978-981-19-6974-4_13
Ogwu MC, Osawaru ME (2022) Traditional methods of plant conservation for sustainable utilization and development. In: Izah SC (ed) Biodiversity in Africa: potentials, threats and conservation. Sustainable development and biodiversity, vol 29. Springer, Singapore, pp 451–472. https://doi.org/10.1007/978-981-19-3326-4_17
Ogwu MC, Iyiola AO, Izah SC (2023) Overview of African biological resources and environment. In: Izah SC, Ogwu MC (eds) Sustainable utilization and conservation of Africa's biological resources and environment. Sustainable development and biodiversity, vol 888. Springer, Singapore, pp 1–31. https://doi.org/10.1007/978-981-19-6974-4_1
Ogwu MC, Izah SC, Iyiola AO (2022) An overview of the potentials, threats and conservation of biodiversity in Africa. In: Izah SC (ed) Biodiversity in Africa: potentials, threats and conservation. Sustainable development and biodiversity, vol 29. Springer, Singapore, pp 3–20. https://doi.org/10.1007/978-981-19-3326-4_1
Okot T (2023) The internet of things (IoT) for sustainability: a framework for costa rica. J Technol Manag Innov 18(4):3–17. https://doi.org/10.4067/s0718-27242023000400003
Oppong S (2023) Promoting global ECD top-down and bottom-up. Ethos 51(3):321–325. https://doi.org/10.1111/etho.12393
Oskam I, Bossink B, Man A (2018) The interaction between network ties and business modeling: case studies of sustainability-oriented innovations. J Clean Prod 177:555–566. https://doi.org/10.1016/j.jclepro.2017.12.202
Ostrom E (2009) A general framework for analyzing sustainability of social-ecological systems. Science 325(5939):419–422. https://doi.org/10.1126/science.1172133
Pan F (2023) Selection and application of Bohai Sea environmental governance policy instrumental: a quantitative analysis based on policy text (1996–2022). Sustainability 15(18):13454. https://doi.org/10.3390/su151813454
Parmentola A, Petrillo A, Tutore I, Felice F (2021) Is blockchain able to enhance environmental sustainability? a systematic review and research agenda from the perspective of sustainable development goals (SDGs). Bus Strateg Environ 31(1):194–217. https://doi.org/10.1002/bse.2882
Pavlovskaia E (2014) Sustainability criteria: their indicators, control, and monitoring (with examples from the biofuel sector). Environ Sci Eur 26(1). https://doi.org/10.1186/s12302-014-0017-2
Pei W, Dong C, Chen N, Qi M, Yang S, Nnenna A, Li W (2021) Environmental regulation, government subsidies, and green technology innovation—a provincial panel data analysis from China. Int J Environ Res Public Health 18(22):11991. https://doi.org/10.3390/ijerph182211991
Politis C, Halligan M, Keen D, Kerner J (2014) Supporting the diffusion of healthy public policy in Canada: the prevention policies directory. Online J Public Health Inform 6(2). https://doi.org/10.5210/ojphi.v6i2.5372
Pospelova V, López-Baldominos I, Sanz L, Castillo-Martinez A, Misra S (2023) User and professional aspects for sustainable computing based on the internet of things in Europe. Sensors 23(1):529. https://doi.org/10.3390/s23010529
Poto M, Murray E (2023) The new horizons of law and science through the lens of 2030 sustainable development agenda: some emerging issues. Environ Policy Law 53(1):35–48. https://doi.org/10.3233/epl-220039
Pramudianto A (2019) Paris agreement agreement 2015 and its impact on indonesian national law. https://doi.org/10.4108/eai.20-1-2018.2281915

Qin L, Ying Z, Yang W, Wang X (2022) Research on the impact of environmental regulation on green technology innovation from the perspective of regional differences: a Quasi-natural experiment based on China's new environmental protection law. Sustainability 14(3):1714. https://doi.org/10.3390/su14031714

Quesada G, Klenke T, Mejía-Ortíz L (2018) Regulatory challenges in realizing integrated coastal management—lessons from Germany, Costa Rica, Mexico and South Africa. Sustainability 10(10):3772. https://doi.org/10.3390/su10103772

Roelfsema M, van Soest HL, Harmsen M, van Vuuren DP, Bertram C, den Elzen M, Höhne N, Iacobuta G, Krey V, Kriegler E, Luderer G (2020) Taking stock of national climate policies to evaluate implementation of the Paris Agreement. Nat Commun 11(1). https://doi.org/10.1038/s41467-020-15414-6

Roşca M, Nicolae C, Sanda E, Madan A (2021) Internet of things (IoT) and sustainability, 346–352. https://doi.org/10.24818/basiq/2021/07/044

Routray S, Tengshe R, Javali A, Sarkar S, Sharma L, Ghosh A (2019) Satellite based IoT for mission critical applications, 1–6. https://doi.org/10.1109/icondsc.2019.8817030

Rowe B, McCann W, Hemmens C (2016) Persona non grata: the marginalization of legal scholarship in criminology and criminal justice journals. J Crim Justice Educ 28(4):514–541

Sachs JD, Schmidt-Traub G, Kroll C, Lafortune G, Fuller G (2019) Sustainable development report 2019: transformations to achieve the sustainable development goals. Bertelsmann Stiftung and sustainable development solutions network (SDSN). Retrieved from https://s3.amazonaws.com/sustainabledevelopment.report/2019/2019_sustainable_development_report.pdf

Saunders F, Gilek M, Linke S (2017) Knowledge for environmental governance: probing science–policy theory in the cases of eutrophication and fisheries in the Baltic Sea. J Environ Planning Policy Manage 19(6):769–782. https://doi.org/10.1080/1523908x.2017.1286575

Seddon N, Daniels E, Davis R, Harris R, Hou-Jones X, Huq S, Kapos V, Mace GM, Rizvi AR, Reid H, Roe D (2018) Global recognition that ecosystems are key to human resilience in a warming world. https://doi.org/10.20944/preprints201810.0203.v1

Silva K, Yapa P, Vesty G (2020) The impact of accountability mechanisms on public sector environmental sustainability performance: a case study of Sri Lanka. Australas Account Bus Financ J 14(3):38–55. https://doi.org/10.14453/aabfj.v14i3.4

Sjögren T (2010) Optimal taxation and environmental policy in a decentralized economic federation with environmental and labor market externalities. Public Financ Manag 10(2):284–330. https://doi.org/10.1177/152397211001000205

Stadelmann-Steffen I, Dermont C (2018) The unpopularity of incentive-based instruments: what improves the cost–benefit ratio? Public Choice 175(1–2):37–62. https://doi.org/10.1007/s11127-018-0513-9

Steffen W, Richardson K, Rockström J, Cornell SE, Fetzer I, Bennett EM, Biggs R, Carpenter SR, De Vries W, De Wit CA, Folke C (2015) Planetary boundaries: guiding human development on a changing planet. Science 347(6223):1259855. https://doi.org/10.1126/science.1259855

Sun Y, Yu B (2023) Greening China's belt and road initiative: from norm localization to norm subsidiarity? Glob Environ Polit 23(1):91–116. https://doi.org/10.1162/glep_a_00685

Suwandayani B, Ekowati D, Darmawan S, Haryono A (2021) Analysis of planning, implementation, assessment of learning from home strategies during the covid-19 pandemic in private elementary schools. Al-Bidayah J Pendidik Dasar Islam 13(1):107–124. https://doi.org/10.14421/al-bidayah.v13i1.623

Switzer D (2017) Citizen partisanship, local government, and environmental policy implementation. Urban Aff Rev 55(3):675–702. https://doi.org/10.1177/1078087417722863

Tan S, Yeo S, Ling G, Chong S, Cheah C (2014) Factors that influence green purchase behaviour among young consumers. Adv Mater Res 1051:1035–1039. https://doi.org/10.4028/www.scientific.net/amr.1051.1035

Tan Y, Shen L, Yao H (2011) Sustainable construction practice and contractors' competitiveness: a preliminary study. Habitat Int 35(2):225–230. https://doi.org/10.1016/j.habitatint.2010.09.008

Tawiah V (2022) Does political ideology affect the stringency of environmental policy? Polit Policy 50(3):631–653. https://doi.org/10.1111/polp.12465
Томашук I, Baldynyuk V (2023) Eco-innovation as a basis for sustainable development. Three Seas Econ J 4(1):71–87. https://doi.org/10.30525/2661-5150/2023-1-9
United Nations (2015) Transforming our world: the 2030 agenda for sustainable development. United Nations General Assembly. Retrieved from https://sdgs.un.org/2030agenda
Uwaoma P (2023) Strategies for sustainable economic growth: the role of green taxation and renewable energy investment in the U.S. World J Adv Res Rev 20(3):924–940. https://doi.org/10.30574/wjarr.2023.20.3.2542
Vaka M, Walvekar R, Rasheed A, Khalid M (2020) A review on Malaysia's solar energy pathway towards carbon-neutral malaysia beyond Covid'19 pandemic. J Clean Prod 273:122834. https://doi.org/10.1016/j.jclepro.2020.122834
Vorontsova E, Vorontsov A, Balashova T (2019) Ecological and legal support for the sustainable development of the Russian Federation. https://doi.org/10.2991/ispcbc-19.2019.150
Walsh-Führing M (2018) The Brazilian Federal Government's role in the prioritization of eu foreign direct investment and its environmental agenda. Braz Polit Sci Rev 12(3). https://doi.org/10.1590/1981-3821201800030004
Wang J, Tong C, Hu X (2021) Policy zoning method for innovation districts to sustainably develop the knowledge-economy: a case study in Hangzhou, China. Sustainability 13(6):3503. https://doi.org/10.3390/su13063503
Wang M, Li Y, Wang Z (2022) A nonlinear relationship between corporate environmental performance and economic performance of green technology innovation: moderating effect of government market-based regulations. Bus Strateg Environ 32(6):3119–3138. https://doi.org/10.1002/bse.3290
Ward H, Cao X (2012) Domestic and international influences on green taxation. Comp Pol Stud 45(9):1075–1103. https://doi.org/10.1177/0010414011434007
Werts A, Brewer C (2014) Reframing the study of policy implementation. Educ Policy 29(1):206–229. https://doi.org/10.1177/0895904814559247
Willett K, Caplanova A, Sivák R (2013) Pricing mechanisms for cap and trade policies: computer-assisted smart markets for air quality. J Environ Planning Manage 57(8):1240–1251. https://doi.org/10.1080/09640568.2013.795887
Williams J (2023) Roadmap for action on the environmental risk assessment of chemicals for insect pollinators (IPol-Era). EFSA Support Publ 20(11). https://doi.org/10.2903/sp.efsa.2023.en-8431
Wilson C, Williams I, Kemp S (2011) An evaluation of the impact and effectiveness of environmental legislation in small and medium-sized enterprises: experiences from the UK. Bus Strateg Environ 21(3):141–156. https://doi.org/10.1002/bse.720
Xu Z, Liu H, Lin S (2022) The influence of government green development policy on a firm's disruptive innovation. Sustainability 14(23):16040. https://doi.org/10.3390/su142316040
Yang S, Shen Q, Jahanger A, Ye P, Zhang H, Balsalobre-Lorente D (2022) The impact of carbon emission trading scheme on export: firm-level evidence from China. Front Environ Sci 10:1035650. https://doi.org/10.3389/fenvs.2022.1035650
Yuan Q (2021) How to restrain regulatory capture and promote green innovation in China: an analysis based on evolutionary game theory. Sustainability 13(17):9752. https://doi.org/10.3390/su13179752
Zanti S, Thomas ML (2021) Evidence-based policymaking: what human service agencies can learn from implementation science and integrated data systems. Glob Implement Res Appl 1(4):304–314. https://doi.org/10.1007/s43477-021-00028-x
Zheng G, Siddik A, Masukujjaman M, Fatema N, Alam S (2021) Green finance development in Bangladesh: the role of private commercial banks (PCBs). Sustainability 13(2):795. https://doi.org/10.3390/su13020795

Zhou B (2023) How does emissions trading affect the efficiency of enterprise resource allocation? evidence from China. Sage Open 13(4). https://doi.org/10.1177/21582440231212040

Zhu Y, Sun Z, Zhang S, Wang X (2021) Economic policy uncertainty, environmental regulation, and green innovation—an empirical study based on Chinese high-tech enterprises. Int J Environ Res Public Health 18(18):9503. https://doi.org/10.3390/ijerph18189503

Correction to: Carbon Capture, Storage, and Utilization Technologies

Amit Kumar Chaturwedi, Nand Kumar Kashyap, Archana Shukla, Shailendra Yadav, Gourisankar Roymahapatra, and Milan Hait

Correction to:
Chapter 11 in: M. C. Ogwu and S. Chibueze Izah (eds.), *Evaluating Environmental Processes and Technologies*, Environmental Science and Engineering, https://doi.org/10.1007/978-3-031-85327-2_11

In the original version of the book, the author name in Chapter 11 has been updated as Gourisankar Roymahapatra. The book and the chapter have been updated with the changes.

The updated version of this chapter can be found at
https://doi.org/10.1007/978-3-031-85327-2_11

M. C. Ogwu and S. Chibueze Izah (eds.), *Evaluating Environmental Processes and Technologies*, Environmental Science and Engineering,
https://doi.org/10.1007/978-3-031-85327-2_17

Index

M. C. Ogwu and S. Chibueze Izah (eds.), *Evaluating Environmental Processes and Technologies*, Environmental Science and Engineering, https://doi.org/10.1007/978-3-031-85327-2

D

J

K

L

M

N

O

P

S

T

MIX
Papier aus verantwortungsvollen Quellen
Paper from responsible sources
FSC® C105338

If you have any concerns about our products,
you can contact us on
ProductSafety@springernature.com

In case Publisher is established outside the EU,
the EU authorized representative is:
Springer Nature Customer Service Center GmbH
Europaplatz 3, 69115 Heidelberg, Germany

Printed by Libri Plureos GmbH
in Hamburg, Germany